Long Noncoding RNAs in Plants: Roles in Development and Stress

Long Noncoding RNAs in Plants: Roles in Development and Stress

Edited by

Santosh Kumar Upadhyay
Assistant Professor, Department of Botany,
Panjab University, Chandigarh, India

ELSEVIER

ACADEMIC PRESS
An imprint of Elsevier

Academic Press is an imprint of Elsevier
125 London Wall, London EC2Y 5AS, United Kingdom
525 B Street, Suite 1650, San Diego, CA 92101, United States
50 Hampshire Street, 5th Floor, Cambridge, MA 02139, United States
The Boulevard, Langford Lane, Kidlington, Oxford OX5 1GB, United Kingdom

Notices
Knowledge and best practice in this field are constantly changing. As new research and experience broaden our understanding, changes in research methods, professional practices, or medical treatment may become necessary.

Practitioners and researchers must always rely on their own experience and knowledge in evaluating and using any information, methods, compounds, or experiments described herein. In using such information or methods they should be mindful of their own safety and the safety of others, including parties for whom they have a professional responsibility.

To the fullest extent of the law, neither the Publisher nor the authors, contributors, or editors, assume any liability for any injury and/or damage to persons or property as a matter of products liability, negligence or otherwise, or from any use or operation of any methods, products, instructions, or ideas contained in the material herein.

Library of Congress Cataloging-in-Publication Data
A catalog record for this book is available from the Library of Congress

British Library Cataloguing-in-Publication Data
A catalogue record for this book is available from the British Library

ISBN 978-0-12-821452-7

Publisher: Charlotte Cockle
Editorial Project Manager: Lena Sparks
Production Project Manager: Maria Bernard
Cover Designer: Mark Rogers

Typeset by SPi Global, India

Dedication

This book is dedicated to my parents, my wife Silky, and lovely daughter Sristy for being a source of inspiration and strength.

Contents

Contributors

Numbers in parenthesis indicate the pages on which the authors' contributions begin.

Muhammad Zohaib Afzal (279), College of Agriculture, Fujian Agriculture and Forestry University, Fuzhou, Fujian, China

Heena Ambreen (151), National Institute of Plant Genome Research, New Delhi, India

Mehtab Muhammad Aslam (279), College of Life Sciences, Fujian Agriculture and Forestry University, Fuzhou, Fujian, China

Purva Bhalothia (257), Dr. B. Lal Institute of Biotechnology, Jaipur, India

Garima Bhatia (293), Department of Biotechnology, Panjab University, Chandigarh, India

Sabhyata Bhatia (151), National Institute of Plant Genome Research, New Delhi, India

Debasis Chakrabarty (93), Genetics and Molecular Biology Division, CSIR-National Botanical Research Institute, Lucknow, India

Anirban Chakraborty (151), National Institute of Plant Genome Research, New Delhi, India

Priyanka Dhakate (151), National Institute of Plant Genome Research, New Delhi, India

Mahpara Fatima (279), College of Agriculture, Fujian Agriculture and Forestry University, Fuzhou, Fujian, China

Kaushik Ghose (133,243), Fiber and Biopolymer Research Institute, Department of Plant and Soil Science, Texas Tech University, Lubbock, TX, United States

Neetu Goyal (293), Department of Biotechnology, Panjab University, Chandigarh, India

Aradhana Lucky Hans (209), Department of Biotechnology, Babasaheb Bhimrao Ambedkar University, Lucknow, Uttar Pradesh, India

Zhiqiang Hao (189), College of Life Science, Shaanxi Normal University, Xi'an, China

Jyoti Singh Jadaun (313), Botany Department, Dayanand Girls Postgraduate College, Kanpur, Uttar Pradesh, India

Shalini Tiwari (63), CSIR-National Botanical Research Institute, Lucknow, India

Elif Karlik (107), Department of Molecular Biology and Genetics, Istinye University, Istanbul, Turkey

Amandeep Kaur (339), Department of Botany, Panjab University, Chandigarh, India

Girija Kaushal (313), Center of Innovative and Applied Bioprocessing (CIAB), Mohali, Punjab, India

Vinod Kumar (225), Molecular Biology and Biotechnology Division, CSIR-National Botanical Research Institute, Lucknow, India

Deepika Lakhwani (35), University of Angers, France

Charu Lata (63), CSIR-National Institute of Science Communication and Information Resources, New Delhi, India

Guanglin Li (189), College of Life Science, Shaanxi Normal University, Xi'an, China

Madhu (339), Department of Botany, Panjab University, Chandigarh, India

Adity Majee (225), Molecular Biology and Biotechnology Division, CSIR-National Botanical Research Institute, Lucknow, India

Nikhil Malhotra (75), ICAR-National Bureau of Plant Genetic Resources Regional Station, Phagli, Shimla, Himachal Pradesh, India

Lokesh Kumar Narnoliya (313), Center of Innovative and Applied Bioprocessing (CIAB), Mohali, Punjab, India

Vimal Pandey (151), National Institute of Plant Genome Research, New Delhi, India

Muhammad Qasim (279), Key Lab of Molecular Biology of Crop Pathogens and Insects, Institute of Insect Sciences, College of Agriculture & Biotechnology, Zhejiang University, Hangzhou, PR China

Anshulika Rai (243), Department of Plant and Microbial Biology, University of Minnesota, Saint Paul, MN, United States

Krishan Mohan Rai (133,243), Fiber and Biopolymer Research Institute, Department of Plant and Soil Science, Texas Tech University, Lubbock, TX; Department of Plant and Microbial Biology, University of Minnesota, Saint Paul, MN, United States

Nelam Sajjad (279), College of Life Sciences, Fujian Agriculture and Forestry University, Fuzhou, Fujian, China

Sangeeta Saxena (209), Department of Biotechnology, Babasaheb Bhimrao Ambedkar University, Lucknow, Uttar Pradesh, India

Yasir Sharif (279), College of Plant Protection, Fujian Agriculture and Forestry University, Fuzhou, Fujian, China

Alok Sharma (15), Department of Botany, Panjab University, Chandigarh, India

Himanshu Sharma (15), Department of Botany, Panjab University, Chandigarh; I.K. Gujral Punjab Technical University, Jalandhar, India

Neha Sharma (197), Division of Crop Improvement, ICAR-Central Potato Research Institute, Shimla, India

Shivani Sharma (75), Himalayan Forest Research Institute, Panthaghati, Shimla, Himachal Pradesh, India

Shumayla (1,49), Department of Botany, Panjab University, Chandigarh, India

Sukhjeet Sidhu (15), Department of Biotechnology, SUSCET, Mohali, India

Baljinder Singh (151), National Institute of Plant Genome Research, New Delhi, India

Gourav Singh (151), National Institute of Plant Genome Research, New Delhi, India

Kashmir Singh (293), Department of Biotechnology, Panjab University, Chandigarh, India

Rahul Singh (243), Department of Basic and Translational Sciences, School of Dental Medicine, University of Pennsylvania, Philadelphia, PA, United States

Sudhir P. Singh (313), Center of Innovative and Applied Bioprocessing (CIAB), Mohali, Punjab, India

Mehak Taneja (1), Department of Botany, Panjab University, Chandigarh, India

Manish Tiwari (151), National Institute of Plant Genome Research, New Delhi, India

Shivi Tyagi (49), Department of Botany, Panjab University, Chandigarh, India

Santosh Kumar Upadhyay (1,15,49,293,339), Department of Botany, Panjab University, Chandigarh, India

Ira Vashisht (197), Crop Genetics and Information Group, School of Computational and Integrative Sciences, Jawaharlal Nehru University, New Delhi, India

Pankaj Kumar Verma (93), Department of Botany, University of Lucknow, Lucknow, India

Shikha Verma (93), Genetics and Molecular Biology Division, CSIR-National Botanical Research Institute, Lucknow, India

Yongjun Wu (177), College of Life Sciences, College of Science, College of Horticulture, Northwest A&F University, Xianyang, China

Manisha Yadav (151), National Institute of Plant Genome Research, New Delhi, India

Kalpesh Yajnik (257), Dr. B. Lal Institute of Biotechnology, Jaipur, India

Chengcheng Yang (177), College of Life Sciences, College of Science, College of Horticulture, Northwest A&F University, Xianyang, China

Zhao Yang (177), College of Life Sciences, College of Science, College of Horticulture, Northwest A&F University, Xianyang, China

Zhenchao Yang (177), College of Life Sciences, College of Science, College of Horticulture, Northwest A&F University, Xianyang, China

Madiha Zaynab (279), College of Life Sciences and Oceanography, Shenzhen University, Shenzhen, PR China

About the editor

Dr. Santosh Kumar Upadhyay is currently working as an Assistant Professor in the Department of Botany, Punjab University, Chandigarh, India. Prior to this, Dr. Upadhyay was DST-INSPIRE faculty at the National Agri-Food Biotechnology Institute, Mohali, Punjab, India. He did his doctoral work at the CSIR-National Botanical Research Institute, Lucknow and received his PhD in Biotechnology from UP Technical University, Lucknow, India. He has been working in the field of Plant Biotechnology for more than 12 years. His present research focuses on the area of functional genomics. He is involved in the characterization of various insect toxic proteins from plant biodiversity, and defense and stress signaling genes in bread wheat. His research group at PU has characterized numerous important gene families and long noncoding RNAs related to the abiotic and biotic stress tolerance and signaling in bread wheat. He has also established the method for genome editing in bread wheat using CRISPR-Cas system and developed a tool SSinder for CRISPR target site prediction. His research contribution led the publication of more than 50 research papers in leading journals of international repute. Further, he has in his credit more than five national and international patents, 13 book chapters, and two books. In recognition of his strong research record he has been awarded NAAS Young Scientist Award (2017–18) and NAAS-Associate (2018) from the National Academy of Agricultural Sciences, India, INSA Medal for Young Scientist (2013) from the Indian National Science Academy, India, NASI-Young Scientist Platinum Jubilee Award (2012) from the National Academy of Sciences, India, and Altech Young Scientist Award (2011). He has also been the recipient of the prestigious DST-INSPIRE Faculty Fellowship (2012), and SERB-Early Career Research Award (2016) from the Ministry of Science and Technology, Government of India. Dr. Upadhyay also serves as a member of the editorial board and reviewer of several peer-reviewed international journals.

Preface

Abiotic stresses such as heat, cold, drought, heavy metals, UV radiations, etc. and biotic stresses like bacterial and fungal pathogens and insect pests largely affect the productivity of various agricultural and horticultural crops. However, the ever-increasing human population regularly demands an increase in the yield and quality of products. Earlier, various methods including transgenic technology have been utilized to overcome these issues. The majority of them have used the expression of various protein-coding genes. However, in recent decade researchers have identified long noncoding RNAs (lncRNAs) in plants which are associated with various regulatory and defense-related pathways. They are more than 200 bp RNA which do not code for proteins, and they may be present in nucleus or cytoplasm. They are transcribed in similar fashion as the coding mRNA, though a few plant-specific RNA Pol IV and V are also known to be associated with lncRNAs' transcription.

Earlier lncRNAs have been extensively studied in animals for numerous functions, but rarely explored in plant system. However, in recent years, they have been identified and characterized in numerous agricultural and horticultural plants including rice, wheat, cucumber, legumes, grapes, etc. They are found to be associated with various developmental pathways, regulatory systems, abiotic and biotic stress response, and signaling as well. Nowadays, thousands of important lncRNAs are reported in plants and their catalogue is being prepared in various dedicated databases such as GreeNC, CANTATAdb, EVLncRNAs, etc.

Development of various new strategies to cope up with the adverse effects of various abiotic and biotic stresses has always been a top priority in recent years. Agronomical practices, chemical applications, and transgenic expression of protein-coding genes have been explored to alleviate these problems up to some extent. However, new and specific technology is always in demand. The lncRNAs can be an alternative to improve the stress tolerance in cash crops. Further, it will be highly plant-specific due to its nonprotein-coding nature. Taking together into the account of progress made in managing both abiotic and biotic stresses in plants, this book has compiled all the recent knowledge and development in identification, characterization, cataloguing in various databases, and their involvement and applications in stress management.

Further, due to the absence of any alternative book on lncRNAs of plant on the above-discussed aspects, it is presumed that this book will serve as an important repository for students, academicians, researchers, and scientists to augment their understanding and may facilitate further in-depth research. Though a lot of efforts have been made to improve this book, it is well known that the first version has always the scope for improvement. I shall be very happy to receive important suggestions for further improvement of this book.

Santosh Kumar Upadhyay,
Department of Botany, Panjab University, Chandigarh, India

Acknowledgments

I am thankful to the Punjab University, Chandigarh for providing facility to complete this book. I am grateful to all the esteemed authors for their exceptional contributions and reviewers for their critical evaluation and suggestions for the improvement of the quality of book.

I would like to thank Charlotte Cockle (Acquisitions Editor), Lena Sparks (Editorial Project Manager), Narmatha Mohan (Copyrights Coordinator), Maria Bernard (Production Project Manager) and Mark Rogers (Cover Designer) from Elsevier for their excellent management of this project and anonymous reviewers for their recommendation for this book.

I also appreciate the support of my students, whose discussion and comments helped to shape this book. I thank Dr. Sudhir P. Singh, Dr. Kashmir Singh, Dr. Ashutosh Pandey, and Dr. Prashant Mishra for direct or indirect help in shaping this project. I wish to express my gratitude to my parents and my beloved wife for her endless support, patience, and inspiration. I thank my daughter, who missed me during this project. I would like to warmly thank faculties and staffs of the department for providing a great working environment. Last but not least, my sincere thanks to Lord Krishna for endowing me to live with joy and success in the form of this book.

Chapter 1

An overview of long noncoding RNA in plants

Mehak Taneja*, Shumayla*, and Santosh Kumar Upadhyay
Department of Botany, Panjab University, Chandigarh, India

1.1 Introduction

Long noncoding RNAs (lncRNAs) came into light with the beginning of genomics, bioinformatics, and high-throughput sequencing technologies. A large number of transcriptional units with very little or negligible protein-coding potential were discovered (Wang et al., 2017). Such RNA molecules were referred to as the non-protein-coding RNAs or ncRNAs. The ncRNAs belong to two broad categories, i.e., housekeeping and regulatory ncRNAs. The housekeeping ncRNAs are denoted as structural ncRNAs consisting of tRNA, rRNA, snRNA, and snoRNA, which constitutively are expressed in all the cells (Ponting et al., 2009). The regulatory ncRNAs are further of various types based on their length, small ncRNAs that include micro RNAs (miRNAs; 22–23 bp), small interfering RNAs (siRNAs; 19–25 bp), and Piwi-associated RNAs (piRNA; 26–31 bp) and long noncoding RNAs (lncRNAs). The lncRNAs are more than or equal to 200 nt long and are the most prevalent and functionally diverse class of ncRNAs. LncRNAs are reported to occur in a large diversity of species including plants, animals, yeast, prokaryotes, and viruses as well (Clemson et al., 1996; Brown et al., 1992; Bernstein et al., 1993; Reeves et al., 2007; Houseley et al., 2008; Swiezewski et al., 2009). Depending on the origin from the genome, lncRNAs could be classified as: (i) intergenic lncRNAs (lincRNAs), (ii) intronic lncRNAs (incRNAs), and (iii) natural antisense transcripts (NATs). The latter is transcribed from the complementary DNA strand of the associated coding genes. NATs are further of two types, i.e., *cis* and *trans*. *Cis*-NATs are transcribed from the similar genomic loci as their sense transcripts, but on the opposite strand of DNA. *Trans*-NATs, on the contrary, are transcribed from genomic regions other than those encoding their sense transcripts (Wang et al., 2005). The lncRNAs are considered as an important regulator of various crucial biological processes including gene transcription, translation, splicing, protein localization, cell cycle, epigenomic regulation, plant flowering,

* Contributed equally.

Long Noncoding RNAs in Plants: Roles in development and stress
https://doi.org/10.1016/B978-0-12-821452-7.00001-5

reproductive development, and heat shock response (Yu and Zhu, 2019). They have been widely reported across the genomes of both higher and lower organisms including plants, human, mouse, and fly (Tupy et al., 2005; Dinger et al., 2008; Khalil et al., 2009; Gupta et al., 2010). Through genome-wide studies, more than 50,000 lncRNAs have been discovered in the human genome (Khalil et al., 2009; Iyer et al., 2015; Cabili et al., 2011). The lncRNAs in mammals are known to be transcribed by RNA polymerase II, processed through both 5′-capping and 3′ poly(A) addition, and possess introns as well (Guttman et al., 2009; Managadze et al., 2011; Ulitsky et al., 2011). The genome of plants is known to produce numerous amounts of lncRNAs from intergenic, intronic, and coding regions. Majority of the lncRNAs are known to be transcribed by RNA Pol II or III. However, plants also have Pol IV and Pol V that are plant-specific and produce lncRNAs (Wang and Chekanova, 2017). They are processed through splicing or non-splicing, polyadenylation or non-polyadenylation, and could be variably located, i.e., in the nucleus or cytoplasm (Liu et al., 2015a). The lncRNAs regulate gene expression by a range of mechanisms including mediation of chromatin-modifying activity and as potent cis- and trans-regulators of gene transcription. As gene regulators, lncRNAs are expressed in a tissue-specific manner and the majority of these are stress-regulated (Dinger et al., 2008; Cabili et al., 2011). The classification of lncRNAs is majorly based on four prominent features that include genomic location, function exerted on DNA or RNA, mechanism of function, and target mechanism (Ma et al., 2013). This chapter covers the overview of their origin, mechanism of action, discovery, and various biological roles.

1.2 Classification of lncRNAs

A vast amount of noncoding region interspersed among the coding region. Based on their genomic location and orientation with respect to the neighboring genes, lncRNAs could be of five types (Ma et al., 2013; Bhatia et al., 2017). *Intronic lncRNAs* are those transcribed solely from introns of protein-coding genes. *Intergenic lncRNAs* include those transcribed from intergenic regions, i.e., the genomic interval between the two genes. Owing to their mRNA-like features, the lincRNAs are the most abundant class of eukaryotic lncRNAs present in the poly(A)+ RNA seq data (Ulitsky and Bartel, 2013). Further, there are sense, antisense, and bidirectional lncRNAs. *Sense lncRNAs* are transcribed from the sense strand and harboring exons from the protein-coding genes. They might show overlapping with a part of the entire sequence or cover the entire sequence of a protein-coding gene. *Antisense lncRNAs,* on contrary, are transcribed from the antisense strand of protein-coding genes. *Bidirectional lncRNAs* are transcribed from the antisense strand of protein-coding genes whose transcription initiation site is located less than 1000 bp away (Ponting et al., 2009).

1.3 Mode of action of lncRNAs

The action mechanism of lncRNAs divides them into three group, i.e., transcriptional regulation or transcription only, posttranscriptional regulation or *cis* acting, and *trans* acting (Tov-Perry and Ulitsky, 2016). The lncRNAs involved in transcriptional regulation could be further subdivided according to their mode of functioning: (i) transcriptional interference and (ii) chromatin remodelling. Besides this, there are enhancer RNAs (eRNAs) that are a set of lncRNAs transcribed from enhancers and act to positively regulate the transcription of genes. Posttranscriptional regulation mechanisms of lncRNAs include splicing regulation and translational control. LncRNAs are also implicated in posttranscriptional regulation through siRNA mechanism and interaction with miRNAs. Additionally, the lncRNAs might also function through other mechanisms including protein localization, telomere replication, and RNA interference. Ma et al. (2013) have extensively reviewed the mechanism of lncRNAs' functioning. These are also classified based on their mode of action and targeting mechanisms into: (1) signal: showing cell type-specific expression and respond to diverse stimuli, e.g., *Xist*, *COLDAIR*, (2) decoy: that obstructs the binding of protein target to DNA or miRNA binding to mRNAs, (3) guide: binding and directing the localization of ribonucleoprotein complex to specific targets, e.g., *HOTAIR*, (4) scaffold: act as a platform to bring multiple proteins together to form a discrete complex, e.g., *HOTAIR* (Fig. 1.1) (Wang and Chang, 2011).

1.4 Discovery of lncRNAs

The discovery of ncRNAs can be done through computational and experimental procedures. An initial step in the determination of the roles of lncRNA in the cells is its identification. However, tissue specificity and minor expression levels had made their discovery difficult. Hence, it is crucial that an appropriate choice in sample preparation and experimental methodology should be carefully made. Various genome-wide approaches employed for transcriptome analyses such as microarrays and RNA sequencing carried out in model organisms have revealed the non-protein-coding transcripts which occupy a majority of the eukaryotic transcriptome (Liu et al., 2015a). Further, the role played by the next-generation sequencing (NGS) technologies is also very crucial which works by producing high-throughput sequencing data in a time- and cost-efficient manner.

A traditional method of lncRNA identification involves creating a list of transcripts after generation and sequencing of cDNA libraries. For instance, the noncoding transcripts of mouse were catalogued using the FANTOM project at RIKEN, wherein 60,770 full-length cDNAs were subjected to Sanger sequencing. This resulted in identification of 15,000 novel transcripts, of which more than 70% were noncoding. Subsequently, the number identified had risen to 23,000 in FANTOM3 project (Okazaki et al., 2002; Carninci et al., 2005). Another classical method of discovery includes probing transcripts on tilling arrays. These arrays work by hybridizing cDNAs to microarray slides carrying tiled oligonucleotides that can be designed in a manner that they cover non-repetitive sequences of specific chromosomes or, potentially, the whole genome. For instance, HOTAIR lncRNA was discovered in similar manner

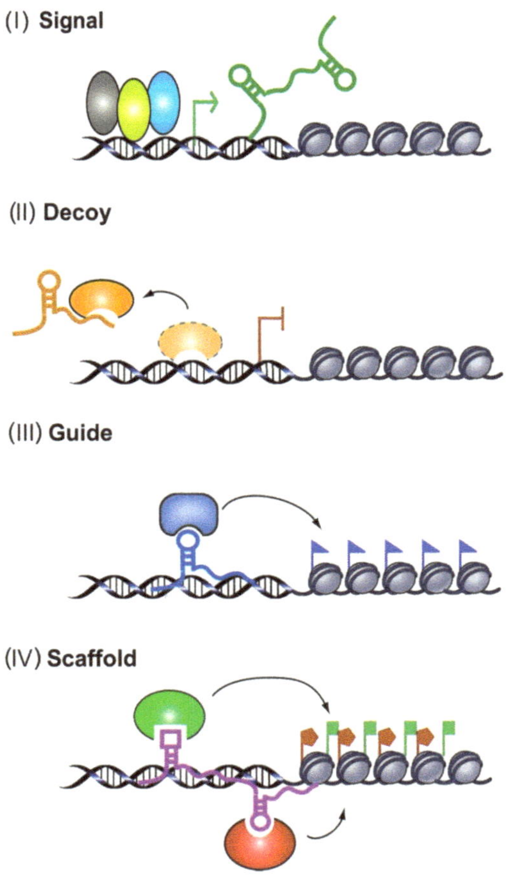

FIG. 1.1 The figure is depicting the mechanism of action of lncRNAs. (I) As signal: the interaction between transcription factors (oval-shaped) and signalling pathway, (II) as decoys: target proteins or miRNA bind to DNA or mRNA, (III) as guides: recruitment of chromatin-modifying enzymes to direct the localization of specific target gene, and (IV) as scaffolds: multiple proteins gather to form discrete complexes. *Figure is adapted from Wang, K.C., Chang, H.Y., 2011. Molecular mechanisms of long noncoding RNAs. Mol. Cell 43(6), 904–14 by permission from Elsevier with License Number 4754720773442.*

by employing a high-resolution tilling array with 400,000 overlapping probes (Rinn et al., 2007). RNA sequencing (RNA seq) is presently the most widely used method for RNA expression detection and for novel lncRNA discovery as well (Tyagi et al., 2018). The technique of RNA seq is based on the conversion of transcripts into a pool of cDNAs using random hexamers that would constitute the sequencing library. The Cap analysis of gene expression (CAGE) is another high-throughput screening method based on NGS that works by mapping and quantifying the expression of 5′-capped RNAs (Shiraki et al., 2003). Majority of the lncRNAs derived from the sense or antisense strand of protein-coding genes can be obtained using CAGE (Ma et al., 2013). As compared to

SAGE and RNA seq, this technique is an effective way to identify transcriptionally active promoter regions and Pol II-derived transcription start sites (Kashi et al., 2016). Some other methods of lncRNAs' discovery have been discussed in detail in Chapter 2.

1.5 Identification of lncRNAs in plants and their role in stress response

When compared with the other eukaryotic species, including yeast, model animal species, and humans, the reports of genome-wide identification of plant lncRNAs are not much comprehensive and date back to only recent times. In plants, genome-wide identification of ncRNAs has been done in *Arabidopsis thaliana* (Song et al., 2009; Wang et al., 2014a), *Medicago truncatula* (Wen et al., 2007), *Zea mays* (Boerner and McGinnis, 2012), *Triticum aestivum*, etc. (Xin et al., 2011; Shumayla et al., 2017). In Arabidopsis, 76 ncRNAs were identified, of which 14 are NATs, 34 lncRNAs considered as potential precursors of sRNAs, and 22 lncRNAs showing altered expression in response to various abiotic stress conditions (Ben Amor et al., 2009). Liu et al. (2012) reported ~ 6480 lncRNAs out of 13,230 intergenic transcripts in *Arabidopsis thaliana* using 200 transcriptomic data sets derived from tilling array experiments conducted using 14 Arabidopsis mutants, 18 variable stress treatments, and six different developmental stages. Further, the modulated expression of 1832 lncRNAs was observed upon exposure to various kinds of abiotic stress conditions like drought, salt, and ABA treatments (Liu et al., 2012). Genome-wide analysis revealed 37,238 NAT pairs in Arabidopsis, of which 626 concordant (sense and antisense transcripts expressing coordinately) and 766 discordant (induction of antisense transcript upon suppression of the sense transcript) NAT pairs were reported to be light-responsive (Wang et al., 2014a). In another study, 303 lncRNAs including 245 poly(A)+ and 58 poly (A)- lncRNA transcripts of Arabidopsis showed differential expression in response to various stress stimuli (Di et al., 2014). Through a strand-specific RNA sequencing approach, Zhu et al. (2014) reported 15 lncNATs and 20 novel intergenic TARs (transcriptionally active regions) or lincRNAs in Arabidopsis in response to *Fusarium oxysporum* infection. Results showed that antisense transcription is a prevalent phenomenon in Arabidopsis. Several lncNATs responsive to *F. oxysporum* infection are located in genes that had role in defense response and 80% of TARs are contiguous to or overlap with the annotated genes (Zhu et al., 2014). In rice, several novel lncRNAs are involved in reproduction and act as ceRNAs (competing endogenous RNAs) and have role in flowering and fertility process (Zhang et al., 2014). Wang et al. (2015) through both nondirectional and strand-specific RNA-sequencing experiments analyzed the noncoding transcriptome of rice and maize organs at different developmental stages. Various tissue and developmental stage-specific lincRNAs were identified in rice and maize in mature seeds, flower buds, leaves, and roots. In *Populus trichocarpa*, high-throughput RNA seq data reported

2542 lncRNAs. Of the total, 504 lncRNAs were drought-responsive and eight of which were experimentally validated. Also, 20 lncRNAs were predicted to be target mimics of miRNAs, while 51 lncRNAs were predicted as putative targets of miRNAs (Shuai et al., 2014). Another study involving identification and characterization of lncRNAs having role in osmotic and salt stress is reported in *Medicago truncatula*. A total of 23,324 putative lncRNAs are reported, of which 13,424 lncRNAs are identified to be specific for osmotic stress and 15,235 lncRNAs are salt stress-responsive in roots and leaves of *M. truncatula* (Wang et al., 2015). In *Brassica rapa*s sp. chinensis, strand-specific RNA sequencing revealed a total of 4594 putative lncRNAs in response to heat stress (Wang et al., 2019). Bhatia et al. (2020) through *in silico* approaches identified 7613 putative lncRNAs in the genome of *B. juncea,* of which 1614 lncRNAs are found to be differentially expressing under heat and drought stress conditions. Apart from this, in various other plants, the lncRNAs such as *MtENOD40*, *TPS11*, *OsPI1*, *AtIPS1*, *COOLAIR*, *COLDAIR*, *LDMAR*, etc. are known for their biological significance (Crespi et al., 1994; Liu et al., 1997; Wasaki et al., 2003; Franco-Zorrilla et al., 2007; Swiezewski et al., 2009; Heo and Sung, 2011; Ding et al., 2012; Zhou et al., 2012). Further, the details of lncRNAs in several other plant species are discussed in subsequent chapters.

1.6 Databases for plant lncRNA identification

An extensive review of the databases for plant ncRNAs is done by Liao et al. (2018) and Bhatia et al. (2017), some of which are discussed herein. As of now, more than 100 plant genomes have been sequenced, which include both crop and model plants. Majority of them are provided with genome databases, such as Arabidopsis, wheat, rice, soybean, and so on. Most of the ncRNA of these crops or model plants could be retrieved using these databases. The genome databases are also useful for surveying the functions of ncRNAs and to search their genomic composition. Thus, they serve as widely used references in ncRNA studies, e.g., TAIR-Arabidopsis gene structure and function annotation (Swarbreck et al., 2008). The major web resource present for the lncRNAs includes lncRNAdb, which is a database of lncRNAs that are speculated to have biological functions in eukaryotes (Amaral et al., 2011). The lncRNAdb v2.0 (Quek et al., 2015) serves as a reference database for more than 280 eukaryotic lncRNAs from scientific literature. This database aims to summarize the information about eukaryotic lncRNAs in a simple accessible format. Further, NONCODE (Xie et al., 2014; Zhao et al., 2016) is an integrated and interactive database that has data collected from the recently published literature and incorporates the latest version of several public databases including RefSeq, lncRNAdb, and GENCODE. Coding Potential Calculator (CPC) is a user-friendly web server that could be used for assessing the potential protein-coding transcripts. Further, CPC2 additionally has a feature of downloadable standalone package and is about 1000 times faster and exhibits greater accuracy than CPC1

(Kong et al., 2007; Kang et al., 2017). NPInter is a database that assembles the experimentally proven functional interactions between noncoding specifically lncRNAs and RNAs, except tRNAs and rRNAs, and biomolecules (proteins, RNAs, and genomic DNAs) and so on. These databases enable the user to have a better understanding of the ncRNAs and the details of the databases have been discussed in chapter 3.

1.7 Biological roles of lncRNAs in plants

The role of lncRNAs has been documented in wide array of important biological processes like regulation of flowering time, gene silencing, root organogenesis, seedling photo morphogenesis, and reproduction (Liu et al., 2015b). Few biological roles of plant lncRNAs have been discussed below.

1.7.1 Role of plant lncRNAs in the regulation of flowering

Owing to the significance of flowering time regulation in the life cycle of a plant, the lncRNAs involved in regulating flowering are the most widely studied class of plant lncRNAs. LncRNAs including COLD-INDUCED LONG ANTISENSE INTRAGENIC RNAs (COOLAIR) and COLD-ASSISTED INTRONIC NONCODING RNA (COLDAIR) are known to regulate the expression of the FLOWERING LOCUS C (FLC) at the time of vernalization. Sheldon et al. (2000) reported that, in Arabidopsis, FLC encodes a MADS-box transcription factor and is known to negatively regulate flowering during vernalization (Sheldon et al., 2000). A high expression of FLC results in delayed flowering; however, flowering is promoted upon the repression of FLC by vernalization. Both COLDAIR and COLDWRAP are necessary to retain Polycomb at the *FLC* promoter during vernalization through the formation of a repressive intragenic chromatin loop (Kim and Sung, 2017).

1.7.2 Role of plant lncRNAs in RNA-directed DNA methylation

A distinct pathway in plants employing noncoding RNA in epigenetic regulation was recently discovered. The pathway is referred to as the RNA-directed DNA methylation (RdDM). Two RNA polymerases that are unique to plants, i.e., Pol IV and Pol V, are involved in the production of 24-nt siRNAs and lncRNAs, respectively (Zhang et al., 2013; Matzke et al., 2014). The RdDM pathway is essential for maintaining silencing of transposons and repetitive elements to regulate biological processes such as allelic crosstalk and antiviral defense. This pathway is known to regulate gene expression under stress conditions in Arabidopsis. NRPD1 (Nuclear RNA Polymerase D) and NRPE1 (Nuclear RNA Polymerase E) correspond to the distinct largest subunits of Pol IV and Pol V, respectively. Under the conditions of heat stress, the expression of *NRPD1* and *NRPE1* is upregulated, resulting in increased methylation of AtSIN1 (SINE-like retrotransposon) (Naydenov et al., 2015). In maize, RNA-mediated transcriptional silencing has been implicated in paramutation. The mutant *mediator of paramutation 1* (*mop1*) is known to be orthologous to the Arabidopsis RNA-dependent RNA polymerase that acts in RdDM and is necessary in the gene silencing (Quan et al., 2015; Alleman et al., 2006; Dorweiler et al., 2000).

1.7.3 Role of plant lncRNAs as miRNA target mimics

Plant lncRNAs are shown to interact with miRNA competitively and act as miRNA target mimics. Target mimicry is a mechanism towards regulating the functions of miRNA, during which interaction between miRNA and their authentic targets is blocked through the binding of decoy RNAs to miRNAs by partially complementary sequence. A similar mechanism operating through competing endogenous RNAs (ceRNAs) was also identified in human and animal cells, indicating that target mimicry is a widespread phenomenon (Franco-Zorrilla et al., 2007). In Arabidopsis, upon phosphate starvation, miR399 is induced to mediate cleavage of *PHO2* mRNA that encodes an E2 ubiquitin conjugase-related enzyme. Low activity of *PHO2* resulted in higher expression of root-specific phosphate transporter genes, which led to an increase in phosphate uptake. Phosphate starvation induces the activity of lncRNA IPS1 (Induced by Phosphate Starvation 1), which has a conserved motif with sequence complementarity to miR399, except a mismatched central loop that is the site of cleavage. This mismatch blocks the miR399-mediated cleavage of *IPS1,* and at the same time, enables *IPS1* to function as miRNA target mimic and weaken the miR399-mediated repression of *PHO2*. Hence, the overexpression of *IPS1* relieves *PHO2* mRNA degradation which results in lower shoot Pi content in response to phosphate starvation.

1.7.4 Role of plant lncRNAs in photo morphogenesis

Wang et al. (2014b) identified and characterized a novel 236 nt lncRNA, named *HIDDEN TREASURE 1(HID1),* which acts as a positive regulator of photo morphogenesis. Loss-of-function of *HID1* yields a mutant named *hid1,* exhibiting hypo-photomorphogenic phenotype under continuous red light. *HID1* functions by negatively regulating the expression of *PHYTOCHROME-INTERACTING FACTOR 3*(*PIF3*). *PIF3* encodes a transcription factor that is found to be a significant repressor of photo morphogenesis. *HID1* is located in the nucleus, and upon binding to the promoter region of *PIF3*, represses the expression of latter. Further, it is seen that *HID1* is both structurally and functionally conserved across dicots and monocot species, as *OsHID1* rescued the elongated hypocotyl phenotype of the *hid1* mutant in Arabidopsis (Wang et al., 2014b).

1.7.5 Role of plant lncRNAs in alternative splicing

Alternative splicing is an important mechanism for gene expression in plants and cueing developmental plasticity responses to environmental signals (Bardou et al., 2014; James et al., 2012; Syed et al., 2012). It plays crucial role in regulating the structure and function of proteins. Alternative splicing of pre-mRNA is an important mechanism contributing to the increased complexity of transcriptome and proteome. The involvement of lncRNA in alternative splicing in Arabidopsis was reported by Bardou et al. (2014). They identified that nuclear speckle RNA-binding proteins (NSRs) are a family of RNA-binding proteins (RBPs) that act as alternative splicing regulators in Arabidopsis. They focussed on the Arabidopsis alternative splicing regulators, AtNSR a and b, originally reported in the legume, *Medicago truncatula* (Bardou et al., 2014). The NSRs localize within nuclear compartments referred to as speckles. They exist in both animal and plant cells and act as splicing factor reservoirs.

1.7.6 Role of plant lncRNAs in modulation of chromatin loop dynamics

Eukaryotic gene expression pattern depends on the interaction between transcription factors and the DNA, which adopts diverse 3D chromatin structural conformations (Cantone and Fisher, 2013). The pattern of gene expression is determined by the dynamics of chromatin structure and topology (Cavalli and Misteli, 2013). The lncRNAs reported in Arabidopsis from full-length cDNA databases through employing bioinformatic approaches consist of one such lncRNA, npc34 (Ben Amor et al., 2009). The latter was renamed as AUXIN REGULATED PROMOTER LOOP (APOLO) (Ariel et al., 2014). This lincRNA is transcribed by RNA Pol II and Pol V. It works by inducing the silencing of neighboring gene *PID*, an important regulator of polar auxin transport and referred to as *PINOID*. *APOLO* mediates the silencing of *PINOID* through chromatin folding, loop formation, and promoter methylation.

Further, the detailed biological roles of lncRNAs in different plant species have been discussed in the subsequent chapters.

1.8 Conclusions and future perspectives

Considerable progress has been made in the discovery of lncRNAs with the advent of high-throughput sequencing approaches and computational tools. The latter have paved the way for predicting the function, location, and type of lncRNAs. Also, genome-wide identification and transcriptome analyses in plants have revealed the role of lncRNAs in various stress conditions including both biotic and abiotic stresses. LncRNAs are also known to play several key molecular functions in plants including their role as scaffolds to cause chromatin modifications, in mediating protein relocalization, and in target mimicry. Besides, the role of lncRNAs in secondary metabolism and fruit ripening is also emerging, which requires detailed study to establish their mode of action in regulating these processes. Thus, lncRNAs serve myriad of functions in plants. However, studies unveiling the mechanism of lncRNA functioning in plants are still lagging behind when compared to the studies done in case of animals. Therefore, in future studies, identification of novel lncRNAs and their functional validation in plants needs to be performed. This, in turn, would enable the use of lncRNAs in expanding agricultural production and understanding other complex phenomenon.

Acknowledgments

Authors are thankful to Panjab University, Chandigarh, for providing facility. MT and S are thankful to UGC and CSIR for research fellowship. SKU is thankful to the Department of Science and Technology, Government of India, for partial financial support under Promotion of University Research and Scientific Excellence (PURSE) grant scheme.

References

Alleman, M., Sidorenko, L., McGinnis, K., Seshadri, V., Dorweiler, J.E., White, J., et al., 2006. An RNA-dependent RNA polymerase is required for paramutation in maize. Nature 442, 295–298.

Amaral, P.P., Clark, M.B., Gascoigne, D.K., Dinger, M.E., Mattick, J.S., 2011. LncRNAdb: a reference database for long noncoding RNAs. Nucleic Acids Res. 39, D146–D151.

Ariel, F., Jegu, T., Latrasse, D., Romero-Barrios, N., Christ, A., Benhamed, M., et al., 2014. Noncoding transcription by alternative RNA polymerases dynamically regulates an auxin-driven chromatin loop. Mol. Cell 55 (3), 383–396.

Bardou, F., Ariel, F., Simpson, C.G., Romero-Barrios, N., Laporte, P., Balzergue, S., et al., 2014. Long noncoding RNA modulates alternative splicing regulators in Arabidopsis. Dev. Cell 30 (2), 166–176.

Ben Amor, B., Wirth, S., Merchan, F., Laporte, P., d'Aubenton-Carafa, Y., Hirsch, J., et al., 2009. Novel long non-protein coding RNAs involved in Arabidopsis differentiation and stress responses. Genome Res. 19 (1), 57–69.

Bernstein, H.D., Zopf, D., Freymann, D.M., Walter, P., 1993. Functional substitution of the signal recognition particle 54-kDa subunit by its *Escherichia coli* homolog. Proc. Natl. Acad. Sci. U. S. A. 90, 5229–5233.

Bhatia, G., Goyal, N., Sharma, S., Upadhyay, S.K., Singh, K., 2017. Present scenario of long noncoding RNAs in plants. Noncoding RNA 3 (2), E16.

Bhatia, G., Singh, A., Verma, D., Sharma, S., Singh, K., 2020. Genome-wide investigation of regulatory roles of lncRNAs in response to heat and drought stress in *Brassica juncea* (Indian mustard). Environ. Exp. Bot. 171, 103922.

Boerner, S., McGinnis, K.M., 2012. Computational identification and functional predictions of long noncoding RNA in *Zea mays*. PLoS One 7, e43047.

Brown, C.J., Hendrich, B.D., Rupert, J.L., Lafrenière, R.G., Xing, Y., Lawrence, J., et al., 1992. The human XIST gene: analysis of a 17 kb inactive X-specific RNA that contains conserved repeats and is highly localized within the nucleus. Cell 71, 527–542.

Cabili, M.N., Trapnell, C., Goff, L., Koziol, M., Tazon-Vega, B., Regev, A., et al., 2011. Integrative annotation of human large intergenic noncoding RNAs reveals global properties and specific subclasses. Genes Dev. 25, 1915–1927.

Cantone, I., Fisher, A.G., 2013. Epigenetic programming and reprogramming during development. Nat. Struct. Mol. Biol. 20 (3), 282–289.

Carninci, P., Kasukawa, T., Katayama, S., Gough, J., Frith, M.C., Maeda, N., et al., 2005. The transcriptional landscape of the mammalian genome. Science 309 (5740), 1559–1563.

Cavalli, G., Misteli, T., 2013. Functional implications of genome topology. Nat. Struct. Mol. Biol. 20, 290–299.

Clemson, C.M., McNeil, J.A., Willard, H.F., Lawrence, J.B., 1996. XIST RNA paints the inactive X chromosome at inter-phase: evidence for a novel RNA involved in nuclear/chromosome structure. J. Cell Biol. 132, 259–275.

Crespi, M.D., Jurkevitch, E., Poiret, M., d'Aubenton-Carafa, Y., Petrovics, G., Kondorosi, E., et al., 1994. *Enod40*, a gene expressed during nodule organogenesis, codes for a non-translatable RNA involved in plant growth. EMBO J. 13 (21), 5099–5112.

Di, C., Yuan, J., Wu, Y., Li, J., Lin, H., Hu, L., et al., 2014. Characterization of stress-responsive lncRNAs in *Arabidopsis thaliana* by integrating expression, epigenetic and structural features. Plant J. 80, 848–861.

Ding, J., Lu, Q., Ouyang, Y., Mao, H., Zhang, P., Yao, J., et al., 2012. A long noncoding RNA regulates photoperiod-sensitive male sterility, an essential component of hybrid rice. Proc. Natl. Acad. Sci. U. S. A. 109, 2654–2659.

Dinger, M.E., Amaral, P.P., Mercer, T.R., Pang, K.C., Bruce, S.J., Gardiner, B.B., et al., 2008. Long noncoding RNAs in mouse embryonic stem cell pluripotency and differentiation. Genome Res. 18 (9), 1433–1445.

Dorweiler, J.E., Carey, C.C., Kubo, K.M., Hollick, J.B., Kermicle, J.L., Chandler, V.L., 2000. Mediator of paramutation1 is required for establishment and maintenance of paramutation at multiple maize loci. Plant Cell 12 (11), 2101–2118.

Franco-Zorrilla, J.M., Valli, A., Todesco, M., Mateos, I., Puga, M.I., Rubio-Somoza, I., et al., 2007. Target mimicry provides a new mechanism for regulation of microRNA activity. Nat. Genet. 39, 1033–1037.

Gupta, R.A., Shah, N., Wang, K.C., Kim, J., Horlings, H.M., Wong, D.J., et al., 2010. Long noncoding RNA HOTAIR reprograms chromatin state to promote cancer metastasis. Nature 464 (7291), 1071–1076.

Guttman, M., Amit, I., Garber, M., French, C., Lin, M.F., Feldser, D., et al., 2009. Chromatin signature reveals over a thousand highly conserved large non-coding RNAs in mammals. Nature 458 (7235), 223–227.

Heo, J.B., Sung, S., 2011. Vernalization-mediated epigenetic silencing by a long intronic noncoding RNA. Science 331 (6013), 76–79.

Houseley, J., Rubbi, L., Grunstein, M., Tollervey, D., Vogelauer, M., 2008. A ncRNA modulates histone modification and mRNA induction in the yeast GAL gene cluster. Mol. Cell 32, 685–695.

Iyer, M.K., Niknafs, Y.S., Malik, R., Singhal, U., Sahu, A., Hosono, Y., et al., 2015. The landscape of long noncoding RNAs in the human transcriptome. Nat. Genet. 47, 199–208.

James, A.B., Syed, N.H., Bordage, S., Marshall, J., Nimmo, G.A., Jenkins, G.I., et al., 2012. Alternative splicing mediates responses of the arabidopsis circadian clock to temperature changes. Plant Cell 24 (3), 961–981.

Kang, Y.J., Yang, D.C., Kong, L., Hou, M., Meng, Y.Q., Wei, L., et al., 2017. CPC2: a fast and accurate coding potential calculator based on sequence intrinsic features. Nucleic Acids Res. 45 (W1), W12–W16.

Kashi, K., Henderson, L., Bonetti, A., Carninci, P., 2016. Discovery and functional analysis of lncRNAs: methodologies to investigate an uncharacterized transcriptome. Biochim. Biophys. Acta 1859 (1), 3–15.

Khalil, A.M., Guttman, M., Huarte, M., Garber, M., Raj, A., Rivea, M.D., et al., 2009. Many human large intergenic noncoding RNAs associate with chromatin-modifying complexes and affect gene expression. Proc. Natl. Acad. Sci. U. S. A. 106, 11667–11672.

Kim, D.H., Sung, S., 2017. Vernalization-triggered intragenic chromatin loop formation by long noncoding RNAs. Dev. Cell 40 (3), 302–312.

Kong, L., Zhang, Y., Ye, Z.Q., Liu, X.Q., Zhao, S.Q., Wei, L., et al., 2007. CPC: assess the protein-coding potential of transcripts using sequence features and support vector machine. Nucleic Acids Res. 35, W345–W349.

Liao, P., Li, S., Cui, X., Zheng, Y., 2018. A comprehensive review of web-based resources of non-coding RNAs for plant science research. Int. J. Biol. Sci. 14 (8), 819–832.

Liu, C., Muchhal, U.S., Raghothama, K.G., 1997. Differential expression of *TPS11*, a phosphate starvation-induced gene in tomato. Plant Mol. Biol. 33 (5), 867–874.

Liu, J., Jung, C., Xu, J., Wang, H., Deng, S., Bernad, L., et al., 2012. Genome-wide analysis uncovers regulation of long intergenic noncoding RNAs in Arabidopsis. Plant Cell 24, 4333–4345.

Liu, J., Wang, H., Chua, N.H., 2015a. Long noncoding RNA transcriptome of plants. Plant Biotechnol. J. 13 (3), 319–328.

Liu, X., Hao, L., Li, D., Zhu, L., Hu, S., 2015b. Long non-coding RNAs and their biological roles in plants. Genomics Proteomics Bioinformatics 13 (3), 137–147.

Ma, L., Bajic, V.B., Zhang, Z., 2013. On the classification of long non-coding RNAs. RNA Biol. 10 (6), 925–933.

Managadze, D., Rogozin, I.B., Chernikova, D., Shabalina, S.A., Koonin, E.V., 2011. Negative correlation between expression level and evolutionary rate of long intergenic noncoding RNAs. Genome Biol. Evol. 3, 1390–1404.

Matzke, M.A., Kanno, T., Matzke, A.J., 2014. RNA-directed DNA methylation: the evolution of a complex epigenetic pathway in flowering plants. Annu. Rev. Plant Biol. 66, 243–267.

Naydenov, M., Baev, V., Apostolova, E., Gospodinova, N., Sablok, G., Gozmanova, M., et al., 2015. High-temperature effect on genes engaged in DNA methylation and affected by DNA methylation in Arabidopsis. Plant Physiol. Biochem. 87, 102–108.

Okazaki, Y., Furuno, M., Kasukawa, T., Adachi, J., Bono, H., Kondo, S., et al., 2002. Analysis of the mouse transcriptome based on functional annotation of 60,770 full-length cDNAs. Nature 420, 563–573.

Ponting, C.P., Oliver, P.L., Reik, W., 2009. Evolution and functions of long noncoding RNAs. Cell 136, 629–641.

Quan, M., Chen, J., Zhang, D., 2015. Exploring the secrets of long noncoding RNAs. Int. J. Mol. Sci. 16, 5467–5496.

Quek, X.C., Thomson, D.W., Maag, J.L., Bartonicek, N., Signal, B., Clark, M.B., et al., 2015. lncRNAdb v2.0: expanding the reference database for functional long noncoding RNAs. Nucleic Acids Res. 43, D168–D173.

Reeves, M.B., Davies, A.A., McSharry, B.P., Wilkinson, G.W., Sinclair, J.H., 2007. Complex I binding by a virally encoded RNA regulates mitochondria-induced cell death. Science 316, 1345–1348.

Rinn, J.L., Kertesz, M., Wang, J.K., Squazzo, S.L., Xu, X., Brugmann, S.A., et al., 2007. Functional demarcation of active and silent chromatin domains in human HOX loci by noncoding RNAs. Cell 129 (7), 1311–1323.

Sheldon, C.C., Rouse, D.T., Finnegan, E.J., Peacock, W.J., Dennis, E.S., 2000. The molecular basis of vernalization: the central role of flowering locus C (FLC). Proc. Natl. Acad. Sci. U. S. A. 97 (7), 3753–3758.

Shiraki, T., Kondo, S., Katayama, S., Waki, K., Kasukawa, T., Kawaji, H., et al., 2003. Cap analysis gene expression for high-throughput analysis of transcriptional starting point and identification of promoter usage. Proc. Natl. Acad. Sci. U. S. A. 100 (26), 15776–15781.

Shuai, P., Liang, D., Tang, S., Zhang, Z., Ye, C.Y., Su, Y., et al., 2014. Genome-wide identification and functional prediction of novel and drought-responsive lincRNAs in *Populus trichocarpa*. J. Exp. Bot. 65, 4975–4983.

Shumayla, Sharma, S., Taneja, M., Tyagi, S., Singh, K., Upadhyay, S.K., 2017. Survey of high throughput RNA-seq data reveals potential roles for lncRNAs during development and stress response in bread wheat. Front. Plant Sci. 8, 1019.

Song, D., Yang, Y., Yu, B., Zheng, B., Deng, Z., Lu, B.L., et al., 2009. Computational prediction of novel non-coding RNAs in Arabidopsis thaliana. BMC Bioinformat. 30 (10). 1:S36.

Swarbreck, D., Wilks, C., Lamesch, P., Berardini, T.Z., Garcia-Hernandez, M., Foerster, H., et al., 2008. The arabidopsis information resource (TAIR): gene structure and function annotation. Nucleic Acids Res. 36, D1009–D1014.

Swiezewski, S., Liu, F., Magusin, A., Dean, C., 2009. Cold-induced silencing by long antisense transcripts of an Arabidopsis Polycomb target. Nature 462 (7274), 799–802.

Syed, N.H., Kalyna, M., Marquez, Y., Barta, A., Brown, J.W., 2012. Alternative splicing in plants—coming of age. Trends Plant Sci. 17 (10), 616–623.

Tov-Perry, R.B., Ulitsky, I., 2016. The functions of long noncoding RNAs in development and stem cells. Development 143, 3882–3894.

Tupy, J.L., Bailey, A.M., Dailey, G., Evans-Holm, M., Siebel, C.W., Misra, S., et al., 2005. Identification of putative noncoding polyadenylated transcripts in *Drosophila melanogaster*. Proc. Natl. Acad. Sci. U. S. A. 102 (15), 5495–5500.

Tyagi, S., Sharma, A., Upadhyay, S.K., 2018. Role of next-generation RNA-Seq data in discovery and characterization of long non-coding RNA in plants. In: Çiftçi, Y.Ö. (Ed.), IntechOpen. Next Generation Plant Breeding, https://doi.org/10.5772/intechopen.72773.

Ulitsky, I., Bartel, D.P., 2013. lincRNAs: genomics, evolution, and mechanisms. Cell 154, 26–46.

Ulitsky, I., Shkumatava, A., Jan, C.H., Sive, H., Bartel, D.P., 2011. Conserved function of lincRNAs in vertebrate embryonic development despite rapid sequence evolution. Cell 147 (7), 1537–1550.

Wang, K.C., Chang, H.Y., 2011. Molecular mechanisms of long noncoding RNAs. Mol. Cell 43 (6), 904–914.

Wang, H.V., Chekanova, J.A., 2017. Long noncoding RNAs in plants. Adv. Exp. Med. Biol. 1008, 133–154.

Wang, X.J., Gaasterland, T., Chua, N.H., 2005. Genome-wide prediction and identification of cis-natural antisense transcripts in *Arabidopsis thaliana*. Genome Biol. 6, R30.

Wang, H., Chung, P.J., Liu, J., Jang, I.C., Kean, M.J., Xu, J., et al., 2014a. Genome-wide identification of long noncoding natural antisense transcripts and their responses to light in Arabidopsis. Genome Res. 24 (3), 444–453.

Wang, Y., Fan, X., Lin, F., He, G., Terzaghi, W., Zhu, D., et al., 2014b. Arabidopsis noncoding RNA mediates control of photomorphogenesis by red light. Proc. Natl. Acad. Sci. U. S. A. 111 (28), 10359–10364.

Wang, T.Z., Liu, M., Zhao, M.G., Chen, R., Zhang, W.H., 2015. Identification and characterization of long non-coding RNAs involved in osmotic and salt stress in *Medicago truncatula* using genome-wide high-throughput sequencing. BMC Plant Biol. 15, 131.

Wang, J., Meng, X., Dobrovolskaya, O.B., Orlov, Y.L., Chen, M., 2017. Non-coding RNAs and their roles in stress response in plants. Genomics Proteomics Bioinformatics 15 (5), 301–312.

Wang, A., Hu, J., Gao, C., Chen, G., Wang, B., Lin, C., et al., 2019. Genome-wide analysis of long non-coding RNAs unveils the regulatory roles in the heat tolerance of Chinese cabbage (*Brassica rapa* ssp. chinensis). Sci. Rep. 9 (1), 5002.

Wasaki, J., Yonetani, R., Shinano, T., Kai, M., Osaki, M., 2003. Expression of the *OsPI1* gene, cloned from rice roots using cDNA microarray, rapidly responds to phosphorus status. New Phytol. 158 (2), 239–248.

Wen, J., Parker, B.J., Weiller, G.F., 2007. In silico identification and characterization of mRNA-like noncoding transcripts in *Medicago truncatula*. In Silico Biol 7, 485–505.

Xie, C., Yuan, J., Li, H., Li, M., Zhao, G., Bu, D., et al., 2014. NONCODEv4: exploring the world of long non-coding RNA genes. Nucleic Acids Res. 42, D98–D103.

Xin, M., Wang, Y., Yao, Y., Song, N., Hu, Z., Qin, D., et al., 2011. Identification and characterization of wheat long non-protein coding RNAs responsive to powdery mildew infection and heat stress by using microarray analysis and SBS sequencing. BMC Plant Biol. 11, 61.

Yu, T., Zhu, H., 2019. Long non-coding RNAs: rising regulators of plant reproductive development. Agronomy 9, 53.

Zhang, H., He, X., Zhu, J.K., 2013. RNA-directed DNA methylation in plants: Where to start? RNA Biol. 10, 1593–1596.

Zhang, Y.C., Liao, J.Y., Li, Z.Y., Yu, Y., Zhang, J.P., Li, Q.F., et al., 2014. Genome-wide screening and functional analysis identify a large number of long noncoding RNAs involved in the sexual reproduction of rice. Genome Biol. 15 (12), 512.

Zhao, Y., Li, H., Fang, S., Kang, Y., Wu, W., Hao, Y., et al., 2016. NONCODE 2016: an informative and valuable data source of long non-coding RNAs. Nucleic Acids Res. 44 (D1), D203–D208.

Zhou, H., Liu, Q., Li, J., Jiang, D., Zhou, L., Wu, P., et al., 2012. Photoperiod- and thermo-sensitive genic male sterility in rice are caused by a point mutation in a novel noncoding RNA that produces a small RNA. Cell Res. 22, 649–660.

Zhu, Q.H., Stephen, S., Taylor, J., Helliwell, C.A., Wang, M.B., 2014. Long noncoding RNAs responsive to *Fusarium oxysporum* infection in *Arabidopsis thaliana*. New Phytol. 201, 574–584.

Chapter 2

Discovery and history of long noncoding RNAs in plants

Alok Sharma[a,*], Himanshu Sharma[a,b,*], Sukhjeet Sidhu[c], and Santosh Kumar Upadhyay[a]
[a]*Department of Botany, Panjab University, Chandigarh, India,* [b]*I.K. Gujral Punjab Technical University, Jalandhar, India,* [c]*Department of Biotechnology, SUSCET, Mohali, India*

2.1 Introduction

Bioinformatics and genomics have been known to be a boon for the society. The increasing advancements have opened up the doors to look deeply inside the genome of an organism. Conventionally, scientific societies have been focused on protein-coding part of the genome. For the world, noncoding genes were "junk" until the discovery of various roles of these noncoding genes. Noncoding genes, generally, are unable to get translated into a functional protein; however, their transcripts may have various regulatory roles (Bhatia et al., 2017; Tyagi et al., 2018). The noncoding genes, with the size greater than 200 nucleotides, are considered as long noncoding RNAs (lncRNAs). Further, lncRNAs can be classified into two ways, i.e., based on their function and their genomic locations (Ma et al., 2013; Rai et al., 2018). Function-based classification of lncRNAs includes *cis*-acting lncRNAs, target mimicry lncRNAs, lncRNAs as scaffold and guides, and lncRNAs as mediators of DNA methylation (Ma et al., 2013; Rai et al., 2018). On the basis of their genomic locations, lncRNAs are classified into sense and intronic lncRNAs, antisense lncRNAs, and intergenic lncRNAs (Ma et al., 2013; Rai et al., 2018). A number of studies have suggested the possible important roles of these lncRNAs in plants. In this chapter, we have compiled the reports of lncRNAs in various plants (Table 1).

2.2 First breakthrough in the discovery of lncRNAs

Major advancements in the experimental and computational approaches made it possible to find out the several noncoding RNAs in different plant species (Huttenhofer et al., 2002). Furthermore, other genome-wide study approaches started working towards the identification of lncRNA(s) in the variety of plant

* Equal contribution.

Long Noncoding RNAs in Plants: Roles in development and stress
https://doi.org/10.1016/B978-0-12-821452-7.00002-7

TABLE 1 Some of the identified lncRNAs in various plant groups.

Organism	Number of lncRNAs	Year of discovery	References
Algae			
Chlamydomonas reinhardtii	1440	2016	Li et al. (2016)
Bryophytes			
Physcomitrella patens	4648	2019	Simopoulos et al. (2019)
Pteridophytes			
Selaginella moellendorffii	4422	2017	Zhu et al. (2017)
Pteris vitata	381	2019	Yan et al. (2019)
Gymnosperms			
Ginkgo biloba	1323	2018	Wang et al. (2018)
Angiosperms			
Arabidopsis thaliana	6510	2018	Zhao et al. (2018)
Pisum sativum	47,322	2017	Kerr et al. (2017)
Brassica rapa	2040	2018	Liu et al. (2018)
Zea mays	1704	2014	Li et al. (2014)
Triticum aestivum	44,698	2017	Shumayla et al. (2017)

species. Next-generation sequencing has become a powerful tool for extracting the knowledge about the lncRNA.

In eukaryotes, *H19* was the first lncRNA gene to be discovered (Pachnis et al., 1984; Pachnis and Tilghman, 1988). The non-translational nature of *H19*, despite having the small open reading frames, made it odd; furthermore, the function of this gene remained unknown until the late 1990s (Jarroux et al., 2017). In 1990, Brannan and his team suggested that, despite having similarity with other RNAs in terms of transcription, these *H19* were not behaving like the typical mRNA and an RNA molecule might be the product of this odd gene (Brannan et al., 1990). In independent studies, two genes, *Igf2r* and *H19*, present in the *H19/Igf2r* cluster were found to be localized at mouse chromosome 7 (Bartolomei et al., 1991; Barlow et al., 1991). Overexpression of *H19* in transgenic mice was found to be lethal, which suggested strict regulation of this lncRNA (Brunkow and Tilghman, 1991). The function of *H19* was revealed after the functional characterization of another lncRNA involved in dosage compensation in mammals, i.e., X-inactive-specific transcript (*Xist*) (Borsani et al., 1991). The genomic imprinting techniques and various characterization studies suggested the important role of *H19* in dosage compensation. After these novel discoveries, the study for lncRNAs in different organisms was revolutionized.

LncRNAs have been widely studied in animals; however, in recent times, plant lncRNAs have been in the light of study. In plants, a *Glycine max* (soybean) gene, *GmENOD40*, of size 700 nucleotides is known to be the first reported lncRNA (Yang et al., 1993). Yang et al. found that there was the absence of 'AUG' in the ORF sequence of the cloned gene, which made the study of this lncRNA more intriguing. Additionally, the *GmENOD40* gene has been found to be involved in soybean nodule development. Moreover, this gene regulates the translocation of proteins from the nucleus to the cytoplasm (Yang et al., 1993). The discovery of *GmENOD40* made other researchers curious about the different roles of lncRNAs and lead to the discovery of many other lncRNAs till now. Introduction of high-throughput sequencing technologies boosted the studies related to the identification and characterization of lncRNA in plants (Tyagi et al., 2018).

2.3 LncRNAs in algae

Algae are primitive organisms having both unicellular and multicellular members. A unicellular member, *Chlamydomonas reinhardtii*, is known to be useful in understanding the mechanism of photosynthesis, flagella, and various other physiological processes (Merchant et al., 2007). In the year 2016, Li et al. reported a genome-wide identification of lncRNAs in *C. reinhardtii*. RNA sequencing resulted in the identification of a total of 1440 lncRNAs consisting of 936 intergenic, 310 intronic, and 194 antisense lncRNAs (Li et al., 2016). Genomic studies suggested that the average size of these lncRNAs was 509 nucleotides with 1-2 exons. Additionally, functional characterization resulted in the identification of 367 sulfur-deficit-responsive lncRNAs consisting of 36 lncRNAs related to photosynthesis. The study concluded the possible role of lncRNAs in *C. reinhardtii* under sulfur-deprived conditions (Li et al., 2016).

2.4 LncRNAs in bryophytes

In the lower group of plants such as bryophytes, comparatively less study has been carried out on lncRNAs. This may be attributed to the lack of availability of sequenced genome. However, a genome-wide study on bryophyte *Physcomitrella patens* has been reported. Recently, Simopoulos et al. (2019) have studied 11 plants including members of algae, bryophytes, and angiosperms. Only one bryophyte, *P. patens*, has been analyzed using 'classifying RNA by ensemble machine learning algorithm (CREMA)' approach. A total of 4648 lncRNAs were predicted, which was increased by 6.4% more lncRNA when CREMA approach was used (Simopoulos et al., 2019).

2.5 LncRNAs in pteridophytes

Despite the popularity of lncRNAs in the scientific world, only a few studies have been reported for lncRNAs' identification in pteridophytes. In a pioneer study, a lycophyte, '*Selaginella moellendorffii*,' has been studied comprehensively to unveil the coding and noncoding part of its genome. RNA-seq data from three tissues including root, stem, and leaf have been used for transcriptome analysis. The results indicate the presence of a total of 4422 lncRNAs in the transcriptome of *S. moellendorffii* (Zhu et al., 2017). They performed an expression analysis to understand their roles during tissue development. A total of 144, 39, and 17 lncRNAs were found to be expressed specifically in root, stem, and leaf tissues, respectively. These results indicated the intriguing roles of lncRNAs in tissue development. Further, another team of scientists has reported a total of 2269 lncRNAs in the genome of *S. moellendorfii* in their genome-wide analysis (Simopoulos et al., 2019). The difference in the predicted lncRNAs in *S. moellendorfii* may be due to the difference in used approaches and the removal of redundancy. Future genome-wide studies may clarify the actual number of lncRNAs in *S. moellendorfii*. Another pteridophyte, *Pteris vitata*, has been studied for its lncRNAs and their roles. *P. vitata* is an important pteridophyte, mainly known for its phytoremediating properties. Moreover, it is a hyperaccumulator of arsenic. In 2019, Yan and his team created a *de novo* full-length transcriptomic tonoplast proteomic database for *P. vitata* under arsenic stress. In their study, they have found 381 lncRNAs, which were suggested to be involved in glutathione regulation in root tissues. Thus, the *P. vitata* lncRNAs may have roles in arsenic hyperaccumulation for phytoremediation. The diverse roles of lncRNAs in plants make it more interesting to explore their roles in different plants under discrete stress conditions (Yan et al., 2019).

2.6 LncRNAs in gymnosperms

Gymnosperm is a group of nonflowering plants possessing naked seeds. *Ginkgo biloba*, also known as living fossil, is an important gymnosperm owing to its ornamental, economical, and medicinal properties. In 2018, Wang and his team carried out the identification and characterization of lncRNAs in *G. biloba*. They have reported 1323 novel lncRNAs using deep strand-specific RNA-sequencing method. Features including a lesser number of exons and shorter gene length as compared to mRNA genes have been found in these lncRNAs. All major developmental stages of *G. biloba* leaves have been used for obtaining data for lncRNA identification. The results suggest their important roles in the regulation of mRNAs involved in photosynthesis and secondary metabolite processes (Wang et al., 2018).

Recently, paired-end sequencing has been used by another research group for the identification of novel lncRNAs in the control and yellow mutant leaves of *G. biloba* (Wu et al., 2019). They have screened 2044 lncRNA sequences with an average length of 702 nucleotides. Out of these lncRNAs, 238 were found to be differentially expressed lncRNAs (DELs). Further, 48 *cis*-acting DELs and 31 *trans*-acting DELs were regulating 72 and 31 target genes, respectively. The study suggests that lncRNAs of *G. biloba* may have important roles in photosynthesis, oxidative phosphorylation, and leaf coloration mechanisms (Wu et al., 2019). Both the studies were focused on *G. biloba*; however, diversity of their roles motivates more lncRNAs-based researches in gymnosperms.

2.7 LncRNAs in angiosperms

Angiosperms are the most evolved and diverse group of plants. These are vascular seed-bearing plants, classified into the two groups, i.e., dicots and monocots, based on the occurrence of number of cotyledons. Here, we have included some researches focused on dicotyledonous and monocotyledonous lncRNAs.

2.7.1 Dicotyledonous lncRNAs

Arabidopsis, a model dicotyledonous plant, is known to be one of the most important plants in the context of research. There have been many long noncoding RNAs-based studies reported in *Arabidopsis thaliana* by various research groups. Thousands of lncRNAs have been identified in Arabidopsis till date. Further, various Arabidopsis lncRNA genes have been functionally characterized, including *IPS1*, *COLDAIR*, *COOLAIR*, *HID1*, etc.

Induced by phosphate starvation 1 (*IPS1*) gene is a noncoding RNA gene characterized in the year 2007 (Franco-Zorrilla et al., 2007). IPS1 contains a motif complementary to the sequence of phosphate (Pi) starvation-induced miRNA, miR-399. IPS1 inhibits the activity of miR-399 through the mechanism called 'target mimicry.' Furthermore, IPS1 is reported to be involved in phosphate homeostasis (Franco-Zorrilla et al., 2007). Another lncRNA, cold-induced long antisense intragenic RNA (COOLAIR), has been discovered and examined in Arabidopsis (Swiezewski et al., 2009). The researchers have tried to find out the link between noncoding RNA and regulation of chromatin through *FLC* (*Flowering locus C*) analysis in Arabidopsis. COOLAIR lncRNA was found to be involved in the regulation of flowering through the mechanism of epigenetic silencing and histone modifications (Swiezewski et al., 2009). Similar research includes the discovery of cold-assisted intronic noncoding RNA (COLDAIR) in Arabidopsis (Heo and Sung, 2011). The biological function of COLDAIR lncRNA is the regulation of flowering through promoter interference and epigenetic changes (Heo and Sung, 2011).

In the year 2014, Wang et al. discovered *Hidden Treasure 1* (*HID1*), a new lncRNA gene from Arabidopsis. *HID1* acts as a positive regulator of photomorphogenesis. It represses the transcription of phytochrome interaction factor 3 (*PIF3*) and modulates the red light-mediated photomorphogenesis (Wang et al., 2014). In the same year, a heat-inducible long noncoding antisense RNA, asHSFB2a, was reported to control the expression of heat shock factor HSFB2a in *A. thaliana* (Wunderlich et al., 2014). This lncRNA is ultimately involved in the regulation of gametophytic and vegetative development of plant.

In another study, alternative splicing competitor long noncoding RNA (ASCO-lncRNA) of Arabidopsis was discovered (Bardou et al., 2014). This lncRNA takes control over nuclear alternative splicing regulators to alter the pattern of alternative splicing during development. Besides, it regulates the development of lateral roots (Bardou et al., 2014). Furthermore, Ariel et al. suggested that Arabidopsis long intergenic noncoding RNA (lincRNA), APOLO, is transcribed by RNA polymerase II and V in response to the auxin hormone (Ariel et al., 2014). This transcription influences the chromatin topology and regulates the expression of its neighboring genes. Moreover, APOLO lncRNA is involved in the development through auxin signaling in Arabidopsis (Ariel et al., 2014).

ELF18-induced long noncoding RNA1 (ELENA1) is another lncRNA of Arabidopsis discovered in the year 2017 (Seo et al., 2017). ELENA1 has been found to be involved in resistance against *Pseudomonas syringe* pv tomato DC300. It increases the expression of genes involved in plant immunity including *pathogenesis-related gene 1* (*PR1*) and *pathogenesis-related gene 2* (*PR2*), β-1,3-glucanase encoding genes, and salicylic acid genes (Seo et al., 2017; Mach, 2017). Qin et al. (2017) discovered another Arabidopsis lncRNA, i.e., drought-induced lncRNA (DRIR), using RNA sequencing of salt-treated Arabidopsis seedlings. DRIR is mainly localized in the nucleus and its size is 755 nucleotides. Its expression has been found to be induced by salt and drought stress. *DRIR* modulates the expression of stress-responsive genes, such as ABA response, and water response; hence, it plays an important role in response to abiotic stress conditions (Qin et al., 2017). In the year 2017, Henriques et al. identified *CDF5 LONG NONCODING RNA* (*FLORE*), a circadian-regulated lncRNA. *FLORE* is a natural antisense transcript (NAT) of cycling DOF factor 5 (CDF5). FLORE promotes the flowering of Arabidopsis and is also known to be involved in photomorphogenesis (Henriques et al., 2017). In 2018, Zhao and his team claimed to identify 6510 lncRNAs in normal and stress conditions in Arabidopsis (Zhao et al., 2018).

Recently, 695 lncRNAs (215 novel lncRNAs) have been identified in the roots of Arabidopsis treated with nitrate (Liu et al., 2019). In addition, *T5120* lncRNA showed the highest expression, thus was further characterized. Overexpression of *T5120* resulted in an increase in nitrate assimilation and root growth (Liu et al., 2019). The study concluded the regulatory roles of *T5120* lncRNA in nitrate assimilation and response.

Along with the Arabidopsis, there have been a number of studies reported on lncRNAs in various angiosperms including pea, sea buckthorn, sunflower, potato, tomato, etc. In a study, Kerr et al. (2017) assembled a transcriptome of *Pisum sativum* (pea) axillary bud into 81,774 transcripts comprising 194,067 isoforms. Approximately, one quarter, i.e., 47,322, isoforms were found to be putative lncRNAs. The average length of these lncRNAs was predicted to be 293 nucleotides and in a range from 201 to 2781 nucleotides. Additionally, ten differentially expressed lncRNAs were predicted to be controlled by pseudo-response regulator 5, PRR5, which ultimately suggested their probable involvement in diurnal functions (Kerr et al., 2017).

Further, genome-wide analysis of lncRNAs at the mature stage of sea buckthorn (*Hippophae rhamnoides* L.) has been done in the year 2017 using RNA sequencing (Zhang et al., 2017). In this study, they have identified 3428 lncRNAs classified into 2498 intergenic, 593 antisense, and 337 intronic lncRNAs. Moreover, all three subtypes of lncRNAs have been analyzed based on their target genes. Among these lncRNAs, 23 showed differential expression in red fruit and 22 showed expression in yellow fruit. The study also suggests that some differentially expressed lncRNAs of sea buckthorn fruit play regulatory roles in genes involved in carotenoid biosynthesis, ascorbate and aldarate metabolism, and fatty acid metabolism pathways (Zhang et al., 2017).

A total of 34,304 transcripts were found to be unknown in sunflower during transcriptome analysis (Florez-Zapata et al., 2016). About 23,960 and 1367 unknown transcripts with or without genomic hits were identified as lncRNAs, respectively. Further, 18,128 of the lncRNAs with a genome hit showed evidence of expression in the three tested sunflower genotypes and 441 of the lncRNAs without a genome hit showed expression in all studied sunflower genotypes (Florez-Zapata et al., 2016). Moreover, 6895 lncRNAs were found to be exclusively expressed in meiocytes, which suggested the possible role of these lncRNAs in plant meiosis and chromatin modification processes along with other regulatory roles (Florez-Zapata et al., 2016).

Brassica rapa is an important part of family *Brassicaceae*. It includes a variety of subspecies cultivated worldwide as a vegetable, oil source, fodder, etc. A total of 2040 lncRNAs have been identified in *Brassica rapa* transcriptome using RNA-seq (Liu et al., 2018). 1961 lncRNAs have been found to be differentially expressed during vernalization. Furthermore, 254 lncRNAs were differentially expressed between the control and vernalization-treated groups of *B. rapa*. In addition, two target genes, *2C2-Dof* (Bra004396) and *BraZF-HD21* (Bra026812), have been found to be regulated by lncRNAs during vernalization (Liu et al., 2018). These vernalization-induced lncRNAs may have possible role in plant development such as flowering.

Non-heading Chinese cabbage (NHCC, *Brassica rapa ssp. chinensis*) is a subspecies of *Brassica rapa* and an important leafy vegetable. NHCC is one of the important vegetables in China and now it is cultivated worldwide. Song et al., in the year 2016, identified 10,001 lncRNAs using RNA-seq under cold and heat treatments in NHCC. Of these identified lncRNAs, 9687 were the novel lncRNAs. Additionally, a co-expression network for lncRNAs and miRNAs revealed that 192 and 67 genes were regulated by lncRNAs under heat and cold treatments, respectively, and thus, suggested their roles in abiotic stress conditions (Song et al., 2016).

Another angiosperm, *Brassica napus* (Canola), is an important oilseed crop cultivated worldwide. *Sclerotinia sclerotiorum* is a necrotrophic phytopathogen, which attacks the *Brassica napus* and causes sclerotinia stem rot (Del Rio et al., 2007). In the year 2016, Joshi et al. have performed the genome-wide identification and functional prediction of lncRNAs, which are responsive to the *S. sclerotiorum* infection in *B. napus*. A total of 3181 lncRNAs have been identified including 2821 intergenic lncRNAs, 111 natural antisense transcripts, 76 having exonic overlap with reference transcripts, and the other 173 novel lnc-isoforms. Further, 931 lncRNAs have been found to be differentially expressed in response to *S. sclerotiorum* infection; 12 out of them have been further validated with qRT-PCR (Joshi et al., 2016). The study has suggested the role of lncRNAs in response to biotic stress in *B. napus*. Further information regarding lncRNAs in brassicaceae can be found in the subsequent chapter entitled 'Role of Long noncoding RNAs in Brassica family.'

Kiwifruit (*Actinidia chinensis*) is an economically important fruit crop with great nutritional value and unique taste. A total of 7051 potential lncRNAs have been identified in kiwifruit transcriptome dataset using RNA-seq (Tang et al., 2016). These lncRNAs include 6009 intergenic, 597 overlapping, 169 natural antisense, 881 intronic, and 347 other lncRNAs. In another study, 14,845 lncRNA transcripts have been identified in kiwifruit plants infected by *Pseudomonas syringae pv. actinidiae* (Wang et al., 2017a). In addition, they have been classified as 10,238 intergenic lncRNAs, 1874 antisense lncRNAs, 75 intronic lncRNAs, 1434 overlapping lncRNAs, and 1805 other lncRNAs. Furthermore, only 2049 lncRNAs have been found to be overlapped with the lncRNAs reported by Tang et al. (2016). Besides, the mean length of these lncRNAs was found to be 774 bp with an average GC content of 36.78% (Wang et al., 2017a).

Orange fruit is famous worldwide for its economic and health values. In 2017, Wang et al. have screened and characterized long-noncoding RNAs in trifoliate orange (*Poncirus trifoliata* L. *Raf.*) (Wang et al., 2017b). Paired-end strand-specific RNA sequencing has been used for the identification of 6584 potential lncRNAs in precocious trifoliate orange and its wild-type. The size of these lncRNAs is in the range of 200–4026 nucleotides. Furthermore, 555 and 276 lncRNAs have been found to be upregulated and downregulated, respectively, in trifoliate orange, indicating their possible roles in flowering. Additionally, seven lncRNAs act as target mimics of citrus miRNAs and 59 lncRNAs have been identified as putative targets. The findings of this study provided the idea of possible roles of lncRNAs in orange fruit. Further detail about the lncRNAs in fruits is given in the chapter entitled 'Long noncoding RNAs in fruit crops.'

In 2016, Kwenda et al. identified 1113 lincRNAs in potato (*Solanum tuberosum*) using strand-specific RNA sequencing. Further, 559 lincRNAs were found to be responsive to *Pectobacterium carotovorum* subsp. *brasiliense* using a time-course RNA sequence analysis between a susceptible potato cultivar and a tolerant potato cultivar. Additionally, the co-expression analysis indicated that

17 lincRNAs were highly associated with 12 potato-defense-related genes. The whole study indicated that the potato lincRNAs have possible roles in plant defense mechanisms (Kwenda et al., 2016). Another plant, *Prunus persica* (peach), is an important deciduous fruit tree grown worldwide. Transcriptome analysis of peach revealed the presence of 2140 novel transcribed regions, out of which around 1416 were long noncoding RNAs (Wang et al., 2013).

Tomato is the most consuming vegetable after potato having characteristic features of fleshy fruit ripening, which is not present in any other plant (Bergougnoux, 2014). To understand the role of lncRNAs in the fleshy fruit ripening, Zhu et al. (2015) identified a total of 3679 lncRNAs from wild- and a mutant-type tomato. Expression profiling revealed that 490 lncRNAs were upregulated and 187 were downregulated in ripening mutant tomato. Moreover, silencing of lncRNA1459 and lncRNA1840 resulted in the delayed ripening of wild-type tomato, which suggested that the lncRNAs have some role in fruit ripening of tomato (Zhu et al., 2015). In another study, genome-wide identification of lncRNAs has been carried out in the tomato yellow leaf curl virus (TYLCV)-infected tomato plants. A total of 1565 lncRNAs were identified. A few identified lncRNAs were found to be working as endogenous target mimics (eTMs) for miRNAs involving in the regulation of TYLCV infection (Wang et al., 2015c). After analyzing both the studies, we can say that the lncRNAs have diverse roles including their role in fruit ripening as well as in the defense mechanism. With the history of diversity in the roles of lncRNAs, it will be interesting to see specific lncRNAs working in specific stress treatments.

In another such study, a tomato plant infected by *Phytophthora infestans* has been studied. *P. infestans* is a fungal pathogen, which causes the late blight disease in the tomato plant. To understand the role of lncRNAs in late blight disease resistance of tomato, Cui et al. (2017) analyzed the comparative transcriptome of resistant and susceptible tomato. A total of 688 lncRNAs were found to be differentially expressed in resistant variety as compared to the susceptible variety. Additionally, *lncRNA16367* has been reported to induce *glutaredoxin (GRX)*, which in turn reduces the level of Reactive oxygen species (ROS) to ultimately play a role in disease resistance (Cui et al., 2017). More comprehensive studies are required to further get the knowledge about lncRNAs to unveil the different pending questions regarding long noncoding RNAs. The bioinformatic revolution made it nearly possible to explore this area with more interest. Next-generation sequencing is such another advanced approach, which allows us to look into the genome more comprehensively.

Illumina-sequencing method was used by Scarano et al. (2017) for transcriptome analysis of tomato. In their analysis, they have found a total of 39,095 transcripts. Out of 39,095 transcripts, 3514 new lncRNAs have been reported to be novel and were not annotated earlier.

To understand the functions of a single gene in plants, virus-induced gene silencing (VIGS) is a well-known method. It is a molecular tool that uses the cleaving property of plant RNA silencing machinery to modulate the expression of the desired gene. In year the 2017, Zheng et al. observed the changes in the transcriptome of tomato using a VIGS vector based upon the tobacco rattle virus (TRV). This VIGS vector-mediated approach resulted in the change in expression of coding and noncoding genes. A total of 22 lncRNAs with altered expression were reported in this study (Zheng et al., 2017). Further research is expected for better knowledge about other different intriguing roles of lncRNAs in tomato and other plants. The detailed information about tomato lncRNAs is discussed in the chapter entitled "Long noncoding RNAs in Tomato: roles in development and stress response."

Cotton (*Gossypium* spp.) is a well-known crop for its worldwide fiber production. We can also say that fiber is the lifeline for textile industries. With the idea that lncRNAs may play an important role in fiber development, some scientists have given important researches in this area. For instance, in the year 2015, Wang and his research group studied *Gossypium barbadense* and found 30,550 and 4718 lincRNAs and NAT, respectively. Their findings based upon the expression and co-expression analysis concluded that some cotton lncRNAs might be involved in the initiation and elongation of fiber growth (Wang et al., 2015d). Another species of cotton, i.e., *G. arboreum,* has been thoroughly studied (Zou et al., 2016). They have used a strand-specific RNA-sequencing approach on RNA samples isolated from cotton fibers and leaves. A total of 5996 lncRNAs have been identified including 3510 lincRNAs and 2486 NATs. Modulated expression of lncRNAs in fiber developmental stages depicts their role in fiber development (Zou et al., 2016). Further information is provided in the chapter entitled "Long noncoding RNAs' involvement in comprehensive development of Gossypium species."

Chickpea, a legume, has been studied comprehensively using RNA seq data from three tissues and eight consecutive stages of flower development (Khemka et al., 2016). Genome-wide analysis identified 2248 lincRNAs in chickpea. Furthermore, expression profiling and co-expression analysis suggested the role of these lincRNAs in flower development (Khemka et al., 2016). Another dicotyledonous plant, *Cucumis sativus,* is known to be an important model plant to study fruit development and sex determination processes. A genome-wide study on *Cucumis sativus* reported a total of 3274 lincRNAs. In addition, co-expression analysis suggested the role of these lincRNAs in growth and reproductive processes (Hao et al., 2015). Further detail is given in chapter "Genome-wide identification and functional analysis of lincRNAs in Cucumber."

Furthermore, 2542 lincRNAs were identified in populus tree under control and drought stress conditions. Out of these, 504 lincRNAs showed differential expression under drought response (Shuai et al., 2014). In 2015, Chen et al. carried out genome-wide identification of lncRNAs in *Populus tomentosa.* They have reported 1377 putative lncRNAs in the xylem tissues of normal, opposite, and tension wood. Functional prediction and target gene analysis have predicted their roles in various physiological processes and wood development (Chen et al., 2015). *P. tomentosa* has also been analyzed for finding out the role of lncRNAs in hormonal responses. RNAs from leaves of control and gibberellic acid (GA)-treated plants were used for the high-throughput RNA sequencing. A total of 7655 lncRNAs have been identified including 410 differentially expressed lncRNAs in response to GA treatment. The functional prediction has suggested their role in growth and hormonal signal transduction (Tian et al., 2016). In addition, lncRNAs have also been found to be involved in nitrogen deficiency response in *P. tomentosa* (Chen et al., 2016). A total of 388 *P. tomentosa* lncRNAs have been identified in response to low nitrogen conditions. Out of these lncRNAs, 126 have shown differential expression in nitrogen deficiency. Furthermore, functional annotation suggested their role in regulatory mechanisms as well as in miRNAs precursors (Chen et al., 2016). In a recent conducted experiment, salt stress treatment in two populus species, *P. euphratica* and *P. alba* var. *pyramidalis,* led to the identification of 10,646 and 10,531 lncRNAs, respectively. Further, the differential expression of some lncRNAs in salt stress suggested their role in salt stress resistance (Ma et al., 2019).

2.7.2 Monocotyledonous lncRNAs

Zea mays (maize) is an important cereal crop, which was studied by various groups to identify their lncRNAs. Ma et al. (2008) found that *Zm401,* a possible lncRNA, is involved in the development of male gametophyte and anther development. In a study, Boerner and McGinnis (2012) used the dual approach of python language and SVM tool for the prediction of noncoding RNAs in the genome of maize. A total of 1101 lncRNAs were predicted in maize (Boerner and McGinnis, 2012). In the year 2014, Li et al. carried out genome-wide identification and characterization of lncRNAs and identified 1704 high confidence lncRNAs, which were the possible precursors of small RNAs. Furthermore, a detailed exploration of transcriptome under the drought conditions in maize was done by a group of scientists in 2014 (Zhang et al., 2014a). Drought stress treatments resulted in the identification of 8449 transcripts responsive to drought stress. Moreover, 664 transcripts were the true lncRNAs. Comparative and expression analysis found that, out of 664 transcripts, 538 lncRNAs were novel. Most of these novel lncRNAs were found to be upregulated in response to drought stress. Functional annotation analysis suggested their role as the precursors of miRNAs (Zhang et al., 2014a). In 2015, Wang and his team analyzed the noncoding transcriptome of two important cereal crops, maize and rice (Wang et al., 2015b). Their combined approach of using both nondirectional and strand-specific RNA-sequencing method resulted in the identification of a total of 22,334 and 6673 lincRNAs and NATs, respectively. The coding genes, which are related to these lncRNAs, were mostly found to be involved in stress and developmental processes depicting their similar roles (Wang et al., 2015b).

The difference in timing, as well as the diversity of genes expressed in various tissues, depicts their numerous roles in the developmental processes of plants. In 2017, Kim et al. carried out the transcription profiling of three different types of cells in endosperm at 8, 12, and 16 days after pollination in maize plants. A total of 1540 new lncRNAs were identified, out of which 1312 lncRNAs showed differential expression in these cells at different times. Spatio-temporal expression of these transcripts suggested their roles in plant development (Kim et al., 2017). Although there have been many studies in the context of maize lncRNAs, there is a lot of scope in exploring the transcriptome under various conditions. Additional information for maize lncRNAs can be found in chapter "Roles of Long noncoding RNA during stress response in Maize."

In another such study, *Setaria italica* (a member of the grass family) has been studied for its lncRNAs under drought stress condition. Genome-wide analysis of lncRNAs using *S. italica* transcriptome (under drought stress condition) has resulted in the identification of 584 lncRNAs. On the basis of expression pattern, 17 lincRNAs and two NATs have been identified to be differentially expressed during the drought stress condition. The results suggest the possible involvement of these lncRNAs in combating the water-deficit conditions (Qi et al., 2013). Further, in the case of Barley, a lncRNA named *HvCesA6* lnc-NAT is reported to act as a precursor of siRNAs and is involved in the cell wall formation of barley (Held et al., 2008).

Another intriguing result of a lncRNAs-based study was the involvement of lncRNAs in photoperiodism of rice plants. Photoperiod is the time duration of light, which plants require for their appropriate growth and development. Photoperiodism is an important factor in the regulation of flowering in plants; however, the role of photoperiodism in other developmental processes is studied lesser than in flowering. In a study of hybrid rice, Ding et al. (2012) have reported that the transcription level of a lncRNA, long-day-specific male-fertility-associated RNA (LDMAR), is critical for normal pollen development in long-day conditions of plants. It is reported that a single nucleotide mutation in the promoter region of LDMAR decreased its transcript level. Furthermore, this causes apoptosis in premature anthers, leading to the photoperiod-sensitive male sterility (PSMS) in plants. This study suggests the role of LDMAR lncRNA in the fertility of rice plants (Ding et al., 2012). A range of functions about lncRNAs has been discussed and proved in different research publications. Another such study includes the role of rice lncRNAs in the tolerance of cadmium stress. Cadmium (Cd) is a nonessential metal for plants and it is highly toxic to both plants and animals. Cadmium accumulation affects the development of plants by various mechanisms like ROS formation, enzyme activation, etc. (Wang et al., 2015a; Hall, 2002; Boominathan and Doran, 2003). RNA seq data generated from cadmium-treated roots of rice were analyzed to understand the role of lncRNAs in cadmium stress (He et al., 2015). Nine lncRNAs showed the relation with cadmium stress depicting their role in cadmium stress tolerance in rice plant.

Zhang et al. (2014b) analyzed the rice genome to identify the lncRNAs and their association with sexual reproduction. In their experiment, they have carried out the strand-specific RNA seq of the whole transcriptome of rice using four tissues. They have used RNA samples from anthers, pistils, five days old post-pollinated seeds, and 14 days post-germinated shoots. A total of 2224 lncRNAs, including the 1624 lincRNAs and 700 lncNATs, were identified in this genome-wide study (Zhang et al., 2014b). Further, the functional annotation analysis showed the role of lncRNAs in target mimicry. Most importantly, *XLOC_057324*, a lncRNA, is revealed to be involved in the panicle development. In other studies, rice lncRNAs were also found to regulate the phosphate homeostasis in rice (Wasaki et al., 2003; Jabnoune et al., 2013). Chapter 'Recent advancements in long noncoding RNA-mediated abiotic stress tolerance in rice' provides further information about rice lncRNAs.

Triticum aestivum is an important cereal consumed globally. It is a staple crop and has a complex genome attributed to its allohexaploid nature. In 2011, Xin et al. identified lncRNAs in wheat, in response to powdery mildew and heat stress. Out of 125 lncRNAs, 23 lncRNAs were found to be responsive to both stresses. Further, functional prediction suggested some of these lncRNAs to be precursor of small RNAs (Xin et al., 2011). In 2016, Zhang et al. conducted a study in *T. aestivum*, for the identification and functional characterization of novel lncRNAs involved in fungal resistance. Two main fungal pathogens of wheat, *Puccinia striiformis* (*Pst*) and *Blumeria graminis* (*Bgt*), responsible for stripe rust and powdery mildew disease, respectively, were used in this study. In their findings, they have reported the 58,218 putative lincRNAs. Further, expression profiling found the differential expression of 283 lincRNAs, including 254 and 52 lincRNAs expressing against *Bgt* and *Pst* infection, respectively. Additionally, 23 lincRNAs were found to be expressed in both *Bgt* and *Pst* infections (Zhang et al., 2016). A total of 44,698 lncRNAs have been identified in the genome of *T. aestivum* (Shumayla et al., 2017). Expression profiling in different stress conditions, including heat, drought, and salt stresses, suggested the possible roles of lncRNAs in abiotic stresses. In addition, co-expression studies depicted their role in stress resistance. Furthermore, a few lncRNAs were predicted to be precursor of miRNAs (Shumayla et al., 2017). Chapter "Long noncoding RNAs in wheat and wild relatives" covered the further details about lncRNAs in *Triticum*.

2.8 Conclusions and perspectives

LncRNAs, once known to be junk of genome, have now become an interesting topic of research for scientists worldwide. They are reported to be involved in a range of functions. The lncRNAs are classified according to their locations and their working methods. With the advancements in bioinformatics and genomics, it has become possible to magnify the hidden information about the genome of plants and animals.

Since the discovery of first lncRNAs, i.e., *H19* and *Xist*, there has been a hustle to discover more and more lncRNAs in different animals and plants. In plants, first reported lncRNA is *GmENOD40* in *Glycine max* (soybean) in the year 1993. Thereafter, many different research groups reported a variety of lncRNAs responsible for diverse functions. A number of genome-wide studies with the help of advanced bioinformatics have reported thousands of lncRNAs in various plants. LncRNAs of plants have been studied less in comparison to the animals. Moreover, we were able to find lncRNAs-based researches in almost every group of plants, including algae, bryophytes, pteridophytes, gymnosperms, and angiosperms. The chapter indicates the history and scope of plants' lncRNAs. It also gives crisp information about the diversity of functions and the number of lncRNAs discovered in a diverse group of plants. Although there have been many publications on lncRNAs in plants, there is a huge scope in this field for the enhancement of knowledge.

Acknowledgments

Authors are thankful to Panjab University, Chandigarh, India, for providing the research facility and infrastructure. AS and HS are thankful to CSIR for senior research fellowships. HS is also thankful to IKGPTU, Jalandhar for Ph.D. registration. SKU is thankful to the Department of Science and Technology, Government of India, for partial financial support under Promotion of University Research and Scientific Excellence (PURSE) grant scheme.

References

Ariel, F., Jegu, T., Latrasse, D., Romero-Barrios, N., Christ, A., Benhamed, M., et al., 2014. Noncoding transcription by alternative RNA polymerases dynamically regulates an auxin-driven chromatin loop. Mol. Cell 55, 383–396.

Bardou, F., Ariel, F., Simpson, C.G., Romero-Barrios, N., Laporte, P., Balzergue, S., et al., 2014. Long noncoding RNA modulates alternative splicing regulators in Arabidopsis. Dev. Cell 30, 166–176.

Barlow, D.P., Stöger, R., Herrmann, B.G., Saito, K., Schweifer, N., 1991. The mouse insulin-like growth factor type-2 receptor is imprinted and closely linked to the Tme locus. Nature 349, 84–87.

Bartolomei, M.S., Zemel, S., Tilghman, S.M., 1991. Parental imprinting of the mouse *H19* gene. Nature 351, 153–155.

Bergougnoux, V., 2014. The history of tomato: From domestication to biopharming. Biotechnol Adv. 32, 170–189. https://doi.org/10.1016/j.biotechadv.2013.11.003.

Bhatia, G., Goyal, N., Sharma, S., Upadhyay, S.K., Singh, K., 2017. Present scenario of long noncoding RNAs in plants. Noncoding RNA 3 (2), E16.

Boerner, S., McGinnis, K.M., 2012. Computational Identification and Functional Predictions of Long Noncoding RNA in *Zea mays*. PLoS One 7 (8), e43047https://doi.org/10.1371/journal.pone.0043047.

Boominathan, R., Doran, P.M., 2003. Cadmium tolerance and antioxidative defenses in hairy roots of the cadmium hyperaccumulator, Thlaspicaerulescens. Biotechnol. Bioeng. 83 (2), 158–167. https://doi.org/10.1002/bit.10656.

Borsani, G., Tonlorenzi, R., Simmler, M.C., Dandolo, L., Arnaud, D., Capra, V., et al., 1991. Characterization of a murine gene expressed from the inactive X chromosome. Nature 351, 325–329.

Brannan, C.I., Dees, E.C., Ingram, R.S., Tilghman, S.M., 1990. The product of the H19 gene may function as an RNA. Mol. Cell. Biol. 10, 28–36.

Brunkow, M.E., Tilghman, S.M., 1991. Ectopic expression of the *H19* gene in mice causes prenatal lethality. Genes Dev. 5 (6), 1092–1101. https://doi.org/10.1101/gad.5.6.1092.

Chen, J., Quan, M., Zhang, D., 2015. Genome-wide identification of novel long non-coding RNAs in Populus tomentosa tension wood, opposite wood and normal wood xylem by RNA-seq. Planta 241 (1), 125–143. https://doi.org/10.1007/s00425-014-2168-1.

Chen, M., Wang, C., Bao, H., Chen, H., Wang, Y., 2016. Genome-wide identification and characterization of novel lnc RNAs in Populus under nitrogen deficiency. Mol. Genet. Genomics 291 (4), 1663–1680. https://doi.org/10.1007/s00438-016-1210-3.

Cui, J., Luan, Y., Jiang, N., Bao, H., Meng, J., 2017. Comparative transcriptome analysis between resistant and susceptible tomato allows the identification of lnc RNA16397 conferring resistance to *Phytophthora infestans* by co-expressing glutaredoxin. Plant J. 89 (3), 577–589. https://doi.org/10.1111/tpj.13408.

Del Rio, L.E., Bradley, C.A., Henson, R.A., Endres, G.J., Hanson, B.K., McKay, K., et al., 2007. Impact of Sclerotinia stem rot on yield of canola. Plant Dis. 91, 191–194.

Ding, J., Lu, Q., Ouyang, Y., Mao, H., Zhang, P., Yao, J., et al., 2012. A long noncoding RNA regulates photoperiod-sensitive male sterility an essential component of hybrid rice. PNAS USA 109 (7), 2654–2659. https://doi.org/10.1073/pnas.1121374109.

Florez-Zapata, N.M., Reyes-Valdes, M.H., Martinez, O., 2016. Long non-coding RNAs are major contributors to transcriptome changes in sunflower meiocytes with different recombination rates. BMC Genomics 17, 490.

Franco-Zorrilla, J.M., Valli, A., Todesco, M., Mateos, I., Puga, M.I., Rubio-Somoza, I., et al., 2007. Target mimicry provides a new mechanism for regulation of micro RNA activity. Nat. Genet. 39, 1033–1037.

Hall, J.L., 2002. Cellular mechanisms for heavy metal detoxification and tolerance. J. Exp. Bot. 53 (366), 1–11. https://doi.org/10.1093/jexbot/53.366.1.

Hao, Z., Fan, C., Cheng, T., Su, Y., Wei, Q., Li, G., 2015. Genome-wide identification, characterization and evolutionary analysis of long intergenic noncoding rnase in cucumber. PLoS One 10 (3), https://doi.org/10.1371/journal.pone.0121800.

He, F., Liu, Q., Zheng, L., Cui, Y., Shen, Z., Zheng, L., 2015. RNA-Seq analysis of rice roots reveals the involvement of post-transcriptional regulation in response to cadmium stress. Front. Plant Sci. 6, 1136. https://doi.org/10.3389/fpls.2015.01136.

Held, M.A., Penning, B., Brandt, A.S., Kessans, S.A., Yong, W., Scofield, S.R., et al., 2008. Small-interfering RNAs from natural antisense transcripts derived from a cellulose synthase gene modulate cell wall biosynthesis in barley. PNAS USA 105 (51), 20534–20539. https://doi.org/10.1073/pnas.0809408105.

Henriques, R., Wang, H., Liu, J., Boix, M., Huang, L.F., Chua, N.H., 2017. The antiphasic regulatory module comprising CDF5 and its antisense RNA FLORE links the circadian clock to photoperiodic flowering. New Phytol. 216, 854–867.

Heo, J.B., Sung, S., 2011. Vernalization-mediated epigenetic silencing by a long intronic noncoding RNA. Science 331, 76–79.

Huttenhofer, A., Brosius, J., Bachellerie, J.P., 2002. RNomics: identification and function of small, non-messenger RNAs. Curr. Opin. Chem. Biol. 6, 835–843.

Jabnoune, M., Secco, D., Lecampion, C., Robaglia, C., Shu, Q., Poirier, Y., 2013. A rice cis-natural antisense RNA acts as a translational enhancer for its cognate mRNA and contributes to phosphate homeostasis and plant fitness. Plant Cell 25 (10), 4166–4182. https://doi.org/10.1105/tpc.113.116251.

Jarroux, J., Morillon, A., Pinskaya, M., 2017. History, discovery, and classification of lncRNAs. Adv. Exp. Med. Biol. 1008, 1–46. https://doi.org/10.1007/978-981-10-5203-3_1.

Joshi, R.K., Megha, S., Basu, U., Rahman, M.H., Kav, N.N., 2016. Genome wide identification and functional prediction of long non-coding RNAs responsive to *Sclerotinia sclerotiorum* infection in *Brassica napus*. PLoS One 11, e0158784https://doi.org/10.1371/journal.pone.0158784.

Kerr, S.C., Gaiti, F., Beveridge, C.A., Tanurdzic, M., 2017. De novo transcriptome assembly reveals high transcriptional complexity in *Pisum sativum* axillary buds and shows rapid changes in expression of diurnally regulated genes. BMC Genomics 18, 221.

Khemka, N., Singh, V.K., Garg, R., Jain, M., 2016. Genome-wide analysis of long intergenic noncoding RNAs in chickpea and their potential role in flower development. Sci. Rep. 6, https://doi.org/10.1038/srep33297.

Kim, E.D., Xiong, Y., Pyo, Y., Kim, D.H., Kang, B.H., Sung, S., 2017. Spatio-temporal analysis of coding and long noncoding transcripts during maize endosperm development. Sci. Rep. 7 (1), https://doi.org/10.1038/s41598-017-03878-4.

Kwenda, S., Birch, P.R., Moleleki, L.N., 2016. Genome-wide identification of potato long intergenic noncoding RNAs responsive to *pectobacterium carotovorum* subspecies *brasiliense* infection. BMC Genomics 17, 614.

Li, L., Eichten, S.R., Shimizu, R., Petsch, K., Yeh, C.-T., Wu, W., et al., 2014. Genome-wide discovery and characterization of maize long non-coding RNAs. Genome Biol. 15 (2), R40. https://doi.org/10.1186/gb-2014-15-2-r40.

Li, H., Wang, Y., Chen, M., et al., 2016. Genome-wide long non-coding RNA screening, identification and characterization in a model microorganism *Chlamydomonas reinhardtii*. Sci. Rep. 6, 34109. https://doi.org/10.1038/srep34109.

Liu, T., Wu, P., Wang, Q., Wang, W., Zhang, C., Sun, F., et al., 2018. Comparative transcriptome discovery and elucidation of the mechanism of long noncoding RNAs during vernalization in *Brassica rapa*. Plant Growth Regul. 85, 27–39.

Liu, F., Xu, Y., Chang, K., Li, S., Liu, Z., Qi, S., et al., 2019. The long noncoding RNA T5120 regulates nitrate response and assimilation in Arabidopsis. New Phytol. 224 (1), 117–131. https://doi.org/10.1111/nph.16038.

Ma, J., Yan, B., Qu, Y., Qin, F., Yang, Y., Hao, X., et al., 2008. Zm401, a short-open reading-frame mRNA or noncoding RNA, is essential for tapetum and microspore development and can regulate the floret formation in maize. J. Cell. Biochem. 105 (1), 136–146. https://doi.org/10.1002/jcb.21807.

Ma, L., Bajic, V.B., Zhang, Z., 2013. On the classification of long non-coding RNAs. RNA Biol. 10, 925–933.

Ma, J., Bai, X., Luo, W., Feng, Y., Shao, X., Bai, Q., et al., 2019. Genome-wide identification of long noncoding RNAs and their responses to salt stress in two closely related poplars. Front. Genet.,10. https://doi.org/10.3389/fgene.2019.00777.

Mach, J., 2017. The long-noncoding RNA ELENA1 functions in plant immunity. Plant Cell 29, 916.

Merchant, S.S., Prochnik, S.E., Vallon, O., Harris, E.H., Karpowicz, S.J., Witman, G.B., 2007. The *Chlamydomonas* genome reveals the evolution of key animal and plant functions. Science 318 (5848), 245–250.

Pachnis, V., Brannan, C.I, Tilghman, S.M., 1988. The structure and expression of a novel gene activated in early mouse embryogenesis. EMBO J. 7 (3), 673-681.

Pachnis, V., Belayew, A., Tilghman, S.M., 1984. Locus unlinked to alpha-fetoprotein under the control of the murine *raf* and *Rif* genes. PNAS USA 81, 5523–5527.

Qi, X., Xie, S., Liu, Y., Yi, F., Yu, J., 2013. Genome-wide annotation of genes and noncoding RNAs of foxtail millet in response to simulated drought stress by deep sequencing. Plant Mol. Biol. 83 (4–5), 459–473. https://doi.org/10.1007/s11103-013-0104-6.

Qin, T., Zhao, H., Cui, P., Albesher, N., Xiong, L., 2017. A nucleus-localized long non coding RNA enhances drought and salt stress tolerance. Plant Physiol. 175, 1321–1336.

Rai, M.I., Alam, M., Lightfoot, D.A., Gurha, P., Afzal, A.J., 2018. Classification and experimental identification of plant long non-coding RNAs. Genomics 111 (5), 997–1005. https://doi.org/10.1016/j.ygeno.2018.04.014.

Scarano, D., Rao, R., Corrado, G., 2017. In silico identification and annotation of noncoding RNAs by RNA-seq and de Novo assembly of the transcriptome of Tomato Fruits. PLoS One 12 (2), https://doi.org/10.1371/journal.pone.0171504.

Seo, J.S., Sun, H.X., Park, B.S., Huang, C.H., Yeh, S.D., Jung, C., et al., 2017. ELF18-induced long-noncoding RNA associates with mediator to enhance expression of innate immune response genes in arabidopsis. Plant Cell 29 (5), 1024–1038. https://doi.org/10.1105/tpc.16.00886.

Shuai, P., Liang, D., Tang, S., Zhang, Z., Ye, C.Y., Su, Y., 2014. Genome-wide identification and functional prediction of novel and drought-responsive lincRNAs in *Populus trichocarpa*. J. Exp. Bot. 65 (17), 4975–4983. https://doi.org/10.1093/jxb/eru256.

Shumayla, Sharma, S., Taneja, M., Tyagi, S., Singh, K., Upadhyay, S.K., 2017. Survey of high throughput RNA-seq data reveals potential roles for lncrnas during development and stress response in bread wheat. Front. Plant Sci. 8, https://doi.org/10.3389/fpls.2017.01019.

Simopoulos, C.M.A., Weretilnyk, E.A., Golding, G.B., 2019. Molecular traits of long non protein coding RNAs from diverse plant species show little evidence of phylogenetic relationships. G3 (Bethesda) 9 (8), 2511–2520. https://doi.org/10.1534/g3.119.400201.

Song, X., Liu, G., Huang, Z., Duan, W., Tan, H., Li, Y., et al., 2016. Temperature expression patterns of genes and their coexpression with LncRNAs revealed by RNA-Seq in non-heading Chinese cabbage. BMC Genomics 17, 297.

Swiezewski, S., Liu, F., Magusin, A., Dean, C., 2009. Cold-induced silencing by long antisense transcripts of an Arabidopsis polycomb target. Nature 462, 799–802.

Tang, W., Zheng, Y., Dong, J., Yu, J., Yue, J., Liu, F., et al., 2016. Comprehensive transcriptome profiling reveals long noncoding RNA expression and alternative splicing regulation during fruit development and ripening in kiwifruit (*Actinidia chinensis*). Front. Plant Sci. 7, 1–15.

Tian, J., Song, Y., Du, Q., Yang, X., Ci, D., Chen, J., et al., 2016. Population genomic analysis of gibberellin-responsive long non-coding RNAs in Populus. J. Exp. Bot. 67 (8), 2467–2482. https://doi.org/10.1093/jxb/erw057.

Tyagi, S., Sharma, A., Upadhyay, S.K., 2018. Role of next-generation RNA-seq data in discovery and characterization of long non-coding RNA in plants. Next Gen Plant Breed, Intech Open https://doi.org/10.5772/intechopen.72773.

Wang, L., Zhao, S., Gu, C., Zhou, Y., Zhou, H., Ma, J., et al., 2013. Deep RNA-Seq uncovers the peach transcriptome landscape. Plant Mol. Biol. 83, 365–377.

Wang, Y., Fan, X., Lin, F., He, G., Terzaghi, W., Zhu, D., et al., 2014. Arabidopsis noncoding RNA mediates control of photomorphogenesis by red light. PNAS USA 111, 10359–10364.

Wang, M., Chen, W., Peng, C., 2015a. Risk assessment of Cd polluted paddy soils in the industrial and township areas in Hunan, Southern China. Chemosphere 144, 346–351. https://doi.org/10.1016/j.chemosphere.2015.09.001.

Wang, H., Niu, Q.W., Wu, H.W., Liu, J., Ye, J., Yu, N., et al., 2015b. Analysis of non-coding transcriptome in rice and maize uncovers roles of conserved lncRNAs associated with agriculture traits. Plant J. 84 (2), 404–416. https://doi.org/10.1111/tpj.13018.

Wang, J., Yu, W., Yang, Y., Li, X., Chen, T., Liu, T., et al., 2015c. Genome-wide analysis of tomato long non-coding RNAs and identification as endogenous target mimic for microRNA in response to TYLCV infection. Sci. Rep. 5, https://doi.org/10.1038/srep16946.

Wang, M., Yuan, D., Tu, L., Gao, W., He, Y., Hu, H., et al., 2015d. Long noncoding RNAs and their proposed functions in fibre development of cotton (*Gossypium* spp.). New Phytol. 207 (4), 1181–1197. https://doi.org/10.1111/nph.13429.
Wang, Z., Liu, Y., Li, L., Li, D., Zhang, Q., Guo, Y., et al., 2017a. Whole transcriptome sequencing of Pseudomonas syringae pv. actinidiae-infected kiwifruit plants reveals species-specific interaction between long non-coding RNA and coding genes. Sci. Rep. 7, 4910.
Wang, C.Y., Liu, S.R., Zhang, X.Y., Ma, Y.J., Hu, C.G., Zhang, J.Z., 2017b. Genome-wide screening and characterization of long non-coding RNAs involved in flowering development of trifoliate orange (*Poncirus trifoliata* L. Raf.). Sci. Rep. 7, 43226.
Wang, L., Xia, X., Jiang, H., Lu, Z., Cui, J., Cao, F., et al., 2018. Genome-wide identification and characterization of novel lncRNAs in *Ginkgo biloba*. Trees-Struct Funct. 32 (5), 1429–1442. https://doi.org/10.1007/s00468-018-1724-x.
Wasaki, J., Yonetani, R., Shinano, T., Kai, M., Osaki, M., 2003. Expression of the OsPI1 gene, cloned from rice roots using cDNA microarray, rapidly responds to phosphorus status. New Phytol. 158 (2), 239–248. https://doi.org/10.1046/j.1469-8137.2003.00748.x.
Wu, Y., Guo, J., Wang, T., Cao, F., Wang, G., 2019. Transcriptional profiling of long noncoding RNAs associated with leaf-color mutation in *Ginkgo biloba* L. BMC Plant Biol. 19 (1), 527. https://doi.org/10.1186/s12870-019-2141-z.
Wunderlich, M., Gross-Hardt, R., Schoffl, F., 2014. Heat shock factor HSFB2a involved in gametophyte development of *Arabidopsis thaliana* and its expression is controlled by a heat-inducible long non-coding antisense RNA. Plant Mol. Biol. 85, 541–550.
Xin, M., Wang, Y., Yao, Y., Song, N., Hu, Z., Qin, D., et al., 2011. Identification and characterization of wheat long nonprotein coding RNAs responsive to powdery mildew infection and heat stress by using microarray analysis and SBS sequencing. BMC Plant Biol. 11 (1), 61. https://doi.org/10.1186/1471-2229-11-61.
Yan, H., Gao, Y., Wu, L., Wang, L., Zhang, T., Dai, C., et al., 2019. Potential use of the Pteris vittata arsenic hyper accumulation-regulation network for phytoremediation. J. Hazard. Mater. 368, 386–396. https://doi.org/10.1016/j.jhazmat.2019.01.072.
Yang, W.C., Katinakis, P., Hendriks, P., Smolders, A., Vries, F., Spee, J., et al., 1993. Characterization of *GmENOD40*, a gene showing novel patterns of cell-specific expression during soybean nodule development. Plant J. 3, 573–585.
Zhang, W., Han, Z., Guo, Q., Liu, Y., Zheng, Y., Wu, F., et al., 2014a. Identification of maize long non-coding RNAs responsive to drought stress. PLoS One 9 (6), e98958. https://doi.org/10.1371/journal.pone.0098958.
Zhang, Y.-C., Liao, J.-Y., Li, Z.-Y., Yu, Y., Zhang, J.-P., Li, Q.-F., et al., 2014b. Genome-wide screening and functional analysis identify a large number of long noncoding RNAs involved in the sexual reproduction of rice. Genome Biol. 15 (12), 512. https://doi.org/10.1186/s13059-014-0512-1.
Zhang, H., Hu, W., Hao, J., Lv, S., Wang, C., Tong, W., et al., 2016. Genome-wide identification and functional prediction of novel and fungi-responsive lincRNAs in *Triticum aestivum*. BMC Genomics 17 (1), https://doi.org/10.1186/s12864-016-2570-0.
Zhang, G., Duan, A., Zhang, J., He, C., 2017. Genome-wide analysis of long non-coding RNAs at the mature stage of sea buckthorn (*Hippophae rhamnoides* Linn) fruit. Gene 596, 130–136.
Zhao, X., Li, J., Lian, B., Gu, H., Li, Y., Qi, Y., 2018. Global identification of Arabidopsis lncRNAs reveals the regulation of MAF4 by a natural antisense RNA. Nature Comm. 9, 5056. https://doi.org/10.1038/s41467-018-07500-7.
Zheng, Y., Ding, B., Fei, Z., Wang, Y., 2017. Comprehensive transcriptome analyses reveal tomato plant responses to tobacco rattle virus-based gene silencing vectors. Sci. Rep. 7 (1), https://doi.org/10.1038/s41598-017-10143-1.

Zhu, B., Yang, Y., Li, R., Fu, D., Wen, L., Luo, Y., et al., 2015. RNA sequencing and functional analysis implicate the regulatory role of long non-coding RNAs in tomato fruit ripening. J. Exp. Bot. 66 (15), 4483–4495. https://doi.org/10.1093/jxb/erv203.

Zhu, Y., Chen, L., Zhang, C., Hao, P., Jing, X., Li, X., 2017. Global transcriptome analysis reveals extensive gene remodeling, alternative splicing and differential transcription profiles in non-seed vascular plant *Selaginella moellendorffii*. BMC Genomics 18 (1), 1042. https://doi.org/10.1186/s12864-016-3266-1.

Zou, C., Wang, Q., Lu, C., Yang, W., Zhang, Y., Cheng, H., et al., 2016. Transcriptome analysis reveals long noncoding RNAs involved in fiber development in cotton (*Gossypium arboreum*). Sci. China Life Sci. 59 (2), 164–171. https://doi.org/10.1007/s11427-016-5000-2.

Chapter 3

Databases and tools for long noncoding RNAs

Deepika Lakhwani
University of Angers, France

3.1 Introduction

Regulation of thousands of genes across the eukaryote genomes requires different regulatory molecules with roles in diverse biological processes. In 2003, it was found that 80% bases of sequenced human genome were transcribed in noncoding RNAs and these noncoding RNAs of human genome participate in numerous biological processes such as histone modification and RNA transcription, etc. (van Bakel et al., 2010; Yotsukura et al., 2017). Later in several studies, the role of noncoding transcripts has been found in cell differentiation and development, epigenetic regulation, transcriptional and posttranscriptional regulation in animals as well as in plants (Yotsukura et al., 2017). Since then, research on noncoding RNAs is rapidly increased to understand the role in eukaryotes genome. These noncoding RNAs are classified as housekeeping RNAs and regulatory RNAs (Bhatia et al., 2017; Mattick and Rinn, 2015). Housekeeping RNAs include tRNA, rRNA, and snRNA, while regulatory RNAs are classified into two categories: (1) small noncoding RNA (sRNA) with <200 nt including microRNAs (miRNAs), small interfering RNAs (siRNAs), Piwi-interacting RNAs (piRNAs); and (2) Long noncoding RNAs (lncRNAs). Long noncoding RNAs are 200 bp in length and found their role in various stress conditions and diverse biological processes involving different regulatory mechanisms (Liao et al., 2018). The biogenesis of lncRNAs is mainly processed by RNA Pol-II in eukaryotes genome, while RNA Pol IV/V are transcribed plant-specific class of lncRNAs which mediate RNA-directed DNA methylation (rdDM) gene silencing (Wang et al., 2014b; Wierzbicki et al., 2008; Wierzbicki et al., 2009). They are also mediated by splicing or non-splicing, polyadenylation or non-polyadenylation, and can be located in the nucleus or cytoplasm (Bhatia et al., 2017). There are different types of biotypes of LncRNAs on the basis of genomic location with respect to nearby genes such as intergenic lncRNA (present between two genes), intronic lncRNA (present in intron), and sense (derived from same strand) and antisense lncRNA (derived

Long Noncoding RNAs in Plants: Roles in development and stress
https://doi.org/10.1016/B978-0-12-821452-7.00003-9

from opposite strand) (Mattick and Rinn, 2015; Ponting et al., 2009). The role of LncRNAs has been extensively studied and characterized in plants. In 1993, first lncRNA (~700bp) was identified in *Glycine max* involved in changing the subcellular localization of a protein (Yang et al., 1993). Later, several lncRNAs have been identified and their role discovered in numerous plant species using experimental as well as computational approaches (Shumayla et al., 2017; Tyagi et al., 2018). For instance, *OsENOD40* was identified in *Oryza sativa* involved in nodule formation (Kouchi et al., 1999). Furthermore, COLDAIR (COLD-ASSISTED INTRONIC NONCODING RNA) and COOLAIR (COLD-INDUCED LONG-ANTISENSE INTRAGENIC RNA) were involved in vernalization in *Arabidopsis thaliana* (Heo and Sung, 2011; Wang et al., 2014a). In different studies, lncRNAs also displayed tissue-specific expression pattern in several plants. For example, the role of lncRNAs in flower and fruit development in *Fragaria vesca* (Kang and Liu, 2015), fiber development in *Gossypium arboretum* (Zou et al., 2016), wood formation in poplar, and photomorphogenesis in *Arabidopsis thaliana* (Wang et al., 2014a) has been identified. Somewhere, it shows that plants have complex regulatory system. So, it's essential to understand various roles and regulation of lncRNAs which can boost our knowledge about the role of lncRNAs in plants.

3.2 Overview of some plant-specific tools and programs to predict lncRNAs

The revolution in the high-throughput sequencing technology has unlocked the door for sequencing platforms to discover novel transcripts at genome-wide level and understand regulatory mechanism at molecular level in model and non-model plants. Transcriptomic approaches such as microarray and RNA sequencing are emerged as powerful techniques to explore novel transcripts and their expression analysis at large scale. Transcriptome studies have been done in several plants to identify long noncoding RNAs at genome-wide level such as *Medicago truncatula*, *Oryza sativa*, *Prunus mume*, *Solanum lycopersicum*, *Gossypium arboreum*, *Zea mays*, etc. (Wang et al., 2018; Wang et al., 2015; Wu et al., 2019; Zhang et al., 2014; Zhu et al., 2017; Zou et al., 2016). A total of 125 putative stress-responsive lncRNAs have been identified in wheat using both computational and experimental approaches (Xin et al., 2011). Li et al. identified 20,163 putative lncRNAs in maize by integrating the available EST data, annotation information of maize genome, and RNA-seq datasets obtained from 30 different experiments (Li et al., 2014b). In plants, different methodologies with different strategies and multiple parameters have been designed to predict lncRNAs for functional characterization which is summarized in Fig. 3.1.

In the progress of technology era, transcriptome generates millions of reads which requires computational facilities as well as programmes with better algorithms to do faster analyses. Computational detection of long noncoding RNA is one of the preferred methods to study the role of lncRNA in the plant genome.

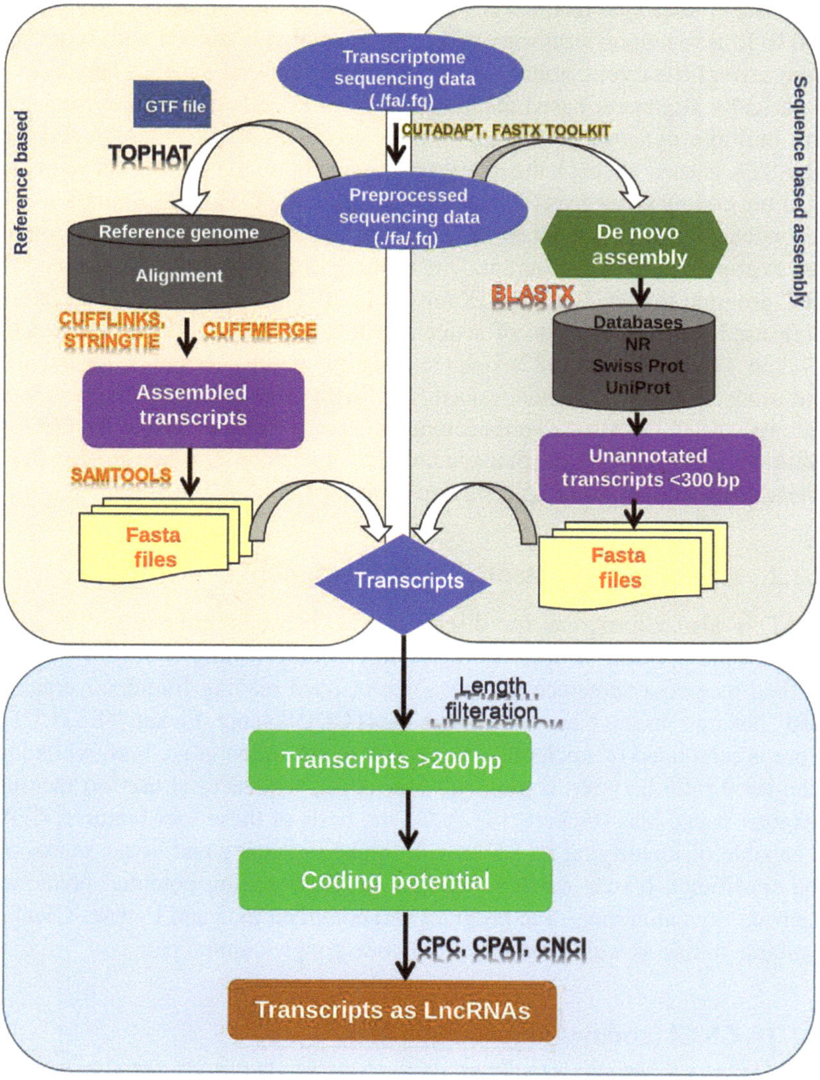

FIG. 3.1 A computational pipeline for the identification of lncRNAs in plants.

Therefore, some tools and programs have been developed to discriminate coding and noncoding sequences on the basis of sequence features and largely used in numerous transcriptome studies of different plants which provide a way to gain insights into the evolution. Basic introduction of these tools is represented here.

3.2.1 The coding potential calculator (CPC)

The Coding Potential Calculator (CPC) is based on machine learning program which classifies potential lncRNA sequences from coding sequences on the basis of sequence features (ORF length, sequence similarity) (Kong et al., 2007). It uses support vector machine (SVM) which is trained with sequences from several eukaryotic noncoding RNA and protein-coding RNA databases. It is basically alignment-based method that uses log-odds of hexamer frequency and multiple dynamic frame shift model to calculate ORF length. Further, BLASTX is used to check the quality of alignment on the basis of hit score and filter out coding sequences (Altschul et al., 1990). CPC also has limitations like dependency on previous set of sequences and it is a time-consuming program due to alignment-based program. Due to these limitations, authors updated the CPC program to CPC2 in 2017 (Kang et al., 2017). In the CPC2, four features were used for classification of sequences that are Fickett TESTCODE score (Fickett, 1982) (Fickett, 1982), open reading frame (ORF) length, ORF integrity and isoelectric point (pI), and trained SVM with these features. It increases not only the speed, but also overall accuracy of result. The web server for CPC2 is available on http://cpc2.cbi.pku.edu.cn/ and standalone program is also freely available on http://cpc2.cbi.pku.edu.cn/download.php/.

3.2.2 CPAT (coding potential assessment tool)

CPAT is also a program for discriminating between protein and noncoding sequences based on logistic regression method (Wang et al., 2013). This method uses four sequence features: size of open reading frame, coverage of ORF, hexamer usage bias, and Fickett TESTCODE score. Fickett TESTCODE score is calculated by nucleotide composition and codon usage bias, while log-likelihood ratio between coding and noncoding sequences is used to measure hexamer usage bias (Fickett, 1982). On the basis of these four features, CPAT is capable of identifying lncRNAs with higher accuracy and better sensitivity and specificity. It runs much faster than previous coding potential prediction methods. It is alignment-free program and compiled in C and Python. CPAT is available online as well as offline http://code.google.com/p/cpat/.

3.2.3 CNCI (coding-noncoding index)

CNCI is a powerful tool to classify coding and noncoding sequences by analyzing nucleotide composition of their sequence (Sun et al., 2013). It shows better performance than existing tools like CPC and phyloCSF because CNCI has capability to determine lncRNAs from many newly evolved proteins that do not contain a conserved ORF because both utilize conserved ORF region to classify sequences. CNCI distinguishes incomplete transcripts and sense–antisense transcript pairs. CNCI uses adjoining nucleotide triplets (ANT) in the sequences and construct score matrix. First, it calculates the frequency of ANT in coding and noncoding sequences separately, then calculates the log ratio of frequency for each codon. According to this, it makes a scoring matrix and calculates S-score according to matrix. Further, according to the cumulative S-score of the combined windows, most-like CDS (MLCDS) is identified for each reading frame using dynamic programming algorithm. MLCDS region with high S-score incorporates into the SVM and it discriminates the protein-coding and noncoding sequences based on selected features. In 2019, the author updated the CNCI to CNIT (Coding–Noncoding Identifying Tool) because CNCI ran poorly for plant species (Guo et al., 2019). CNIT runs ~200 times faster with improved accuracy, especially for plants. CNIT is user-friendly web server and available on http://cnit.noncode.org/CNIT/.

3.2.4 PLIT (plant LncRNA and identification tool)

PLIT is an alignment-free tool for prediction of lncRNAs in plants using RNA-Seq data (Deshpande et al., 2019). It utilizes the 31 feature set that includes ORF, sequence- and codon-based features such as maximum ORF length, ORF coverage, mean ORF coverage, transcript length, GC content, codon usage bias, Relative Codon Bias, Frequency of optimal codons, Synonymous Codon Usage Order, etc. These features are extracted from noncoding and coding transcripts of eight plant species. To identify lncRNAs from the RNA-Seq data, reads are used for trimming and quality filtering using cutadapt (Martin, 2011). After that, filtered reads are mapped to genome using tophat (Trapnell et al., 2009). Further, Samtool mpileup is used to call variants from mapping file (BAM) and provide the output in VCF (variant call format) file (Li et al., 2009). On the basis of genome sequence and VCF output file, Bcftools is used to create consensus sequences which are considered as transcripts and used for classification as coding or long noncoding transcripts. Random Forest model is randomly classified sequences into training and test datasets. Iterative Random Forest Feature Selection (LiRFFS) algorithm is used to select optimal feature sets to provide high accuracy result (Deshpande et al., 2019). To evaluate the performance of classification, Accuracy (ACC), Sensitivity (SENS), and Specificity (SPEC) are calculated which will define quality of result for plants. Comparative study with other existing tools like CPC2, PLEK, and lncScore shows higher specificity but much lower sensitivity, which represents higher false negatives result (Kang et al., 2017; Li et al., 2014a; Zhao et al., 2016). It also shows that PLIT is more stable and accurate tool to identify long noncoding RNAs in plants. PLIT will provide valuable resources for plants to identify novel lncRNAs in plants. It is an offline tool and GitHub is available on https://github.com/deshpan4/PLIT/.

3.2.5 PLncPRO (plant long noncoding RNA prediction by random forests)

PLncPRO is widely used to predict lncRNAs from RNA-seq data with higher accuracy in plant species (Singh et al., 2017). It supports random forest model which is an efficient computational method to classify transcripts. In this method, training model is constructed from different plant species using a set of known lncRNAs and coding RNAs. In this model, frame finder and BLASTX tools are used to extract features for each transcript from the training dataset. These features are ORF score and ORF coverage, Number of hits, Significance Score, Total bit score, Frame entropy, codon usage bias, and length of the transcript. Further, trained model is used to classify noncoding and coding transcript from set of unknown transcripts. When PLncPRO tool is compared with other available tools, PLncPRO shows better performance and higher accuracy in the prediction of lncRNAs from plant species. PLncPRO is python-based standalone program which is multi-threaded and memory-efficient. PLncPRO is freely available and can be downloaded from http://ccbb.jnu.ac.in/plncpro/.

3.3 Plant-specific long noncoding RNA databases

After identification and characterization of the role of lncRNA in plants, it's important to store this meaningful data and utilize for further analysis and updates. In order to manage this huge amount of information and further help in the mapping and profiling of putative lncRNAs, several databases have been developed for lncRNAs which provide not only the information about sequences and genome coordinates of lncRNAs, but also their expression profiles in different conditions. Here, we will cover some lncRNA-specific databases which manage the information specifically for single or numerous plant species. These databases are summarized in Table 3.1 and basic description about these databases is mentioned here:

3.3.1 NONCODEv5

NONCODEv5 is an integrated database of eukaryotic noncoding RNA for 17 species such as human, mouse, pig, cow, rat, chicken, zebrafish, chimpanzee, gorilla, platypus, fruit fly, *Caenorhabditis elegans*, yeast, *Arabidopsis,* etc. (Fang et al., 2017). It provides information about RNA secondary structure, expression of exosome, and association between lncRNA and disease. The source of raw data in NONCODE database includes those from public literature and lncRNA databases. It retrieves the information from PubMed and supplementary material or websites of articles using different keywords like 'noncoding', 'noncoding', lncRNA, etc. Ensembl, RefSeq, lncRNAdb, LNCipedia, and the old versions of the NONCODE databases are used to obtain sequencing data of lncRNAs (Cunningham et al., 2015; Pruitt et al., 2014; Quek et al., 2015; Volders et al., 2019). The updated NONCODE (v5) includes two different databases, Arabidopsis Information Resource (TAIR) and FlyBase databases. The sequencing data and literature data are first preprocessed on the basis of redundancy. Further, the preprocessed data go through a pipeline which includes different software/tools to fetch out final result. This pipeline has six following steps: (1) Normalization (data is converted into GTF or bed format); (2) Combination

TABLE 3.1 Summarization of lncRNA databases for plants.

S. no.	Name of the database	Plant species	Number of lncRNAs for plants	References
1	NONCODE v5	*A. thaliana*	3853 lncRNAs	Fang et al. (2017)
2	EVLncRNAs	*Arabidopsis*, Tomato, *Zea mays*, and *Foxtail millet*	428 lncRNAs	Zhou et al. (2018)
3	CANTATAdb	36 plant species: *Amborella trichopoda, A.thaliana, Chlamydomonasreinhardtii, G. max, O.sativa, Physcomitrella patens,Selaginella moellendorffii, Solanum tuberosum,Vitis vinifera*,and *Z. mays etc*	239,631 lncRNAs	Szczesniak et al. (2016)
4	PLncDB	*Arabidopsis thaliana*	13,466 lncRNAs	Jin et al. (2013)
5	GreeNC	37 Plant species and 6 algae	>190,000 lncRNAs	(Paytuvi Gallart et al. (2016)
6	PLncRNAdb	Four plant species: *A. thaliana, A. lyrata, P.trichocarpa*, and *Z. mays*	5100 lncRNAs	Chen (2016)
7	PNRD	Four plant species: *A. thaliana, Oryza sativa, Populus trichocarpa*, and *Zea mays*	28214 lncRNAs	Yi et al. (2015)

(combine the normalized data using cuffcompare); (3) Filtering protein-coding RNA (filter out the coding transcript using Coding-Noncoding Index (CNCI) or compared with the coding RNA in RefSeq and Ensemble); (4) information retrieval (assign the name to transcript and retrieve basic information such as location, exons, length, assembly sequence, source, etc.); (5) Advanced annotation (provide the expression profiles, functions, conservation, and disease information); and (6) Web presence (new web page availability of transcript). The information of analyzed data is stored and accessed by users according to their requirements because of their user-friendly interface. This database is based on MySQL and the website is built by an Apache server. It has a number of convenient browse and search options so that users can ensure rapid access to the expression profiles or to information on potential functions of the ncRNAs.

3.3.2 EVLncRNAs

EVLncRNAs is also an integrated database of experimentally validated lncRNAs (Zhou et al., 2018). It provides information about functional and disease-related lncRNAs. It contains 1543 lncRNAs from 77 species that include 869 lncRNAs for humans, 220 for animal, 428 for plant, and 26 for microbe. In plants, it mainly contains *Arabidopsis*, Tomato, *Zea mays,* and *Foxtail millet*. For plants, they mentioned functional information of lncRNAs such as gene regulation, cell differentiation, or embryonic development. This database also integrates CPC (coding potential calculator) as the lncRNA prediction tool and user-friendly interface allows the users to browse, search, and download (Kong et al., 2007). This is MySQL-based database and powered by Apache Tomcat.

3.3.3 CANTATAdb 2.0

CANTATAdb 2.0 database is a collection of computationally identified lncRNAs using publically available RNA-Seq datasets which contain 36 plant species and 3 algae (Szczesniak et al., 2016). A computational pipeline is used in the background of database, which identified 239,631 lncRNAs in 39 species and also calculated expression value, lncRNA–RNA interactions. In the pipeline, short reads are mapped to genome using tophat and bowtie software which generates output file in the bam format (Langmead et al., 2009; Trapnell et al., 2009). Tophat is mainly used as splice junction mapper and bowtie is used for small reads. Further, StringTie is used to assemble mapped alignment into the transcript on the basis of reference GTF file (Pertea et al., 2015). It is run for one file per RNA-Seq library. The output files from StringTie are applied in cuffmerge to find out the single set of transcripts and their annotation (Trapnell et al., 2010). BLAST and CNCI are applied to predicted protein-coding or other type of RNAs and discarded them for further analysis. Transcripts with > 200nt are considered for estimation of expression value as RPKM (reads per kilobase of transcript per million mapped reads) using custom python script. Finally, LAST package is used to identify G:U (wobble) pairs to identify lncRNA-RNA interaction with defined parameters. CANTATAdb 2.0 stores all these information and provides the output in a meaningful manner. It is very user-friendly and is constructed using Hypertext Markup Language (HTML), Cascading Style Sheets (CSS), PHP 5.4 **http://www.php.net/**, MySQL 5.5 **http://www.mysql.com/**, and Bootstrap 3 **http://getbootstrap.com/** framework. Users can access species-specific information on the basis of some filtration like coding potential, functionality, expression value, etc. Till date, a total of 239,631 lncRNAs are available in this database.

3.3.4 PLncDB (plant long noncoding RNA database)

PLncDB is an online database of lncRNAs collected from diverse resources and provides comprehensive information about genomic and siRNA information, expression profiles, and associated epigenetic markers of related lncRNAs (Jin et al., 2013). It specifically provides genomic view of *Arabidopsis* that contains 13,466 lncRNAs using Tiling array Analysis Strategy (RepTAS) method. Oligonucleotide microarray and RNA-seq datasets are used to detect the expression profile of lncRNAs and integrated the processed signal intensities into the genome browser. This online genome browser can allow the users to download and deposit the information related to lncRNAs.

3.3.5 GREENC (green noncoding database)

GreeNC database is an online database which provides the annotation of lncRNAs from 37 plant species and six algae. Users can extract the information about the genomic sequences and their coordinates, folding energy of lncRNAs (Paytuvi Gallart et al., 2016). To annotate lncRNAs, two bash scripts are used in the background of database. In the first script, transcripts sequences of different plant species (downloaded from phytozome) are utilized to determine coding potentiality and transcripts with >200bp and ORF <120aa. In the second script, transcripts are discriminated as lncRNAs from other noncoding RNAs, transposable elements, and miRNA precursors. For this, different databases are used such as mirbase, Rfam, MIReNA, and RepBase. Further, transcripts are divided into two categories (low and high confidence lncRNAs) based on the results of BLASTX and CPC software. GreeNC is a MySQL relational database which includes information about $>190{,}000$ annotated lncRNA transcripts in approximately 200,000 pages. In future, new information regarding lncRNA expression patterns, phylogeny, and conservation across different species will be added soon. GreeNC database is utilized as a powerful tool for plant researchers to extend their knowledge on this transcript.

3.3.6 PlncRNAdb (plant long noncoding RNA database)

PlncRNAdb consists approx. 5100 lncRNAs from four plant species (*P. trichocarpa*, *Z. mays*, *A. thaliana*, and *A. lyrata*) (Chen, 2016). Authors developed a pipeline to identify lncRNAs from the sequences which are obtained either from annotation based on reference-guided transcriptome assembly or by the literature and classified these identified RNAs using training dataset (*Arabidopsis* coding sequences and LncRNAs from PLcDB). It is a very simple and valuable database that provides the relationship between lncRNAs and various RNA-binding proteins (RBPs). Due to user-friendly interface, lncRNA-protein networks can be visualized on web interface.

3.3.7 PNRD (a plant noncoding RNA database)

PNRD is a useful resource for noncoding RNAs across various plant species. It contains information about lncRNAs, tRNA, rRNA, tasiRNA, snRNA and snoRNA, etc. PNRD is basically an updated version of PMRD (plant microRNA database), which is a plant miRNA database (Yi et al., 2015). Due to the importance of ncRNAs in plants field, authors designed comprehensive integrated and updated database for noncoding RNAs which include different types of RNAs and more useful toolkits such as CPC toolkit which allow to calculate coding potential of sequences (Kong et al., 2007). PNRD also provides two genome browsers to determine the location of ncRNA and their relationship with epigenetic modifications. PNRD is a comprehensive analysis platform that includes information about 28214 noncoding RNAs from 166 plant species using public domain (literature and high-throughput sequencing data). However, it provides information about lncRNAs from 18 species. PNRD has significantly 11 ncRNAs categories from 166 plant species, which shows high data coverage as compared to other databases. The PNRD database is based on a standard Linux, Mysql, Apache, and PHP system. It has many functional search and analysis tools for browsing, studying, searching, downloading, predicting, and visualizing different types of ncRNAs.

Apart from it, there are several other repositories of lncRNAs such as lncRNAdb, TAIR10, and RNAcentral which are comprehensive resources of lncRNAs. These databases help us in deciphering biologically meaningful information which might be the key in understanding the regulation of biological processes in plants. Most of the databases are mainly focused on expression and interaction of lncRNAs, but there is a need to focus on some other features like splicing pattern, binding interaction of lncRNAs, etc. Advancement in databases will occur based on the requirement of users and also updated on a regular basis.

3.4 Conclusions

The number of lncRNAs has been increased across different plant species using bioinformatics approaches and deciphered their involvement in several biological processes. To utilize this information, several databases and web tools have been developed for plants as well as for animals. Here, we summarized some important databases and tools specific for plant lncRNAs with their description. These databases and tools will provide significant information about lncRNAs.

Acknowledgments

The author acknowledges Prof. Santosh K. Upadhyay for copyediting this chapter to refine the language.

References

Altschul, S.F., Gish, W., Miller, W., Myers, E.W., Lipman, D.J., 1990. Basic local alignment search tool. J. Mol. Biol. 215, 403–410.

Bhatia, G., Goyal, N., Sharma, S., Upadhyay, S.K., Singh, K., 2017. Present scenario of long non-coding RNAs in plants. Noncoding RNA 3, 16.

Chen, M., 2016. PLncRNAdb Ming Chen's Lab. [Accessed on 25 December 2016].

Cunningham, F., Amode, M.R., Barrell, D., Beal, K., Billis, K., Brent, S., et al., 2015. Ensembl 2015. Nucleic Acids Res. 43, D662–D669.

Deshpande, S., Shuttleworth, J., Yang, J., Taramonli, S., England, M., 2019. PLIT: an alignment-free computational tool for identification of long non-coding RNAs in plant transcriptomic datasets. Comput. Biol. Med. 105, 169–181.

Fang, S., Zhang, L., Guo, J., Niu, Y., Wu, Y., Li, H., et al., 2017. NONCODEV5: a comprehensive annotation database for long non-coding RNAs. Nucleic Acids Res. 46, D308–D314.

Fickett, J.W., 1982. Recognition of protein coding regions in DNA sequences. Nucleic Acids Res. 10, 5303–5318.

Guo, J.C., Fang, S.S., Wu, Y., Zhang, J.H., Chen, Y., Liu, J., Wu, B., et al., 2019. CNIT: a fast and accurate web tool for identifying protein-coding and long non-coding transcripts based on intrinsic sequence composition. Nucleic Acids Res. 47, W516–W522.

Heo, J.B., Sung, S., 2011. Vernalization-mediated epigenetic silencing by a long intronic noncoding RNA. Science 331, 76–79.

Jin, J., Liu, J., Wang, H., Wong, L., Chua, N.H., 2013. PLncDB: plant long non-coding RNA database. Bioinformatics 29, 1068–1071.

Kang, C., Liu, Z., 2015. Global identification and analysis of long non-coding RNAs in diploid strawberry *Fragaria vesca* during flower and fruit development. BMC Genomics 16, 815.

Kang, Y.J., Yang, D.C., Kong, L., Hou, M., Meng, Y.Q., Wei, L., et al., 2017. CPC2: a fast and accurate coding potential calculator based on sequence intrinsic features. Nucleic Acids Res. 45, W12–W16.

Kong, L., Zhang, Y., Ye, Z.Q., Liu, X.Q., Zhao, S.Q., Wei, L., et al., 2007. CPC: assess the protein-coding potential of transcripts using sequence features and support vector machine. Nucleic Acids Res. 35, W345–W349.

Kouchi, H., Takane, K., So, R.B., Ladha, J.K., Reddy, P.M., 1999. Rice *ENOD40*: isolation and expression analysis in rice and transgenic soybean root nodules. Plant J. 18, 121–129.

Langmead, B., Trapnell, C., Pop, M., Salzberg, S.L., 2009. Ultrafast and memory-efficient alignment of short DNA sequences to the human genome. Genome Biol. 10, R25.

Li, H., Handsaker, B., Wysoker, A., Fennell, T., Ruan, J., Homer, N., Marth, G., Abecasis, G., Durbin, R., 2009. The sequence alignment/map format and SAMtools. Bioinformatics 25, 2078–2079.

Li, A., Zhang, J., Zhou, Z., 2014a. PLEK: a tool for predicting long non-coding RNAs and messenger RNAs based on an improved k-mer scheme. BMC Bioinform. 15, 311.

Li, L., Eichten, S.R., Shimizu, R., Petsch, K., Yeh, C.T., Wu, W., et al., 2014b. Genome-wide discovery and characterization of maize long non-coding RNAs. Genome Biol. 15, R40.

Liao, P., Li, S., Cui, X., Zheng, Y., 2018. A comprehensive review of web-based resources of non-coding RNAs for plant science research. Int. J. Biol. Sci. 14, 819–832.

Martin, M., 2011. Cutadapt removes adapter sequences from high-throughput sequencing reads. EMBnet J. 17, 10–12.

Mattick, J.S., Rinn, J.L., 2015. Discovery and annotation of long noncoding RNAs. Nat. Struct. Mol. Biol. 22, 5–7.

Paytuvi Gallart, A., Hermoso Pulido, A., de Lagran, I.A.M., Sanseverino, W., Aiese Cigliano, R., 2016. GREENC: a Wiki-based database of plant lncRNAs. Nucleic Acids Res. 44, D1161–D1166.

Pertea, M., Pertea, G.M., Antonescu, C.M., Chang, T.C., Mendell, J.T., Salzberg, S.L., 2015. StringTie enables improved reconstruction of a transcriptome from RNA-seq reads. Nat. Biotechnol. 33, 290–295.

Ponting, C.P., Oliver, P.L., Reik, W., 2009. Evolution and functions of long noncoding RNAs. Cell 136, 629–641.

Pruitt, K.D., Brown, G.R., Hiatt, S.M., Thibaud-Nissen, F., Astashyn, A., Ermolaeva, O., et al., 2014. RefSeq: an update on mammalian reference sequences. Nucleic Acids Res. 42, D756–D763.

Quek, X.C., Thomson, D.W., Maag, J.L., Bartonicek, N., Signal, B., Clark, M.B., et al., 2015. lncRNAdb v2.0: expanding the reference database for functional long noncoding RNAs. Nucleic Acids Res. 43, D168–D173.

Singh, U., Khemka, N., Rajkumar, M.S., Garg, R., Jain, M., 2017. PLncPRO for prediction of long non-coding RNAs (lncRNAs) in plants and its application for discovery of abiotic stress-responsive lncRNAs in rice and chickpeas. Nucleic Acids Res. 45, e183.

Shumayla, Sharma, S., Taneja, M., Tyagi, S., Singh, K., Upadhyay, S.K., 2017. Survey of high throughput RNA-seq data reveals potential roles for lncRNAs during development and stress response in bread wheat. Front. Plant Sci. 8, 1019.

Sun, L., Luo, H., Bu, D., Zhao, G., Yu, K., Zhang, C., et al., 2013. Utilizing sequence intrinsic composition to classify protein-coding and long non-coding transcripts. Nucleic Acids Res. 41, e166.

Szczesniak, M.W., Rosikiewicz, W., Makalowska, I., 2016. CANTATAdb: a collection of plant long non-coding RNAs. Plant Cell Physiol. 57, e8.

Trapnell, C., Pachter, L., Salzberg, S.L., 2009. TopHat: discovering splice junctions with RNA-Seq. Bioinformatics 25, 1105–1111.

Trapnell, C., Williams, B.A., Pertea, G., Mortazavi, A., Kwan, G., van Baren, M.J., et al., 2010. Transcript assembly and quantification by RNA-Seq reveals unannotated transcripts and isoform switching during cell differentiation. Nat. Biotechnol. 28, 511–515.

Tyagi, S., Sharma, A., Upadhyay, S.K., 2018. Role of next-generation RNA-Seq data in discovery and characterization of long non-coding RNA in plants. In: Çiftçi, Y.Ö. (Ed.), Next Generation Plant Breeding. IntechOpen, https://doi.org/10.5772/intechopen.72773.

van Bakel, H., Nislow, C., Blencowe, B.J., Hughes, T.R., 2010. Most "dark matter" transcripts are associated with known genes. PLoS Biol. 8, e1000371.

Volders, P.J., Anckaert, J., Verheggen, K., Nuytens, J., Martens, L., Mestdagh, P., et al., 2019. LNCipedia 5: towards a reference set of human long non-coding RNAs. Nucleic Acids Res. 47, D135–D139.

Wang, L., Park, H.J., Dasari, S., Wang, S., Kocher, J.P., Li, W., 2013. CPAT: Coding-Potential Assessment Tool using an alignment-free logistic regression model. Nucleic Acids Res. 41, e74.

Wang, Y., Fan, X., Lin, F., He, G., Terzaghi, W., Zhu, D., et al., 2014a. *Arabidopsis* noncoding RNA mediates control of photomorphogenesis by red light. Proc. Natl. Acad. Sci. U. S. A. 111, 10359–10364.

Wang, Y., Wang, X., Deng, W., Fan, X., Liu, T.T., He, G., et al., 2014b. Genomic features and regulatory roles of intermediate-sized non-coding RNAs in *Arabidopsis*. Mol. Plant 7, 514–527.

Wang, T.Z., Liu, M., Zhao, M.G., Chen, R., Zhang, W.H., 2015. Identification and characterization of long non-coding RNAs involved in osmotic and salt stress in *Medicago truncatula* using genome-wide high-throughput sequencing. BMC Plant Biol. 15, 131.

Wang, M., Zhao, W., Gao, L., Zhao, L., 2018. Genome-wide profiling of long non-coding RNAs from tomato and a comparison with mRNAs associated with the regulation of fruit ripening. BMC Plant Biol. 18, 75.

Wierzbicki, A.T., Haag, J.R., Pikaard, C.S., 2008. Noncoding transcription by RNA polymerase Pol IVb/Pol V mediates transcriptional silencing of overlapping and adjacent genes. Cell 135, 635–648.

Wierzbicki, A.T., Ream, T.S., Haag, J.R., Pikaard, C.S., 2009. RNA polymerase V transcription guides ARGONAUTE4 to chromatin. Nat. Genet. 41, 630–634.

Wu, X., Shi, T., Iqbal, S., Zhang, Y., Liu, L., Gao, Z., 2019. Genome-wide discovery and characterization of flower development related long non-coding RNAs in *Prunus mume*. BMC Plant Biol. 19, 64.

Xin, M., Wang, Y., Yao, Y., Song, N., Hu, Z., Qin, D., et al., 2011. Identification and characterization of wheat long non-protein coding RNAs responsive to powdery mildew infection and heat stress by using microarray analysis and SBS sequencing. BMC Plant Biol. 11, 61.

Yang, W.C., Katinakis, P., Hendriks, P., Smolders, A., de Vries, F., Spee, J., et al., 1993. Characterization of *GmENOD40*, a gene showing novel patterns of cell-specific expression during soybean nodule development. Plant J. 3, 573–585.

Yi, X., Zhang, Z., Ling, Y., Xu, W., Su, Z., 2015. PNRD: a plant non-coding RNA database. Nucleic Acids Res. 43, D982–D989.

Yotsukura, S., duVerle, D., Hancock, T., Natsume-Kitatani, Y., Mamitsuka, H., 2017. Computational recognition for long non-coding RNA (lncRNA): Software and databases. Brief. Bioinform. 18, 9–27.

Zhang, Y.C., Liao, J.Y., Li, Z.Y., Yu, Y., Zhang, J.P., Li, Q.F., et al., 2014. Genome-wide screening and functional analysis identify a large number of long noncoding RNAs involved in the sexual reproduction of rice. Genome Biol. 15, 512.

Zhao, J., Song, X., Wang, K., 2016.lncScore: alignment-free identification of long noncoding RNA from assembled novel transcripts. Sci Rep 6, 34838.

Zhou, B., Zhao, H., Yu, J., Guo, C., Dou, X., Song, F., et al., 2018.EVLncRNAs: a manually curated database for long non-coding RNAs validated by low-throughput experiments. Nucleic Acids Res 46, D100-D105.

Zhu, M., Zhang, M., Xing, L., Li, W., Jiang, H., Wang, L., et al., 2017. Transcriptomic analysis of long non-coding RNAs and coding genes uncovers a complex regulatory network that is involved in maize seed development. Genes (Basel) 8 (10), 274.

Zou, C., Wang, Q., Lu, C., Yang, W., Zhang, Y., Cheng, H., et al., 2016. Transcriptome analysis reveals long noncoding RNAs involved in fiber development in cotton (*Gossypium arboreum*). Sci. China Life Sci. 59, 164–171.

Chapter 4

Role of lncRNAs in wheat and its wild relatives

Shumayla*, Shivi Tyagi*, and Santosh Kumar Upadhyay
Department of Botany, Panjab University, Chandigarh, India

4.1 Introduction

Bread wheat (*Triticum aestivum* L.) is the most important staple food crop which accounts for approximately 30% of the global cereal production. Among all other cereals, it is the richest source of protein and calorie for more than 7 billion people (www.wheatinitiative.org). During the process of domestication, two spontaneous hybridization events occurred in the genome of three wheat progenitors which led to the evolution of allohexaploid (6x = 2n = 42, AABBDD) modern wheat, i.e., *T. aestivum* L. (Gale and Devos, 1998). The proposed contributors of A and D sub-genomes of *T. aestivum* were *T. urartu* and *Aegilops tauschii*, respectively, but the progenitor of B sub-genome is still a matter of debate (Kilian et al., 2007). The first event of hybridization occurred between *T. urartu* (2n = 2x = 14, AA) and an *Aegilops* sp., forming the allotetraploid *T. turgidum* (2n = 4x = 28, AABB) (Marcussen et al., 2014). However, hybridization between *T. turgidum* and *Ae. tauschii* (DD) led to the formation of *T. aestivum* (Marcussen et al., 2014). *T. aestivum* genome size is approximately 17-Gb and highly repetitive with > 80% transposable elements, making it the most complex species (Pingault et al., 2015).

Wheat and its wild relatives belong to Poaceae, a grass family. Wild wheat grasses are found in different parts of the world and comprise various species like *Ae. speltoides*, *Ae. umbellulata*, *Ae. neglecta*, *Ae. caudate*, *Ae. triuncialis*, *Ae. cylindrical*, *Ae. tauschii*, *T. compactum*, *T. dicoccum*, *T. dicoccoides*, *T. carthilicum*, *T. durum*, *T. monococcum*, etc. (Ahmadi et al., 2018). Most of them are diploid and tetraploid with a few hexaploid species (van Slageren, 1994). These wild relatives are an excellent source of genetic diversity which can be utilized for trait improvement and stress tolerance in cultivated wheat (Schneider et al., 2008; Ahmadi et al., 2018). For the first time, a disease resistance gene was transferred into wheat from a wild relative (McFadden, 1930).

* Contributed equally.

Long Noncoding RNAs in Plants: Roles in development and stress
https://doi.org/10.1016/B978-0-12-821452-7.00004-0

Later, leaf rust resistance from *Ae. umbellulata* was incorporated into the wheat genome (Sears, 1956), and so on. However, after the discovery of long noncoding RNAs (lncRNAs), a new class of noncoding RNAs, their diverse biological roles emerged in the plants. These untranslated RNAs have been associated with the regulation of gene expression, epigenetic modification of chromatin structures, RNA modification, etc. (Bhatia et al., 2017). They regulate the function of the majority of coding genes directly or indirectly through miRNAs (Wang and Chekanova, 2017; Tyagi et al., 2018). The information regarding the lncRNA and their role in development and stress tolerance is limited in wheat and its wild grasses. The increasing environmental stresses and changing climatic conditions affect the growth and productivity of staple wheat crop. It is necessary to explore the importance of lncRNA in wheat and its relative grasses for high agronomic performance and better adaptation to the changing environment.

The advancement in high-throughput sequencing techniques has opened up the scope of surveying the lncRNAs at the genomic level in any species. Further, the deep transcriptome sequencing data have allowed the spatio-temporal gene expression study of lncRNAs in different plant tissues and stress conditions (Tyagi et al., 2018). Numerous lncRNAs have been identified and studied in various plant species such as *Arabidopsis thaliana*, *Oryza sativa*, *Populus trichocarpa*, *Zea mays*, *Setaria italic*, etc. (Bhatia et al., 2017). Here, we are focusing on the lncRNAs in wheat and its wild relatives. The available findings suggested the occurrence of a total 44,698 lncRNAs from 52 RNA-seq data in *T. aestivum* genome (Shumayla et al., 2017). Besides this, a total of 6309 lncRNAs were identified in *T. aestivum* in response to calcium channel blocker, out of which 177 lncRNAs were differentially expressed (Ma et al., 2018). In draft genomes of *Ae. tauschii* and *T. urartu*, a total of 20,908 and 14,515 lncRNAs, respectively, were identified from 68 RNA-seq libraries. Further, a total of 20,523 and 14,131 lncRNAs were classified as long intervening/intergenic noncoding RNAs (lincRNA) according to their genomic location in *Ae. tauschii* and *T. urartu*, respectively (Pieri et al., 2018). In the case of *T. turgidum* ssp. *Dicoccoides*, TR39477 and TTD-22 (two wild emmer wheat genotypes), and modern durum wheat, i.e., *T. turgidum* ssp. *durum var.* Kiziltan, total 61,823, 43,932, and 63,773 putative lncRNAs were identified, respectively, during drought stress (Cagirici et al., 2017). It indicates the variability in the number of lncRNAs in different species.

The lncRNAs differ from mRNAs in various aspects like the coding potential, length, AU content, number of exons, localization, etc. For instance, the average length of lncRNAs and mRNAs is 453 and 832 nucleotides (nt), respectively, in *T. aestivum* (Shumayla et al., 2017). Similarly, in *Ae. tauschii* and *T. urartu*, the lncRNAs have an average length of 464 and 987 nt, respectively, and mRNAs have 465 and 861 nucleotides, respectively (Pieri et al., 2018). Apart from this, they also share similarities with mRNAs in terms of chromosomal distribution, gene structure, gene duplication, transcription, etc. (Shumayla et al., 2017; Pieri et al., 2018). For instance, approximately 28%, 42%, and 30% of lncRNAs and

32%, 35%, and 33% of mRNAs were localized on A, B, and D sub-genomes, respectively (Shumayla et al., 2017). In *T. urartu* and *Ae. tauschii*, the lncRNAs comprise fewer exons in comparison to mRNAs. Further, their genome has approximately 60% of the transposable elements (TE), which overlap with 1376 Tu-lncRNAs and 4156 Aet-lncRNAs with the abundance of *Copia* and *Gypsy* long terminal repeat (LTR) retro-transposons, respectively (Ling et al., 2013; Jia et al., 2013). These TEs have made significant contributions to the genome size and its complexity. Therefore, the study of molecular and structural features is important for understanding the genome evolution in wheat.

The lncRNAs are known to regulate various biological processes such as vernalization, fertility, photomorphogenesis, phosphate homeostasis, cell wall synthesis, auxin-controlled development, biotic and abiotic stress responses, etc. (Liu et al., 2015). They regulate these processes through diverse mechanisms, for instance, histone modification, promoter methylation, chromatin association, translational enhancer, target mimicry, protein re-localization, alternative splicing regulators, etc. (Liu et al., 2015). In wheat and relative wild grasses, precursor and decoy lncRNAs have been identified and the mechanism of target mimicry has been explored to date (Huang et al., 2017; Ma et al., 2018). A lot of research is still needed to be done to unfold the magnificence of lncRNAs in cultivated and wild wheat to utilize them for crop improvement.

4.2 Conservation of lncRNAs among the wheat and its relative

During the process of evolution, lncRNAs were poorly conserved in the context of the primary structure and nucleotide sequence. However, the secondary structure of RNA was found somewhat conserved (Diederichs, 2014). Approximately, 5400 unique hits were found between *T. urartu* and *Ae. Tauschii*. Most of them showed conservation at the genomic level, while the remaining 28% showed non-conserveness in terms of exon-intron structure. To study the lncRNAs' conservation during evolution and domestication, the conserved patterns were searched using the whole genome alignment method among the emmer wheat (tetraploid) and the bread wheat (hexaploid). The genomic sequences of *T. urartu* lncRNAs were compared with that of A sub-genome of *T. aestivum* and *T. turgidum* ssp. dicoccoides, while the *Ae. tauschii* lncRNAs sequences were compared with the D sub-genome of *T. aestivum*. A total of 2074 lncRNAs were conserved among *Ae. tauschii* and *T. aestivum*, while 1850 and 926 lncRNAs were conserved between *T. urartu* and *T. turgidum* and *T. urartu* and *T. aestivum*, respectively. Contrary to the non-conserveness at the sequence level, the *cis*-acting regions were more conserved (Amaral et al., 2016). Further investigation in the promoter region of lncRNAs revealed the conservation of 2022 regulatory regions in *T. urartu* and A sub-genome of bread wheat and 3581 regulatory regions in *T. urartu* and emmer wheat genome. Additionally, a total of 4227 promoters were conserved in *Ae. tauschii* and D sub-genome of bread wheat. Interestingly, during the process of evolution and domestication, the comparison of A genome of three species resulted in the conservation of 557 lncRNA and 1888 regulatory regions.

4.3 LncRNAs in growth and development

Wheat encounters climatic variations at different stages of its life cycle, which are more harmful at the reproductive phase. The stresses at the reproductive phase affect the number, grain setting, dry weight, and size of the wheat kernel (Kajla et al., 2015). The advancement in sequencing technologies has uncovered a wide range of lncRNAs in various plants (Chung et al., 2016; Niu et al., 2018; Pingault et al., 2015). Few of these lncRNAs are involved in plant growth and development; however, the role of most of the lncRNAs is still not known. Expression profiling is an approach to determine the biological function of a gene. Various lncRNAs are known to modulate the expression of a gene involved in multiple pathways and a wide range of development mechanisms (Ulitsky, 2016; Chekanova, 2015). Some of the reports have elucidated that lncRNAs control the expression of genes through the chromatin modification mechanism in plants. For instance, in Arabidopsis, the *AUXIN REGULATED PROMOTER LOOP RNA* controls the expression of *PINOID* (auxin polar transporter) in the root (Ariel et al., 2014). In rice, normal pollen development is mediated by the *LONG DAY SPECIFIC MALE FERTILITY ASSOCIATED RNA* (*LDMAR*) (Ding et al., 2012). However, only a few studies have been done in wheat and its progenitors which suggest the involvement of lncRNAs in reproductive processes. Pieri et al. (2018) examined the organ-specific expression level of lncRNAs in *Ae. tauschii* and *T. urartu*. Thousands of lncRNAs showed significantly modulated expression in various organs, the majority of them were found in spike tissue. However, the expression level of these lncRNAs was very low across all the tissues in both the species, but they have strong organ specificity. Despite the low expression level, they showed evident-modulated expression in different organs such as root, seed, and spike (Pieri et al., 2018). This has developed the hypothesis of their important role in reproductive organs. It was also found that organ-specific lncRNAs of both the species showed association with some transposable elements, indicating their diverse possible role in origin, differentiation, and function (Pieri et al., 2018). These associated lncRNAs were also highly organ-specific with predominant accumulation in the immature and young pistil, stamen, and spike, suggesting their role in the reproductive processes. Further, a total of 22,549 lncRNAs in *T. aestivum* showed involvement in growth and development (Shumayla et al., 2017). Their role was explored by expression profiling in three developmental stages of five different tissues such as root, leaf, stem, spike, and grain. A majority of lncRNAs exhibited differential expression in these tissues, suggesting their role in tissue development. It was further stated that rather than being tissue-specific, these lncRNAs were also developmental stage-specific. A total of 485 lncRNAs were found to be specific to a particular development stage. Majority of lncRNAs were expressed in the later stage of tissue development like 135, 55, and 102 lncRNAs were specific to grain_Z85, spike_Z65, and leaf_Z71 stages, respectively. For instance, TR9409 and TR8446 were specific to leaf_Z71, while TR1756 and TR1873 were specifically expressed in grain_Z85 (Shumayla et al., 2017). This suggested the major role of wheat lncRNAs in growth and development.

Further, Ma et al. (2018) studied the function of lncRNAs in the growth of wheat roots in response to Ca^{2+}-channel blocking. The GO enrichment analysis showed the involvement of 23 lncRNAs in the gene regulation because of the transcription factor activity of their target gene. They also showed that, out of 23, eight lncRNAs were upregulated and 15 were downregulated. However, 18 and five lncRNAs were positive and negative to their targets, respectively (Ma et al., 2018). The developmental roles of lncRNAs in wheat and its wild progenitors are still limited which needs to be further explored in future studies.

4.4 Biotic stress-responsive lncRNAs

Complex traits regulate the biotic stress responses in plants, which comprise plant-pathogen interaction and host specificity. Various proteins, lncRNAs, transcription factors, etc., are involved in stress tolerance in plants. In the case of wheat and its wild relatives, several genes are known for providing disease resistance against specific pathogens (Zhang et al., 2013, 2016). However, they come across various pathogenic fungi and viruses due to the loss of disease resistance genes in cultivars with time or due to the blooming of resistant pathogens. The addition of disease resistance traits along with the other desirable traits is necessary to enhance productivity without yield loss. The role of lncRNAs in providing the biotic stress tolerance and regulating the other disease-resistant genes is necessary for understanding their mechanism in response to the different pathogens.

Blumeria graminis f. sp. *tritici* (Bgt) and *Puccinia striiformis* f. sp. *tritici* (Pst) are two major fungal pathogens of wheat causing powdery mildew and stripe rust disease, respectively (Li et al., 2014; Chen et al., 2014). They cause significant damage to the crop and affect their growth and yield. In response to powdery mildew, the expression of 71 lncRNAs of wheat was affected, out of which only 48 were specifically responsive to *Bgt*. The upregulated expression of *TapmlnRNA19* and *TapmlnRNA30* was also confirmed through qRT PCR after powdery mildew inoculation (Xin et al., 2011). Apart from this, three lncRNAs (*TalnRNA5*, *TapmlnRNA8*, and *TapmlnRNA19*) were responsive to *Bgt* infection and also act as the precursors of miRNAs. *TapmlnRNA8* acts as the precursor of miR2066, while *TalnRNA5* and *TapmlnRNA19* are precursors of miR2004. The small RNA high-throughput sequencing data indicate the upregulated expression of miR2004 after *Bgt* infestation, suggesting the regulatory role of these lncRNAs in biotic stress (Xin et al., 2011).

In response to defense against stripe rust, Zhang et al. (2013) worked on elucidating the molecular regulatory mechanisms of lncRNAs in wheat. They identified four intron-less genes encoding *TalncRNA18*, *TalncRNA73*, *TalncRNA106*, and *TalncRNA108*, which were 1393, 667, 449, and 647 nucleotides, respectively. The homology of *TalncRNA18* with E3 ubiquitin-protein ligase UPL1-like gene and *TalncRNA108* with rRNA intron-encoded homing endonuclease suggests their involvement in many biological processes. However, *TalncRNA73* represents an antisense transcript of hypothetical protein and *TalncRNA106* shares no similarity with a known sequence. The expression of these lncRNAs was upregulated at 1 and 3 dpi (days postinoculation) contrarily, the expression of *TalncRNA108* showed downregulation at early stages (Zhang et al., 2013). The modulated expression of these lncRNAs suggested their role in regulating the expression of other resistance genes.

The transcriptome analysis of wheat line N9134 inoculated with *Pst* and *Bgt* at 1, 2, and 3 dpi suggested the involvement of long intergenic ncRNAs (*lincRNAs*) against these pathogens. A total of 254 and 52 *lincRNAs* were differentially expressed after *Bgt* and *Pst* infestation, respectively, of which 23 *lincRNAs* were affected during both the stress conditions (Zhang et al., 2016). These lncRNAs also showed temporal expression after pathogen attack. For instance, after 1, 2, and 3 dpi of *Pst* infestation, 30, 18, and 14 *lincRNAs* were differentially expressed, respectively, while 98, 112, and 144 *lincRNAs* showed differential expression after *Bgt* infestation at corresponding time points. In addition to this, 25%–38% *lincRNA* were differentially expressed at different time points during *Bgt* stress in contrast to *Pst* stress, where a few lincRNA shared different time points of stress. In response to different pathogens, the gene expression profile of *lncRNAs* proposed their role in pathogen-defense mechanisms (Zhang et al., 2016). During fungal stress, differentially expressed *lincRNAs* also play an important role in spliceosome and regulate alternative splicing events (Zhang et al., 2016).

In the case of *T. dicoccoides*, a wild progenitor of wheat, the functional annotation of miRNA targets revealed their role in diverse biological processes and stress conditions. For example, the members of miR2118 family target *Tsn1* gene (involved in response to tan spot disease) and control the biotic stress (Akpinar et al., 2018). The lncRNAs may act as a precursor of miRNAs involved in biotic stress. So far, a detailed functional characterization of lncRNAs is limited in the wheat and its wild relatives. Moreover, the wild relatives might be targeted as an exotic source for disease resistance genes which might get lost in bread wheat during the domestication and cultivation process.

4.5 Abiotic stress-responsive lncRNAs

The environmental stress often decreases growth and yield; therefore, the research is focused on improving the stress tolerance of most of the crop plants (Tran and Mochida, 2010; Prado et al., 2014; Ronald, 2014; Nemali et al.,

2015). The unpredictable climatic change in recent years has affected wheat development and its productivity (Wang and Xia, 2018). Most of the reduction in crop yield is due to the various abiotic stresses such as extreme temperature, salinity, drought, and various other kinds of stresses.

Several studies have shown the involvement of long noncoding RNAs in various abiotic stresses (Zhang et al., 2014; Chung et al., 2016; Shumayla et al., 2017). Shumayla et al. (2017) have reported the role of lncRNAs under various abiotic stresses such as heat stress (HS), drought stress (DS), heat-drought combined stress (HD), and salt stress. A total of 2064 lncRNAs were upregulated after 1 h of HS, DS, and HD treatments, for instance, TR25521, TR22764, TR23279, etc. In contrast, 2278 lncRNAs showed downregulation after 1 h of the above treatments, for instance, TR24190, TR25702, and TR23279. However, after 6 h treatment of HS, DS, and HD, 2024 lncRNAs were upregulated, for instance, TR22764 and TR893. However, 1970 lncRNAs including TR20814, TR34104, and TR24754 were downregulated. Cagirici et al. (2017) performed the comparative study of the differentially expressed lncRNAs and coding genes in response to drought stress for understanding the regulation of gene expression. Two lncRNAs (lncRNA_c118446_g1_i1 and lncRNA_c47700_g1_i1) of Kiziltan, (moderate drought-tolerant wheat variety) showed enhanced expression in response to drought stress (Cagirici et al., 2017). In response to heat stress, Xin et al. (2011) proposed the function of 54 lncRNAs in wheat. Further, the upregulated expression of *TahlnRNA27* and *TalnRNA5* suggested their role in heat stress.

The functionality of lncRNAs has been described in response to salinity stress, another major abiotic factor affecting wheat yield globally (Zhang et al., 2016; Shumayla et al., 2017). The temporal expression of lncRNAs at 6, 12, 24, and 48 h of salt stress suggested their putative role at different time intervals. For instance, TR2912, TR2728, TR17935, and TR15627 were highly upregulated lncRNAs after 6, 12, 24, and 48 h of salt stress, respectively (Shumayla et al., 2017). Also, numerous lncRNAs were early responsive (6 and 12 h) and late responsive (24 and 48 h) to the salinity stress, for instance, TR2556 and TR5273, respectively (Shumayla et al., 2017).

During abiotic stress, abscisic acid (ABA) is involved in the plant signalling through the activation of various transcription factors (TFs) and regulates the stress-responsive pathways (Saibo et al., 2009; Lindemose et al., 2013). The putative role of lncRNAs in ABA biosynthesis in wheat has been discussed by Shumayla et al. (2017). They identified 12 development-related, 14 heat and drought stress-related, and 13 salt stress-related lncRNAs co-expressed with mRNAs encoding BCH, ZEP, NCED, and AAO enzymes involved in ABA biosynthesis (Shumayla et al., 2017).

4.6 Role of co-expressing lncRNAs

The amount of lncRNAs in the genome directly corresponds to the genome complexity which further promotes their regulatory roles (Guleria et al., 2011;

Budak et al., 2016). The gene ontology (GO) mapping and annotation of mRNAs predict the function of co-expressing lncRNAs (Li et al., 2017). The co-expression analysis of lncRNAs with mRNAs during growth and development and various abiotic stress conditions was performed to study the involvement of lncRNAs in these conditions (Shumayla et al., 2017). During plant development, a total of 185, 68, and 29 lncRNAs were found co-expressing with 75, 39, and 12 mRNAs in grain, leaf, and spike, respectively. These co-expressing mRNAs were annotated as leucine-rich repeat receptor-like kinases, cysteine-rich receptor-like kinases, disease-resistant protein RPM1, etc. These genes are known to be involved in numerous developmental and biological processes which further suggested the similar function of co-expressing lncRNAs in wheat (Ljungberg et al., 1986; Nordin et al., 1993; Hubert et al., 2003; Okuda et al., 2009; Saito and Ueda, 2009; Gu et al., 2016; Xing et al., 2017).

During HS, DS, and HD stress conditions, 509 lncRNAs showed co-expression with respective mRNAs. However, under salt stress, 1853 lncRNAs were co-expressed with respective mRNAs, out of which 381 were also co-expressing during developmental stages and 32 were co-expressing during heat-drought stress. It suggests the overlapping role of numerous lncRNAs in the developmental process and different stress conditions. Further, the co-expressed mRNAs function in response to abiotic and biotic stimulus, transport, biological process regulation and the myriad biosynthetic processes (Shumayla et al., 2017).

The transcription factors (TFs) are reported to be involved in the regulation of various pathways related to development and stress. Therefore, the co-expression analysis of lncRNAs with TFs was performed (Saibo et al., 2009; Lindemose et al., 2013; Wang et al., 2016; Shumayla et al., 2017). A total of 121 TFs were reported as co-expressing with lncRNAs during the developmental stages of wheat. In these developmental stages, the TFs of BZIP, HSF, MYB NAC, and WRKY families were highly enriched. However, during various abiotic stresses like heat, drought, heat-drought, and salt stresses, a total of 181 TFs of 34 TF families and 828 TFs of 48 different families were found to be co-expressing with lncRNAs. During the above stress conditions, the TFs of WRKY, ERF, C2H2, Bhlh, C3H, MYB-related, etc. families were highly enriched. Further, co-expression analysis during developmental and various abiotic stresses showed the enrichment of several Auxin-related TFs such as HD-Zip, bHLH, BZIP, ERF, ARF, etc. The TFs related to hormones are reported to be involved in various functions of plants. (Chapmen and Estelle, 2009; Chew et al., 2013; Chen et al., 2017). The co-expression analysis of TFs with lncRNAs showed their involvement in myriad regulatory processes. Further, the co-expression analysis of few lncRNAs during the heat, drought, and salt stresses with ABA-related TFs like MYB, WRKY, bZIP, C2H2, ERF, etc. predicted their involvement in ABA-dependent signalling. Few lncRNAs were found co-expressing with ABA-independent TFs such as NAC and ZF-HD, predicting their involvement in ABA-independent signalling.

Further, Ma et al. (2018) found 32 lncRNAs related to 63 miRNAs based on co-expression and correlation in wheat roots which were responsible for Ca^{2+}-channel blocking. Out of 63, 21 lncRNAs were found to be positively correlated with 58 miRNAs, and 12 lncRNAs were found to be negatively correlated with 16 miRNAs. Additionally, 165 lncRNAs were found co-expressing with 1626 mRNAs, 157 lncRNAs of them were positively correlated with 1253 mRNAs and 82 were negatively correlated with 738 mRNAs. GO enrichment analysis was performed to study the function of these co-expressing lncRNAs. These differentially expressed lncRNAs are reported to have roles in cellular components, biological processes, and molecular functions (Ma et al., 2018).

4.7 LncRNA-miRNA-mediated gene regulation

The current knowledge of gene regulatory mechanisms of lncRNAs is infantry in plants, especially in wheat. However, lncRNAs are known to function as miRNA decoy or precursor and regulate numerous biological pathways. The first well-known example of target mimicry or miRNA decoy was reported in Arabidopsis for lncRNA gene '*INDUCED BY PHOSPHATE STARVATION 1* (*IPS1*),' which mimics the target of miR399. Subsequently, the inactivation of miR399 leads to the upregulation of its primary target, i.e., *PHOSPHATE 2* (*PHO2*) gene (Franco-Zorrilla et al., 2007). The miRNA triggers gene expression through target mRNA cleavage and translational inhibition process (Llave et al., 2002; Brodersen et al., 2008). The stem-loop structure of precursor lncRNAs is processed for the formation of mature miRNAs by Dicer-like enzyme, and together with the silencing complex, they direct complementary mRNA cleavage (Jones-Rhoades et al., 2006).

In the wheat root, eight lncRNAs were identified as a putative precursor of four miRNAs in response to Ca^{2+}-channel blocker. Intriguingly, different miRNAs target the same lincRNAs, for instance, two sites of lncRNA_033754 matched with each ssp-miR444b.2_R15-1L21 and tae-miR1121_R3-21L22. Further, each lncRNAs has multiple target sites for single miRNA, for instance, lncRNA_051551 consists of four sites similar to bdimiR1127_R18-4L21, lncRNA_042235 have two sites of ssp-miR444b.2_R15-1L21, and one site of lncRNA_054399 showed similarity with mtr-miR7701-5p_R18-4L21 (Ma et al., 2018). These lncRNAs have a role in wheat root growth in response to the blocking of Ca^{2+}-channel such as smo-miR159_R2-21L21 plays a role in cell proliferation (Palatnik et al., 2007). Further, lncRNA_061738, lncRNA_074658, and lncRNA_047461 might act as decoys for miRNAs and regulate cell-cycle and growth process (Ma et al., 2018). In the case of fungal-responsive lincRNAs in wheat, a total of 101 lincRNAs were identified as putative targets of miRNA. Out of this, 39 lincRNAs were triggered by miRNA for translational inhibition. Apart from this, miRNA targets numerous functional coding genes, for instance, five functional genes were regulated by tae-miR1137a (Zhang et al., 2016).

These sequence-dependent interactions of lncRNA-miRNA-mRNA were identified in various tissue developmental stages of wheat and during abiotic stress conditions like heat, drought, a combination of heat and drought, salt stresses, etc. A total of 222 miRNAs have putative target sites in both 1047 lncRNAs and 209 mRNAs. The gene ontology (GO) mapping of these mRNAs suggested their role in cellular, metabolic, biosynthetic, and biological processes (Shumayla et al., 2017). Further, 19 lncRNAs act as a putative precursor of 28 miRNAs. Here also, lncRNAs were identified with putative multiple target sites or single target site in multiple lncRNAs (Shumayla et al., 2017).

In durum wheat, *INHIBITOR of WAX1* (*Iw1*) gene (a lncRNA) has been described to control the glaucousness and β-diketone waxes through miRNA. These waxes form the major composition of the epicuticular layer of Triticeae, and during the reproductive-age, provide bluish-white glaucous trait to the crop. *Iw1* acts as a precursor of miRW1 (a miRNA) which suppresses a *carboxylesterase-like protein* (*W1-COE*) gene in WAX1 (W1) multigene locus (Huang et al., 2017). *Iw1* was also identified in W1 locus on chromosome 2B; however, in *Ae. Tauschii*, the paralog *Iw2* gene was present on W2 locus on chromosome 2D. Therefore, the introduced *Iw1* in glaucous bread wheat resulted in the upregulation of miRW1 and downregulation of *W1-COE* and *W2-COE* and phenotype was nonglaucous and lacked β-diketone waxes (Huang et al., 2017). It suggests that lncRNA may regulate the functional genes through miRNAs.

4.8 Conclusions

The chapter revealed that wheat lncRNAs function as important regulators and exert crucial roles in various biological processes. Various lncRNAs have been identified from Arabidopsis, maize and rice, and several have been functionally characterized. However, the study is still limited in wheat and its wild progenitors. The regulatory role of these lncRNAs in development and various stress responses not only increases our understanding of lncRNAs interaction but also reveals their involvement in various signalling pathways. The co-expression analysis of wheat lncRNAs uncovers its myriad role in growth and development, various stress responses, and hormonal pathways. In the future study, it would be indispensable to further explore various functional motifs and their secondary or tertiary structure to understand the different gene regulatory mechanisms and to develop effective approaches to investigate their target genes. These events would provide a new way into the lncRNAs-mediated gene regulatory network during the entire plant life cycle.

Acknowledgments

Authors are thankful to Panjab University, Chandigarh, India, for providing the research facility and infrastructure. S and ST are thankful to the CSIR and the Science and Engineering Research Board (SERB) for research fellowship, respectively. SKU is thankful to SERB,

Government of India, for Early Career Research Award (ECR/2016/001270) and CSIR for research grant (38(1489)/19/EMR-II). We are thankful to the Department of Science and Technology, Government of India, for partial financial support under Promotion of University Research and Scientific Excellence (PURSE) grant scheme.

References

Ahmadi, J., Pour-Aboughadareh, A., Ourang, S.F., Mehrabi, A.A., Siddique, K.H.M., et al., 2018. Wild relatives of wheat: *Aegilops–Triticum* accessions disclose differential antioxidative and physiological responses to water stress. Acta Physiol. Plant. 40, 90.

Akpinar, B.A., Biyiklioglu, S., Alptekin, B., Havrankova, M., Vrana, J., Dolezel, J., et al., 2018. Chromosome-based survey sequencing reveals the genome organization of wild wheat progenitor Triticum dicoccoides. Plant Biotechnol. J. 16, 2077–2087.

Amaral, P.P., Leonardi, T., Han, N., Vire, E., Gascoigne, D.K., Arias-Carrasco, R., et al., 2016. Genomic positional conservation identifies topological anchor point (tap) RNAs linked to developmental loci. bioRxiv. http://biorxiv.org/content/early/2016/05/04/051052.

Ariel, F., Jegu, T., Latrasse, D., Romero-Barrios, N., Christ, A., Benhamed, M., et al., 2014. Non-coding transcription by alternative RNA polymerases dynamically regulates an auxin-driven chromatin loop. Mol. Cell 55, 383–396.

Bhatia, G., Goyal, N., Sharma, S., Upadhyay, S.K., Singh, K., 2017. Present scenario of long non-coding RNAs in plants. Non Coding RNA 3, 16.

Brodersen, P., Sakvarelidze-Achard, L., Bruun-Rasmussen, M., Dunoyer, P., Yamamoto, Y.Y., Sieburth, L., et al., 2008. Widespread translational inhibition by plant miRNAs and siRNAs. Science 320, 1185–1190.

Budak, H., Bulut, R., Kantar, M., Alptekin, B., 2016. MicroRNA nomenclature and the need for a revised naming prescription. Brief. Funct. Genomics 15, 65–71.

Cagirici, H.B., Alptekin, B., Budak, H., 2017. RNA sequencing and co-expressed long non-coding RNA in modern and wild wheats. Sci. Rep. 7 (1), 10670.

Chapmen, E.J., Estelle, M., 2009. Mechanism of Auxin regulated gene expression in plants. Annu. Rev. Genet. 43, 265–285. https://doi.org/10.1146/annurev-genet-102108-134148.

Chekanova, J.A., 2015. Long non-coding RNAs and their functions in plants. Curr. Opin. Plant Biol. 27, 207–216.

Chen, W., Wellings, C., Chen, X., Kang, Z., Liu, T., 2014. Wheat stripe (yellow) rust caused by *Puccinia striiformis* f. sp. *tritici*. Mol. Plant Pathol. 15 (5), 433–446.

Chen, Z., Yuan, Y., Fu, D., Shen, C., Yang, Y., 2017. Identification and expression profiling of the auxin response factors in dendrobium officinale under abiotic stresses. Int. J. Mol. Sci. 18, 927. https://doi.org/10.3390/ijms18050927.

Chew, W., Hrmova, M., Lopato, S., 2013. Role of homeodomain leucine zipper (HD-Zip) IV transcription factors in plant development and plant protection from deleterious environmental factors. Int. J. Mol. Sci. 14, 8122–8147. https://doi.org/10.3390/ijms14048122.

Chung, P.J., Jung, H., Jeong, D.H., Ha, S.H., Choi, Y.D., Kim, J.K., 2016. Transcriptome profiling of drought responsive non-coding RNAs and their target genes in rice. BMC Genomics 17, 563.

Diederichs, S., 2014. The four dimensions of noncoding RNA conservation. Trends Genet. 30, 121–123.

Ding, J., Lu, Q., Ouyang, Y., Mao, H., Zhang, P., Yao, J., et al., 2012. A long noncoding RNA regulates photoperiod-sensitive male sterility, an essential component of hybrid rice. Proc. Natl. Acad. Sci. U. S. A. 109, 2654–2659.

Franco-Zorrilla, J.M., Valli, A., Todesco, M., Mateos, I., Puga, M.I., Rubio-Somouza, I., et al., 2007. Target mimicry provides a new mechanism for regulation of microRNA activity. Nat. Genet. 39, 1033–1037.

Gale, M.D., Devos, K.M., 1998. Comparative genetics in the grasses. Proc. Natl. Acad. Sci. 95, 1971–1974.

Gu, Q., Chen, M., Huang, J., Wei, Y.D., Hsiang, T., Zheng, L., 2016. Multifaceted roles of the Ras guanine-nucleotide exchange factor ChRgf in development, pathogenesis, and stress responses of Colletotrichum higginsianum. Phytopathology 107, 433–443. https://doi.org/10.1094/PHYTO-03-16-0137-R.

Guleria, P., Mahajan, M., Bhardwaj, J., Yadav, S.K., 2011. Plant small RNAs: biogenesis, mode of action and their roles in abiotic stresses. Genomics Proteomics Bioinformatics 9, 183–199.

Huang, D., Feurtado, J.A., Smith, M.A., Flatman, L.K., Koh, C., Cutler, A.J., 2017. Long noncoding miRNA gene represses wheat β-diketone waxes. Proc. Natl. Acad. Sci. 114 (15), E3149–E3158.

Hubert, D.A., Tornero, P., Belkhadir, Y., Krishna, P., Takahashi, A., Shirasu, K., et al., 2003. Cytosolic HSP90 associates with and modulates the Arabidopsis RPM1 disease resistance protein. EMBO J. 22, 5679–5689. https://doi.org/10.1093/emboj/cdg547.

Jia, J., Zhao, S., Kong, X., Li, Y., Zhao, G., He, W., et al., 2013. *Aegilops tauschii* draft genome sequence reveals a gene repertoire for wheat adaptation. Nature 496, 91–95.

Jones-Rhoades, M.W., Bartel, D.P., Bartel, B., 2006. MicroRNAs and their regulatory roles in plants. Annu. Rev. Plant Biol. 57 (1), 19–53.

Kajla, M., Yadav, V., Chhokar, R.S., Sharma, R.K., 2015. Management practices to mitigate the impact of high temperature on wheat. J. Wheat Res 7 (1), 1–12.

Kilian, B., Ozkan, H., Deusch, O., Effgen, S., Brandolini, A., Kohl, J., et al., 2007. Independent wheat B and G genome origins in outcrossing Aegilops progenitor haplotypes. Mol. Biol. Evol. 24, 217–227.

Li, Z., Lan, C., He, Z., Singh, R.P., Rosewarne, G.M., Chen, X., et al., 2014. Overview and application of QTL for adult plant resistance to leaf rust and powdery mildew in wheat. Crop. Sci. 54 (5), 1907–1925.

Li, J., Li, C., Wang, J., Song, G., Zhao, Z., Wang, H., et al., 2017. Genome-wide analysis of differentially expressed lncRNAs and mRNAs in primary gonadotrophin adenomas by RNA-seq. Oncotarget 8, 4585–4606. https://doi.org/10.18632/oncotarget.13948.

Lindemose, S., O'Shea, C., Jensen, M., Skriver, K., 2013. Structure, function and networks of transcription factors involved in abiotic stress responses. Int. J. Mol. Sci. 14, 5842–5878. https://doi.org/10.3390/ijms14035842.

Ling, H.Q., Zhao, S., Liu, D., Wang, J., Sun, H., Zhang, C., et al., 2013. Draft genome of the wheat A-genome progenitor *Triticum urartu*. Nature 496, 87–90.

Liu, X., Hao, L., Li, D., Zhu, L., Hu, S., 2015. Long non-coding RNAs and their biological roles in plants. Genomics Proteomics Bioinformatics 13 (3), 137–147.

Ljungberg, U., Akerlund, H.E., Andersson, B., 1986. Isolation and characterization of the 10-kDa and 22-kDa polypeptides of higher plant photosystem 2. Eur. J. Biochem. 158, 477–482. https://doi.org/10.1111/j.1432-1033.1986.tb09779.x.

Llave, C., Xie, Z., Kasschau, K.D., Carrington, J.C., 2002. Cleavage of Scarecrow-like mRNA targets directed by a class of Arabidopsis miRNA. Science 297, 2053–2056.

Ma, K., Shi, W., Xu, M., Liu, J., Zhang, F., 2018. Genome-wide identification and characterization of long non-coding RNA in wheat roots in response to Ca^{2+} channel blocker. Font. Plant Sci. 9, 244. https://doi.org/10.3389/fpls.2018.00244.

Marcussen, T., Sandve, S.R., Heier, L., Spannagl, M., Pfeifer, M., Jakobsen, K.S., et al., 2014. Ancient hybridizations among the ancestral genomes of bread wheat. Science 345, 1250092.

McFadden, E.S., 1930. A successful transfer of emmer characters to vulgare wheat. J. Am. Soc. Agron. 22, 1020–1034.

Nemali, K.S., Bonin, C., Dohleman, F.G., Stephens, M., Reeves, W.R., Nelson, D.E., et al., 2015. Physiological responses related to increased grain yield under drought in the first biotechnology-derived drought-tolerant maize. Plant Cell Environ. 38, 1866–1880.

Niu, D.D., Zhang, X., Song, X.O., Wang, Z.H., Li, Y.Q., Qiao, L.L., et al., 2018. Deep sequencing uncovers rice long siRNAs and its involvement in immunity against *Rhizoctonia solani*. Phytopathology 108, 60–69.

Nordin, K., Vahala, T., Palva, E.T., 1993. Differential expression of two related, low-temperature-induced genes in Arabidopsis thaliana (L.) Heynh. Plant Mol. Biol. 21, 641. https://doi.org/10.1007/BF00014547.

Okuda, S., Tsutsui, H., Shiina, K., Sprunck, S., Takeuchi, H., Yui, R., et al., 2009. Defensin-like polypeptide LUREs are pollen tube attractants secreted from synergid cells. Nature 458, 357–361. https://doi.org/10.1038/nature07882.

Palatnik, J.F., Wollmann, H., Schommer, C., Schwab, R., Boisbouvier, J., Rodriguez, R., et al., 2007. Sequence and expression differences underlie functional specialization of Arabidopsis, microRNAs miR159 and miR319. Dev. Cell 13, 115–125.

Pieri, A., Pè, M.E., Bertolini, E., 2018. Long non-coding RNAs in wild wheat progenitors. https://doi.org/10.1101/301804.

Pingault, L., Choulet, F., Alberti, A., Glover, N., Wincker, P., Feuillet, C., et al., 2015. Deep transcriptome sequencing provides new insights into the structural and functional organization of the wheat genome. Genome Biol. 16, 29.

Prado, J.R., Segers, G., Voelker, T., et al., 2014. Genetically engineered crops: from idea to product. Annu. Rev. Plant Biol. 65, 769–790.

Ronald, P.C., 2014. Lab to farm: applying research on plant genetics and genomics to crop improvement (S Kamoun, Ed.). PLoS Biol. 12, e1001878.

Saibo, N.J.M., Lourenço, T., Oliveira, M.M., 2009. Transcription factors and regulation of photosynthetic and related metabolism under environmental stresses. Ann. Bot. 103, 609–623. https://doi.org/10.1093/aob/mcn227.

Saito, C., Ueda, T., 2009. Chapter 4 functions of RAB and SNARE proteins in plant life. Int. Rev. Cell Mol. Biol. 274, 183–233. https://doi.org/10.1016/S1937-6448(08)02004-2.

Schneider, A., Molnár, I., Molnár-Láng, M., 2008. Utilisation of *Aegilops* (goatgrass) species to widen the genetic diversity of cultivated wheat. Euphytica 163, 1–19.

Sears, E.R., 1956. The transfer of leaf rust resistance from *Aegilops umbellulata* to wheat. Brookhaven Symp. Biol. 9, 1–22.

Shumayla, Sharma, S., Taneja, M., Tyagi, S., Singh, K., Upadhyay, S.K., 2017. Survey of high throughput RNA-seq data reveals potential roles for lncRNAs during development and stress response in bread wheat. Front. Plant Sci. 8, 1019.

Tran, L.-S.P., Mochida, K., 2010. Functional genomics of soybean for improvement of productivity in adverse conditions. Funct. Integr. Genomics 10, 447–462.

Tyagi, S., Sharma, A., Upadhyay, S.K., 2018. Role of next-generation RNA-seq data in discovery and characterization of long non-coding RNA in plants, next generation plant breeding, Yelda Özden Çiftçi. IntechOpen. https://doi.org/10.5772/intechopen.72773.

Ulitsky, I., 2016. Evolution to the rescue: using comparative genomics to understand long non-coding RNAs. Nat. Rev. Genet. 17, 601–614.

van Slageren, M., 1994. Wild wheats: a monograph of Aegilops L. and Amblyopyrum (Jaub. & Spach) Eig (Poaceae). In: ICARDA/Wageningen Agricultural University Papers. vol. 94(7). i–xiv, 1–512.

Wang, H.L.V., Chekanova, J.A., 2017. Long noncoding RNAs in plants. In: Rao, M. (Ed.), Long Non Coding RNA Biology. Adv. Exp. Med. Biol. 1008. Springer, Singapore.

Wang, M., Xia, G., 2018. The landscape of molecular mechanisms for salt tolerance in wheat. Crop J. 6, 42–47.

Wang, H., Wang, H., Shao, H., Tang, X., 2016. Recent advances in utilizing transcription factors to improve plant abiotic stress tolerance by transgenic technology. Front. Plant Sci. 7, 67. https://doi.org/10.3389/fpls.2016.00067.

Xin, M., Wang, Y., Yao, Y., Song, N., Hu, Z., Qin, D., et al., 2011. Identification and characterization of wheat long non-protein coding RNAs responsive to powdery mildew infection and heat stress by using microarray analysis and SBS sequencing. BMC Plant Biol. 11, 61.

Xing, X., Li, X., Zhang, M., Wang, Y., Liu, B., Xi, Q., et al., 2017. Transcriptome analysis of resistant and susceptible tobacco (Nicotiana tabacum) in response to root-knot nematode Meloidogyne incognita infection. Biochem. Biophys. Res. Commun. 482, 1114–1121. https://doi.org/10.1016/j.bbrc.2016.11.167.

Zhang, H., Chen, X., Wang, C., Xu, Z., Wang, Y., Liu, X., et al., 2013. Long non-coding genes implicated in response to stripe rust pathogen stress in wheat (*Triticum aestivum* L.). Mol. Biol. Rep. 40, 6245–6253.

Zhang, J. Y., Liao, Z. Y., Li, Y., Yu, J. P., Zhang, Q. F., Li, et al., 2014. Genome-wide screening and functional analysis identify a large number of long noncoding RNAs involved in the sexual reproduction of rice. Genome Biol., 15, 512.

Zhang, H., Hu, W., Hao, J., Lv, S., Wang, C., Tong, W., et al., 2016. Genome-wide identification and functional prediction of novel and fungi-responsive lincRNAs in *Triticum aestivum*. BMC Genomics 17, 238.

Chapter 5

Recent advancements in long noncoding RNA-mediated stress responses in rice

Shalini Tiwari[a] and Charu Lata[b]
[a]*CSIR-National Botanical Research Institute, Lucknow, India,* [b]*CSIR-National Institute of Science Communication and Information Resources, New Delhi, India*

5.1 Introduction

Nowadays, a reduction in crop productivity due to adverse environmental conditions is gaining major concern globally. Numerous defense mechanisms have been evolved by plants to withstand stress conditions. Various omics approaches are commonly used, such as transcriptomics, proteomics, and metabolomics, to decipher the regulatory networks participating in this defense mechanism (Tiwari and Lata, 2018). Such various functional genomic methods help to understand molecular networking better and grow better varieties with increased stress tolerance (Mosa et al., 2017). Among various crop plants, rice (*Oryza sativa* L.) is one of the world's most prominent grain crops. Different adverse conditions including both abiotic and biotic stress limit its production worldwide (Li et al., 2018; Tiwari et al., 2017a). Like other plants, rice has also evolved complex networks to protect itself from exposure to environmental stresses. Several researches on underlying genetic mechanisms against numerous stresses in rice have been studied till date; however, role of noncoding RNAs (ncRNAs) during stress conditions is gaining momentum nowadays. Currently, a progressive increase in research discovered that long noncoding RNAs (lncRNAs), a class of ncRNAs, play a role against multiple adverse conditions (Golicz et al., 2018). LncRNAs are defined as transcripts with a length of more than 200 bp and have no potential for coding (Golicz et al., 2018; Li et al., 2018). In modern molecular biology, studies related to lncRNA have become one of the most recent hotspots.

Nevertheless, research in plants is still premature compared to human and animal studies. LncRNAs exhibit significant functions in different biological processes and transcriptional regulation in higher eukaryotes including plants (Guha et al., 2019; Golicz et al., 2018). LncRNAs are further divided into three

Long Noncoding RNAs in Plants: Roles in development and stress
https://doi.org/10.1016/B978-0-12-821452-7.00005-2

subgroups based on the relative position of lncRNA loci to protein-coding genes: (1) long intergenic noncoding RNAs (lincRNAs); (2) natural antisense transcripts (NATs); and (3) intronic lncRNAs. Noticeably, lncRNAs can synthesize small nuclear RNAs (snRNAs), small nucleolar RNAs (snoRNAs), and miRNAs (Amor et al., 2009). Since the discovery of lncRNA in mammals, many lncRNAs have been reported in various crops, regulating important agricultural characteristics (Joshi et al., 2016; Liu et al., 2012). Recent discoveries also unravel the association of lncRNAs in the remodeling of chromatin via histone-modifications (Campos and Reinberg, 2009). Therefore, this article mainly summarizes the recent advances in lncRNAs-mediated stress tolerance in rice crops.

5.2 Identification of long noncoding RNA from rice

Over the past few decades, due to diverse functionality, the lncRNAs have gained interest in different biological processes. Identification of lncRNAs is still challenging; however, with the help of computational method for lncRNAs, identification in many species is now possible due to their complete genome sequencing and annotations. But in a wide range of species, identification remains challenging due to lack of well-established sequences, annotations, and prior knowledge. A flowchart displaying arrays of computational methods for the identification of lncRNA is illustrated in Fig. 5.1. Due to the well-annotated genome sequence of rice, the identification of lncRNAs in rice is comparatively easier than other crop plants. For example, to annotate lncRNAs, Singh et al. (2017) developed new tool PLncPRO and input transcriptomes of rice (42,064) in tool with default parameters and annotated a total of 7345 transcripts in rice as lncRNAs. Comparing these data to other previous studies on lncRNAs in rice revealed the identification of at least 65% novel lncRNAs in rice. Another tool, PNRD, makes available lncRNAs of 150 plant species, including *Arabidopsis thaliana*, *Oryza sativa*, *Zea mays*, and *Populus trichocarpa* (Liu et al., 2015). Recently, Deshpande et al. (2019) also develop a tool named PLIT for identification of lncRNA in plants including rice. Overall, it indicates that the computational method imparts a significant role for in-depth study of transcripts and in identifying lncRNAs in different plants.

5.3 Diverse functions of long noncoding RNA in rice

The abundant amount of lncRNAs has been identified using genome-wide transcriptome sequencing technology in rice which showed its role in several diverse functions. In a study by Zhang et al. (2014), strand-specific RNA sequencing of rice pistils, anthers, seeds, and shoots identified 2224 lncRNAs, and with a combination of other available RNA-Seq datasets of rice, demonstrated that lncRNAs are highly stage- or tissue-specific in rice. Some of the biological functions with their example in rice have been described in this section. In plants, lncRNAs act as a precursor to ncRNAs such as miRNAs and siRNAs (Liu et al., 2015). In similarity to coding genes, some lncRNAs genes are reported to be transcribed

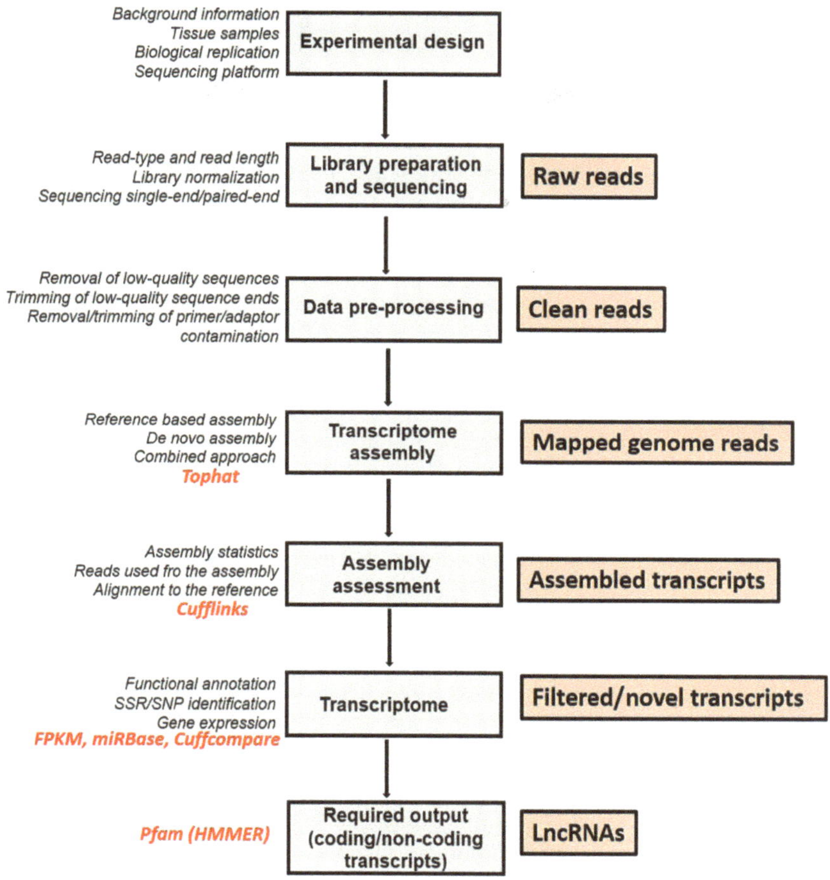

FIG. 5.1 A flowchart illustrating a stepwise computational method for the identification of long noncoding RNA.

by RNA polymerase II (Lee et al., 2004). However, the Pol IV/Pol V-dependent siRNA biogenesis pathway also gives rise to a plant-specific lncRNA class known as Pol IV/V-dependent lncRNAs, which are required for RNA-directed DNA-methylation (RdDM) (Kim and Sung, 2012). As an example, in rice, NATs have been reported to form double-stranded RNAs that can be used as templates for the generation of endogenous siRNA, cis-nat-siRNA (Kim and Sung, 2012; Li et al., 2007). Besides this, the potent role of lncRNA in plant growth, development, and response to stress has been described below with their example in rice.

5.3.1 Long noncoding RNA association in rice development

LncRNAs function in the regulation of various biological processes, including plant height, grain size, and floral development in rice (Yu and Zhu, 2019). During rice crop development progression including vegetative to gametophytic transition, several researchers reported the role of lncRNAs (Liu et al., 2019; Wang et al., 2018). In plants, COOLAIR is the first well-characterized lncRNA that regulates vernalization in *Arabidopsis*. Further, COLDAIR, another long intronic ncRNA, was identified in *Arabidopsis*. Besides COOLAIR and COLDAIR, several lncRNAs are found to be involved in plant reproduction, particularly in male fertility (Li et al., 2015). Though the regulatory genes and molecular mechanisms that underlie male sterility in the plant are not widely known, the role of lncRNA in plant, especially in rice, has not been reported till date. However, Ding et al. (2012a) termed adequate amount of the lncRNA as long day (LD)-specific male-fertility-associated RNA (LDMAR) and reported them as necessary for the fertility of rice under LD conditions. Identification of lncRNAs, including LTCONS_00035676, LTCONS_00042496, and LTCONS_00042583, by genome-wide screening showed its involvement in rice ovule development and female gametophyte abortion (Liu et al., 2019). Wang et al. (2018) overexpressed a lncRNA in rice that ultimately led to increased grain yield. In another study, Zhang et al. (2014) reported another rice lncRNA XLOC_057324 that plays a role in panicle development and sexual reproduction. Similarly, Luan et al. (2019) revealed that ESP encodes a putative pseudogene-derived lncRNA that helps in the formation of the panicle. Alternatively, spliced LncRNAs are also identified in alteration of flowering time in *Arabidopsis* and fertility in rice (Kiegle et al., 2018). In this study, it was showed that lncRNAs' alternative splicing takes place at the time of seed development in rice by comparing alternative splicing in immature seeds vs. embryo and endosperm of mature seeds. In another aspect of growth and development in plants, light plays an essential role in approximately all the related processes (Liu et al., 2015). The contribution of lncRNAs in this process of photo-morphogenesis is yet to be explored in detail. The rice homolog, OsHID1, is capable of rescuing hid1 mutant phenotype in A. thaliana (Wang et al., 2014). HID1 is the first known photo-morphogenesis lncRNA that sheds light on the relationship of ncRNAs and light response in plants (Wang et al., 2014). LncRNAs are also known to have an essential role in molecular processes. Genetic and molecular analysis showed that an increasing number of lncRNAs are involved in regulating gene expression and the structure of chromatin (Wang et al., 2015). These lncRNAs mediate epigenetic modification by recruiting chromatin remodeling complexes to a particular locus of chromatin (Bhat et al., 2016). Genome-wide association studies (GWAS) unravel the genetic basis of trait variability, establishing causal loci associated with phenotypic plant diversity. Additionally, GWAS executes small haplotype blocks associated with trait dissimilarity via single-nucleotide polymorphism (SNP) typing (Brachi et al., 2011). These SNPs, within intergenic regions, state potential functions of lncRNAs in plants. A study by Ding et al. (2012b) in rice showed that an SNP regulates the corresponding lncRNA gene expression that results in photoperiod-sensitive male sterility, indicating its role in modulation of chromatin loop dynamics in rice. Together with the abovementioned examples, other lncRNAs are listed in Table 5.1. These examples suggest the diverse role of lncRNAs in various biological functions of rice.

TABLE 5.1 Diverse functions of long noncoding RNA in rice.

LncRNA_ID	Biological function		References
TCONS_00028567	Plant hormone signal transduction		Li et al. (2019)
OS02T0626200-01	Phenylalanine metabolism		Li et al. (2019)
OS04T0412225-00	Photosynthesis		Li et al. (2019)
LTCONS_00035676, LTCONS_00042496, LTCONS_00042583, LTCONS_00078014	Ovule development		Liu et al. (2019)
ESP	Panicle development		Luan et al. (2019)
LAIR	Panicle development		Wang et al. (2018)
XLOC_086119, XLOC_066284	Carotenoid biosynthesis		Chen et al. (2018)
LOC107279416, LOC107279928	Seed development		Kiegle et al. (2018)
osa-eTM160	Anther development		Wang et al. (2017)
LNC_Os03 g44325	Seed color		Wang et al. (2015)
LNC_Os05 g27795	Leaf pubescence		Wang et al. (2015)
XLOC_057324, XLOC_010670, XLOC_004275, XLOC_053418	Reproductive organ development	(Panicle)	Zhang et al. (2014)
		(Sperms)	
		(Sperms)	
		(Ovule)	
Cis-NATPHO1;2	Phosphate homeostasis		Jabnoune et al. (2013)
LDMAR	Fertility/Pollen development		Ding et al. (2012a,b)
OsPI1	Phosphate homeostasis		Wasaki et al. (2003)
OsENOD40	Nodule formation		Kouchi et al. (1999)

5.3.2 Long noncoding RNA functioning under various stresses

Abiotic and biotic stresses are major constrains that adversely affect rice growth and productivity across the globe. Rice evolved numerous complex networks to withstand adverse conditions. This complex mechanism involves the functioning of several genes and their transcripts. Several studies reported the role of lncRNAs under stress conditions in different plants. With the help of genome-wide transcriptome sequencing, a large amount of stress-responsive lncRNAs have been identified in plants (Li et al., 2018; Shumayla et al., 2017). However, in rice, the biological functions of most lncRNAs under stress remain unclear. Only a few lncRNAs were studied for their role during environmental stress conditions (Table 5.2.).

5.3.2.1 Role of long noncoding RNA under abiotic stress

Abiotic stress such as drought, salinity, heavy metal toxicity, etc. are the common environmental constraints that affect crop development and limit its productivity globally (Tiwari et al., 2017a). Out of these, drought and salinity responses are one of the most important complex traits governed by several factors (Tiwari et al., 2017b; Tiwari et al., 2016). Chung et al. (2016) performed deep sequencing under drought stress in rice and obtained a large number of lncRNAs, i.e., 98 drought-responsive lncRNAs (31 drought-inducible and 67 drought-repressible), and their sense or antisense transcripts were detected. Similarly, Singh et al. (2017) develop a PLncPRO tool and detect a total of 3714 lncRNAs under the effect of drought and salinity and validated it via RT-qPCR. On the other hand, heavy metals accumulation also adversely affects various physiological and molecular processes of plants that lead to a reduction in plant growth including rice (Tiwari and Lata, 2018). In a recent study, Tang et al. (2019) deciphered several ncRNAs in response to arsenic (As) stress in rice, out of which a total of nine AGO1-enriched sRNAs were extracted that target 56 lncRNAs, indicating the potential role of these lncRNAs in As signaling. In another study, He et al. (2015) treated rice seedlings with two different concentrations of cadmium (Cd) and performed deep RNA sequencing. Comprehensive analysis of RNA-Seq in rice roots revealed a total of 1169 Cd toxicity-responsive genes. Out of total toxicity-responsive genes, 93 were predicted as lncRNA, out of which nine were detected as specifically Cd-responsive lncRNAs. Similarly, in response to Cd stress, Chen et al. (2018) documented the characterization and expression of lncRNAs at an early stage of development in rice root. These abovementioned abiotic stresses indirectly lead to limited or less nutrient availability conditions to plant roots in the soil (Bista et al., 2018; Hu and Schmidhalter, 2005). For example, nitrogen is a crucial macronutrient, and its availability has a strong influence on crop growth and production. In recent decades, the use of synthetic nitrogen fertilizers on crops has increased substantially; however, the applied nitrogen is not entirely used due to the low nitrogen use efficiency of crops. To resolve this limitation, a comprehensive study of the

TABLE 5.2 Role of long noncoding RNA under stress in rice.

LncRNA_ID	Stress	References
Abiotic stress		
TCONS_00028567, OS02T0626200-01, OS04T0412225-00	Drought	Li et al. (2019)
XLOC_086119, XLOC_066284	Metal (Cadmium)	Chen et al. (2018)
Chr03G0008, Chr04G0017	Nitrogen starvation	Shin et al. (2018)
GRNC_osa_lcl | Osativa_LOC_ Os12g19080.1	Nitrogen	Borah et al. (2018)
TCONS_00067186, TCONS_00012580	Drought and Salinity	Singh et al. (2017)
XLOC_054416/TCONS_00110660, XLOC_011965/TCONS_00026082, XLOC_029597/TCONS_00062817	Metal (Cadmium)	He et al. (2015)
LNC_Os10g18830, LNC_Os02g06640	–	Wang et al. (2015)
GRNC_osa_lcl | Osativa_LOC_ Os12g19080.1	Low Nitrogen	Yan et al. (2014)
Biotic stress		
ALEX1	*Xanthomonas oryzae*	Yu et al. (2020)
TCONS_00054645, TCONS_00055598, TCONS_00011946	*Dickeya zeae*	Li et al. (2018)
ChrSy25, ChrSy28, ChrSy29	*Maganaporthe oryzae*	Jain et al. (2017)
Osa-lsiR51031, osa-lsiR118183, osa-lsiR194568, osa-lsiR73750	*Rhizoctonia solani*	Niu et al. (2017)

functions of potential genes and regulatory factors in nitrogen metabolism is needed. Therefore, Shin et al. (2018) characterized alteration in the rice (*Oryza sativa*) transcriptome and identified 2588 novel putative lncRNA encoding loci. This study delivers thorough information about the regulatory pathways modified by nitrogen starvation-responsive lncRNAs. Similarly, Borah et al. (2018) summarized several lncRNAs under low nutrient stress tolerance in plants and reported GRNC_osa_lcl|Osativa_LOC_Os12g19080.1 lncRNA in rice under low nitrogen concentration. Phosphate, another mineral nutrient in plants, also plays a role in plant growth and development (Lv et al., 2014). Numerous studies suggested the involvement of lncRNAs in the phosphate homeostasis

(Jabnoune et al., 2013; Franco-Zorrilla et al., 2007). Because miRNAs are predominantly regulated by lncRNAs, the participation of miRNAs in phosphate homeostasis suggests the contribution of lncRNAs in phosphate homeostasis (Liu et al., 2015). Jabnoune et al. (2013) reported that the cis-natural antisense RNA, cis-NATPHO1;2, can assist as a translational enhancer for the expression of its sensory gene PHOSPHATE1;2 (PHO1;2) in rice.

5.3.2.2 Role of long noncoding RNA under biotic stress

Comparably, biotic stress also imposes huge loss to crop plants, and it is well-known that noncoding RNA imparts a major role in tolerance to biotic stress. However, in the case of rice crop, to date, very few studies have reported the involvement of lncRNA against the adverse effect of biotic stress. Recently, a vast amount of differentially expressed intergenic lncRNAs were detected in resistant rice lines after infection with the rice blast fungal pathogens, *Magnaporthe oryzae,* and their role to alter the resistance mechanism during *M. oryzae* invasion in rice was also implied (Jain et al., 2017). Similarly, Yu et al. (2020) studied the expression of *Xanthomonas oryzae*-responsive lncRNAs in rice and shed light on the link between lncRNAs and jasmonic acid pathway for the regulation of disease resistance in rice. Li et al. (2018) also revealed that lncRNAs might show a response to the infection of *Dickeya zeae* in rice by regulating the transcript levels of their targets and miRNAs. Moreover, deep sequencing profiling by Niu et al. (2017) also uncovered a large number of long siRNAs in rice under *Rhizoctonia solani* infection and studied its involvement in immunity against *R. solani*. The overall study concludes the potential role of lncRNAs under stress conditions in rice. In rice, functions of very few lncRNAs are known under stress; hence, information of some lncRNAs along with other related lncRNAs are listed in Table 5.2.

5.4 Conclusion and future prospects

Rice is both an important crop and a monocot model organism. Nevertheless, less is known about the genes involved in the regulation and molecular mechanisms of the plant stress tolerance. To develop improved stress tolerance lines of rice, an in-depth study is needed. The increasing development of RNA-Seq and associated bioinformatics strategies offers innovative ways to discover novel lncRNAs in model plants including rice (Singh et al., 2017; Tyagi et al., 2018). In recent years, research based on lncRNA for stress tolerance in plants is gaining momentum. As described in this chapter, the abovementioned examples give new insights into the lncRNAs' functions in tolerance of abiotic and biotic stress and highlight selected candidate lncRNAs for further studies in plants. Functional validation of putative lncRNAs will improve the overall understanding of lncRNAs' function in stress tolerance in plants. Apart from these findings, no study was conducted to date about the role of lncRNA in plant association with beneficial microbes. A study by Chauhan et al. (2019) performed in-depth sequencing in rice in response to plant growth-promoting rhizobacteria (PGPR), *Bacillus amyloliquefaciens,* under salt stress; however, this study did not provide information about noncoding RNAs involved in this study. However, another study by Jatan et al. (2019) in chickpea under drought stress revealed the alteration in expression of *Pseudomonas putida*-responsive miRNAs. Therefore, this study indicates that there may be a possibility of involvement of lncRNAs in response to plant-PGPR interaction. This attempt of functional validation of lncRNAs and to study the role of lncRNA in plant-rhizobia association may be a promising tool for the development of improved crop plants and enhanced agricultural production. Besides, the traditional reverse genetics, including overexpression and RNAi, genome engineering by CRISPR/cas9 may also be used for functional validation of candidate lncRNAs in plants.

Acknowledgments

ST acknowledges Senior Research Fellowship from Council of Scientific & Industrial Research (CSIR), New Delhi, India. CL acknowledges "Start-up Research Grant for INSA Young Scientist" (Grant No. INSA/SP/YSP/148/2017/1582) by the Indian National Science Academy (INSA), New Delhi, India. Authors acknowledge the Director, CSIR-National Botanical Research Institute, Lucknow, India, for providing facilities and support for the study.

References

Amor, B.B., Wirth, S., Merchan, F., Laporte, P., d'Aubenton-Carafa, Y., Hirsch, J., Maizel, A., Mallory, A., Lucas, A., Deragon, J.M., Vaucheret, H., 2009. Novel long non-protein coding RNAs involved in Arabidopsis differentiation and stress responses. Genome Res. 19 (1), 57–69.

Bhat, S.A., Ahmad, S.M., Mumtaz, P.T., Malik, A.A., Dar, M.A., Urwat, U., Shah, R.A., Ganai, N.A., 2016. Long non-coding RNAs: mechanism of action and functional utility. Non-coding RNA Res. 1 (1), 43–50.

Bista, D., Heckathorn, S., Jayawardena, D., Mishra, S., Boldt, J., 2018. Effects of drought on nutrient uptake and the levels of nutrient-uptake proteins in roots of drought-sensitive and-tolerant grasses. Plants 7 (2), 28.

Borah, P., Das, A., Milner, M.J., Ali, A., Bentley, A.R., Pandey, R., 2018. Long non-coding RNAs as endogenous target mimics and exploration of their role in low nutrient stress tolerance in plants. Genes *9* (9), 459.

Brachi, B., Morris, G.P., Borevitz, J.O., 2011. Genome-wide association studies in plants: the missing heritability is in the field. Genome Biol. *12* (10), 232.

Campos, E.I., Reinberg, D., 2009. Histones: annotating chromatin. Annu. Rev. Genet. 43, 559–599.

Chauhan, P.S., Lata, C., Tiwari, S., Chauhan, A.S., Mishra, S.K., Agrawal, L., Chakrabarty, D., Nautiyal, C.S., 2019. Transcriptional alterations reveal *Bacillus amyloliquefaciens*-rice cooperation under salt stress. Sci. Rep. 9 (1), 1–13.

Chen, L., Shi, S., Jiang, N., Khanzada, H., Wassan, G.M., Zhu, C., Peng, X., Xu, J., Chen, Y., Yu, Q., He, X., 2018. Genome-wide analysis of long non-coding RNAs affecting roots development at an early stage in the rice response to cadmium stress. BMC Genomics 19 (1), 460.

Chung, P.J., Jung, H., Jeong, D.H., Ha, S.H., Do Choi, Y., Kim, J.K., 2016. Transcriptome profiling of drought responsive noncoding RNAs and their target genes in rice. BMC Genomics *17* (1), 563.

Deshpande, S., Shuttleworth, J., Yang, J., Taramonli, S., England, M., 2019. PLIT: an alignment-free computational tool for identification of long non-coding RNAs in plant transcriptomic datasets. Comput. Biol. Med. 105, 169–181.

Ding, J., Lu, Q., Ouyang, Y., Mao, H., Zhang, P., Yao, J., Xu, C., Li, X., Xiao, J., Zhang, Q., 2012a. A long noncoding RNA regulates photoperiod-sensitive male sterility, an essential component of hybrid rice. Proc. Natl. Acad. Sci. 109 (7), 2654–2659.

Ding, J., Shen, J., Mao, H., Xie, W., Li, X., Zhang, Q., 2012b. RNA-directed DNA methylation is involved in regulating photoperiod-sensitive male sterility in rice. Mol. Plant 5 (6), 1210–1216.

Franco-Zorrilla, J.M., Valli, A., Todesco, M., Mateos, I., Puga, M.I., Rubio-Somoza, I., Leyva, A., Weigel, D., García, J.A., Paz-Ares, J., 2007. Target mimicry provides a new mechanism for regulation of microRNA activity. Nat. Genet. 39 (8), 1033.

Golicz, A.A., Singh, M.B., Bhalla, P.L., 2018. The long intergenic noncoding RNA (LincRNA) landscape of the soybean genome. Plant Physiol. 176 (3), 2133–2147.

Guha, P.K., Mazumder, A., Das, A., Pani, D.R., Mondal, T.K., 2019. In silico identification of long non-coding RNA based simple sequence repeat markers and their application in diversity analysis in rice. Gene Rep. 16, 100418.

He, F., Liu, Q., Zheng, L., Cui, Y., Shen, Z., Zheng, L., 2015. RNA-Seq analysis of rice roots reveals the involvement of post-transcriptional regulation in response to cadmium stress. Front. Plant Sci. 6, 1136.

Hu, Y., Schmidhalter, U., 2005. Drought and salinity: a comparison of their effects on mineral nutrition of plants. J. Plant Nutr. Soil Sci. 168 (4), 541–549.

Jabnoune, M., Secco, D., Lecampion, C., Robaglia, C., Shu, Q., Poirier, Y., 2013. A rice cis-natural antisense RNA acts as a translational enhancer for its cognate mRNA and contributes to phosphate homeostasis and plant fitness. Plant Cell 25 (10), 4166–4182.

Jain, P., Sharma, V., Dubey, H., Singh, P.K., Kapoor, R., Kumari, M., Singh, J., Pawar, D.V., Bisht, D., Solanke, A.U., Mondal, T.K., 2017. Identification of long non-coding RNA in rice lines resistant to Rice blast pathogen *Maganaporthe oryzae*. Bioinformation 13 (8), 249.

Jatan, R., Tiwari, S., Asif, M.H., Lata, C., 2019. Genome-wide profiling reveals extensive alterations in Pseudomonas putida-mediated miRNAs expression during drought stress in chickpea (*Cicer arietinum* L.). Environ. Exp. Bot. 157, 217–227.

Joshi, R.K., Megha, S., Basu, U., Rahman, M.H., Kav, N.N., 2016. Genome wide identification and functional prediction of long non-coding RNAs responsive to *Sclerotinia sclerotiorum* infection in *Brassica napus*. PLoS One 11 (7), e0158784.

Kiegle, E.A., Garden, A., Lacchini, E., Kater, M.M., 2018. A genomic view of alternative splicing of long non-coding RNAs during rice seed development reveals extensive splicing and lncRNA gene families. Front. Plant Sci. 9, 115.

Kim, E.D., Sung, S., 2012. Long noncoding RNA: unveiling hidden layer of gene regulatory networks. Trends Plant Sci. 17 (1), 16–21.

Kouchi, H., Takane, K.I., So, R.B., Ladha, J.K., Reddy, P.M., 1999. Rice ENOD40: isolation and expression analysis in rice and transgenic soybean root nodules. Plant J. 18 (2), 121–129.

Lee, Y., Kim, M., Han, J., Yeom, K.H., Lee, S., Baek, S.H., Kim, V.N., 2004. MicroRNA genes are transcribed by RNA polymerase II. EMBO J. 23 (20), 4051–4060.

Li, L., Wang, X., Sasidharan, R., Stolc, V., Deng, W., He, H., Korbel, J., Chen, X., Tongprasit, W., Ronald, P., Chen, R., 2007. Global identification and characterization of transcriptionally active regions in the rice genome. PLoS One 2 (3), e294.

Li, Z.F., Zhang, Y.C., Chen, Y.Q., 2015. miRNAs and lncRNAs in reproductive development. Plant Sci. 238, 46–52.

Li, W.Q., Jia, Y.L., Liu, F.Q., Wang, F.Q., Fan, F.J., Wang, J., Zhu, J.Y., Xu, Y., Zhong, W.G., Yang, J., 2018. Genome-wide identification and characterization of long non-coding RNAs responsive to *Dickeya zeae* in rice. RSC Adv. 8 (60), 34408–34417.

Li, P., Yang, H., Wang, L., Liu, H., Huo, A.H., Zhang, C., Liu, A., Zhu, A., Hu, J., Lin, Y., Liu, L., 2019. Physiological and transcriptome analyses reveal short-term responses and formation of memory under drought stress in rice. Front. Genet. 10, 55.

Liu, J., Jung, C., Xu, J., Wang, H., Deng, S., Bernad, L., Arenas-Huertero, C., Chua, N.H., 2012. Genome-wide analysis uncovers regulation of long intergenic noncoding RNAs in Arabidopsis. Plant Cell 24 (11), 4333–4345.

Liu, X., Hao, L., Li, D., Zhu, L., Hu, S., 2015. Long non-coding RNAs and their biological roles in plants. Genom. Proteom. Bioinform. 13 (3), 137–147.

Liu, H., Wang, R., Mao, B., Zhao, B., Wang, J., 2019. Identification of lncRNAs involved in rice ovule development and female gametophyte abortion by genome-wide screening and functional analysis. BMC Genomics 20 (1), 90.

Luan, X., Liu, S., Ke, S., Dai, H., Xie, X.M., Hsieh, T.F., Zhang, X.Q., 2019. Epigenetic modification of ESP, encoding a putative long noncoding RNA, affects panicle architecture in rice. Rice 12 (1), 20.

Lv, Q., Zhong, Y., Wang, Y., Wang, Z., Zhang, L., Shi, J., Wu, Z., Liu, Y., Mao, C., Yi, K., Wu, P., 2014. SPX4 negatively regulates phosphate signaling and homeostasis through its interaction with PHR2 in rice. Plant Cell 26 (4), 1586–1597.

Mosa, K.A., Ismail, A., Helmy, M., 2017. Functional genomics combined with other omics approaches for better understanding abiotic stress tolerance in plants. In: Plant Stress Tolerance. Springer, Cham, pp. 55–73.

Niu, D., Zhang, X., Song, X., Wang, Z., Li, Y., Qiao, L., Wang, Z., Liu, J., Deng, Y., He, Z., Yang, D., 2017. Deep sequencing uncovers rice long siRNAs and its involvement in immunity against *Rhizoctonia solani*. Phytopathology 108 (1), 60–69.

Shin, S.Y., Jeong, J.S., Lim, J.Y., Kim, T., Park, J.H., Kim, J.K., Shin, C., 2018. Transcriptomic analyses of rice (Oryza sativa) genes and non-coding RNAs under nitrogen starvation using multiple omics technologies. BMC Genomics 19 (1), 532.

Shumayla, Sharma, S., Taneja, M., Tyagi, S., Singh, K., Upadhyay, S.K.., 2017. Survey of high throughput RNA-seq data reveals potential roles for lncRNAs during development and stress response in bread wheat. Front. Plant Sci. 8, 1019.

Singh, U., Khemka, N., Rajkumar, M.S., Garg, R., Jain, M., 2017. PLncPRO for prediction of long non-coding RNAs (lncRNAs) in plants and its application for discovery of abiotic stress-responsive lncRNAs in rice and chickpea. Nucleic Acids Res. 45 (22), e183.

Tang, Z., Xu, M., Ito, H., Cai, J., Ma, X., Qin, J., Yu, D., Meng, Y., 2019. Deciphering the non-coding RNA-level response to arsenic stress in rice (*Oryza sativa*). Plant Signal. Behav. 14 (9), 1–11.

Tiwari, S., Lata, C., 2018. Heavy metal stress, signaling, and tolerance due to plant-associated microbes: an overview. Front. Plant Sci. 9, 452.

Tiwari, S., Lata, C., Chauhan, P.S., Nautiyal, C.S., 2016. Pseudomonas putida attunes morphophysiological, biochemical and molecular responses in *Cicer arietinum* L. during drought stress and recovery. Plant Physiol. Biochem. 99, 108–117.

Tiwari, S., Prasad, V., Chauhan, P.S., Lata, C., 2017a. *Bacillus amyloliquefaciens* confers tolerance to various abiotic stresses and modulates plant response to phytohormones through osmoprotection and gene expression regulation in rice. Front. Plant Sci. 8, 1510.

Tiwari, S., Lata, C., Singh Chauhan, P., Prasad, V., Prasad, M., 2017b. A functional genomic perspective on drought signalling and its crosstalk with phytohormone-mediated signalling pathways in plants. Curr. Genom. 18 (6), 469–482.

Tyagi, S., Sharma, A., Upadhyay, S.K., 2018. Role of next-generation RNA-Seq data in discovery and characterization of long non-coding RNA in plants. In: Çiftçi, Y.Ö. (Ed.), Next Generation Plant Breeding. IntechOpen, https://doi.org/10.5772/intechopen.72773.

Wang, Y., Fan, X., Lin, F., He, G., Terzaghi, W., Zhu, D., Deng, X.W., 2014. Arabidopsis noncoding RNA mediates control of photomorphogenesis by red light. Proc. Natl. Acad. Sci. 111 (28), 10359–10364.

Wang, H., Niu, Q.W., Wu, H.W., Liu, J., Ye, J., Yu, N., Chua, N.H., 2015. Analysis of non-coding transcriptome in rice and maize uncovers roles of conserved lncRNA s associated with agriculture traits. Plant J. 84 (2), 404–416.

Wang, M., Wu, H.J., Fang, J., Chu, C., Wang, X.J., 2017. A long noncoding RNA involved in rice reproductive development by negatively regulating osa-miR160. Sci. Bull. 62 (7), 470–475.

Wang, Y., Luo, X., Sun, F., Hu, J., Zha, X., Su, W., Yang, J., 2018. Overexpressing lncRNA LAIR increases grain yield and regulates neighbouring gene cluster expression in rice. Nat. Commun. 9.

Wasaki, J., Yonetani, R., Shinano, T., Kai, M., Osaki, M., 2003. Expression of the OsPI1 gene, cloned from rice roots using cDNA microarray, rapidly responds to phosphorus status. New Phytol. 158 (2), 239–248.

Yan, Y., Wang, H., Hamera, S., Chen, X., Fang, R., 2014. miR444a has multiple functions in the rice nitrate-signaling pathway. Plant J. 78 (1), 44–55.

Yu, T., Zhu, H., 2019. Long non-coding RNAs: rising regulators of plant reproductive development. Agronomy 9 (2), 53.

Yu, Y., Zhou, Y.F., Feng, Y.Z., He, H., Lian, J.P., Yang, Y.W., Lei, M.Q., Zhang, Y.C., Chen, Y.Q., 2020. Transcriptional landscape of pathogen-responsive LncRNAs in rice unveils the role of ALEX1 in jasmonate pathway and disease resistance. Plant Biotechnol. J. 18, 679–690.

Zhang, Y.C., Liao, J.Y., Li, Z.Y., Yu, Y., Zhang, J.P., Li, Q.F., Qu, L.H., Shu, W.S., Chen, Y.Q., 2014. Genome-wide screening and functional analysis identify a large number of long noncoding RNAs involved in the sexual reproduction of rice. Genome Biol. 15 (12), 512.

Chapter 6

Genome-wide analysis of long noncoding RNAs in Sorghum and their roles in development and stress

Nikhil Malhotra[a] and Shivani Sharma[b]
[a]*ICAR-National Bureau of Plant Genetic Resources Regional Station, Phagli, Shimla, Himachal Pradesh, India,* [b]*Himalayan Forest Research Institute, Panthaghati, Shimla, Himachal Pradesh, India*

6.1 Introduction

Long noncoding RNAs (lncRNAs) utilize diverse functions in controlling various biological processes. These are noncoding RNAs of >200 nucleotides, which play a major role in regulating gene transcription and translation (Swiezewski et al., 2007), and can be categorized in different ways according to their length, location, and biological functions (Costa, 2005; Ben Amor et al., 2009; Liu et al., 2012; Zhu and Wang, 2012; Jin et al., 2013). They are transcribed by RNA polymerase II or III, along with polymerase IV/V in plants (Dinger et al., 2008, 2009; Wierzbicki et al., 2008). Although more attention has been paid to lncRNAs in recent years, research in this area is still in its infancy. Till date, only a few lncRNAs have been identified adequately (Heo and Sung, 2011; Swiezewski et al., 2009). Plant research in this area in particular is far below than that of humans and animals (Zhu and Wang, 2012; Zhang and Chen, 2013; Bai et al., 2015). However, available studies suggest that plant lncRNAs have similar regulatory functions to animals (Zhu and Wang, 2012; Zhang and Chen, 2013).

LncRNAs are classified by utilizing genome-wide transcriptomics (Liu et al., 2015), where RNA sequencing (RNA-Seq) has been used prominently. The identified lncRNAs reveal substantial changes in various organs and stresses, indicating that they are robustly regulated and can act in response to plant development and stress conditions (Liu et al., 2012). Approximately 6480 lncRNAs associated with either organ-specific or stress-induced expression profiles have been identified from the 200 *Arabidopsis thaliana* transcriptomic

Long Noncoding RNAs in Plants: Roles in development and stress
https://doi.org/10.1016/B978-0-12-821452-7.00006-4

datasets (Liu et al., 2012). Wang et al. (2014a) found 37,238 long noncoding natural antisense transcripts in *A. thaliana*, with 70 % of annotated mRNAs aligned with antisense transcripts. In maize, 38 lncRNAs were found to be marked from cDNA sequences that were significantly classified between maternal and paternal alleles by their different relative expression (Zhang et al., 2011). In *Populus trichocarpa*, RNA-Seq was used in the identification of 2542 drought-responsive long intergenic noncoding RNAs (Shuai et al., 2014). To recognize 2224 tissue-specific lncRNAs in rice, RNA-Seq combined with transcriptome analysis was used (Shuai et al., 2014), while 5515 lncRNAs were found during the transition stage (Kiegle et al., 2018). In *Solanum commersonii*, computational modeling and RNA-Seq were used to identify 18,882 lncRNAs (Carbonell and Rodriguezescudero, 2015). Similarly, 125 putative stress-responsive lncRNAs were identified by bioinformatics and wet lab analysis in wheat (Liu et al., 2018a). A total of 254 differentially expressed lncRNAs and 300 differentially expressed genes were identified between wild and vernalized *Brassica rapa* by comparative transcriptome analysis (Liu et al., 2018b). In *Solanum lycopersicum*, RNA-Seq revealed 3679 lncRNAs, including 677 lncRNAs, which were significantly found to be differentially expressed in mature mutant fruits, thereby providing the evidence for the involvement of lncRNAs in regulating fruit ripening (Zhu et al., 2015).

High tissue-specific expression patterns are the most evident resemblances among the known lncRNAs from different plant species, which thus signify the specific regulatory role of lncRNAs in plant development. Even though several studies have documented the functions of lncRNAs, the investigations pertaining to the exacting mechanism behind the functions of lncRNAs need to be understood (Cho et al., 2006; Miao-Chih et al., 2010; Nakamoto et al., 2010; Juan et al., 2012; Shuai et al., 2014; Shumayla et al., 2017; Tyagi et al., 2018; Wang et al., 2014a; Wunderlich et al., 2014). The regulation of plant growth and development along with stress responses is shown to be complicated with the research advancements in lncRNAs (Jack, 2004; Filipe and Martienssen, 2015) (Table 6.1).

6.2 Sorghum

Sorghum (*Sorghum bicolor*) is an annual ($2n=2x=20$ chromosomes) smart food, feed, and bioenergy crop, belonging to the family Poaceae and adapted worldwide to temperate and tropical climates. It is the fifth most important cereal grain after barley, maize, rice, and wheat (Kamara et al., 2011). It has a genome size of ~730 Mb and is widely considered as a rich crop in genomic resources (Paterson et al., 2009). Sorghum grains are gluten-free and, thus, an alternative food for people with celiac disease (Dahlberg et al., 2011; Stefoska-Needham et al., 2015). Sorghum is a significant resource of minerals, vitamins, proteins, antioxidants, and energy (Moraes et al., 2012) and is nutritionally comparable with other major cereal grains. It is grown worldwide including

TABLE 6.1 Summary of plant lncRNAs.

Plant species	lncRNA	Regulation mechanism	Biological significance	Reference
Arabidopsis thaliana	asHSFB2a ASCO-lncRNA APOLO COLDAIR COOLAIR DRIR ELENA1 FLORE HID1 IPS1	Antisense transcription Alternative splicing Chromatin loop dynamics Histone modification Promoter interference Stomatal closure Plant immunity Antisense transcription Chromatin association Target mimicry	Gametophytic development Lateral root development Auxin-controlled development Flowering time Flowering time Drought and salt stress *Pseudomonas syringae* infection Photomorphogenesis Photomorphogenesis Phosphate homeostasis	Wunderlich et al. (2014), Bardou et al. (2014), Ariel et al. (2014), Heo and Sung (2011), Swiezewski et al. (2009), Qin et al. (2017), Mach (2017), Henriques et al. (2017), Wang et al. (2014b), Franco-Zorrilla et al. (2007)
Glycine max	GmENOD40	Protein relocalization	Nodule formation	Yang et al. (1993)
Hordeum vulgare	HvCesA6 lnc-NAT	siRNA precursor	Cell-wall synthesis	Held et al. (2008)
Medicago truncatula	MtENOD40	Protein relocalization	Nodule formation	Sousa et al. (2001)
Oryza sativa	Cis-$NAT_{PHO1;2}$ OsPI1 OsENOD40 LDMAR (P/TMS12-1)	Translational enhancer Unknown Unknown Promoter methylation	Phosphate homeostasis Phosphate homeostasis Nodule formation Fertility	Jabnoune et al. (2013), Wasaki et al. (2003), Kouchi et al. (1999), Ding et al. (2012)
Petunia hybrida	SHO lnc-NAT	dsRNA degradation	Cytokinin synthesis	Zubko and Meyer (2007)
Populus trichocarpa	lincRNA2752	Target mimicry	Drought stress	Shuai et al. (2014)
Solanum lycopersicum	TPS11	Unknown	Phosphate homeostasis	Liu et al. (1997)

Africa, America, Asia, Australia, and Europe with global production touching ~64.5 million tons in 2016 (Motuma et al., 2016). More than 90% of total global sorghum harvested areas are in Africa and Asia (Deb et al., 2004; Bhagavatula et al., 2013), with Africa counting for 61% of the area and 41% of the production, and Asia accounting for 22% of area and 18% of production (Bhagavatula et al., 2013) (Table 6.2). Sub-Saharan Africa, and more particularly, Ethiopia, Sudan, and Chad are regarded as the primary centers of origin and domestication of sorghum species, while India and China are secondary centers of diversity (de Wet, 1978; Doggett, 1988).

In recent decades, food production per capita has risen globally with a rise in crop demand, and a further 59%–98% increase in demand is expected by 2050 (Elferink and Schierhorn, 2016). Yield losses are a problem due to extreme weather conditions, plant diversification, quantity of inputs used, farm management, pests, and diseases (Penning de Vries et al., 1997). Many abiotic and biotic stresses restrict the production of sorghum worldwide. The major abiotic stresses are drought, heat, salinity, and aluminum toxicity. Biotic stresses include anthracnose [*Colletotrichum graminicola* (Ces.) G.W. Wilson], charcoal rot and/or stalk rot [*Macrophomina phaseolina* (Tassi) Goid], downy mildew (*Peronosclerospora sorghi* [Wetson and Uppal (Shaw)]), grain mold (caused by a complex of several fungal diseases), leaf blight [*Exserohilum turcicum* (Pass.) K.J. Leonard and E.G. Suggs], rust (*Puccinea purpurea* Cooke), and several viral diseases, while head bug (*Calocoris angustatus* Leth), midge (*Contarinia sorghicola* Coq.), stem borer [*Chilo partellus* (Swinhoe) and *Sesamia inferens* (Walker)], and shoot fly [*Atherigona soccata* (Rondani)] are the major pests of sorghum. These stresses often occur in combinations that cause substantial losses in the production of sorghum (House, 1985; Sharma, 1993; Kumar et al.,

TABLE 6.2 Global production of sorghum from 2000 to 2015 (million tons).

Region	2000	2005	2010	2013	2015
Africa	18.4	24.8	24.7	25.7	24.8
Asia	11.3	18.2	10.5	8.7	8.0
Australia	2.1	2.0	1.6	2.2	2.1
Europe	0.7	0.5	0.7	1.1	0.9
North America	18.0	15.7	15.9	16.4	17.2
South America	4.9	5.3	6.2	6.8	5.5
World	55.8	59.6	60.0	61.3	63.5

Source: FAOSTAT Database Collections, FAO, 2016. Available online: http://www.fao.org/faostat/en/#data (Accessed 22 September 2019); USDA, 2016. World Agricultural Production; Circular Series WAP 1–16, Washington DC, USA.

2011). Furthermore, the evidence, to date, suggests that global warming has an overall negative impact on agricultural biodiversity, food production, and food quality (Dwivedi et al., 2013). The discovery of climate-resilient traits in germplasm collections and their use in crop improvement should be used with utmost necessity to tailor climate change and variability-adapted crop cultivars.

In the last decade, genomics science in sorghum has advanced significantly in terms of assays and marker technologies, genetic maps, quantitative trait loci (QTL) mapping and cloning associated with agronomically beneficial traits, and genome sequencing. At the same time, substantial investment has been made in developing high-throughput phenotyping systems for rapid and reliable evaluation of phenotypic characteristics, including tolerance to abiotic stress. Resequencing diverse germplasms and their sequence comparisons with a reference genome provide opportunities for researchers to discover markers and whether such differences in phenotypes relate to differences in sequence variations between accessions. Therefore, discovering climate-resilient traits by using available germplasm, understanding population structure and diversity, marker–trait associations, and identifying multi-trait genetically diverse and agronomically beneficial germplasm has long-term implications on breeding and genomics of sorghum.

6.2.1 Phenotypic diversity

Landraces form the substantial number of sorghum germplasm and previous reports have demonstrated a high level of morpho-agronomic characteristics among sorghum landraces, highlighting their central role in subsistence farming (Ayana and Bekele, 1998; Grenier et al., 2004; Geleta and Labuschagne, 2005; Manzelli et al., 2005). In addition to edaphic, temporal, and spatial factors, abundant diversity among sorghum landraces is shaped by ethnic, linguistic, customary, end-use, and sociological factors. A detailed evaluation of a sorghum reference set (384 accessions) for five seasonal locations showed high variance (as measured by Shannon–Weaver diversity index) for plant height, 100-seed weight, grain yield, panicle weight, and for chlorophyll content as measured by chlorophyll meter reading for soil crop study (Billot et al., 2013). In addition, qualitative traits such as mid rib color (white), plant pigmentation, glume color (black and purple), and seed color (white, purple, and brown), which scored in greater extent in the reference set, could be used as markers for selecting accessions of desired end-use types as they were associated with fodder quality, resistance to grain mold, or for seed preference to food and industrial uses (Seetharam, 2011). Drought-tolerant accessions identified were early flowering types, had shortest plant height, and high panicle weight and grain yield (Seetharam, 2011).

The primary centers of diversity in sorghum are Ethiopia and Sudan. On-farm sorghum diversity in Ethiopia is stated to be determined by field height, field size, soil pH, P, available N_2, organic matter, exchangeable K, and clay content (Mekbib et al., 2009). Likewise, a higher level of diversity was reported in the intermediate altitudes than in the lowlands and highlands of Ethiopia. On the other hand, Sudanese landraces possess a high level of within-region diversity (Grenier et al., 2004). In India, Upadhyaya et al. (2014) evaluated a sorghum mini core collection for Brix (a measure of sugar concentration) content for two seasons under irrigated and terminal drought-stressed conditions. Accessions IS 13294, 13549, 23216, 23684, 24139, 24939, and 24953 recorded 14%–15% higher Brix concentration than the best control, IS 33844 (Brix, 12.4%); however, the yield was low. IS 1004, 4698, 23891, and 28141 produced 12%–23% greater seed yield over IS 33844 (35–42 g/plant) and had almost the same Brix concentration (~13%). Further, Upadhyaya et al. (2016) reported dense seed mineral (Fe and Zn) accessions from the sorghum mini core collection, some with significantly higher seed yields and specific adaptation to irrigated and/or drought-stressed environments. Clearly, several factors have contributed to phenotypic diversity, requiring further documentation of this diversity on the basis of perceptions of farmers, agro-ecological factors, and end-use patterns from the regions not explored.

6.2.2 Genomic advancements

6.2.2.1 Genotypic diversity

Molecular profiling of sorghum landraces has shown large allelic diversity. A substantial allelic diversity exists among Bostwana's landraces (30 accessions and 10 simple sequence repeats [SSRs]), but these variations could not be linked to diversity in agro-ecological regions, ethnicities, and races (Motlhaodi et al., 2014). Burkina's sub-Sahelian landraces (124 accessions and 29 SSRs) also displayed the highest gene diversity and the color of the seed was the main factor for this diversity (Barro-Kondombo et al., 2010). Burow et al. (2012) grouped the northeastern China landraces (159 accessions and 41 SSRs) into eight clusters and reported a number of genetically distinct accessions with agronomically beneficial traits for use in sorghum breeding. Billot et al. (2013) conducted the largest study ever to analyze population structure and diversity in a composite collection using 41 genomic SSRs (3367 accessions covering 5 races, 10 intermediate races, and wild relatives). This study revealed that accessions from central and eastern Africa had the largest number of alleles, and cultivated races were arranged around geographical regions and race within regions, detecting 13 groups of variable size with peripheral groups in western Africa, southern Africa, and eastern Asia. Ramu et al. (2013) further examined this reference set by using expressed sequence tag (EST)-SSRs to validate the genetic structure deciphered by Billot et al. (2013). Then, using 265,000 simple nucleotide polymorphisms (SNPs) on 971 sorghum accessions, Morris et al. (2013) showed structured populations along both morphological types and geographical origins, offering insights into the historical plant diffusion processes within and throughout the agroclimatic zones of Africa and Asia.

Evans et al. (2013) reported ~2.8 million SNPs and ~300,000 insertions and deletions (INDELS) that distinguished the reference genotype BTx623 from sorghum genotype Tx7000 and/or BTx642. A robust SNP array system has also been reported which enables screening for genome-wide and trait-linked polymorphisms in genetically diverse *S. bicolor* populations (Bekele et al., 2013). The analysis of the genomes of populations growing in contrasting environmental conditions will unravel the genes subjected to natural selection in adaptation to climate variations.

Wild relatives of sorghum harbor genes for adaptation and resistance to pests and diseases (Venkateswaran et al., 2014). In contrast to cultivated germplasm, information on allelic richness and diversity between wild and cultivated species is limited in sorghum. Using 104 accessions including 98 cultivated and 31 wild relatives along with 98 SSRs, Casa et al. (2005) found that landraces retained 86% of the allelic diversity of the wild sorghums, showed weak evidence of population structure between cultivated and wild as well as among cultivated sorghum accessions, and mapped seven SSR loci in or near-genomic regions associated with domestication-related QTLs in sorghum. Mutegi et al. (2012) found a reasonably large number of alleles (257) in cultivated than wild (238 alleles) gene pools by using 24 SSRs; however, wild sorghum gene pools harbored significantly more genetic diversity than their domesticated counterpart, revealing that domestication process was accompanied by a genetic bottleneck in sorghum. Therefore, the sympatric occurrence of wild and cultivated species, coupled with extensive seed-mediated gene flow and high outcrossing rates of wild sorghum, suggests a possible crop-to-wild gene flow through regions and vice versa.

Of late, Targeting-Induced Local Lesions IN Genome (TILLING) has been successfully used to identify both induced and natural variations in pathogenic bacteria and species of plants and animals (Gilchrist and Haughn, 2005). In sorghum, ethyl methane sulfonate-mutanized population consisting of 1600 lines was generated from the inbred line BT×623, and numerous phenotypes with altered morphological and agronomic characteristics were detected when phenotyped (Xin et al., 2008). Sorghum contains the cyanogenic glucoside dhurrin, the ingestion of which can cause a serious problem for animal and human due to its ability to release toxic hydrogen cyanide, often causing the death of grazing animals. Mutant lines with lower or negligible cyanogenic glucoside dhurrin compared to non-mutated parent lines have been identified in sorghum using a TILLING approach (Blomstedt et al., 2012).

6.2.2.2 *Climate change and nutritional security*

Climate models forecast a rise in drought severity and heat stress, as well as a risk to altered pests and diseases due to global warming, which will adversely affect food production and quality (Dwivedi et al., 2013). The way forward for the production of climate-ready crops is to discover and use climate-resilient traits in crop improvement. A number of sorghum germplasm having resistance to abiotic and biotic stresses have been reported. Vadez et al. (2011) recorded a 10-fold difference in seed yield and harvest index, a 2-fold difference in transpiration efficiency, and 1.25-fold difference in water extraction under drought-stressed conditions among 149 reference set accessions. Kapanigowda et al. (2013) recorded a large variation in the exotic germplasm and advanced breeding lines evaluated for post-flowering drought tolerance for drought tolerance-related traits. Likewise, heat stress-tolerant (DK-54-00, Pioneer 84G62, and SC 1019) and responsive (DK 28-E, SC 15, and TX 7078) sorghum germplasm lines were reported by Djanaguiraman et al. (2014). Leiser et al. (2014) identified a genomic region associated with a major Al tolerance gene, SbMATE, on chromosome 3 in sorghum, which showed a very high correlation to grain yield in P-deficient soils. Venkateswaran et al. (2014) compiled a list of wild sorghum species with reported resistance to pests and diseases, including the resistance to downy mildew, shoot fly, spotted stem borer, and midge.

As the staple food in developing countries for people with low economic status, sorghum is quite comparable to the main nutritional crops. The high energy content and high antioxidant capacity of sorghum ensure the improvement in nutritional status, thereby aiding in food security along with nutritional security (Proietti et al., 2015; Gaytán-Martínez et al., 2017; Chhikara et al., 2018; Zhu, 2018). Further, the presence of high concentrations of nutrients in wild sorghum species is to be considered as potential source for Fe and protein biofortification that can be used in wide hybridization for transferring useful genes from wild species into cultivated germplasm (Abdelhalim et al., 2019).

6.3 lncRNAs and Sorghum research

Recent studies in various eukaryotic transcriptomes using high-throughput technologies, especially RNA-Seq, have identified large numbers of lncRNAs. Total RNA extraction, RNA enrichment, library preparation and sequencing are the main steps in a typical RNA-Seq workflow. Either selection/enrichment of polyA RNA using oligo (dT) primers or depletion of ribosomal RNA (rRNA depletion) can be performed for RNA enrichment. RNA is fragmented (200–500 bp) after enrichment of RNA by oligo dT or rRNA depletion and an adapter is ligated to each fragment. RNA fragments are then reverse-transcribed to cDNA. Fragments of RNA are then transcribed back to cDNA. High-throughput sequencing follows this, resulting in short sequences from both ends (pair-end sequencing) or from one end (single-end sequencing). The resulting reads are processed after sequencing to obtain the expression level and/or transcriptional structure for each gene. In order to identify the complete range of lncRNAs within a plant species, many modifications to these RNA-Seq methods are commonly used. lncRNAs are implicated in regulating flowering, immunity, root development, biotic and abiotic stress responses, and several imperative biological processes in plants (Chekanova, 2015; Wang and Chekanova, 2017). Despite the recent accumulation of data revealing gene expression patterns during development in sorghum (Gelli et al., 2014; Shakoor et al., 2014), limited studies have been done studying the transcriptomic landscape in this important crop species. Alternative splicing and other RNA processing mechanisms significantly contribute to improve genome coding, gene regulation, and eukaryotic transcriptome diversity. A total of ~11,000 expressed genes and more than 2100 novel genes have been identified from sorghum transcriptome using single-molecule long reads (Abdel-Ghany et al., 2016). This finding greatly enhanced gene annotations and aided in studying gene regulation in sorghum. In another study, Wang et al. (2018a) studied the alternative splicing mechanism by comparing the transcriptome changes in maize and sorghum by using RNA-Seq and single-molecule long-read sequencing. The study revealed substantial heterogeneity in both protein-coding and noncoding gene expression, as well as differential alternative splicing in both maize and sorghum. These statistics would be crucial to address the role of posttranscriptional stress-regulated splicing and polyadenylation in plants' adaptation to various stresses.

Although these studies have identified large number of previously unstudied transcriptional events, the progress made in the identification and functional characterization of lncRNAs remains a major challenge in sorghum. The discovery of lncRNAs is the primary step in unraveling the role of lncRNAs in plants, and the past decade has seen most studies concerted on identifying plant lncRNAs (Chekanova, 2015; Wang and Chekanova, 2017). With the availability of sorghum genome (Paterson et al., 2009), a total of 2600 lncRNAs have been identified using RNA-Seq data and reference annotations (Paytuvi-Gallart et al., 2019; Szcześniak et al., 2019), but their role in plant development along with various biotic and abiotic stresses remains unclear. The relationship between lncRNAs and plant stresses has been described by previous studies. While miRNAs and siRNAs are small forms of noncoding RNAs, they are derived from lncRNAs and a form of lncRNA is closely linked to miRNAs and their function in regulating gene expression. Due to the roles of miRNAs in development, nutrient acquisition and use, and tolerance to abiotic and biotic stresses in plants, sorghum has been an interest for such investigation (Katiyar et al., 2012). The tolerance to drought and heat stress has been attributed to the miR169 family for adaptation of sorghum to abiotic stresses (Paterson et al., 2009; Sanousi et al., 2016; Hamzaa et al., 2016).

A large number of lncRNAs which have been identified from different plant species, *viz.*, Arabidopsis, maize rice, tomato, wheat, etc., are stored in several databases developed over time. These include CANTATdb, GreeNc, lncRNAdb, NONCODE, Plant NATsDB, PlncDB, PNRD, PLNlnRbase, and TAIR (Bhatia et al., 2017; Nejat and Mantri, 2017). In this context, a computational pipeline has been developed which utilizes RNA-Seq data to identify co-expression networks comprising lncRNA and mRNA genes to provide putative annotation of lncRNAs having unknown biological functions in plants (Qi et al., 2019). Several studies pertaining to identification of co-expression networks for integrative analysis of lncRNA-mRNA expression have been reported in cabbage (Wang et al., 2018b, 2019), ohwi (Yan et al., 2019), orange (Zhou et al., 2019), tomato (Yang et al., 2019), strawberry (Kang and Liu, 2019), and wheat (Cagirici et al., 2017) for development and stress conditions. Taking cues from such findings and using huge amount of information stored in aforementioned databases, the mapping and profiling of putative lncRNAs could be the next big step in sorghum research.

6.4 Conclusion and future perspectives

The untapped complexity unravelled by genome sequencing of a diverse set of germplasm provides a great opportunity for the genetic improvement of sorghum. Intensive efforts are required to screen more sorghum germplasm for climate-resilient and nutritional traits. Although our knowledge of lncRNA-mediated regulation of plant development and stress responses is gradually improving the understanding of the regulatory mechanisms of lncRNAs, considerable progress remains to be made in this field. As perhaps evident from the research advancements done till date on sorghum, there appears to be insufficient knowledge to accurately classify lncRNAs and understand their molecular function in relation to development and stress. Possible lncRNA functions can be first investigated in model plants because of the availability of biological information and genetic and transgenic plant resources associated with them. The identification of lncRNAs located in chromosomal synteny regions between sorghum and its corresponding model plant will help in the identification of homologous lncRNAs in its genome. Developing accurate data on the biological and/or molecular functions of individual lncRNAs will lead to a better understanding of their diversity to develop sorghum as future crop in combating climate change and nutritional security.

References

Abdel-Ghany, S.E., Hamilton, M., Jacobi, J.L., Ngam, P., Devitt, N., Schilkey, F., Ben-Hur, A., Reddy, A.S., 2016. A survey of the sorghum transcriptome using single-molecule long reads. Nat. Commun. 7, 11706.

Abdelhalim, T.S., Kamal, N.M., Hassan, A.B., 2019. Nutritional potential of wild sorghum: grain quality of Sudanese wild sorghum genotypes (*Sorghum bicolor* L. Moench). Food Sci. Nutr. 7, 1529–1539.

Ariel, F., Jegu, T., Latrasse, D., Romero Barrios, N., Christ, A., Benhamed, M., et al., 2014. Noncoding transcription by alternative RNA polymerases dynamically regulates an auxin-driven chromatin loop. Mol. Cell 55, 383–396.

Ayana, A., Bekele, E., 1998. Geographical pattern of morphological variation in sorghum (*Sorghum bicolor* L. Moench) germplasm from Ethiopia and Eritrea: qualitative characters. Hereditas 129, 195–205.

Bai, Y., Dai, X., Harrison, A.P., Chen, M., 2015. RNA regulatory networks in animals and plants: a long noncoding RNA perspective. Brief. Funct. Genomics 14, 91–101.

Bardou, F., Ariel, F., Simpson, C.G., Romero-Barrios, N., Laporte, P., Balzergue, S., et al., 2014. Long noncoding RNA modulates alternative splicing regulators in Arabidopsis. Dev. Cell 30, 166–176.

Barro-Kondombo, C., Sagnard, F., Chantereau, J., Deu, M., Vom Brocke, K., Durand, P., Gozé, E., Zongo, J.D., 2010. Genetic structure among sorghum landraces as revealed by morphological variation and microsatellite markers in three agroclimatic regions of Burkina Faso. Theor. Appl. Genet. 120, 1511–1523.

Bekele, W.A., Wieckhorst, S., Freidt, W., Snowdon, R.J., 2013. High-throughput genomics in sorghum: from whole-genome resequencing to a SNP screening array. Plant Biotechnol. J. 11, 1112–1125.

Ben Amor, B., Wirth, S., Merchan, F., Laporte, P., D'Aubenton-Carafa, Y., Hirsch, J., et al., 2009. Novel long non-protein coding RNAs involved in Arabidopsis differentiation and stress responses. Genome Res. 19, 57–69.

Bhagavatula, S., Rao, P.P., Basavaraj, G., Nagaraj, N., 2013. Sorghum and Millet Economies in Asia—Facts, Trends and Outlook. International Crops Research Institute for the Semi-Arid Tropics, Patancheru, India.

Bhatia, G., Goyal, N., Sharma, S., Upadhyay, S., Singh, K., 2017. Present scenario of long noncoding RNAs in plants. Non-Coding RNA 3, 16.

Billot, C., Ramu, P., Bouchet, S., Chantereau, J., Deu, M., Gardes, L., Noyer, J.L., Rami, J.F., Rivallan, R., Li, Y., Lu, P., Wang, T., Folkertsma, R.T., Arnaud, E., Upadhyaya, H.D., Glaszmann, J.C., Hash, C.T., 2013. Massive sorghum collection genotyped with SSR markers to enhance use of global genetic resources. PLoS One 8, e59714.

Blomstedt, C.K., Gleadow, R.M., Donnell, N.O., Naur, P., Jensen, K., Laursen, T., Olsen, C.E., Stuart, P., Hamill, J.D., Møller, B.L., Neale, A.D., 2012. A combined biochemical screen and TILLING approach identifies mutations in *Sorghum bicolor* L. Moench resulting in acynogenic forage production. Plant Biotechnol. J. 10, 54–66.

Burow, G., Franks, C.D., Xin, Z., Burke, J.J., 2012. Genetic diversity in a collection of Chinese sorghum landraces accessed by microsatellites. Am. J. Plant Sci. 3, 1722–1729.

Cagirici, H.B., Alptekin, B., Budak, H., 2017. RNA sequencing and co-expressed long non-coding RNA in modern and wild wheats. Sci. Rep. 7, 10670.

Carbonell, P., Rodriguezescudero, A.I., 2015. The *Solanum commersonii* genome sequence provides insights into adaptation to stress conditions and genome evolution of wild potato relatives. Plant Cell 27, 954–968.

Casa, A.M., Mitchell, S.E., Hamblin, M.T., Sun, H., Bowers, J.E., Paterson, A.H., Aquadro, C.F., Kresovich, S., 2005. Diversity and selection in sorghum: simultaneous analyses using simple sequence repeats. Theor. Appl. Genet. 111, 23–30.

Chekanova, J.A., 2015. Long non-coding RNAs and their functions in plants. Curr. Opin. Plant Biol. 27, 207–216.

Chhikara, N., Abdulahi, B., Munezero, C., Kaur, R., Singh, G., Panghal, A., 2018. Exploring the nutritional and phytochemical potential of sorghum in food processing for food security. Nutr. Food Sci. 49, 318–332.

Cho, J., Koo, D.H., Nam, Y.W., Han, C.T., Lim, H.T., Bang, J.W., Hur, Y., 2006. Isolation and characterization of cDNA clones expressed under male sex expression conditions in a monoecious cucumber plant (*Cucumis sativus* L. cv. Winter Long). Euphytica 146, 271–281.

Costa, F.F., 2005. Non-coding RNAs: new players in eukaryotic biology. Gene 357, 83–94.

Dahlberg, J.A., Berenji, J., Sikora, V., Latković, D., 2011. Assessing sorghum (*Sorghum bicolor* (L.) Moench) germplasm for new traits: food, fuel and unique uses. Maydica 56, 5–92.

de Wet, J.M., 1978. Systematics and evolution of sorghum sect. *Sorghum* (Gramineae). Am. J. Bot. 65, 477–484.

Deb, U.K., Bantilan, M.C.S., Hash, C.T., Ndjeunga, J., 2004. Adoption of improved Sorghum cultivars. In: Sorghum Genetic Enhancement: Research Process, Dissemination and Impacts. International Crops Research Institute for the Semi-Arid Tropics, Patancheru, India.

Ding, J., Lu, Q., Ouyang, Y., Mao, H., Zhang, P., Yao, J., et al., 2012. A long noncoding RNA regulates photoperiod-sensitive male sterility, an essential component of hybrid rice. Proc. Natl. Acad. Sci. U. S. A. 109, 2654–2659.

Dinger, M.E., Pang, K.C., Mercer, T.R., Mattick, J.S., 2008. Differentiating protein-coding and noncoding RNA: challenges and ambiguities. PLoS Comput. Biol. 4, e1000176.

Dinger, M.E., Pang, K.C., Mercer, T.R., Crowe, M.L., Grimmond, S.M., Mattick, J.S., 2009. NRED: a database of long noncoding RNA expression. Nucleic Acids Res. 37, D122–D126.

Djanaguiraman, M., Vara Prasad, P.V., Murugan, M., Perumal, R., Reddy, U.K., 2014. Physiological differences among sorghum (*Sorghum bicolor* L Moench) genotypes under high temperature stress. Environ. Exp. Bot. 100, 43–54.

Doggett, H., 1988. Sorghum, second ed. Longmans, London.

Dwivedi, S.L., Sahrawat, K., Upadhyaya, H., Ortiz, R., 2013. Food, nutrition and agrobiodiversity under global climate change. Adv. Agron. 120, 1–128.

Elferink, M., Schierhorn, F., 2016. Global Demand for Food Is Rising. Can We Meet It. Harvard Business Review, Boston, MA, USA. Available online, https://hbr.org/2016/04/global-demand-for-food-is-rising-canwe-meet-it. (Accessed August 25, 2019).

Evans, J., McCormick, R.F., Morishige, D., Olson, S.N., Weers, B., Hilley, J., Klein, P., Rooney, W., Mullet, J., 2013. Extensive variation in the density and distribution of DNA polymorphism in sorghum genomes. PLoS One 8, e79192.

Filipe, B., Martienssen, R.A., 2015. The expanding world of small RNAs in plants. Nat. Rev. Mol. Cell Biol. 16, 727–741.

Franco-Zorrilla, J.M., Valli, A., Todesco, M., Mateos, I., Puga, M.I., Rubio-Somoza, I., et al., 2007. Target mimicry provides a new mechanism for regulation of microRNA activity. Nat. Genet. 39, 1033–1037.

Gaytán-Martínez, M., Cabrera-Ramírez, Á.H., Morales-Sánchez, E., Ramírez-Jiménez, A.K., Cruz-Ramírez, J., Campos-Vega, R., Mendoza, S., 2017. Effect of nixtamalization process on the content and composition of phenolic compounds and antioxidant activity of two sorghums varieties. J. Cereal Sci. 77, 1–8.

Geleta, N., Labuschagne, M.T., 2005. Qualitative trait variation in sorghum (*Sorghum bicolor* L. (Moench)) from eastern highlands of Ethiopia. Biodivers. Conserv. 14, 3055–3064.

Gelli, M., Duo, Y., Konda, A.R., Zhang, C., Holding, D., Dweikat, I., 2014. Identification of differentially expressed genes between sorghum genotypes with contrasting nitrogen stress tolerance by genome-wide transcriptional profiling. BMC Genomics 15, 179.

Gilchrist, E.J., Haughn, G.W., 2005. TILLING without a plough: a new method with applications for reverse genetics. Curr. Opin. Plant Biol. 8, 211–215.

Grenier, C., Bramel, P.J., Dahlberg, J.A., El-Ahmadi, A., Mahmoud, M., Peterson, G.C., Rosenow, D.T., Ejeta, G., 2004. Sorghums of the Sudan: analysis of regional diversity and distribution. Genet. Resour. Crop. Evol. 51, 489–500.

Hamzaa, N.B., Sharma, N., Tripathi, A., Sanan-Mishra, N., 2016. MicroRNA expression profiles in response to drought stress in *Sorghum bicolor*. Gene Expr. Patterns 20, 88–98.

Held, M.A., Penning, B., Brandt, A.S., Kessans, S.A., Yong, W., Scofield, S.R., et al., 2008. Small-interfering RNAs from natural antisense transcripts derived from a cellulose synthase gene modulate cell wall biosynthesis in barley. Proc. Natl. Acad. Sci. U. S. A. 105, 20534–20539.

Henriques, R., Wang, H., Liu, J., Boix, M., Huang, L.F., Chua, N.H., 2017. The antiphasic regulatory module comprising CDF5 and its antisense RNA FLORE links the circadian clock to photoperiodic flowering. New Phytol. 216, 854–867.

Heo, J.B., Sung, S., 2011. Vernalization-mediated epigenetic silencing by a long intronic noncoding RNA. Science 331, 76–79.

House, L.R., 1985. A Guide to Sorghum Breeding, second ed. International Crops Research Institute for the Semi-Arid Tropics, Patancheru, India.

Jabnoune, M., Secco, D., Lecampion, C., Robaglia, C., Shu, Q.Y., Poirier, Y., 2013. A rice cis-natural antisense RNA acts as a translational enhancer for its cognate mRNA and contributes to phosphate homeostasis and plant fitness. Plant Cell 25, 4166–4182.

Jack, T., 2004. Molecular and genetic mechanisms of floral control. Plant Cell 16, S1–S17.

Jin, J., Liu, J., Wang, H., Wong, L., Chua, N.H., 2013. PLncDB: plant long non-coding RNA database. Bioinformatics 29, 1068–1071.

Juan, W., Toshihiro, O., Toru, F., Takahiko, T., Masahiro, S., Yasushi, Y.A., 2012. Novel hypoxic stress-responsive long non-coding RNA transcribed by RNA polymerase III in Arabidopsis. RNA Biol. 9, 302–313.

Kamara, A.Y., Ekeleme, F., Omoigui, L., Menkir, A., Chikoye, D., Dugje, I.Y., 2011. Response of exotic sorghum (*Sorghum bicolor* [L.] Moench) cultivars to planting date under natural infestation of *Striga hermonthica* (Del) Benth. in the Sudan savanna zone of northeast Nigeria. Arch. Agron. Soil Sci. 6, 679–692.

Kang, C., Liu, Z., 2019. An easy-to-follow pipeline for long noncoding RNA identification: a case study in diploid strawberry *Fragaria vesca*. In: Chekanova, J.A., Wang, H.L.V. (Eds.), Plant Long Non-Coding RNAs, Methods in Molecular Biology. vol. 1933. Humana Press, New York, NY.

Kapanigowda, M.H., Perumal, R., Djanaguiraman, M., Aiken, R.M., Tesso, T., Vara Prasad, P.V., Little, C.R., 2013. Genotypic variation in sorghum [*Sorghum bicolor* (L.) Moench] exotic germplasm collections for drought and disease resistance. Springerplus 2, 650.

Katiyar, A., Smita, S., Chinnusamy, V., Pandey, D.M., Bansal, K.C., 2012. Identification of miRNAs in sorghum by using bioinformatics approach. Plant Signal. Behav. 7, 246–259.

Kiegle, E.A., Garden, A., Lacchini, E., Kater, M.M., 2018. A genomic view of alternative splicing of long noncoding RNAs during rice seed development reveals extensive splicing and lncRNA gene families. Front. Plant Sci. 9, 115.

Kouchi, H., Takane, K., So, R.B., Ladha, J.K., Reddy, P.M., 1999. Rice ENOD40: isolation and expression analysis in rice and transgenic soybean root nodules. Plant J. 18, 121–129.

Kumar, A.A., Reddy, B.V.S., Sharma, H.C., Hash, C.T., Rao, P.S., Ramaiah, B., Reddy, P.S., 2011. Recent advances in sorghum genetic enhancement research at ICRISAT. Am. J. Plant Sci. 2, 589–600.

Leiser, W.L., Rattunde, H.F.W., Weltzien, E., Cisse, N., Abdou, M., Diallo, A., Tourè, A.O., Magalhaes, J.V., Haussmann, B.I.G., 2014. Two in one sweep: aluminum tolerance and grain yield in P-limited soils are associated to the same genomic region in West African sorghum. BMC Plant Biol. 14, 206.

Liu, C., Muchhal, U.S., Raghothama, K.G., 1997. Differential expression of TPS11, a phosphate starvation-induced gene in tomato. Plant Mol. Biol. 33, 867–874.

Liu, J., Jung, C., Xu, J., Wang, H., Deng, S., Bernad, L., et al., 2012. Genome-wide analysis uncovers regulation of long intergenic noncoding RNAs in Arabidopsis. Plant Cell 24, 4333–4345.

Liu, J., Wang, H., Chua, N.-H., 2015. Long noncoding RNA transcriptome of plants. Plant Biotechnol. J. 13, 319–328.

Liu, S., Sun, Z., Xu, M., 2018a. Identification and characterization of long non-coding RNAs involved in the formation and development of poplar adventitious roots. Ind. Crop. Prod. 118, 334–346.

Liu, T., Wu, P., Wang, Q., Wang, W., Zhang, C., Sun, F., Liu, Z., Li, Y., Hou, X., 2018b. Comparative transcriptome discovery and elucidation of the mechanism of long noncoding RNAs during vernalization in *Brassica rapa*. Plant Growth Regul. 85, 27–39.

Mach, J., 2017. The long-noncoding RNA ELENA1 functions in plant immunity. Plant Cell 29, 916.

Manzelli, M., Benedettelli, S., Vecchio, V., 2005. Agricultural biodiversity in Northwest Somalia—an assessment among selected Somali sorghum (*Sorghum bicolor* (L.) Moench) germplasm. Biodivers. Conserv. 14, 3381–3392.

Mekbib, F., Bjørnstad, A., Sperling, L., Synnevåg, G., 2009. Factors shaping on-farm genetic resources of sorghum (*Sorghum bicolor* L. (Moench)) in the center of diversity, Ethiopia. Int. J. Biodiversity Conserv. 1, 45–59.

Miao-Chih, T., Ohad, M., Yue, W., Nima, M., Wang, J.K., Fei, L., Yang, S., Eran, S., Chang, H.Y., 2010. Long noncoding RNA as modular scaffold of histone modification complexes. Science 329, 689–693.

Moraes, É.A., Queiroz, V.A.V., Shaffert, R.E., Costa, N.M.B., Nelson, J.D., Ribeiro, S.M.R., Martino, H.S.D., 2012. In vivo protein quality of new sorghum genotypes for human consumption. Food Chem. 3, 1549–1555.

Morris, G.P., Ramu, P., Deshpande, S.P., Hash, C.T., Shah, T., Upadhyaya, H.D., Riera-Lizarazu, O., Brown, P.J., Achary, C.B., Mitchell, S.E., Harriman, J., Glaubitz, J.C., Buckler, E.S., Kresovich, S., 2013. Population genomic and genome-wide association studies of agroclimatic traits in sorghum. Proc. Natl. Acad. Sci. U. S. A. 110, 453–458.

Motlhaodi, T., Geleta, M., Bryngelsson, T., Fatih, M., Chite, S., Ortiz, R., 2014. Genetic diversity in ex-situ conserved sorghum accessions of Botswana as estimated by microsatellite markers. Aust. J. Crop. Sci. 8, 35–43.

Motuma, M., Suryabhagavan, K.V., Balakrishnan, M., 2016. Land suitability analysis for wheat and sorghum crops in Wogdie district, South Wollo, Ethiopia, using geospatial tools. Appl. Geomat. 1, 57–66.

Mutegi, E., Sagnard, F., Labuschagne, M., Herselman, L., Semagn, K., Deu, M., de Villers, S., Kanyenji, B.M., Mwongera, C.N., Traore, P.C.S., Kiambi, D., 2012. Local scale patterns of gene flow and genetic diversity in a crop-wild-weedy complex of sorghum (*Sorghum bicolor* (L.) Moench) under traditional agricultural field conditions in Kenya. Conserv. Genet. 13, 1059–1071.

Nakamoto, S., Tashiro, K., Matsumoto, A., Rice ENOD40, 2010. Isolation and expression analysis in rice and transgenic soybean root nodules. Plant J. 18, 121–129.

Nejat, N., Mantri, N., 2017. Emerging roles of long non-coding RNAs in plant response to biotic and abiotic stresses. Crit. Rev. Biotechnol. 1–13.

Paterson, A., Bowers, J., Bruggmann, R., Dubchak, I., Grimwood, J., et al., 2009. The *Sorghum bicolor* genome and the diversification of grasses. Nature 457, 551–556.

Paytuvi-Gallart, A., Sanseverino, W., Aiese Cigliano, R., 2019. A walkthrough to the use of GreeNC: the plant lncRNA database. In: Chekanova, J.A., Wang, H.L.V. (Eds.), Plant Long Non-Coding RNAs. Methods in Molecular Biology. vol. 1933. Humana Press, New York, NY.

Penning de Vries, F.W.T., Rabbinge, R., Groot, J.J.R., 1997. Potential and attainable food production and food security in different regions. Philos. Trans. R. Soc. Lond. Ser. B Biol. Sci. 352, 917–928.

Proietti, I., Frazzoli, C., Mantovani, A., 2015. Exploiting nutritional value of staple foods in the world's semi-arid areas: risks, benefits, challenges and opportunities of sorghum. Healthcare 3, 172–193.

Qi, S., Akter, S., Li, S., 2019. Identification of novel lincRNA and co-expression network analysis using RNA-sequencing data in plants. In: Chekanova, J.A., Wang, H.L.V. (Eds.), Plant Long Non-Coding RNAs, Methods in Molecular Biology. vol. 1933. Humana Press, New York, NY.

Qin, T., Zhao, H., Cui, P., Albesher, N., Xiong, L., 2017. A nucleus-localized long non-coding RNA enhances drought and salt stress tolerance. Plant Physiol. 175, 1321–1336.

Ramu, P., Billot, C., Rami, J.F., Senthilvel, S., Upadhyaya, H.D., Reddy, L.A., Hash, C.T., 2013. Assessment of genetic diversity in the sorghum reference set using EST-SSR markers. Theor. Appl. Genet. 126, 2051–2064.

Sanousi, R.S., Sanan-Mishra, N., Abdelmula, A.A., Mohammed, I.A., Seif, M.G., Hamza, N.B., 2016. Differential expression of miRNAs in *Sorghum bicolor* under drought and salt stress. Am. J. Plant Sci. 7, 870–878.

Seetharam, K., 2011. Phenotypic Assessment of Sorghum (*Sorghum bicolor* L. Moench) Germplasm Reference Set for Yield and Related Traits Under Postflowering Drought Conditions. Ph.D. Thesis Tamil Nadu Agricultural University, Coimbatore, Tamil Nadu, India.

Shakoor, N., Nair, R., Crasta, O., Morris, G., Feltus, A., Kresovich, S., 2014. A *Sorghum bicolor* expression atlas reveals dynamic genotype-specific expression profiles for vegetative tissues of grain, sweet and bioenergy sorghums. BMC Plant Biol. 14, 35.

Sharma, H.C., 1993. Host plant resistance to insects in sorghum and its role in integrated pest management. Crop Prot. 12, 11–34.

Shuai, P., Liang, D., Tang, S., Zhang, Z., Ye, C.Y., Su, Y., Xia, X., Yin, W., 2014. Genome-wide identification and functional prediction of novel and drought-responsive lincRNAs in *Populus trichocarpa*. J. Exp. Bot. 65, 4975–4983.

Shumayla, Sharma S., Taneja, M., Tyagi, S., Singh, K., Upadhyay, S.K., 2017. Survey of high throughput RNA-seq data reveals potential roles for lncRNAs during development and stress response in bread wheat. Front. Plant Sci. 8, 1019.

Sousa, C., Johansson, C., Charon, C., Manyani, H., Sautter, C., Kondorosi, A., et al., 2001. Translational and structural requirements of the early nodulin gene enod40, a short-open reading frame containing RNA, for elicitation of a cell-specific growth response in the alfalfa root cortex. Mol. Cell. Biol. 21, 354–366.

Stefoska-Needham, A., Beck, E.J., Johnson, S.K., Tapsell, L.C., 2015. Sorghum: an underutilized cereal whole grain with the potential to assist in the prevention of chronic disease. Food Rev. Int. 4, 401–437.

Swiezewski, S., Crevillen, P., Liu, F., Ecker, J.R., Jerzmanowski, A., Dean, C., 2007. Small RNA-mediated chromatin silencing directed to the 3′ region of the Arabidopsis gene encoding the developmental regulator, FLC. Proc. Natl. Acad. Sci. U. S. A. 104, 3633–3638.

Swiezewski, S., Liu, F., Magusin, A., Dean, C., 2009. Cold-induced silencing by long antisense transcripts of an Arabidopsis polycomb target. Nature 462, 799–802.

Szcześniak, M.W., Bryzghalov, O., Ciomborowska-Basheer, J., Makałowska, I., 2019. CANTATAdb 2.0: expanding the collection of plant long noncoding RNAs. In: Chekanova, J.A., Wang, H.L.V. (Eds.), Plant Long Non-Coding RNAs, Methods in Molecular Biology. vol. 1933. Humana Press, New York, NY.

Tyagi, S., Sharma, A., Upadhyay, S.K., 2018. Role of next-generation RNA-Seq data in discovery and characterization of long non-coding RNA in plants. In: Çiftçi, Y.Ö. (Ed.), Next Generation Plant Breeding. IntechOpen, https://doi.org/10.5772/intechopen.72773.

Upadhyaya, H.D., Dwivedi, S.L., Ramu, P., Shailesh, S.K., Singh, S., 2014. Genetic variability and effect of post flowering drought on stalk sugar content in sorghum mini core collection. Crop Sci. 54, 2120–2130.

Upadhyaya, H.D., Dwivedi, S.L., Singh, S., Sahrawat, K.L., Singh, S.K., 2016. Genetic variation and post flowering drought effects on seed iron and zinc in ICRISAT sorghum mini core collection. Crop Sci. 56, 374–384.

Vadez, V., Krishnamurthy, L., Hash, C.T., Upadhyaya, H.D., Borrell, A.K., 2011. Yield, transpiration efficiency, and water-use variations and their interrelationships in the sorghum reference collection. Crop Pasture Sci. 62, 645–655.

Venkateswaran, K., Muraya, M., Dwivedi, S.L., Upadhyaya, H.D., 2014. Wild sorghums—their potential use in crop improvement. In: Wang, Y.H., Upadhyaya, H.D., Kole, C. (Eds.), Genetics, Genomics and Breeding of Sorghum. CRC Press, Boca Raton, FL.

Wang, H.L.V., Chekanova, J.A., 2017. Long noncoding RNAs in plants. Adv. Exp. Med. Biol. 1008, 133–154.

Wang, H., Chung, P.J., Liu, J., Jang, I.C., Kean, M.J., Xu, J., et al., 2014a. Genome-wide identification of long noncoding natural antisense transcripts and their responses to light in Arabidopsis. Genome Res. 24, 444–453.

Wang, Y., Fan, X., Lin, F., He, G., Terzaghi, W., Zhu, D., et al., 2014b. Arabidopsis noncoding RNA mediates control of photomorphogenesis by red light. Proc. Natl. Acad. Sci. U. S. A. 111, 10359–10364.

Wang, B., Regulski, M., Tseng, E., Olson, A., Goodwin, S., McCombie, W.R., Ware, D., 2018a. A comparative transcriptional landscape of maize and sorghum obtained by single-molecule sequencing. Genome Res. 28, 921–932.

Wang, R., Zou, J., Meng, J., Wang, J., 2018b. Integrative analysis of genome-wide lncRNA and mRNA expression in newly synthesized *Brassica* hexaploids. Ecol. Evol. 8, 6034–6052.

Wang, A., Hu, J., Gao, C., et al., 2019. Genome-wide analysis of long non-coding RNAs unveils the regulatory roles in the heat tolerance of Chinese cabbage (*Brassica rapa* ssp. chinensis). Sci. Rep. 9, 5002.

Wasaki, J., Yonetani, R., Shinano, T., Kai, M., Osaki, M., 2003. Expression of the OsPI1 gene, cloned from rice roots using cDNA microarray, rapidly responds to phosphorus status. New Phytol. 158, 239–248.

Wierzbicki, A.T., Haag, J.R., Pikaard, C.S., 2008. Noncoding transcription by RNA polymerase Pol IVb/Pol V mediates transcriptional silencing of overlapping and adjacent genes. Cell 135, 635–648.

Wunderlich, M., Gross-Hardt, R., Schoffl, F., 2014. Heat shock factor HSFB2a involved in gametophyte development of *Arabidopsis thaliana* and its expression is controlled by a heat-inducible long non-coding antisense RNA. Plant Mol. Biol. 85, 541–550.

Xin, Z., Wang, M.L., Barkley, N.A., Burow, G., Franks, C., Pederson, G., Burke, J., 2008. Applying genotyping (TILLING) and phenotyping analyses to elucidate gene function in a chemically induced sorghum mutant population. BMC Plant Biol. 8, 103.

Yan, Q., Wu, F., Yan, Z., et al., 2019. Differential co-expression networks of long non-coding RNAs and mRNAs in *Cleistogenes songorica* under water stress and during recovery. BMC Plant Biol. 19, 23.

Yang, W.C., Katinakis, P., Hendriks, P., Smolders, A., de Vries, F., Spee, J., et al., 1993. Characterization of GmENOD40, a gene showing novel patterns of cell-specific expression during soybean nodule development. Plant J. 3, 573–585.

Yang, Z., Yang, C., Wang, Z., Yang, Z., Chen, D., Wu, Y., 2019. LncRNA expression profile and ceRNA analysis in tomato during flowering. PLoS One 14, e0210650.

Zhang, Y.C., Chen, Y.Q., 2013. Long noncoding RNAs: new regulators in plant development. Biochem. Biophys. Res. Commun. 436, 111–114.

Zhang, M., Zhao, H., Xie, S., Chen, J., Xu, Y., Wang, K., Zhao, H., Guan, H., Hu, X., Jiao, Y., 2011. Extensive, clustered parental imprinting of protein-coding and noncoding RNAs in developing maize endosperm. Proc. Natl. Acad. Sci. U. S. A. 108, 20042–20047.

Zhou, G.-F., Zhang, L.-P., Li, B.-X., Sheng, O., Wei, Q.-J., Yao, F.-X., Guan, G., Liu, G.-D., 2019. Genome-wide identification of long non-coding RNA in trifoliate orange (*Poncirus trifoliata* (L.) Raf) leaves in response to boron deficiency. Int. J. Mol. Sci. 20, 5419.

Zhu, F., 2018. Anthocyanins in cereals: composition and health effects. Food Res. Int. 109, 232–249.

Zhu, Q.H., Wang, M.B., 2012. Molecular functions of long non-coding RNAs in plants. Genes (Basel) 3, 176–190.

Zhu, B., Yang, Y., Li, R., Fu, D., Wen, L., Luo, Y., Zhu, H., 2015. RNA sequencing and functional analysis implicate the regulatory role of long non-coding RNAs in tomato fruit ripening. J. Exp. Bot. 66, 4483–4495.

Zubko, E., Meyer, P., 2007. A natural antisense transcript of the *Petunia hybrida* Sho gene suggests a role for an antisense mechanism in cytokinin regulation. Plant J. 52, 1131–1139.

Chapter 7

Roles of long noncoding RNA during stress response in maize

Shikha Verma[a], Pankaj Kumar Verma[b], and Debasis Chakrabarty[a]
[a]*Genetics and Molecular Biology Division, CSIR-National Botanical Research Institute, Lucknow, India,* [b]*Department of Botany, University of Lucknow, Lucknow, India*

7.1 Introduction

Maize (*Zea mays* L.) is a globally cultivated crop having more extensive adaptability under varied agroclimatic conditions (Dass et al., 2018). Maize grows in about 160 countries under a wide diversity of soil, climate, biodiversity, and management practices that contribute 36% to global grain production. Various stresses affect maize production as well as its grain quality (Fischer and Connor, 2018). High soil salinity and drought are major abiotic stresses that can significantly reduce the production of maize crops. To cope up with these abiotic stresses, plants respond in various ways, including changing their metabolism, physiology, and development (Haggag et al., 2015). These changes may occur due to the dynamic reprogramming of gene expression in response to stress. Various mechanisms were identified by researchers, which help to intensify the plant response to stress by modifying their genetic makeup. Several biotechnology strategies have been applied to modify maize crops to generate transgenic plants with a superior reaction to numerous abiotic and biotic stresses (Kamthan et al., 2016). However, the use of transgenic plants is a serious concern due to the versatility in the action of genes and diverse signaling networks, including crosstalk between different signaling pathways (van Esse et al., 2019). Therefore, unintended environmental effects are possible on generating genetically modified organisms (Biricolti et al., 2016). Consequently, it is crucial to identify and alter some regulatory mechanisms that have a positive impact on plants as well as overcome the problems related to transgenic.

For improving the plant defense, earlier research was focused mainly on stress management by the regulation of protein-encoding genes. The advancement in high-throughput sequencing and transcriptomics analysis has gained the knowledge of intergenic region between protein-coding genes named "junk DNA" (Nejat and Mantri, 2018). In the eukaryotic genome, a large part of the genome transcribed into RNAs, but did not encode proteins; that potentially

Long Noncoding RNAs in Plants: Roles in development and stress
https://doi.org/10.1016/B978-0-12-821452-7.00007-6

functional RNAs are called nonprotein-coding RNAs (Ulitsky, 2016). In recent years, nonprotein-coding transcripts, including long noncoding RNAs (lncRNAs) are emerging as critical regulators of gene expression. The lncRNAs represent a distinct class of transcripts longer than 200 nucleotides without or with little protein-coding potential (Wu et al., 2017). The lncRNAs act as precursors of microRNAs and other small RNAs that regulate the biological processes (Catalanotto et al., 2016), alternative splicing of pre-mRNA, and are also involved in epigenetic responses (Ariel et al., 2015; Hanly et al., 2018). Besides, numerous small RNAs are reacted to different environmental stimuli and involved in the modulation of gene expression, adaption, and stress response (Wang and Chekanova, 2016).

Long noncoding RNAs have recently appeared as fundamental regulators of gene expression and have attracted the researcher's attention. The role of lncRNAs in growth, development, and cellular response to various stresses has been explored and reviewed. This chapter focuses on lncRNAs and their role in coordinating maize stress response.

7.2 Discovery and classification of lncRNA

In animal cells, lncRNAs have been identified and their functions investigated for more than 20 years (Liu et al., 2015). The beginning of next-generation sequencing (NGS) accelerated the production of a massive amount of transcriptomic data. The data showed that DNA sequences-encoded proteins involve a small amount (2%–25%) of the genome space depending on the organism in question. Except for housekeeping genes, RNAs and transposons, the majority of genomic sequences are intergenic, which means they are present between protein-coding genes (Liu et al., 2015). However, this conventional view was subsequently challenged by the discovery of discrete transcripts arising from these intergenic regions (Liu et al., 2015). Like mRNAs, these transcripts have capped 5′ end, polyadenylated 3′ end, and many contain spliced introns (Jarroux et al., 2017). In contrast to mRNAs, their expression levels are very low and do not have an evolutionarily conserved long open reading frame (Liu et al., 2015).

This group of RNAs can be classified, according to size, into: (1) small noncoding RNAs; and (2) long noncoding RNAs (lncRNAs) (Rai et al., 2019). The lncRNAs are further classified based on genomic location into: (1) natural antisense transcripts (NATs); (2) long intergenic noncoding RNAs (lincRNAs); and (3) intronic RNAs (incRNAs) (Wang et al., 2015) (Table 7.1). In the past, thousands of lncRNAs have been explored in eukaryotes, including yeast and human, whereas in plants, identification of lncRNAs is recent but not global (Zheng et al., 2015). By coordinating directional tiling array and RNA-sequencing (RNA-seq) data, around 6480 lincRNAs and 37238 NAT pairs were identified in *Arabidopsis thaliana* (Wang et al., 2015). A recent analysis of maize EST and RNA-seq datasets has uncovered many lncRNAs (Li et al., 2014). As these maize lncRNAs were assembled from nondirectional transcriptome data,

TABLE 7.1 List of plant lncRNAs classification based on genome location.

Classification	Name	Species	Function	Reference
Sense and Intronic lncRNAs	COLDAIR	*Arabidopsis thaliana*	Regulates flowering	Heo and Sung (2011)
Antisense lncRNAs	COOLAIR	*Arabidopsis thaliana*	Regulates flowering	Swiezewski et al. (2009)
	cis-NAT*PHO1;2*	*Oryza sativa*	Regulates phosphate homeostasis	Jabnoune et al. (2013)
	HvCesA6 lnc-NAT	*Hordeum vulgare*	Involved in the synthesis of cell wall	Held et al. (2008)
	SHO lnc-NAT	*Petunia hybrid*	Regulates the synthesis of local cytokines	Zubko and Meyer (2007)
	lncNATs: 600	*Oryza sativa*	Regulates sexual reproduction	Zhang et al. (2014b)
	HID1	*Arabidopsis thaliana*	Regulates photomorphogenesis	Wang et al. (2014)
	HSFB2a	*Arabidopsis thaliana*	Regulates the gametophytic and vegetative development	Wunderlich et al. (2014)
Intergenic lncRNAs	APOLO	*Arabidopsis thaliana*	Regulated plant development through auxin signaling	Ariel et al. (2014)
	ASCO	*Arabidopsis thaliana*	Regulates the development of lateral roots	Bardou et al. (2014)
	ENOD40	*Oryza sativa*	Involved in the development of nodules	Kouchi et al. (1999)
	IPS1	*Arabidopsis thaliana*	Regulates phosphate homeostasis	Franco-Zorrilla et al. (2007)
	DRIR	*Arabidopsis thaliana*	Upregulated in drought and salt stress	Qin et al. (2017)
	LDMAR	*Oryza sativa*	Male fertility regulation	Ding et al. (2012a,b)
	ELENA1	*Arabidopsis thaliana*	Upregulated in disease caused by *Pseudomonas syringae*	Mach (2017)

the characterization and description of lncRNAs, particularly for NATs, was constrained for transcription and transcript borders (Wang et al., 2015). A plant lncRNA, APOLO was described to play a role in response to auxin in the chromatin loop formation (Ariel et al., 2014). Therefore, the information involved in epigenetic modification could also be used in the characterization and re-annotation of genes for noncoding transcripts in maize.

Conservation analysis improves the understanding of lncRNAs in numerous aspects: (1) identification of functional sequence elements and structures; (2) potential biological functions; and (3) their mode of action (Ulitsky and Bartel, 2013). Because of the scarcity of prospective data on maize lncRNAs, it is still an open question to what extent maize lncRNAs are conserved at the sequence or genomic position level. A more comprehensive lncRNA catalog from this cereal crop is required to address these issues. Genetic and molecular analysis showed that an increased number of lncRNAs are involved in the gene expression regulation and chromatin structure. LncRNAs can regulate the targets in cis or trans through various mechanisms (Ulitsky and Bartel, 2013). However, whether lncRNAs serve biological functions or are purely transcriptional by-products is still controversial, because mostly they do not have a decipherable feature (Kung et al., 2013). The genome-wide association studies (GWAS) have been successful in unraveling the genetic basis of trait variation and identification of loci linked to phenotypic diversity in plants (Wang et al., 2015).

Moreover, single nucleotide polymorphism (SNP) typing allowed GWAS to find out small haplotype blocks that are associated with trait variation (Brachi et al., 2011). Based on existing GWAS results in animals and plants, $<20\%$ of these causal loci are assigned to coding regions (Hindorff et al., 2009). Nevertheless, trait-associated SNPs embedded in the intergenic areas serve as precious resources to address the potential roles of lncRNAs. For example, genetic variations associated with human diseases have been shown to impact the expression of lincRNA (Kumar et al., 2013).

Additionally, rice SNP regulates the expression of the corresponding lncRNA gene, resulting in photoperiod-sensitive male sterility (Ding et al., 2012a,b). These examples suggested that association analysis of maize lncRNAs and GWAS data recognizes the potential functions of lncRNAs and also categorizes candidate noncoding genes essential for GWAS. Understanding the genetic basis of trait variation is also critical to improve the yield and quality of this cereal crop.

7.3 Maize long noncoding RNAs (lncRNAs)

Maize reference genome was initially released in 2009 and updated in 2017, but the genome-wide annotation of lncRNAs is continuing (Jiao et al., 2017). Boerner and McGinnis (2012) identified hundreds of putative lncRNAs in maize using computational analysis on maize and its wild relative teosinte. The underlying molecular mechanism of maize speciation of one ancient teosinte was scrutinized by genome-wide identification and characterization of conserved lncRNAs and non-conserved lncRNAs among maize and teosinte. Additionally, selection analysis indicated that conserved lncRNAs are more likely than non-conserved lncRNAs to be located in regions subject to the current selection, showing evolutionary differentiation (Deng et al., 2018). However, an evolutionary contribution of conserved and non-conserved lncRNAs during maize civilization and domestication is generally unknown. Following this study, Li et al. (2018) recognized and characterized around 2000 high confidence maize lncRNAs, implying these lncRNAs are widespread in the maize genome. A nearby 664 dehydration-responsive maize lncRNAs were identified (Zhang et al., 2014a; Han et al., 2019b), followed by several nitrogen-responsive intergenic and intron-derived lncRNAs in maize leaves by deep total RNA sequencing (Lv et al., 2016b). Hence, the most advanced genome-wide annotation and maize lncRNAs analysis explore that the possible functional divergence between protein-coding, conserved lncRNA, and non-conserved lncRNA genes corresponds to the high complexity of maize transcriptome. The reports of functional maize lncRNAs have been increased day by day; yet, a new comprehensive annotation of lncRNAs has been required for the maize genome.

7.4 Synthesis and emerging significance of lncRNAs and status in plants

Like mRNAs, the majority of lncRNAs are transcribed by RNA polymerase II and function as noncoding molecules or considered as transcriptional noise (Ulitsky and Bartel, 2013; Ruiz-Orera et al., 2014). However, earlier reports explained that in plants, lncRNAs could also be synthesized by RNA polymerase IV and RNA polymerase V (Wierzbicki et al., 2008).

In plant system, initially reported biologically important plant lncRNA genes were GmENOD40, *Glycine max* (soybean) lncRNA (Yang et al., 1993), MtENOD40 (Charon et al., 1997), TPS11 (Liu et al., 1997), OsPI1 (Wasaki et al., 2003), etc. and recently reported lncRNA genes, AtIPS1 (Meng et al., 2012), COOLAIR (Hawkes et al., 2016), COLDAIR (Kim and Sung, 2012; Kim et al., 2017), and LDMAR (Shafiq et al., 2016), have shown distinct biological roles. For example, during nodule organogenesis, ENOD40 is expressed in plants. Further, *A. thaliana* lncRNA, AtIPS1, was studied and it was found that it interacts with miRNA ath-miR399 (Meng et al., 2012).

7.5 Functional roles of lncRNAs

In 1992, a pioneering study reported primate brain-specific lncRNAs (Lukiw et al., 1992). Till now, numerous attempts have been made to identify and understand the biological roles of these lncRNAs. In order to improve understanding, several plants lncRNAs have been functionally characterized, among them *A. thaliana* and rice lncRNAs are most studied till now. These studies explore that lncRNAs participate in the epigenetic regulation of flowering time in *A. thaliana* by triggering floral repressor *FLOWERING LOCUS C* (*FLC*). Another most studied *Arabidopsis Drought-induced lncRNA* (*DRIR*) provides tolerance to both salt and drought (Qin et al., 2017). In addition, rice antisense lncRNA regulates *PHOSPHATE1;2* (*PHO1;2*) and plays a vital role in phosphate homeostasis. The detailed studies suggested that *PHO1;2* is required for phosphate loading in rice as *cis*-NAT*PHO1;2* is the antisense transcript of *PHO1;2*. Alternatively, *cis*-NAT*PHO1;2* enhances *PHO1;2* translation to influence phosphate homeostasis (Jabnoune et al., 2013; Reis and Deforges, 2017). Similarly, another lncRNA *INDUCED BY PHOSPHATE STARVATION 1* (*IPS1*) serves as a target mimic of miR-399 and represses the activity of miR-399. *IPS1* overexpression increases the expression of miR-399 target *PHO2* and reduces phosphate content in the shoot; thus, antisense lncRNA maintains phosphate homeostasis in rice (Franco-Zorrilla et al., 2007). Although a large number of lncRNAs are discovered in different species (Bhatia et al., 2017), only a few of them have been related to function as precursors of small RNAs, microRNA mimics, and epigenetic complex decoys (Yoon and Abdelmohsen, 2014). Thus, lncRNAs have multiple functions and expected to play significant roles in progression and speciation.

7.6 Functional roles of lncRNAs in maize

With the reference of whole-genome sequencing resources of maize, RNA-seq data, and EST information, Li et al. (2014) systematically interpreted the distribution and expression behavior of lncRNA in the maize genome. Since now, few maize lncRNAs were identified, which are linked to stress response (listed in Table 7.2). Empowered by the advancement of next-generation sequencing technique, large-scale lncRNAs profiles have been described in maize using the full-length cDNA sequences (Boerner and McGinnis, 2012).

Recently, the functional role of lncRNA in response to arbuscular mycorrhizal fungi was identified by strand-specific RNA-sequencing. For this, maize roots were colonized with AM fungus, *Rhizophagus irregularis,* and AM fungi-responsive lncRNAs were identified. After strand-specific RNA-sequencing analysis, a total of 5941 lncRNAs were identified in which more than 3000 were new. A few lncRNAs were differentially expressed, which coordinate in the bidirectional nutrient exchange between maize and AM fungi via miRNA target mimicry. This study might contribute to understand the regulatory network between plants and AM fungi (Han et al., 2019a).

Ma and coworkers in 2008 identified a novel ncRNA or sORF gene zm401, which was expressed mainly in tapetal cells and microspores. The silencing of zm401 dramatically influences stamen growth and anther development. The maize, zm401, is an essential gene that offers a pleiotropic role during the development of the male flower and can be used to generate sterile male plants.

TABLE 7.2 lncRNAs associated with stress response in maize.

lncRNAs	Identification approach	Tissue	Role in	Reference
Zm401	Overexpression and RNAi transgenic	pollen	Tapetum and microspore development	Ma et al. (2008)
ZmMNC-2, ZmMNC-9, ZmMNC-18, ZmMNC-24	Transcriptomic analysis	Seed and Endosperm	Endosperm development	Zhang et al. (2011)
TCONS_00012690, TCONS_00000649	Bioinformatics and Molecular analysis	Leaf	Drought stress	Zhang et al. (2014a)
lncRNA-7245	Ribosomal RNA depletion and ultra-deep total RNA-seq	Leaf	Nitrogen stress	Lv et al. (2016b)
lncRNA-35524, lncRNA-31273, lncRNA-69328, lncRNA-71309, lncRNA-02785, lncRNA-86055, lncRNA-58195	High-throughput Illumina paired-end sequencing and Bioinformatics analysis	Embryo Endosperm	Seed development	Zhu et al. (2017)

A large number of lncRNAs were identified in maize, which regulates parent of origin-dependent epigenetic endosperm development. One of the studies found 38 noncoding imprinted RNAs, of which 25 were maternally expressed transcripts and 13 were paternally expressed transcripts. The four maternal specifically expressed transcripts were intronic noncoding RNAs and complement their associated paternal specifically transcripts, which strongly suggested their role in imprinting. This study provides an extensive analysis of imprinted noncoding RNAs and improves our understanding of the impact of genetic imprinting during maize endosperm development (Zhang et al., 2011).

7.7 Functional roles of lncRNAs in maize under stress

7.7.1 Role of lncRNAs under nitrogen stress

There are only a few studies that explain the involvement of lncRNAs in response to nitrogen in maize. To explore the possible role of lncRNAs under nitrogen stress, Lv et al. (2016a,b) worked on differentially expressed genes analysis in maize leaves using ribosomal RNA depletion and ultra-deep total RNA sequencing. They identified around 2391 differentially expressed transcripts, of which 637 lncRNAs were nitrogen-responsive. Under high nitrogen, 211 lncRNAs were upregulated, while 426 lncRNAs were downregulated. Additionally, GWAS, co-expression network, and differential expression analysis of intergenic/intronic lncRNA presented a global aspect of transcriptional responses of lncRNAs to nitrogen stress in maize (Lv et al., 2016b). Co-expression analysis suggested that lncRNAs are linked to energy metabolic pathways and enriched in the co-expressed module. The genes linked in the enriched module were associated with metabolic processes like oxidative phosphorylation, NADH dehydrogenase activity, and nitrogen stress. This report suggested that lncRNAs could have a putative role in maize under nitrogen stress.

7.7.2 Role of lncRNAs in drought and salt stress

The lncRNAs also participate in drought stress-responsive regulation. Through genome-wide screening, a total of 664 out of 8449 differentially expressed maize drought stress-responsive lncRNA transcripts, including 126 known and 538 novel lncRNAs, were identified. Among the 664 drought-responsive lncRNAs, 315 lncRNAs (10 antisenses and 305 nonantisenses) were identified as other lncRNAs that expected to work as longer molecules (Zhang et al., 2014a).

Recently, Qin et al. (2017) identified DROUGHT-INDUCED lncRNA (DRIR) in *Arabidopsis thaliana* that was induced by drought and salt stress. To reveal the molecular mechanism of DRIR in response to drought stress, overexpression, and mutated (drirD) seedlings were generated and treated with dehydration stress and it was found that expression of FUCOSYLTRANSFERASE4 (FUT4) was significantly upregulated. Apart from this, expression levels of other stress-related genes, like WRKY8, ANNAT7, NIP1, and NAC3, were also markedly increased in overexpression and mutated lines under drought stress.

Thus, DRIR might control maize to drought stress by enhancing the expression of stress-related genes by increasing water transport, and also by normalizing the stomata response to ABA.

Additionally, DRIR also has a functional role in salt stress. During salt stress, the expression levels of RD29B, RD29A, P5CS1, and AtrbohB were increased in the mutant (drirD) and overexpression lines. Besides, the expression of TIP4 and WRKY8 in salt-treated wild-type seedlings was even higher than in drirD and overexpressing lines. These studies suggested that DRIR may differentially regulate stress-related genes through complex mechanisms. DRIR may regulate plant tolerance to salt and drought stress through a different mechanism. Conclusively these reports indicated that DRIR controls the expression of genes involved in ABA-mediated signaling and facilitates the plant to tolerate drought and salt stress. They discovered the expression level using real-time PCR and determined that DRIR accumulates a 168-fold higher transcript in salt-treated seedlings relative to untreated seedlings. Additionally, the expression level of DRIR in dehydrated seedlings was 59-fold higher than in untreated seedlings. This study exhibited that DRIR may play a vital role in drought and salt stress tolerance in maize.

7.8 Concluding remarks and future perspectives

The lncRNAs play crucial roles in plant adaptation and defense during both abiotic and biotic stresses. However, their straight role in defense-related signaling pathways or crosstalk between lncRNAs and components of defined networks is still unexplored. Recently, researchers exhibited multifunctional characters of lncRNAs as regulatory elements in intricate regulatory pathways. Therefore, it is essential to explore that the functional roles of lncRNAs are not just in stress, but also in developmental and cellular processes, hormone signaling, physiology, and pathogenesis. Overall, the identification of stress-related lncRNAs, their transcriptional profiling, functional characterization, and localization studies are now crucial to fully understand the role of these multifunctional elements. Hence, sincere efforts should be attempted to systematically analyze the regulatory functions and pathways of lncRNAs to provide new genetic resources for crop breeding and genetic engineering. Various methods may be applied to analyze the novel functional role, including loss of function and gain of function using knockouts or overexpression of lncRNAs, respectively. In this series, use of recently developed techniques like Transcription Activator-Like Effector Nucleases (TALENs), RNA interference (RNAi), virus-induced gene silencing, and Clustered Regulatory Interspaced Short Palindromic Repeats (CRISPR)/Cas9 systems will become more fruitful to understand their precise functions. Apart from this, the development of lncRNA prediction tools and well-maintained repositories will allow the researchers to extract meaningful data for both novel and already known lncRNAs. However, existing resources for lncRNA failed to embrace most of the newly identified lncRNAs in the recent investigation. Consequently, it would be expected that the current databases are refreshed often and continuously to strengthen the effectiveness of the investigation.

References

Ariel, F., Jegu, T., Latrasse, D., Romero-Barrios, N., Christ, A., Benhamed, M., Crespi, M., 2014. Noncoding transcription by alternative RNA polymerases dynamically regulates an auxin-driven chromatin loop. Mol. Cell 55 (3), 383–396.

Ariel, F., Romero-Barrios, N., Jégu, T., Benhamed, M., Crespi, M., 2015. Battles and hijacks: noncoding transcription in plants. Trends Plant Sci. 20 (6), 362–371.

Bardou, F., Ariel, F., Simpson, C.G., Romero-Barrios, N., Laporte, P., Balzergue, S., Brown, J.W., Crespi, M., 2014. Long noncoding RNA modulates alternative splicing regulators in Arabidopsis. Dev. Cell 30 (2), 166–176.

Bhatia, G., Goyal, N., Sharma, S., Upadhyay, S., Singh, K., 2017. Present scenario of long noncoding RNAs in plants. Non-coding RNA 3 (2), 16.

Biricolti, S., Bartsch, D., Boerjan, W., Fladung, M., Glandorf, D.C., Sweet, J., Fernando, G., 2016. Potential impacts of GM trees on the environment and on plant "omics": questionnaire-based responses. In: Biosafety of Forest Transgenic Trees, pp. 195–205.

Boerner, S., McGinnis, K.M., 2012. Computational identification and functional predictions of long noncoding RNA in Zea mays. PLoS One 7 (8), e43047.

Brachi, B., Morris, G.P., Borevitz, J.O., 2011. Genome-wide association studies in plants: the missing heritability is in the field. Genome Biol. 12 (10), 232.

Catalanotto, C., Cogoni, C., Zardo, G., 2016. MicroRNA in control of gene expression: an overview of nuclear functions. Int. J. Mol. Sci. 17 (10), 1712.

Charon, C., Johansson, C., Kondorosi, E., Kondorosi, A., Crespi, M., 1997. enod40 induces dedifferentiation and division of root cortical cells in legumes. Proc. Natl. Acad. Sci. 94 (16), 8901–8906.

Dass, S., Jat, S., Chikkappa, G.K., Parihar, C., 2018. Crop production management to climate change. In: Food Security and Climate Change, pp. 251–287.

Deng, P., Liu, S., Nie, X., Weining, S., Wu, L., 2018. Conservation analysis of long non-coding RNAs in plants. Sci. China Life Sci. 61 (2), 190–198.

Ding, J., Lu, Q., Ouyang, Y., Mao, H., Zhang, P., Yao, J., Xu, C., Li, X., Xiao, J., Zhang, Q., 2012a. A long noncoding RNA regulates photoperiod-sensitive male sterility, an essential component of hybrid rice. Proc. Natl. Acad. Sci. 109 (7), 2654–2659.

Ding, J., Shen, J., Mao, H., Xie, W., Li, X., Zhang, Q., 2012b. RNA-directed DNA methylation is involved in regulating photoperiod-sensitive male sterility in rice. Mol. Plant 5 (6), 1210–1216.

Fischer, R., Connor, D., 2018. Issues for cropping and agricultural science in the next 20 years. Field Crop Res. 222, 121–142.

Franco-Zorrilla, J.M., Valli, A., Todesco, M., Mateos, I., Puga, M.I., Rubio-Somoza, I., Leyva, A., Weigel, D., García, J.A., Paz-Ares, J., 2007. Target mimicry provides a new mechanism for regulation of microRNA activity. Nat. Genet. 39 (8), 1033.

Haggag, W.M., Abouziena, H., Abd-El-Kreem, F., El Habbasha, S., 2015. Agriculture biotechnology for management of multiple biotic and abiotic environmental stress in crops. J. Chem. Pharm. Res. 7 (10), 882–889.

Han, G., Cheng, C., Zheng, Y., Wang, X., Xu, Y., Wang, W., et al., 2019a. Identification of long noncoding RNAs and the regulatory network responsive to arbuscular mycorrhizal fungi colonization in maize roots. Int. J. Mol. Sci. 20 (18), 4491.

Han, L., Mu, Z., Luo, Z., Pan, Q., Li, L., 2019b. New lncRNA annotation reveals extensive functional divergence of the transcriptome in maize. J. Integr. Plant Biol. 61 (4), 394–405.

Hanly, D.J., Esteller, M., Berdasco, M., 2018. Interplay between long non-coding RNAs and epigenetic machinery: emerging targets in cancer? Philos. Trans. R. Soc. B: Biol. Sci. 373 (1748), 20170074.

Hawkes, E.J., Hennelly, S.P., Novikova, I.V., Irwin, J.A., Dean, C., Sanbonmatsu, K.Y., 2016. COOLAIR antisense RNAs form evolutionarily conserved elaborate secondary structures. Cell Rep. 16 (12), 3087–3096.

Held, M.A., Penning, B., Brandt, A.S., Kessans, S.A., Yong, W., Scofield, S.R., Carpita, N.C., 2008. Small-interfering RNAs from natural antisense transcripts derived from a cellulose synthase gene modulate cell wall biosynthesis in barley. Proc. Natl. Acad. Sci. 105 (51), 20534–20539.

Heo, J.B., Sung, S., 2011. Vernalization-mediated epigenetic silencing by a long intronic noncoding RNA. Science 331 (6013), 76–79.

Hindorff, L.A., Sethupathy, P., Junkins, H.A., Ramos, E.M., Mehta, J.P., Collins, F.S., Manolio, T.A., 2009. Potential etiologic and functional implications of genome-wide association loci for human diseases and traits. Proc. Natl. Acad. Sci. 106 (23), 9362–9367.

Jabnoune, M., Secco, D., Lecampion, C., Robaglia, C., Shu, Q., Poirier, Y., 2013. A rice cis-natural antisense RNA acts as a translational enhancer for its cognate mRNA and contributes to phosphate homeostasis and plant fitness. Plant Cell 25 (10), 4166–4182.

Jarroux, J., Morillon, A., Pinskaya, M., 2017. History, discovery, and classification of lncRNAs. In: Long Non Coding RNA Biology, pp. 1–46.

Jiao, Y., Peluso, P., Shi, J., Liang, T., Stitzer, M.C., Wang, B., Campbell, M.S., Stein, J.C., Wei, X., Chin, C.S., Guill, K., Regulski, M., Kumari, S., Olson, A., Gent, J., Schneider, K.L., Wolfgrube, T.K., May, M.R., Springer, N.M., Antoniou, E., McCombie, W.R., Presting, G.G., McMullen, M., Ross-Ibarra, J., Dawe, R.K., Hastie, A., Rank, D.R., Ware, D., 2017. Improved maize reference genome with single-molecule technologies. Nature 546 (7659), 524.

Kamthan, A., Chaudhuri, A., Kamthan, M., Datta, A., 2016. Genetically modified (GM) crops: milestones and new advances in crop improvement. Theor. Appl. Genet. 129 (9), 1639–1655.

Kim, D.-H., Sung, S., 2012. Environmentally coordinated epigenetic silencing of FLC by protein and long noncoding RNA components. Curr. Opin. Plant Biol. 15 (1), 51–56.

Kim, D.-H., Xi, Y., Sung, S., 2017. Modular function of long noncoding RNA, COLDAIR, in the vernalization response. PLoS Genet. 13 (7), e1006939.

Kouchi, H., Takane, K.I., So, R.B., Ladha, J.K., Reddy, P.M., 1999. Rice ENOD40: isolation and expression analysis in rice and transgenic soybean root nodules. Plant J. 18 (2), 121–129.

Kumar, V., Westra, H.-J., Karjalainen, J., Zhernakova, D.V., Esko, T., Hrdlickova, B., Almeida, R., Zhernakova, A., Reinmaa, E., Võsa, U., Hofker, M.H., Fehrmann, R.S.N., Fu, J., Withoff, S., Metspalu, A., Franke, L., Wijmenga, C., 2013. Human disease-associated genetic variation impacts large intergenic non-coding RNA expression. PLoS Genet. 9 (1), e1003201.

Kung, J.T.Y., Colognori, D., Lee, J.T., 2013. Long noncoding RNAs: past, present, and future. Genetics 193 (3), 651–669. https://doi.org/10.1534/genetics.112.146704.

Li, L., Eichten, S.R., Shimizu, R., Petsch, K., Yeh, C.-T., Wu, W., et al., 2014. Genome-wide discovery and characterization of maize long non-coding RNAs. Genome Biol. 15 (2), R40.

Li, R., Qu, H., Wang, S., Wei, J., Zhang, L., Ma, R., et al., 2018. GDCRNATools: an R/Bioconductor package for integrative analysis of lncRNA, miRNA and mRNA data in GDC. Bioinformatics 34 (14), 2515–2517.

Liu, C., Muchhal, U.S., Raghothama, K., 1997. Differential expression of TPS11, a phosphate starvation-induced gene in tomato. Plant Mol. Biol. 33 (5), 867–874.

Liu, J., Wang, H., Chua, N.H., 2015. Long noncoding RNA transcriptome of plants. Plant Biotechnol. J. 13 (3), 319–328.

Lukiw, W., Handley, P., Wong, L., McLachlan, D.C., 1992. BC200 RNA in normal human neocortex, non-Alzheimer dementia (NAD), and senile dementia of the Alzheimer type (AD). Neurochem. Res. 17 (6), 591–597.

Lv, J., Qiu, M., Xia, W., Liu, C., Xu, Y., Wang, J., et al., 2016a. High expression of long non-coding RNA SBF2-AS1 promotes proliferation in non-small cell lung cancer. J. Exp. Clin. Cancer Res. 35 (1), 75.

Lv, Y., Liang, Z., Ge, M., Qi, W., Zhang, T., Lin, F., et al., 2016b. Genome-wide identification and functional prediction of nitrogen-responsive intergenic and intronic long non-coding RNAs in maize (Zea mays L.). BMC Genomics 17 (1), 350.

Ma, J., Yan, B., Qu, Y., Qin, F., Yang, Y., Hao, X., et al., 2008. Zm401, a short-open reading-frame mRNA or noncoding RNA, is essential for tapetum and microspore development and can regulate the floret formation in maize. J. Cell. Biochem. 105 (1), 136–146.

Mach, J., 2017. The long-noncoding RNA ELENA1 functions in plant immunity. Plant Cell 29, 916.

Meng, P.-H., Zhao, N., Ma, Y.-H., Chen, Z.-L., Deng, Y., Guo, J.-T., et al., 2012. Interaction between AtIPS1 and ATXR5/6 related to programmed cell death. Sci. Agric. Sin. 2012 (20), 2.

Nejat, N., Mantri, N., 2018. Emerging roles of long non-coding RNAs in plant response to biotic and abiotic stresses. Crit. Rev. Biotechnol. 38 (1), 93–105.

Qin, T., Zhao, H., Cui, P., Albesher, N., Xiong, L., 2017. A nucleus-localized long non-coding RNA enhances drought and salt stress tolerance. Plant Physiol. 175 (3), 1321–1336.

Rai, M.I., Alam, M., Lightfoot, D.A., Gurha, P., Afzal, A.J., 2019. Classification and experimental identification of plant long non-coding RNAs. Genomics 111 (5), 997–1005.

Reis, R.S., Deforges, J., 2017. Plant non-coding RNAs and the new paradigms. In: Plant Epigenetics. Springer, pp. 163–182.

Ruiz-Orera, J., Messeguer, X., Subirana, J.A., Alba, M.M., 2014. Long non-coding RNAs as a source of new peptides. elife 3, e03523.

Shafiq, S., Li, J., Sun, Q., 2016. Functions of plants long non-coding RNAs. Biochim. Biophys. Acta 1859 (1), 155–162.

Swiezewski, S., Liu, F., Magusin, A., Dean, C., 2009. Cold-induced silencing by long antisense transcripts of an Arabidopsis Polycomb target. Nature 462 (7274), 799–802.

Ulitsky, I., 2016. Evolution to the rescue: using comparative genomics to understand long non-coding RNAs. Nat. Rev. Genet. 17 (10), 601.

Ulitsky, I., Bartel, D.P., 2013. lincRNAs: genomics, evolution, and mechanisms. Cell 154 (1), 26–46.

van Esse, H.P., Reuber, L., van der Does, D., 2019. GM approaches to improve disease resistance in crops. New Phytol. 225, 70–86.

Wang, H.L.V., Chekanova, J.A., 2016. Small RNAs: essential regulators of gene expression and defenses against environmental stresses in plants. Wiley Interdiscip. Rev. RNA 7 (3), 356–381.

Wang, Y., Fan, X., Lin, F., He, G., Terzaghi, W., Zhu, D., Deng, X.W., 2014. Arabidopsis noncoding RNA mediates control of photomorphogenesis by red light. Proc. Natl. Acad. Sci. 111 (28), 10359–10364.

Wang, H., Niu, Q.W., Wu, H.W., Liu, J., Ye, J., Yu, N., et al., 2015. Analysis of non coding transcriptome in rice and maize uncovers roles of conserved lnc RNA s associated with agriculture traits. Plant J. 84 (2), 404–416.

Wasaki, J., Yonetani, R., Shinano, T., Kai, M., Osaki, M., 2003. Expression of the OsPI1 gene, cloned from rice roots using cDNA microarray, rapidly responds to phosphorus status. New Phytol. 158 (2), 239–248.

Wierzbicki, A.T., Haag, J.R., Pikaard, C.S., 2008. Noncoding transcription by RNA polymerase Pol IVb/Pol V mediates transcriptional silencing of overlapping and adjacent genes. Cell 135 (4), 635–648.

Wu, H., Yang, L., Chen, L.-L., 2017. The diversity of long noncoding RNAs and their generation. Trends Genet. 33 (8), 540–552.

Wunderlich, M., Groß-Hardt, R., Schöffl, F., 2014. Heat shock factor HSFB2a involved in gametophyte development of Arabidopsis thaliana and its expression is controlled by a heat-inducible long non-coding antisense RNA. Plant Mol. Biol. 85 (6), 541–550.

Yang, W.C., Katinakis, P., Hendriks, P., Smolders, A., de Vries, F., Spee, J., et al., 1993. Characterization of GmENOD40, a gene showing novel patterns of cell-specific expression during soybean nodule development. Plant J. 3 (4), 573–585.

Yoon, J.-H., Abdelmohsen, K., Gorospe, M., 2014. Functional interactions among microRNAs and long noncoding RNAs. Semin. Cell Dev. Biol. 34, 9–14.

Zhang, M., Zhao, H., Xie, S., Chen, J., Xu, Y., Wang, K., et al., 2011. Extensive, clustered parental imprinting of protein-coding and noncoding RNAs in developing maize endosperm. Proc. Natl. Acad. Sci. 108 (50), 20042–20047. https://doi.org/10.1073/pnas.1112186108.

Zhang, W., Han, Z., Guo, Q., Liu, Y., Zheng, Y., Wu, F., et al., 2014a. Identification of maize long non-coding RNAs responsive to drought stress. PLoS One 9 (6), e98958.

Zhang, Y.C., Liao, J.Y., Li, Z.Y., Yu, Y., Zhang, J.P., Li, Q.F., Qu, L.H., Shu, W.S., Chen, Y.Q., 2014b. Genome-wide screening and functional analysis identify a large number of long non-coding RNAs involved in the sexual reproduction of rice. Genome Biol. 15 (12), 512.

Zheng, L.-L., Li, J.-H., Wu, J., Sun, W.-J., Liu, S., Wang, Z.-L., et al., 2015. deepBase v2.0: identification, expression, evolution and function of small RNAs, LncRNAs and circular RNAs from deep-sequencing data. Nucleic Acids Res. 44 (D1), D196–D202.

Zhu, M., Zhang, M., Xing, L., Li, W., Jiang, H., Wang, L., et al., 2017. Transcriptomic analysis of long non-coding RNAs and coding genes uncovers a complex regulatory network that is involved in maize seed development. Genes 8 (10), 274.

Zubko, E., Meyer, P., 2007. A natural antisense transcript of the Petunia hybrida Sho gene suggests a role for an antisense mechanism in cytokinin regulation. Plant J. 52 (6), 1131–1139.

Chapter 8

Role of long noncoding RNAs during stress in cereal crops

Elif Karlik
Department of Molecular Biology and Genetics, Istinye University, Istanbul, Turkey

8.1 Introduction

Transcriptome studies, especially in eukaryotes, demonstrated that >90% of the genome is transcribed and a plenty of them covers to noncoding RNAs (ncRNAs) (Chekanova et al., 2007; Kapranov et al., 2007). ncRNAs can be classified according to their lengths, localizations, and functions that long noncoding RNAs (lncRNAs) are classically longer than ≥200 nt, which can exert their functions in epigenomic regulation, including posttranslational modifications, regulation of gene transcription, expression of genes, chromosome structure modulation, splicing, and messenger RNA (mRNA) stability (Swiezewski et al., 2007; Chen et al., 2017; Kopp and Mendell, 2018). lncRNAs' localization is a good marker for their functions that they can be within the nucleus, nucleolus, cytoplasm, and even in the organelles such as mitochondria and chloroplast (Rackham et al., 2011; Cabili et al., 2015). In plants, most lncRNAs are transcribed by RNA Pol II (from intergenic, intronic, or coding regions of sense or antisense strands); however, two plant-specific RNA polymerases, Pol IV and Pol V, are also involved in transcription of some lncRNAs (Wierzbicki et al., 2008; Li et al., 2014a,b). Nowadays, majority of plant lncRNAs are found to be polyadenylated, although many lncRNAs in yeast and mammals are observed to be non-polyadenylated (Andersson et al., 2014). However, well-studied non-polyadenylated plant lncRNAs are found to be abiotic stress-related; in a recent work in *Arabidopsis*, hundreds of non-polyadenylated lncRNAs have been demonstrated to be transcribed under abiotic stress (Di et al., 2014; Yuan et al., 2016; Li et al., 2016).

Most lncRNAs are grouped due to their relationships to protein-coding genes, such as long intergenic ncRNAs (lincRNAs), intronic, sense, and natural antisense lncRNAs (NATs) (Jun et al., 2012; Mattick and Rinn, 2015). The expression levels of numerous lncRNAs differ significantly based on the tissue and/or alter during stress, indicating that lncRNAs are dynamically regulated and may act as important regulator in regulation of development and stress

Long Noncoding RNAs in Plants: Roles in development and stress
https://doi.org/10.1016/B978-0-12-821452-7.00008-8

responses (Jun et al., 2012; Liu et al., 2012). According to reported studies, the identified lncRNAs from distinctive plant species show highly tissue-specific expression patterns, indicating that these lncRNAs play important roles as a specific regulator in plant development, including tissue specification and organogenesis (Wang et al., 2014; Shuai et al., 2014; Tyagi et al., 2018). Studies demonstrated that lncRNAs interact with other biological molecules, including regulatory proteins and RNAs such as miRNAs in these regulatory processes. These regulatory lncRNAs mostly interact with these molecules in many physiological properties, including decoys, scaffolds, guides, and signals (Wang and Chang, 2011). Additionally, the regulation mechanisms of lncRNAs in the plant reproductive development are initiated by competing between lncRNAs and miRNAs for target genes which is termed the target mimicry. Competition between lncRNAs and miRNAs enable to participate in regulating the accumulation of target genes (Wu et al., 2013). In plants, transcription factors, proteins, or miRNAs are shown to be induced as decoys or molecular sponges by target mimicry mechanism, indicating that this mechanism indirectly regulates transcriptionally and posttranscriptionally target genes (Franco-Zorrilla et al., 2007; Salmena et al., 2011; Yvonne et al., 2011; Wu et al., 2013).

LncRNAs also play important roles in regulation of chromatin remodelling by serving as guides for binding to chromatin-modifying enzymes. Guide lncRNA–ribonucleoprotein complexes bind to specific target regions to change chromatin structure in local regions, which is defined as a type of transcriptional regulation (Swiezewski et al., 2009). Additionally, chromatin-bound lncRNAs are the most common regulatory mechanism and they are considered as scaffolds to recruit chromatin-modifying enzymes in the target genomic locations. The cooperative assembly of chromatin-modifying complexes via chromatin-bound lncRNAs occurs in either a small RNA (sRNA)-dependent and forming ribonucleoprotein complexes or a sRNA-independent manner. So far, the recent findings suggest that recruitment of lncRNA scaffolds arises independently from sRNAs. In plants, chromatin-bound lncRNAs interact with these protein complexes by lncRNA-mediated chromatin modifications (lncR2Epi) regulation pathway. However, how chromatin-modifying complexes or lncRNAs recognize, join, or target each other remains still unclear (Jing et al., 2008; Miao-Chih et al., 2010; Wang et al., 2016).

8.2 Functions of LncRNA in stress responses

Plant growth and productivity are extensively limited by various environmental stresses, including biotic (fungi, bacteria, viruses, parasites, weeds, insects, and nematodes) and abiotic stresses (drought, salinity, heat, and cold stress) (Hirayama and Shinozaki, 2010; Newton et al., 2011). Recently, ncRNAs, especially lncRNAs, have been demonstrated to function as regulators in response to biotic and abiotic stress in plants (Zhang and Chen, 2013; Liu et al., 2015). Numerous studies have shown that lncRNAs participated in many regulatory processes such as gene silencing (Franco-Zorrilla et al., 2007; Wu et al., 2013), responses to biotic and abiotic stress (Ben Amor et al., 2009; Zhu et al., 2014; Cui et al., 2017), and other important developmental pathways by regulating gene transcription, mRNA stability, translation, and epigenetic modifications (Gong and Maquat, 2011; Gosai et al., 2015; Bazin et al., 2017). However, the regulatory functions in stress response of lncRNAs are still unclear in plants and under investigation. In this chapter, recent findings of lncRNAs associated with plant stress response mechanism are summarized for cereal crops.

8.2.1 Wheat

Wheat (*Triticum* ssp.), especially bread wheat (*Triticum aestivum*) and durum wheat (*Triticum durum* Desf.), is one of the major sources of increasing food demand and is mostly cultivated in the Mediterranean area (Southern Italy), Argentina, Australia, Canada, India, and the United States (FAO, 2014; Larsen and Jensen, 2014). Despite its cultivation area extensity, the production of wheat is not adequate to provide world food demand because its yield rate is restricted by environmental factors, including biotic and abiotic stress (Budak et al., 2015). Xin et al. (2011) have revealed that 125 putative lncRNAs, among which 48 lncRNAs were only responsive to powdery mildew infection and 54 were only responsive to heat stress, were identified as stress response associated with microRNAs, siRNAs, signal recognition particle (SRP), 7S RNA, and U3 snoRNAs. However, 23 lncRNAs were reported to be expressed in response to both powdery mildew infection and heat stress. Besides, wheat lncRNAs displayed tissue-specific expression patterns related to powdery mildew infection and/or heat stress; their expressions were found to be different in leaf, internode, flag leaf, root, seed, awn, young spike, and glume and were responded to powdery mildew infection and/or heat stress. Their study indicated that wheat lncRNAs may play important roles in regulation of stress responses to biotic and/or abiotic stresses (Xin et al., 2011). In another study, alterations in lncRNA expression patterns suggested that lncRNAs may function in modulation and gene silencing in response to biotic stress such as *Puccinia striiformis* f. sp. *Tritici,* causing stripe or yellow wheat rust (Yang et al., 2003; Zhang et al., 2013). In plants, sophisticated defense mechanisms have been evolved to delay or arrest pathogen growth (Yang et al., 2003). To overcome pathogen attacks, studies mostly focus on protein-coding genes or regulation of gene expression. However, only a few reports have centered upon lncRNAs to understand the plant defense mechanism. Zhang et al. (2013) have defined four lncRNAs as relative modulator for pathogen attacks. These lncRNAs are referred as TalncRNA18, TalncRNA73, TalncRNA106, and TalncRNA108 having the same stem-structure. Alterations of these lncRNAs' gene expression patterns have been suggested to be important in response to pathogen attack to trigger defense response genes and may relate to transcriptional and posttranscriptional gene regulatory mechanisms to perform a diverse range of biological roles (Faghihi and Wahlestedt, 2009; Zhang et al., 2013).

To provide stress tolerance and maintenance of vitality, many lncRNAs have been identified as important regulators under distinct stress conditions; however, few reports have been focused on wheat lncRNAs in response to drought stress (Lu et al., 2016; Shumayla et al., 2017). Drought-responsive mRNAs and lncRNAs have been analyzed in three different tetraploid wheat varieties with two wild emmer wheat genotypes, *T. turgidum* ssp. *dicoccoides* TR39477 and TTD-22, and one modern durum wheat, *T. turgidum* ssp. *durum var.* Kiziltan (Cagirici et al., 2017). More than 200 putative stress-responsive lncRNAs per cultivar have been identified and they tend to sample specific expressions, suggesting they may play roles in distinct molecular functions in response to drought stress. Totally, 28 lncRNAs were established to build 47 miRNA-lncRNA target pairs, indicating that gene regulation of miRNAs on stress-responsive lncRNAs is well-correlated with the stress-tolerance mechanism. Differentially expressed lncRNAs have been suggested to be associated with ABA-responsive transcription factors of ROS scavengers, osmolytes, and secondary metabolite including 'Glutathione' and 'Thiamine' to survive under drought stress (Krasensky and Jonak, 2012; Liang et al., 2013; Boubakri et al., 2013; Cagirici et al., 2017). Cold or chilling stress at low temperatures causes a large spectrum of reprograming of gene expression and metabolism at both vegetative and reproductive stages in plants (Thomashow, 1999; Basualdo, 2013). Díaz et al. (2019) have reported 31 lncRNAs from durum wheat CBW0101 plants exposed to low temperatures, indicating that lncRNAs may play important roles in response to cold or chilling stress. Traes_2BS_7A04BF5D5 lncRNA, which targets the cold acclimation protein-coding gene WCOR413, was found to be upregulated. WCOR413 have been demonstrated to be participated in membrane cryoprotection and dehydration protection (Danyluk et al., 1994). Interestingly, positive correlation was observed between WCOR413 and its lncRNA Traes_2BS_7A04BF5D5, suggesting the existence of a miRNA target mimic mechanism. Similarly, Traes_2DL_ABD08139B was demonstrated to be downregulated in response to low temperature stress. Its target gene was predicted as the flowering-promoting factor 1-like protein 3 gene, causing delay of the progression towards flowering and was also downregulated, indicating the similar target mimicry event as lncRNA Traes_2BS_7A04BF5D5 (Díaz et al., 2019). To understand the sophisticated gene regulation mechanism in response to stress, more lncRNAs may be defined in the future by development of new techniques.

8.2.2 Maize

Maize (*Zea mays* L.) is a major cereal with the highest production worldwide, serving as human consumption, livestock feed, and fuel (Gwirtz and Garcia-Casal, 2014). According to international grains council, maize is harvested to just over 1099.91 million tons in 2018/19. However, maize yield and grain quality are affected seriously by biotic and abiotic stress, limiting maize production.

Plants are evolved by genetic and epigenetic regulatory systems in response to biotic and abiotic stresses. Especially, among the regulatory ncRNAs, lncRNAs are suggested as an important component of genetic and epigenetic regulatory mechanisms, including chromatin modification, genomic imprinting, transcriptional control, as well as pre- and posttranslational mRNA processing involved in stress response and adaptation (Wang et al., 2015). Zhang et al. (2014a,b) have identified 8449 drought-responsive transcripts in maize, and among them, 664 transcripts were defined as drought-responsive lncRNAs, 567 were found to be upregulated, and 97 were downregulated. Based on the homology research, 8 lncRNAs were identified as miRNA precursors; GRMZM2G420571_T01 was an intragenic lncRNA and the TCONS_00054544 was an overlapping lncRNAs. Furthermore, the remaining six were reported as intergenic lncRNAs. However, their expression profile suggested that only two lncRNAs, which were precursors of miR167j and miR172c, may be related to drought stress response. Additionally, 279 drought-responsive lncRNAs were categorized as siRNA precursors and may be involved in regulation of gene expression in different pathways due to their expression patterns within maize leaves and roots (Zhang et al., 2014a,b). Recently, in another study with drought stress response, a total of 1535 out of 3488 differentially expressed transcripts were classified as drought-responsive lncRNAs and exhibited higher tissue- and development-specificity (Pang et al., 2019). They used 94 RNA-seq datasets from public resources and data obtained from four developmental stages (vegetative growth stages V12, V14, V18, and reproductive stage R1) from three main maize tissues—ear, tassel, and leaf—were exposed to drought stress and control conditions (Thatcher et al., 2016). Under drought stress, the amount of differentially expressed lncRNAs in the late developmental stages (V18 and R1) were found to be higher than the early developmental stages (V12 and V14) of maize. Specifically, the most significant expression level alterations were demonstrated at the R1 stage when silks surface from the ear to receive pollen and begin the fertilization. Also, two main tissues—leaves and ear—were observed as the most sensitive tissues to drought stress, indicating that expression levels of drought-responsive lncRNAs in maize could be depended on specific tissue and developmental stages. MSTRG.2834.1 lncRNAs were demonstrated alteration in gene expression pattern at R1 stage, suggesting that these drought-responsive lncRNAs could be involved in the developmental process of ear. Positive correlation between 122 out of the 124 lncRNA-mRNA pairs indicated that they may be regulated in the same direction. According to GO molecular function analysis, *cis*-acting lncRNAs participated in biological processes, including electron transport chain and oxidative phosphorylation; however, trans-acting lncRNAs were extensively involved in DNA topological change, protein ubiquitination, and response to stimulus and homoiothermy. As the targets of lncRNAs, expression of *vpp4* (GRMZM2G028432), which encodes a vacuolar (H^+)-pumping ATPase subunit, was correlated with its adjacent MSTRG.6838.1 lncRNA during drought stress (Seidel et al., 2013; Pang et al., 2019).

Forestan et al. (2016) have applied specific pipeline to identify the leaf-transcriptome of maize B73 wild-type and required to maintain repression 6 mutant (rmr6/rpd1–1, PolIV mutant involved in siRNA biogenesis and in the RNA-directed DNA methylation pathway, introgressed into the B73 background) (Erhard et al., 2009), treated for ten days, to control growth condition, drought stress, salt stress, and combined stress. According to their results, they classified 13,387 lncRNAs with 3012 NAT lncRNAs and 10,375 lincRNAs. To uncover the regulation of stress—and epigenetic—factors of the maize genome, they analyzed the differential expression of these lncRNAs and concluded that the expression of protein coding genes could be regulated by lincRNAs at transcriptional and posttranscriptional and epigenetic level via molecular guide or scaffold for chromatin-modifying enzymes or acting as decoys to DNA-binding proteins. Selected lncRNAs were also investigated for their differential expression levels under stress conditions and GRMZM6G851663_T01 lncRNA, which is a 421 nt long and annotated as intergenic lncRNA, was found to be upregulated in both B73 and rmr6. Based on homology analysis, GRMZM2G320373_T01 is partially matched with a gene encoding for a bifunctional inhibitor/lipid-transfer protein/seed storage 2S albumin protein, a family of proteins (LPT) associated with plant defense against pathogens and environmental stress (Des Marais et al., 2012). Nitrogen (N) is an essential restricting nutrient of the plant growth and development. However, lncRNA's functions in N-deficiency are still unknown in plants. Intergenic and intronic lncRNAs were investigated in maize B73 leaves at the seven-leaf (V7) stage under conditions of N-deficiency and N-sufficiency by Lv et al. (2016). Among the total 7882 lncRNAs, 637 lncRNAs were classified as nitrogen-responsive lncRNA and 620 intergenic and 17 intronic RNAs. According to differential expression analysis, 417 intergenic and 9 intronic lncRNAs were downregulated, although 203 lincRNAs and 8 intronic lncRNAs were upregulated under high nitrogen conditions. These lncRNAs were found to be related to biological process, including oxidative phosphorylation, generation of precursor metabolites, and energy and oxidation reduction such as NADH dehydrogenase activity and oxidoreductase activity. Their results also suggested that lncRNAs could play important roles in C-N-P rebalancing in response to nitrogen. Sequence homology research indicated that sequence similarity of intergenic/intronic lncRNAs was higher than randomly selected intergenic regions (Lv et al., 2016).

In 2018, Huanca-Mamani and colleagues performed combined stress (Salt and Boron) to identify lncRNAs in Lluteño maize, a tolerant maize landrace from Chile. They identified a total of 48,345 different lncRNAs, of which 41.9% were exclusive for Lluteño, indicating a potential spatial–temporal divergence. 1710 lncRNAs were observed to be expressed in response to combined stress, of which 848 stress-responsive lncRNAs were defined as trans-NAT lncRNAs. According to the GO analysis, these trans-NAT lncRNAs play important roles for regulating genes related to regulation of transcription, response to stress, and involving the nicotinamide metabolic process (i.e., 'Nicotinamide synthase activity'). In response to salt and boron stress, relative expressions of lncRNAs on roots and leave tissues were investigated and revealed that some lncRNAs were differentially expressed on roots and leaves simultaneously. 53 lncRNAs were found to be expressed only in roots, while 86 lncRNAs were in leaves, suggesting that there may be a potential mechanism of defense to exceed the prolonged stressed conditions. Determination of a set of lncRNAs in Lluteño maize suggests that, under the high concentration of salinity and boron, lncRNAs may play pivotal roles at a posttranscriptional regulation level, in the adaptation and stress response (Huanca-Mamani et al., 2018).

In recent years, diverse lncRNAs in plants have been identified and classified in response to biotic and abiotic stresses, including nitrogen stresses, phosphate deficiency, osmotic and salt stress, cold stress, and drought stress, against fungi, bacteria, and virus pathogens (Wang et al., 2015; Gao et al., 2016; Lv et al., 2016; Wang et al., 2017; Liu et al., 2018; Li et al., 2018; Pang et al., 2019; Jiang et al., 2019). Additionally, some agriculturally beneficial microorganisms such as arbuscular mycorrhiza (AM) fungi and rhizobia could also facilitate directly or indirectly the crop improvement (van Velzen et al., 2018; Ferguson et al., 2019). To maintain mycorrhizal symbiont, studies indicate that ncRNAs may contribute to the regulation mechanisms of the expression of mycorrhiza-inducible genes. However, bidirectional nutrient exchange must be regulated by both partners (plant and microorganism) and this specific mechanism is still unknown. Han et al. (2019) identified 5941 lncRNAs, and among them, 63 lncRNAs were classified as differentially expressed, mainly associated with phosphate ion transmembrane transport, cellular response to potassium ion starvation, and lipid catabolic processes. Additionally, these lncRNAs may be participated in the regulation of bidirectional nutrient exchange between plant and AM fungus *Rhizophagus irregularis*. Among the 63 differentially expressed lncRNAs, 36 lncRNAs were upregulated, while 27 lncRNAs were found to be downregulated. Additionally, 778 protein-coding genes were also predicted as target of these lncRNAs. They observed that several lncRNAs were upregulated such as TCONS_00083135, TCONS_00161821, TCONS_00153885, TCONS_00010587, TCONS_00165850, TCONS_00113799, and TCONS_00165851, which were considered to use the target mimicry mechanism, and that zma-miR395b-5p, zma-miR395h-5p, and zma-miR397b-5p miRNAs were regulated. Especially, TCONS_00113799, TCONS_00165850, and TCONS_00153885 lncRNAs were demonstrated to be overexpressed in AM fungal-colonized roots. Besides, TCONS_00113799, TCONS_00125081, TCONS_00060090, TCONS_00106194, and TCONS_00071551 were also considered as sponge of miR399 family which could be important in the regulation or maintenance of symbiosis of plants with AM fungi (Han et al., 2019). These results indicate that lncRNAs may be involved in the regulation and/or maintenance of bidirectional nutrient exchange between plant and AM fungi and could contribute to understand the regulatory roles in symbiosis mechanism.

~85% of the maize genomic contents are constituted from transposable elements (TEs) and are usually inactivated by epigenetic modification, including DNA methylations (Schnable et al., 2009; Lisch, 2012). Additionally, TEs have been demonstrated to function in important roles in plant evolution and environmental adaptation (Yang et al., 2013). Lv et al. (2019) detected 23,309 lncRNAs in maize; thus 65% (15,312) of the 23,309 lncRNAs, which were called "TE-lncRNAs", have been identified to overlap with known maize TEs, mostly similar to long-terminal-repeat (LTR) retrotransposons from the *Copia* and *Gypsy* superfamilies. However, only a small portion of TE-lncRNAs were found to be derived from DNA transposons. However, their data indicated that *LTR/Copia* elements do not generate lncRNAs as readily as *LTR/Gypsy* elements. The reason for these differences is still unclear, but according to Lv et al. (2019), it may be associated with TE age and location. Additionally, 1077 lncRNAs among the 13,822 polyA+ lncRNAs were detected as stress-related, including 529 TE-lncRNAs. Besides, TE-lncRNAs were observed to be significantly correlated with the adjacent genes, indicating TE-lncRNAs might act as *cis*-regulatory factor, such as open chromatin structure. However, stress-related lncRNAs may function as *trans*-acting regulatory factor in maize. According to co-expression analysis of mRNAs and lncRNAs, totally 40 co-expression modules were identified, and of the 40 modules, 16 were found to significantly relate to stress conditions. Interestingly, these 16 modules showed obvious patterns through developmental stages, indicating that temporal hierarchies may play important roles in response to stress condition (Lv et al., 2019).

8.2.3 Rice

To prevent biotic and abiotic stress effects, plants have evolved sophisticated response mechanisms. Heavy metal pollution as abiotic stress has been causing serious problems for modern agriculture such as crop growth stress, limiting the yield and quality of crops. Besides, heavy metals, including Cadmium (Cd), affect human health after entering the human body through the food chain (Lei et al., 2015; Zhang et al., 2015). Nine out of 93 differentially expressed lncRNAs were predicted as a Cd-stress-related rice lncRNAs (He et al., 2015). However, biological functions of these lncRNAs in Cd stress are still unclear in rice. To reveal lncRNAs' roles in Cd-stress, lncRNAs were systematically analyzed in rice root development at an early stage by Chen et al. (2018). They identified 144 differentially expressed lnRNAs (69 lncRNAs up- and 75 lncRNAs down-regulated) between control and Cd-stress. GO analysis suggested that four main pathways, including secondary metabolites biosynthesis, phenylpropanoid biosynthesis, phenylalanine metabolism, and photosynthesis signaling pathways, could play crucial role in Cd-stress. Besides, significant changes have been detected in the cysteine and methionine metabolism pathway in the Cd stress which is involved in chelating Cd to transform nontoxic complexes (Tang et al., 2014). Interestingly, XLOC_086307 lncRNA-targeted *OS03G0196600* gene related to the cysteine-rich peptide metabolism in *cis* was found to be significantly upregulated, indicating XLOC_086307 likely is involved in Cd response processes in rice. As a secondary metabolism, carotenoid biosynthesis-related OS03G0184000 gene was targeted by XLOC_086119 and XLOC_066284 lncRNAs in *cis,* suggesting that carotene content may increase by regulation of carotenoid biosynthesis. With the increasing of apocarotenoids, ABA signaling pathway was observed to be activated to cope with Cd-induced oxidative stress (Li et al., 2014a,b). Therefore, XLOC_086119 and XLOC_066284 lncRNAs might also play important roles in ABA signaling through carotenoid biosynthesis in Cd-stress. Also, this study showed that the phenylpropanoids and the phenylalanine-related OS11G0552000 gene were regulated by XLOC_058523, XLOC_104363, and XLOC_059778 lncRNAs in trans to contribute removal of excessive Cd from the cell. Furthermore, lncRNAs, including XLOC_122123, XLOC_125848, and XLOC_098316, targeted OS07G0148900 gene involved in photosynthesis, indicating that the photosynthesis-related genes are associated with Cd-stress response in rice. As a common environmental pollutant, arsenic (AS) is toxic and belongs to class I carcinogen (Neubauer, 1947). Contamination with AS in soil and groundwater causes significant problems for rice seedlings, especially in many South-East Asian countries. AS uptake by rice results in high accumulation in grains, causing food safety problems (Zhu et al., 2008). 56 lncRNAs were detected to be targeted by nine Argonaute 1 (AGO1)-enriched, As-responsive sRNAs, suggesting that these lncRNAs are involved in AS stress signaling pathway (Tang et al., 2019). It was demonstrated that some lncRNAs were regulated by two distinct sRNAs. For example, both root-CK-high-13,603 and root-CK-high-2928, which are repressed in roots under AS stress, targeted to lincRNA257. Another lncRNA CNT0027168 was targeted by shoot-As-high-17,275 induced in shoots under AS stress and root-CK-high-1309 (Table 8.1) (Tang et al., 2019).

The future environmental changes worldwide are growing matter of concern, especially increased drought stress because of changing patterns of aridity that affects the human land-use activities (Lobell and Gourdji, 2012). Chung et al. (2016) have identified that 98 lncRNAs (31 upregulated and 67 downregulated) were responsive to drought stress in rice. As expected, their expression level changes were found to be correlated with those of their putative target genes. Especially, two lncRNAs, Os02g0250700-01 and Os02g0180800-01, were showed to be bidirectional, targeting neighboring genes on their opposite strand. Os02g0250700-01 lncRNAs were estimated to target Os02g0250600-01, which is encoded in late embryogenesis abundant protein (LEA). They share the same bidirectional promoter, and their expressions were shown to be negatively correlated. Additionally, the expression levels of Os02g0180800–01 and its target gene, Os02g0180700-01 (cinnamoyl-CoA reductase), altered and are inversely correlated under drought stress. Singh et al. (2017) have developed a new tool to predict plant lncRNAs that PLncPRO works according to machine learning, using random forest algorithm to identify coding and long noncoding

TABLE 8.1 lncRNAs and their putative roles in several plant stress.

Stress	Transcript	Function	Reference
Wheat			
Cold	Traes_2BS_7A04BF5D5.3	Targeting the cold acclimation protein gene *WCOR413*	Díaz et al. (2019)
	Traes_2DL_ABD08139B	Targeting the flowering-promoting factor 1-like protein 3	
Pathogen attack	TalncRNA73	Trigger defense response genes may relate to transcriptional and posttranscriptional gene regulatory mechanisms	Zhang et al. (2013)
	TalncRNA106		
	TalncRNA18		
	TalncRNA108		
Heat stress	TapmlnRNA19	Putative miRNA precursor of miR2004	Xin et al. (2011)
	TapmlnRNA30	Unclear	
	TahlnRNA27	Putative miRNA precursor of miR2010	
	TalnRNA5	Putative miRNA precursor of miR2004	
Maize			
Drought stress	GRMZM2G420571_T01	Putative miRNA precursor	Zhang et al. (2014a,b)
	TCONS_00054544	Putative miRNA precursor	
	TCONS_00056395	Putative siRNA precursor	
	TCONS_00082174	Putative siRNA precursor	
	TCONS_00037470	Putative siRNA precursor	
	TCONS_00012768	Putative siRNA precursor	
	MSTRG.2834.1	Involved in the developmental process of ear	Pang et al. (2019)
	MSTRG.6838.1	Correlated with *vpp4* (GRMZM2G028432) protein-coding gene a vacuolar (H^+)-pumping ATPase subunit	

TABLE 8.1 lncRNAs and their putative roles in several plant stress—cont'd

Stress	Transcript	Function	Reference
Pathogen attack and environmental stress	GRMZM6G851663_T01	Unclear	Forestan et al. (2016)
	GRMZM2G320373_T01	Unclear	
Pathogen attack	TCONS_00083135	Regulating expression as sponge of zma-miR395b-5p, zma-miR395h-5p, and zma-miR397b-5p	Han et al. (2019)
	TCONS_00161821		
	TCONS_00153885		
	TCONS_00010587		
	TCONS_00165850		
	TCONS_00165851		
	TCONS_00113799	Regulating expression as sponge of miR399	
	TCONS_00125081		
	TCONS_00071551		
	TCONS_00060090		
	TCONS_00106194		
Rice			
Cd stress	XLOC_086307	Targeting *OS03G0196600* protein-coding gene related to cysteine-rich peptide metabolism	Chen et al. (2018)
	XLOC_086119	Regulation of carotenoid biosynthesis	
	XLOC_066284		
	XLOC_058523	Regulation of phenylpropanoids and the phenylalanine-related *OS11G0552000* gene	
	XLOC_104363		
	XLOC_059778		
	XLOC_122123	Targeting *OS07G0148900* protein-coding gene involved in photosynthesis	
	XLOC_125848		
	XLOC_098316		

Continued

TABLE 8.1 lncRNAs and their putative roles in several plant stress—cont'd

Stress	Transcript	Function	Reference
AS stress	CNT0027168	Targeted by shoot-As-high-17,275 and root-CK-high-1309 which are AGO1-enriched, As-responsive sRNAs	Tang et al. (2019)
	lincRNA257	Targeted by root-CK-high-13,603 and root-CK-high-2928 which are AGO1-enriched, As-responsive sRNAs	
Drought stress	Os02g0250700-01	Targeting *Os02g0250600–01* which is encoded LEA protein	Chung et al. (2016)
	Os02g0180800-01	Correlated with Os02g0180700–01 (cinnamoyl-CoA reductase),	
	TCONS_00028567 lncRNA might be the precursor of osa-MIR1428e	Regulating ABA signaling pathway via putative miRNA precursor of osa-MIR1428e	
Cold stress	Chr03G0008	Unclear	Shin et al. (2018)
N and Pi starvation	Chr04G0017	Unclear	
	Chr07G0166	Unclear	
Pathogen attack	XLOC_437338 (ALEX1)	Interacting with 7 JA-related genes, *JAZ8*, *MYC2*, *PR1a*, *PR1b*, *PR10a*, and *RSOsPR10*	Yu et al. (2019)
Barley			
Pathogen attack	XLOC_057704	Pathogen resistance by regulating the cellulose biosynthesis	Bischoff et al. (2010); Huang et al. (2016)
	XLOC_050232	Unclear	Huang et al. (2016)

TABLE 8.1 lncRNAs and their putative roles in several plant stress—cont'd

Stress	Transcript	Function	Reference
Salt stress	AK363461 (CNT20168194)	Unclear	Karlik et al. (2018)
	AK370506	Targeting *HORVU7Hr1G007800* protein-coding gene (LEA protein) as NAT-lncRNA	
	CNT20168342	Targeting *HORVU2Hr1G010890* protein-coding gene (40S ribosomal protein S6) as NAT-lncRNA	Karlik and Gozukirmizi (2018a)
	CNT0031477	Putative sno-lncRNA	
	AK370814	Related to *putative pyridoxal reductase (PLR)* gene (AK372815)	Karlik and Gozukirmizi (2018b)
B stress	TCONS_00043651	Putative miRNA precursor of miR399	Unver and Tombuloglu (2019)

transcripts with high accuracy. They identified 3714 high-confidence expressed lncRNAs in rice under drought stress by using PLncPRO tool. In another study, three different types of RNA libraries, including poly(A)+, poly(A)-, and nuclear RNAs, were sequenced in rice and these RNAs were obtained from four abiotic stresses (cold, heat, drought, and salt) conditions (Yuan et al., 2018). ~500 poly(A)+lncRNAs were found to be differentially expressed under stress and were significantly downregulated; however, 25% of non-poly(A) form were upregulated. Interestingly, downregulation of polyadenylation (DPA) events in lncRNAs was observed and this event tends to be highly conserved and demonstrates significant nuclear retention under stress conditions. Moreover, these DPA lncRNAs are significantly located in quantitative trait loci (QTLs) related to stress tolerance or development in rice, indicating that these DPA lncRNAs may function in growth under stress conditions. The DPA events have been observed differently under the four stress tested and DPA lncRNAs were found to be mostly accumulated in rice exposed to drought and salt stress. Therefore, this event was considered to be consistent with the reduction of RNA 3′-end processing factors, which, under these conditions, were differentially expressed under stress conditions. Remarkably, more than 90% of the tested factors were downregulated by salt stress. However, only approximately 20% of the tested factors were repressed in plants exposed to cold and heat (Yuan et al., 2018). Studies showed that the subcellular localization of lncRNAs is related to their biological functions; the fact remains that the subcellular localizations of lncRNAs can alter due to the interaction with their target proteins (Kino et al., 2010; Yuan et al., 2018). Yuan and colleagues observed that the DPA lncRNAs were mostly located in the nucleus in plants exposed specially to salt and drought stress. According to their results, polyadenylation and subcellular localization of many rice lncRNAs tend to be regulated posttranscriptionally, suggesting that traditional RNA-seq analyses require to be carefully evaluated for the influence of posttranscriptional modifications on lncRNAs (Yuan et al., 2018). Recently, studies showed plants exposed to stress can induce a memory response, which is considered to play significant roles in adaptation to stress (Avramova, 2015). Li et al. (2019) identified 6885 transcripts and 238 lncRNAs associated with the drought stress memory in rice after cycles of mild drought and rewatering treatment. They observed endogenous phytohormones [especially abscisic acid (ABA) and jasmonic acid (JA)], lncRNAs, and DNA methylation are involved in rice short-term drought memory. 12 lncRNAs among 238 lncRNAs were predicted as miRNA precursors and 168 were found as lncNATs. TCONS_00028567 lncRNA might be the precursor of osa-MIR1428e and was upregulated after drought stress. TCONS_00028567 lncRNA may regulate ABA signaling pathway via its critical gene *SAPK10* (Table 8.1) (Li et al., 2019).

For plants, nitrogen (N) is one of the key macronutrients influencing crop development and productivity. During last decades, the application of synthetic N fertilizers has been increased and plants can only use less than half of the applied N based on the low N use efficiency (Raun and Johnson, 1999; Shin et al., 2018). In rice, 2588 novel lncRNAs were identified as N starvation response lncRNA with combination of different sequencing data obtained by strand-specific RNA-Seq and Poly A-primed sequencing (2P-Seq) (Shin et al., 2018). 1255 and 1333 were expressed from intergenic regions and cis-NATs, respectively, and their expression levels were found to be tissue-specific. Interestingly, a few rice lncRNAs were observed to be regulated in both N and Pi starvation. Similarly, salt-, cold-, and drought-responsive lncRNAs were concordantly regulated in response to N starvation. For instance, the expression of Chr03G0008 related to cold stress response was induced by N-starved roots. *IPS1* were demonstrated to be co-regulated in response to both N and Pi starvation, indicating that *IPS1* may function as a bridge for regulation of the two macronutrients. Chr04G0017 lncRNA, which was actively translated as 116 a.a. and 13.64 kDa functioning as both RNAs and peptides, was expressed in both N and Pi starvation, especially in rice roots (Laporte et al., 2010; Secco et al., 2013; Shin et al., 2018). Also, the sequence analysis of Chr04G0017 in different rice varieties indicated that this lncRNA was well-conserved. After 7 days of both N and Pi starvation, alternatively polyadenylated Chr07G0166 found with three isoforms were expressed (Shin et al., 2018). In another study, Yu et al. (2019) defined 567 disease-responsive rice lncRNAs in response to *Xanthomonas oryzae* pv. *Oryzae* (*Xoo*). Most of the lncRNAs were found to be distributed through 12 rice chromosomes. They showed 744 lncRNA-mRNA matches, including 282 lncRNAs and 704 mRNAs, indicating lncRNAs can generate *cis*-regulatory pairs with multiple adjacent genes. The analysis of co-expression profiles revealed the association between lncRNAs and JA-related genes and 39 JA-related protein-coding genes were regulated by 73 lncRNAs, indicating the regulatory roles of lncRNAs in JA pathway. Also, among the 73 lncRNAs, 26 were demonstrated to be significantly associated with JA-related genes. One of these lncRNAs, XLOC_437338, was interacted with 7 JA-related genes, indicating that this lncRNA may play important roles in response to diseases. Additionally, the expression of XLOC_437338, which was called A Leaf Expressed, and *Xoo*-induced lncRNA 1 (ALEX1) were observed especially in *Xoo*-infected leaves. ALEX1 was not able to be detected in wild-types; however, it was dramatically upregulated and only the mutant of XLOC_437338 exhibited rice blight resistance. The length of ALEX1 was 294 nt located on the antisense direction of rice Chromosome 8. According to co-regulation analysis, *JAZ8, MYC2, PR1a, PR1b, PR10a,* and *RSOsPR10* genes, which were related to JA pathway, were significantly upregulated in ALEX1 mutants. Additionally, the overexpression of ALEX1 resulted to be resistant to *Xanthomonas oryzae* pv. *oryzicola* similar to *Xoo*, indicating ALEX1 can provide broad-spectrum resistance to rice bacterial pathogens (Yu et al., 2019). Further investigations are proceeding to discover the functions of ALEX1 in playing regulatory key roles in response to bacterial disease.

8.2.4 Barley

In plants, lncRNAs play regulatory roles in sexual reproduction, tissue development, and in response to external stimuli such as drought, salinity, heat stress, and infections (Xin et al., 2011, Kim and Sung, 2012; Zhang et al., 2014a,b). *Fusarium* head blight (FHB), which is resulted by fungal pathogens of *Fusarium* species such as *Fusarium graminearum*, caused reduction in barley (*Hordeum vulgare* L.) production (McMullen et al., 1997; Steffenson, 2003). To assess the defense response against FHB, Huang et al. (2016) performed RNA-Seq analysis in barley exposed to *F. graminearum*. They identified totally 12,366 lncRNAs, among which 604 were defined as FBH-responsive. Especially, one antisense and five lincRNAs were found to be expressed with their neighboring transcripts after *F. graminearum* infection. The XLOC_057704 and its sense transcript MLOC_70487 both exhibited co-induction by fungal infection. Studies with the MLOC_70847, encoding a trichome birefringence like (TBR-like [TBL]) protein, showed it is required for cellulose biosynthesis, indicating that the XLOC_057704 may play important role in pathogen resistance by regulating the cellulose biosynthesis (Bischoff et al., 2010; Huang et al., 2016). Among the five lincRNAs, XLOC_050232 lincRNA was expressed coordinately with its neighboring gene CRK (MLOC_60954) after fungal infection. However, how both these lincRNA and its gene affect the pathogen attack is still unclear.

Barley represents an important source of variation for biotic and abiotic stress adaptations, especially for drought and salinity stress (Wahbi and Gregory, 1989). Barley was demonstrated to be more tolerant than other crops for salinity stress. In 2017, Mascher and colleagues reported the linear order of barley sequences across the pericentromeric space and the spatial organization of chromatin. Moreover, they predicted 19,908 barley lncRNAs, which are not related to stress. However, two barley lncRNAs, AK363461 and AK370506, were observed to be associated with salinity stress in four barley species (Hasat, Beysehir 99, Konevi 98, and Tarm 92) by Karlik et al. (2018). In our group, we observed that the expression of these two lncRNAs was altered in both shoot and root tissues exposed to salt stress. Additionally, their alteration levels were found to be different between four barley varieties, indicating that expression of these lncRNAs might depend on barley genotype. Sequence analysis of AK363461, which was first known as 980 bp, was revealed to be 2512 bp at chromosome four (Karlik et al., 2018). Therefore, it was renamed as CNT20168194 according to CANTATAdb (Szcześniak et al., 2016). Based on the sequence analysis, this genome region in barley does not contain any annotated protein coding or noncoding genes to suggest the CNT20168194 function in salt stress. Also, CNT20168194 was found to be non-conserved means and that it could be unique for barley. However, AK370506 lncRNA was demonstrated to be at chromosome 7, containing *HORVU7Hr1G007800* protein-coding gene at antisense direction, indicating this lncRNA may be paired with LEA protein as a NAT-lncRNA (Karlik et al., 2018). The *LEA* gene family mainly occurs in seedlings, roots, and other organs throughout the whole developmental stage and has many functions, including protection of cellular structures from the effects of water loss and desiccation (Serrano and Montesinos, 2003), protection of proteins from stress-induced damage (Serrano and Montesinos, 2003; Grelet et al., 2005; Shao et al. 2005), sequestration of ions (Grelet et al., 2005), and folding of denatured proteins (Bray, 1993); additionally, LEA proteins can also act as chaperone proteins for limiting cellular damage (Wise, 2003; Umezawa et al., 2006). Moreover, shoot expression levels of AK370506 lncRNA were downregulated compared to control during salt application; root expression levels of this lncRNA were observed to be upregulated, suggesting function of this lncRNA may be tissue-specific and can be altered according to tissue.

To date, approximately 19,908 lncRNAs have been identified in barley (Mascher et al., 2017) and 7970 lncRNAs have been annotated by CANTATAdb (Szcześniak et al., 2016). *Re*-sequencing of barley lncRNA based on homology with CNT0018772 revealed that our sequence was found to be associated with barley CNT20168342 lncRNA according to CANTATAdb. However, CNT20168342 lncRNA was 443 bp, while our re-sequenced data was 915 bp, indicating that this lncRNA is located in chromosome 2 and is longer than that considered before. According to sequence analysis, CNT0018772 had two exons and one intron on genome sequence of barley, while sequencing data of cDNA did not contain any intron part, indicating that this lncRNA may undergo the excision of introns or alternative splicing (Unpublished data). This barley genome region also contains *HORVU2Hr1G010890* protein-coding gene at antisense orientation, indicating that CNT20168342 lncRNA may be a NAT-lncRNA. *HORVU2Hr1G010890* is predicted to be homolog with 40S ribosomal protein S6 (RPS6) in barley involved in the initiation of translation at eukaryotes (Krieg et al., 1988; Fumagalli and Thomas, 2000; Szcześniak et al., 2016). Former studies indicated that phosphorylation of RPS6 in the cytoplasm is considered to promote the selective translation of a subset of transcripts, especially ribosomal proteins and elongation factor mRNAs (Radimerski et al., 2000), indicating that regulation of its expression and phosphorylation by CNT20168342 lncRNA may be related in response to salt stress. Another re-sequencing study in our lab showed that rice lncRNA (CNT0031477) was found to be longer in barley than rice. Our re-sequencing analysis of CNT0031477 revealed that this barley lncRNA was 1451 bp and located in chromosome five which contains three snoRNAs (ENSRNA049476433, ENSRNA049476461, and ENSRNA049476383), suggesting this lncRNA may be a member of sno-lncRNA class. Sno-lncRNAs are considered to participate in imprinting, regulation of AS, and many other posttranscriptional events in human (Sahoo et al., 2008; Duker et al., 2010). However, our expression studies demonstrated that this lncRNA could be expressed both on shoot and root tissues, indicating that this lncRNA may not be tissue-specific or not directly associated with salt stress response (Karlik and Gozukirmizi, 2018a).

Vitamin B6 is an essential coenzyme for plant growth and development; however, there is only little information to exhibit B6 vitamin biosynthesis pathway in plants (Fitzpatrick et al., 2007). Our salt stress response studies demonstrated the relationship between barley putative pyridoxal reductase (PLR) gene (AK372815), which is predicted to be associated with vitamin B6 salvage pathway, and its sense lncRNA (AK370814) (Karlik and Gozukirmizi, 2018b). According to our results, AK372815 was affected by salt stress. Its expression was downregulated in shoots, while its expression in roots was found to be upregulated. Therefore, its lncRNA AK372814 were upregulated in shoots, although expression level changes in roots were found to be divergent in different barley varieties, suggesting that expression level of sense AK372814 lncRNA may depend on genotype (Karlik and Gozukirmizi, 2018b). Former studies indicate that ion channels are important in stress response and that pyridoxal 5′-phosphate, which is one of the vitamers and a coenzyme involved in a wide diverse of biochemical reactions, is an antagonist of the ATP-gated ion channel in animal cells (Ralevic and Burnstock, 1998).

Boron (B) is an essential microelement affecting the plant growth and productivity (Unver and Akkaya, 2008). Also, the range of B concentrations in soil can determine the plant development and that plants generally suffer from B-deficiency or B-toxicity (Camacho-Cristóbal et al., 2015). In 2019, Unver and Tombuloglu reported ~8000 barley lncRNA associated with boron stress. They observed that most lncRNAs were distributed through all barley chromosomes.

However, chromosome 2 contains the highest number of lncRNAs. The GO enrichment analysis demonstrated that these lncRNAs were related to ion transport and establishment of localization, thus six lncRNAs were found to be strongly associated with ion transport-related genes and 12 lncRNAs were observed to have similar expression patterns and co-localized with the "establishment of localization"-related genes. Moreover, some lncRNAs were found to be expressed in tissue-specific manner and be differentially regulated in both leaf and root tissues. Endogenous target mimicry (eTM) mechanism is the regulation of miRNA activity by lncRNAs (Karakulah et al., 2016). In the article, 32 barley lncRNAs have been predicted to have eTM activity to suppress 18 miRNAs such as conserved barley miRNAs such as miR159a, miR166a, and miR399, indicating that eTM lncRNAs may regulate various cellular and molecular processes. For instance, TCONS_00043651, a potential eTM for miR399, was found to be upregulated in three times more than roots compared to control in response to B stress, while miR399 expression level was downregulated in three times, suggesting that boron stress response may be cooperatively regulated by the interaction of miRNA-eTM (lncRNA)-coding target transcript modules. Additionally, expression analysis-sensitive and -tolerant varieties indicated that some lncRNAs were specific for the tolerant variety, contributing to boron stress response (Unver and Tombuloglu, 2019). This has obvious implications that further studies will improve the barley breeding.

8.3 Conclusions and future perspectives

The acceptance of the RNA world hypothesis altered the understanding of the evolution of life on earth. Especially, the recent revolutions in noncoding RNA identification techniques have led to extensive transcriptomic analysis which has aided in the identification of lncRNAs. However, researches in plants are still in limited range. Further key areas of plant lncRNAs include their roles in growth, development, and especially, in stress response. Discovering biological functions of lncRNAs will bring new opportunities and breakthroughs in lncRNA researches in plants, and the unexplored fields of plant stress response will gradually come to light.

References

Andersson, R., Gebhard, C., Miguel-Escalada, I., Hoof, I., Bornholdt, J., Boyd, M., et al., 2014. An atlas of active enhancers across human cell types and tissues. Nature 507 (7493), 455–461. https://doi.org/10.1038/nature12787.

Avramova, Z., 2015. Transcriptional 'memory' of a stress: Transient chromatin and memory (epigenetic) marks at stress-response genes. Plant J. 83, 149–159. https://doi.org/10.1111/tpj.12832.

Basualdo, J., 2013. Estudio de la variabilidad de la tolerancia a bajas temperaturas en trigo candeal (*Triticum turgidum* L. var. *durum*) y genes asociados. Tesis para optar al grado de doctora en Ciencias Biológicas. Universidad Nacional del Sur. http://repositoriodigital.uns.edu.ar/handle/123456789/607.

Bazin, J., Baerenfaller, K., Gosai, S.J., Gregory, B.D., Crespi, M., Bailey-Serres, J., 2017. Global analysis of ribosome-associated noncoding RNAs unveils new modes of translational regulation. Proc. Natl. Acad. Sci. U. S. A. 114, E10018–E10027. https://doi.org/10.1073/pnas.1708433114.

Ben Amor, B., Wirth, S., Merchan, F., Laporte, P., D'aubenton-Carafa, Y., Hirsch, J., et al., 2009. Novel long non-protein coding RNAs involved in *Arabidopsis* differentiation and stress responses. Genome Res. 19, 57–69. https://doi.org/10.1101/gr.080275.108.

Bischoff, V., Nita, S., Neumetzler, L., Schindelasch, D., Urbain, A., Eshed, R., et al., 2010. Trichome birefringence and its homolog AT5G01360 encode plant-specific DUF231 proteins required for cellulose biosynthesis in *Arabidopsis*. Plant Physiol. 153, 590–602.

Boubakri, H., Poutaraud, A., Wahab, M.A., Clayeux, C., Baltenweck-Guyot, R., Steyer, D., et al., 2013. Thiamine modulates metabolism of the phenylpropanoid pathway leading to enhanced resistance to *Plasmopara viticola* in grapevine. BMC Plant Biol. 13, 31.

Bray, E.A., 1993. Molecular responses to water deficit. Plant Physiol. 103, 1035–1040.

Budak, H., Hussain, B., Khan, Z., Ozturk, N.Z., Ullah, N., 2015. From genetics to functional genomics: improvement in drought signalling and tolerance in wheat. Front. Plant Sci. 6, 1012.

Cabili, M.N., Dunagin, M.C., McClanahan, P.D., Biaesch, A., Padovan-Merhar, O., Regev, A., et al., 2015. Localization and abundance analysis of human lncRNAs at single-cell and single-molecule resolution. Genome Biol. 16, 20.

Cagirici, H.B., Alptekin, B., Budak, H., 2017. RNA sequencing and co-expressed long non-coding RNA in modern and wild wheats. Sci. Rep. 7 (1), 10670. https://doi.org/10.1038/s41598-017-11170-8.

Camacho-Cristóbal, J.J., Martín-Rejano, E.M., Herrera-Rodríguez, M.B., Navarro-Gochicoa, M.T., Rexach, J., González-Fontes, A., 2015. Boron deficiency inhibits root cell elongation via an ethylene/auxin/ROS-dependent pathway in *Arabidopsis* seedlings. J. Exp. Bot. 66, 3831–3840. https://doi.org/10.1093/jxb/erv186.

Chekanova, J.A., Gregory, B.D., Reverdatto, S.V., Chen, H., Kumar, R., Hooker, T., et al., 2007. Genome-wide high-resolution mapping of exosome substrates reveals hidden features in the *Arabidopsis* transcriptome. Cell 131 (7), 1340–1353. https://doi.org/10.1016/j.cell.2007.10.056.

Chen, Y.G., Satpathy, A.T., Chang, H.Y., 2017. Gene regulation in the immune system by long noncoding RNAs. Nat. Immunol. 18, 962–972.

Chen, L., Shi, S., Jiang, N., Khanzada, H., Wassan, G.M., Zhu, C., et al., 2018. Genome-wide analysis of long non-coding RNAs affecting roots development at an early stage in the rice response to cadmium stress. BMC Genomics 19, 460. https://doi.org/10.1186/s12864-018-4807-6.

Chung, P.J., Jung, H., Jeong, D.H., Ha, S.H., Choi, Y.D., Kim, J.K., 2016. Transcriptome profiling of drought responsive noncoding RNAs and their target genes in rice. BMC Genomics 17 (1), 563. https://doi.org/10.1186/s12864-016-2997-3.

Cui, J., Luan, Y.S., Jiang, N., Bao, H., Meng, J., 2017. Comparative transcriptome analysis between resistant and susceptible tomato allows the identification of lncRNA16397 conferring resistance to *Phytophthora infestans* by co-expressing glutaredoxin. Plant J. 89, 577–589. https://doi.org/10.1111/tpj.13408.

Danyluk, J., Houde, M., Rassart, E., Sarhan, F., 1994. Differential expression of a gene encoding an acidic dehydrin in chilling sensitive and freezing tolerant gramineae species. FEBS Lett. 344, 20–24. https://doi.org/10.1016/0014-5793(94)00353-X.

Des Marais, D.L., McKay, J.K., Richards, J.H., Sen, S., Wayne, T., Juenger, T.E., 2012. Physiological genomics of response to soil drying in diverse *Arabidopsis* accessions. Plant Cell 24, 893–914.

Di, C., Yuan, J., Wu, Y., Li, J., Lin, H., Hu, L., et al., 2014. Characterization of stress-responsive lncRNAs in *Arabidopsis thaliana* by integrating expression, epigenetic and structural features. Plant J. 80 (5), 848–861. https://doi.org/10.1111/tpj.12679.

Díaz, M.L., Soresi, D.S., Basualdo, J., Cuppari, S.J., Carrera, A., 2019. Transcriptomic response of durum wheat to cold stress at reproductive stage. Mol. Biol. Rep. 46 (2), 2427–2445. https://doi.org/10.1007/s11033-019-04704-y.

Duker, A.L., Ballif, B.C., Bawle, E.V., Person, R.E., Mahadevan, S., Alliman, S., et al., 2010. Paternally inherited microdeletion at 15q11.2 confirms a significant role for the SNORD116 C/D box snoRNA cluster in Prader–Willi syndrome. Eur. J. Hum. Genet. 18, 1196–1201. https://doi.org/10.1038/ejhg.2010.102.

Erhard, K.F., Stonaker, J.L., Parkinson, S.E., Lim, J.P., Hale, C.J., Hollick, J.B., 2009. RNA polymerase IV functions in Paramutation in *Zea mays*. Science 323, 1201–1205.

Faghihi, M.A., Wahlestedt, C., 2009. Regulatory roles of natural antisense transcripts. Nat. Rev. Mol. Cell Biol. 10 (9), 637–643.

FAO, 2014. FAO Statistics. The Food and Agriculture Organization of the United Nations.

Ferguson, B.J., Mens, C., Hastwell, A.H., Zhang, M., Su, H., Jones, C.H., et al., 2019. Legume nodulation: the host controls the party. Plant Cell Environ. 42, 41–51.

Fitzpatrick, T.B., Amrhein, N., Kappes, B., Macheroux, P., Tews, I., Raschle, T., 2007. Two independent routes of de novo vitamin B6 biosynthesis: Not that different after all. Biochem. J. 407, 1–13.

Forestan, C., Cigliano, R.A., Farinati, S., Lunardon, A., Sanseverino, W., Varotto, S., 2016. Stress-induced and epigenetic-mediated maize transcriptome regulation study by means of transcriptome reannotation and differential expression analysis. Sci. Rep. 6, 30446. https://doi.org/10.1038/srep30446.

Franco-Zorrilla, J.M., Valli, A., Todesco, M., Mateos, I., Puga, M.I., Rubio-Somoza, I., et al., 2007. Target mimicry provides a new mechanism for regulation of microRNA activity. Nat. Genet. 39, 1033–1037.

Fumagalli, S., Thomas, G., 2000. S6 phosphorylation and signal transduction. In: Hershey, J.W.B., Mathews, M.B., Sonenberg, N. (Eds.), Translational Control of Gene Expression. Cold Spring Harbor Lab., Cold Spring Harbor, pp. 695–717.

Gao, R., Liu, P., Irwanto, N., Loh, R., Wong, S.M., 2016. Upregulation of LINC-AP2 is negatively correlated with AP2 gene expression with turnip crinkle virus infection in *Arabidopsis thaliana*. Plant Cell Rep. 35, 2257–2267.

Gong, C.G., Maquat, L.E., 2011. lncRNAs transactivate STAU1-mediated mRNA decay by duplexing with 3' UTRs via Alu elements. Nature 470, 284–288. https://doi.org/10.1038/nature09701.

Gosai, S.J., Foley, S.W., Wang, D.X., Silverman, I.M., Selamoglu, N., Nelson, A.D., et al., 2015. Global analysis of the RNA-protein interaction and RNA secondary structure landscapes of the *Arabidopsis* nucleus. Mol. Cell 57, 376–388. https://doi.org/10.1016/j.molcel.2014.12.004.

Grelet, J., Benamar, A., Teyssier, E., Avelange-Macherel, M.H., Grunwald, D., Macherel, D., 2005. Identification in pea seed mitochondria of a late-embryogenesis abundant protein able to protect enzymes from drying. Plant Physiol. 137, 157–167.

Gwirtz, J.A., Garcia-Casal, M.N., 2014. Processing maize flour and corn meal food products. Ann. N. Y. Acad. Sci. 1312, 66–75. https://doi.org/10.1111/nyas.12299.

Han, G., Cheng, C., Zheng, Y., Wang, X., Xu, Y., Wang, W., et al., 2019. Identification of long non-coding RNAs and the regulatory network responsive to arbuscular mycorrhizal Fungi colonization in maize roots. Int. J. Mol. Sci. 20 (18). https://doi.org/10.3390/ijms20184491. pii: E4491.

He, F., Liu, Q., Zheng, L., Cui, Y., Shen, Z., Zheng, L., 2015. RNA-Seq analysis of rice roots reveal the involvement of post-transcriptional regulation in response to cadmium stress. Front. Plant Sci. 6, 1136. https://doi.org/10.3389/fpls.2015.01136.

Hirayama, T., Shinozaki, K., 2010. Research on plant abiotic stress responses in the postgenome era: past, present and future. Plant J. 61, 1041–1052. https://doi.org/10.1111/j.1365-313X.2010.04124.x.

Huanca-Mamani, W., Arias-Carrasco, R., Cárdenas-Ninasivincha, S., Rojas-Herrera, M., Sepúlveda-Hermosilla, G., Caris-Maldonado, J., et al., 2018. Long non-coding RNAs responsive to salt and boron stress in the hyper-arid lluteño maize from Atacama desert. Genes 9, E170. https://doi.org/10.3390/genes9030170.

Huang, Y., Li, L., Smith, K.P., Muehlbauer, G.J., 2016. Differential transcriptomic responses to *Fusarium graminearum* infection in two barley quantitative trait loci associated with *Fusarium* head blight resistance. BMC Genomics 17, 387. https://doi.org/10.1186/s12864-016-2716-0.

Jiang, N., Cui, J., Shi, Y., Yang, G., Zhou, X., Hou, X., et al., 2019. Tomato lncRNA23468 functions as a competing endogenous RNA to modulate NBS-LRR genes by decoying miR482b in the tomato-*Phytophthora infestans* interaction. Hortic. Res. 6, 28.

Jing, Z., Bryan, K.S., Jennifer, A.E., Ji-Joon, S., Jeannie, T.L., 2008. Polycomb proteins targeted by a short repeat RNA to the mouse X chromosome. Science 322, 750–756.

Jun, L., Choonkyun, J., Jun, X., Huan, W., Shulin, D., Lucia, B., et al., 2012. Genome-wide analysis uncovers regulation of long intergenic noncoding RNAs in *Arabidopsis*. Plant Cell 24, 4333–4345.

Kapranov, P., Cheng, J., Dike, S., Nix, D.A., Duttagupta, R., Willingham, A.T., et al., 2007. RNA maps reveal new RNA classes and a possible function for pervasive transcription. Science 316 (5830), 1484–1488. https://doi.org/10.1126/science.1138341.

Karakulah, G., Yucebilgili Kurtoglu, K., Unver, T., 2016. PeTMbase: a database of plant endogenous target mimics (eTMs). PLoS One 11, e016769.

Karlik, E., Gozukirmizi, N., 2018a. Evaluation of barley LncRNA expression analysis in salinity stress. Russ. J. Genet. 54 (2), 198–204.

Karlik, E., Gozukirmizi, N., 2018b. Expression analysis of lncRNA AK370814 involved in the barley vitamin B6 salvage pathway under salinity. Mol. Biol. Rep. 45, 1597–1609.

Karlik, E., Marakli, S., Gozukirmizi, N., 2018. Two lncRNAs expression profiles in salt stressed barley (*Hordeum vulgare* L.) roots. Cytologia 83, 37–43.

Kim, E.D., Sung, S., 2012. Long noncoding RNA: unveiling hidden layer of gene regulatory networks. Trends Plant Sci. 17, 16–21.

Kino, T., Hurt, D.E., Ichijo, T., Nader, N., Chrousos, G.P., 2010. Noncoding RNA gas5 is a growth arrest- and starvation-associated repressor of the glucocorticoid receptor. Sci. Signal. 3, ra8.

Kopp, F., Mendell, J.T., 2018. Functional classification and experimental dissection of long noncoding RNAs. Cell 172, 393–407.

Krasensky, J., Jonak, C., 2012. Drought, salt, and temperature stress-induced metabolic rearrangements and regulatory networks. J. Exp. Bot. 63, 1593–1608.

Krieg, J., Olivier, A.R., Thomas, G., 1988. Analysis of 40S ribosomal protein S6 phosphorylation during the mitogenic response. Methods Enzymol. 164, 575–581.

Laporte, P., Satiat-Jeunemaitre, B., Velasco, I., Csorba, T., van de Velde, W., Campalans, A., et al., 2010. A novel RNA binding peptide regulates the establishment of the *Medicago truncatula-Sinorhizobium meliloti* nitrogen-fixing symbiosis. Plant J. 62 (1), 24–38.

Larsen, A.O., Jensen, C.A., 2014. Evaluación de cultivares de trigo candeal en Barrow—Campaña 2013–2014. EEAI INTA Barrow. http://inta.gob.ar/docum entos /evalu acion -de-culti vares - de-trigo cande al-en-barro w-campa na-2013-2014.

Lei, M., Tie, B.Q., Song, Z.G., Liao, B.H., Lepo, J.E., Huang, Y.Z., 2015. Heavy metal pollution and potential health risk assessment of white rice around mine areas in Hunan Province. China. Food Secur. 7, 45–54. https://doi.org/10.1007/s12571-014-0414-9.

Li, L., Eichten, S.R., Shimizu, R., Petsch, K., Yeh, C.T., Wu, W., et al., 2014a. Genome-wide discovery and characterization of maize long noncoding RNAs. Genome Biol. 15 (2), R40. https://doi.org/10.1186/gb-2014-15-2-r40.

Li, S.W., Leng, Y., Feng, L., Zeng, X.Y., 2014b. Involvement of abscisic acid in regulating antioxidative defense systems and IAA-oxidase activity and improving adventitious rooting in mung bean [*Vigna radiata* (L.) *Wilczek*] seedlings under cadmium stress. Environ. Sci. Pollut. Res. 21 (1), 525–537.

Li, S., Yamada, M., Han, X., Ohler, U., Benfey, P.N., 2016. High-resolution expression map of the *Arabidopsis* root reveals alternative splicing and lincRNA regulation. Dev. Cell 39 (4), 508–522. https://doi.org/10.1016/j.devcel.2016.10.012.

Li, W.Q., Jia, Y.L., Liu, F.Q., Wang, F.Q., Fan, F.J., Wang, J., et al., 2018. Genome-wide identification and characterization of long non-coding RNAs responsive to *Dickeya zeae* in rice. RSC Adv. 8, 34408–34417.

Li, P., Yang, H., Wang, L., Liu, H., Huo, H., Zhang, C., et al., 2019. Physiological and transcriptome analyses reveal short-term responses and formation of memory under drought stress in rice. Front. Genet. 10, 55. https://doi.org/10.3389/fgene.2019.00055.

Liang, X., Zhang, L., Natarajan, S.K., Becker, D.F., 2013. Proline mechanisms of stress survival. Antioxid. Redox Signal. 19, 998–1011.

Lisch, D., 2012. Regulation of transposable elements in maize. Curr. Opin. Plant Biol. 15, 511–516.

Liu, J., Jung, C., Xu, J., Wang, H., Deng, S., Bernad, L., et al., 2012. Genome-wide analysis uncovers regulation of long intergenic noncoding RNAs in *Arabidopsis*. Plant Cell 24 (11), 4333–4345. https://doi.org/10.1105/tpc.112.102855.

Liu, J., Wang, H., Chua, N.H., 2015. Long noncoding RNA transcriptome of plants. Plant Biotechnol. J. 13, 319–328. https://doi.org/10.1111/pbi.12336.

Liu, W., Cheng, C., Lin, Y., Xu Han, X., Lai, Z., 2018. Genome-wide identification and characterization of mRNAs and lncRNAs involved in cold stress in the wild banana (*Musa itinerans*). PLoS One 13, e0200002.

Lobell, D.B., Gourdji, S.M., 2012. The influence of climate change on global crop productivity. Plant Physiol. 160 (4), 1686–1697.

Lu, X., Chen, X., Mu, M., Wang, J., Wang, X., Wang, D., et al., 2016. Genome-wide analysis of long noncoding RNAs and their responses to drought stress in cotton (*Gossypium hirsutum* L.). PLoS One 11 (6). e0156723.

Lv, Y., Liang, Z., Ge, M., Qi, W., Zhang, T., Lin, F., et al., 2016. Genome-wide identification and functional prediction of nitrogen-responsive intergenic and intronic long non-coding RNAs in maize (*Zea mays* L.). BMC Genomics 17, 350.

Lv, Y., Hu, F., Zhou, Y., Wu, F., Gaut, B.S., 2019. Maize transposable elements contribute to long non-coding RNAs that are regulatory hubs for abiotic stress response. BMC Genomics 20 (1), 864. https://doi.org/10.1186/s12864-019-6245-5.

Mascher, M., Gundlach, H., Himmelbach, A., Beier, S., Twardziok, S.O., Wicker, T., et al., 2017. A chromosome conformation capture ordered sequence of the barley genome. Nature 26, 427–433.

Mattick, J.S., Rinn, J.L., 2015. Discovery and annotation of long noncoding RNAs. Nat. Struct. Mol. Biol. 22 (1), 5–7. https://doi.org/10.1038/nsmb.2942.

McMullen, M., Jones, R., Gallenberg, D., 1997. Scab of wheat and barley: a re-emerging disease of devastating impact. Plant Dis. 81, 1340–1348.

Miao-Chih, T., Ohad, M., Yue, W., Nima, M., Wang, J.K., Fei, L., et al., 2010. Long noncoding RNA as modular scaffold of histone modification complexes. Science 329, 689–693.

Neubauer, O., 1947. Arsenical cancer: a review. Br. J. Cancer 1, 192–251. https://doi.org/10.1038/bjc.1947.22.

Newton, A.C., Johnson, S.N., Gregory, P.J., 2011. Implications of climate change for diseases, crop yields and food security. Euphytica 179, 3–18.

Pang, J., Zhang, X., Ma, X., Zhao, J., 2019. Spatio-temporal transcriptional dynamics of maize long non-coding RNAs responsive to drought stress. Genes (Basel) 10 (2), 138. https://doi.org/10.3390/genes10020138.

Rackham, O., Shearwood, A.M.J., Mercer, T.R., Davies, S.M.K., Mattick, J.S., Filipovska, A., 2011. Long noncoding RNAs are generated from the mitochondrial genome and regulated by nuclear-encoded proteins. RNA 17, 2085–2093.

Radimerski, T., Mini, T., Schneider, U., Wettenhall, R.E., Thomas, G., Jenö, P., 2000. Identification of insulin-induced sites of ribosomal protein S6 phosphorylation in *Drosophila melanogaster*. Biochemistry 39, 5766–5774.

Ralevic, V., Burnstock, G., 1998. Receptors for purines and pyrimidines. Pharmacol. Rev. 50, 413–492.

Raun, W.R., Johnson, G.V., 1999. Improving nitrogen use efficiency for cereal production. Agron. J. 91 (3), 357–363.

Sahoo, T., del Gaudio, D., German, J.R., Shinawi, M., Peters, S.U., Person, R.E., et al., 2008. Prader–Willi phenotype caused by paternal deficiency for the HBII-85 C/D box small nucleolar RNA cluster. Nat. Genet. 40, 719–721. https://doi.org/10.1038/ng.158.

Salmena, L., Poliseno, L., Tay, Y., Kats, L., Pandolfi, P.P., 2011. A ceRNA hypothesis: The Rosetta stone of a hidden RNA language? Cell 146, 353–358.

Schnable, P.S., Ware, D., Fulton, R.S., Stein, J.C., Wei, F., Pasternak, S., et al., 2009. The B73 maize genome: complexity, diversity, and dynamics. Science 326, 1112–1115.

Secco, D., Jabnoune, M., Walker, H., Shou, H., Wu, P., Poirier, Y., et al., 2013. Spatiotemporal transcript profiling of rice roots and shoots in response to phosphate starvation and recovery. Plant Cell 25 (11), 4285–4304.

Seidel, T., Siek, M., Marg, B., Dietz, K.J., 2013. Energization of vacuolar transport in plant cells and its significance under stress. Int. Rev. Cell Mol. Biol. 304, 57–131.

Serrano, R., Montesinos, C., 2003. Molecular bases of desiccation tolerance in plant cells and potential applications in food dehydration. Food Sci. Technol. Int. 9, 157–161.

Shin, S.Y., Jeong, J.S., Lim, J.Y., Kim, T., Park, J.H., Kim, J.K., et al., 2018. Transcriptomic analyses of rice (*Oryza sativa*) genes and non-coding RNAs under nitrogen starvation using multiple omics technologies. BMC Genomics 19, 532. https://doi.org/10.1186/s12864-018-4897-1.

Shuai, P., Liang, D., Tang, S., Zhang, Z., Ye, C.Y., Su, Y., et al., 2014. Genome-wide identification and functional prediction of novel and drought-responsive lincRNAs in *Populus trichocarpa*. J. Exp. Bot. 65, 4975–4983.

Shumayla, Sharma, S., Taneja, M., Tyagi, S., Singh, K., Upadhyay, S.K., 2017. Survey of high throughput RNA-seq data reveals potential roles for lncRNAs during development and stress response in bread wheat. Front. Plant Sci. 8, 1019.

Singh, U., Khemka, N., Rajkumar, M.S., Garg, R., Jain, M., 2017. PLncPRO for prediction of long non-coding RNAs (lncRNAs) in plants and its application for discovery of abiotic stress-responsive lncRNAs in rice and chickpea. Nucleic Acids Res. 45, e183https://doi.org/10.1093/nar/gkx866.

Steffenson, B.J., 2003. *Fusarium* head blight of barley: Impact, epidemics, management, and strategies for identifying and utilizing genetic resistance. In: Leonard, K.J., Bushnell, W.R. (Eds.), *Fusarium* Head Blight of Wheat and Barley. St. Paul, The American Phytopathological Society, pp. 241–295.

Swiezewski, S., Crevillen, P., Liu, F., Ecker, J.R., Jerzmanowski, A., Dean, C., 2007. Small RNA-mediated chromatin silencing directed to the 30 region of the *Arabidopsis* gene encoding the developmental regulator. FLC. Proc. Natl. Acad. Sci. USA 104, 3633–3638.

Swiezewski, S., Liu, F., Magusin, A., Dean, C., 2009. Cold-induced silencing by long antisense transcripts of an *Arabidopsis* Polycomb target. Nature 462, 799–802.

Szcześniak, M.W., Rosikiewicz, W., Makałowska, I., 2016. CANTATAdb: a collection of plant long non-coding RNAs. Plant Cell Physiol. 57. 1.e8.

Tang, M., Mao, D., Xu, L., Li, D., Song, S., Chen, C., 2014. Integrated analysis of miRNA and mRNA expression profiles in response to cd exposure in rice seedlings. BMC Genomics 15, 835.

Tang, Z., Xu, M., Ito, H., Cai, J., Ma, X., Qin, J., et al., 2019. Deciphering the non-coding RNA-level response to arsenic stress in rice (*Oryza sativa*). Plant Signal. Behav. 14 (9), 1629268. https://doi.org/10.1080/15592324.2019.1629268.

Thatcher, S.R., Danilevskaya, O.N., Meng, X., Beatty, M., Zastrow-Hayes, G., Harris, C., et al., 2016. Genome-wide analysis of alternative splicing during development and drought stress in maize. Plant Physiol. 170, 586–599.

Thomashow, M.F., 1999. Plant cold acclimation: Freezing tolerance genes and regulatory mechanisms. Annu. Rev. Plant Physiol. 50, 571–599.

Tyagi, S., Sharma, A., Upadhyay, S.K., 2018. Role of next-generation RNA-Seq data in discovery and characterization of long non-coding RNA in plants. In: Çiftçi, Y.Ö. (Ed.), Next Generation Plant Breeding. IntechOpen, https://doi.org/10.5772/intechopen.72773.

Umezawa, T., Fujita, M., Fujita, Y., Yamaguchi-Shinozaki, K., Shinozaki, K., 2006. Engineering drought tolerance in plants: discovering and tailoring genes unlock the future. Curr. Opin. Biotechnol. 17, 113–122.

Unver, T., Bozkurt,O., Akkaya, M.S., 2008. Identification of differentially expressed transcripts from leaves of the boron tolerant plant *Gypsophila perfoliata* L. Plant Cell Rep. 27, 1411–1422, doi:https://doi.org/10.1007/s00299-008-0560-7.

Unver, T., Tombuloglu, H., 2019. Barley long non-coding RNAs (lncRNA) responsive to excess boron. Genomics 7543 (19). https://doi.org/10.1016/j.ygeno.2019.11.007. 30714-1.

van Velzen, R., Holmer, R., Bu, F.J., Rutten, L., van Zeijl, A., Liu, W., et al., 2018. Comparative genomics of the nonlegume *Parasponia* reveals insights into evolution of nitrogen-fixing rhizobium symbioses. Proc. Natl. Acad. Sci. U. S. A. 115, E4700–E4709.

Wahbi, A., Gregory, P.J., 1989. Genotypic differences in root and shoot growth of barley (*Hordeum vulgare*). I. Glasshouse studies of young plants and effects of rooting medium. Exp. Agric. 25, 375–387.

Wang, K.C., Chang, H.Y., 2011. Molecular mechanisms of long noncoding RNAs. Mol. Cell 43, 904–914.

Wang, Y., Wang, X., Deng, W., Fan, X., Liu, T.T., He, G., et al., 2014. Genomic features and regulatory roles of intermediate-sized non-coding RNAs in *Arabidopsis*. Mol. Plant 7, 514–527.

Wang, H., Niu, Q.W., Wu, H.W., Liu, J., Ye, J., Yu, N., et al., 2015. Analysis of non-coding transcriptome in rice and maize uncovers roles of conserved lncRNAs associated with agriculture traits. Plant J. 84, 404–416.

Wang, J., Meng, X., Yuan, C., Harrison, A.P., Chen, M., 2016. The roles of cross-talk epigenetic patterns in *Arabidopsis thaliana*. Brief. Funct. Genomics 15, 278–287.

Wang, T., Zhao, M., Zhang, X., Liu, M., Yang, C., Chen, Y., et al., 2017. Novel phosphate deficiency-responsive long non-coding RNAs in the legume model plant *Medicago truncatula*. J. Exp. Bot. 68, 59375–59948.

Wierzbicki, A.T., Haag, J.R., Pikaard, C.S., 2008. Noncoding transcription by RNA polymerase pol IVb/pol V mediates transcriptional silencing of overlapping and adjacent genes. Cell 135 (4), 635–648. https://doi.org/10.1016/j.cell.2008.09.035.

Wise, M.J., 2003. LEA ping to conclusions: a computational reanalysis of late embryogenesis abundant proteins and their possible roles. BMC Bioinform. 4, 52.

Wu, H.J., Wang, Z.M., Wang, M., Wang, X.J., 2013. Widespread long noncoding RNAs as endogenous target mimics for microRNAs in plants. Plant Physiol. 161, 1875–1884.

Xin, M., Wang, Y., Yao, Y., Song, N., Hu, Z., Qin, D., et al., 2011. Identification and characterization of wheat long non-protein coding RNAs responsive to powdery mildew infection and heat stress by using microarray analysis and SBS sequencing. BMC Plant Biol. 11, 61.

Yang, Z.M., Xie, C.J., Sun, Q.X., 2003. Situation of the sources of stripe rust resistance of wheat in the post-CY32 era in China. Acta Agron. Sin. 29 (2), 161–168.

Yang, Q., Li, Z., Li, W., Ku, L., Wang, C., Ye, J., Li, K., et al., 2013. CACTA-like transposable element in ZmCCT attenuated photoperiod sensitivity and accelerated the postdomestication spread of maize. Proc. Natl. Acad. Sci. U. S. A. 110, 16969–16974.

Yu, Y., Zhou, Y.F., Feng, Y.Z., He, H., Lian, J.P., Yang, Y.W., et al., 2019. Transcriptional landscape of pathogen-responsive lncRNAs in rice unveils the role of ALEX1 in jasmonate pathway and disease resistance. Plant Biotechnol. J. https://doi.org/10.1111/pbi.13234.

Yuan, J., Zhang, Y., Dong, J., Sun, Y., Lim, B.L., Liu, D., et al., 2016. Systematic characterization of novel lncRNAs responding to phosphate starvation in *Arabidopsis thaliana*. BMC Genomics 17 (1), 655. https://doi.org/10.1186/s12864-016-2929-2.

Yuan, J., Li, J., Yang, Y., Tan, C., Zhu, Y., Hu, L., et al., 2018. Stress-responsive regulation of long noncoding RNAs' polyadenylation in *Oryza sativa*. Plant J. 93, 814–827. https://doi.org/10.1111/tpj.13804.

Yvonne, T., Lev, K., Leonardo, S., Dror, W., Mynn, T.S., Ugo, A., et al., 2011. Coding-independent regulation of the tumor suppressor PTEN by competing endogenous mRNAs. Cell 147, 344–357.

Zhang, Y.C., Chen, Y.Q., 2013. Long noncoding RNAs: new regulators in plant development. Biochem. Biophys. Res. Commun. 436, 111–114. https://doi.org/10.1016/j.bbrc.2013.05.086.

Zhang, H., Chen, X., Wang, C., Xu, Z., Wang, Y., Liu, X., et al., 2013. Long non- coding genes implicated in response to stripe rust pathogen stress in wheat (*Triticum aestivum* L.). Mol. Biol. Rep. 40, 6245–6253. https://doi.org/10.1007/s11033-013-2736-7.

Zhang, W., Han, Z., Guo, Q., Liu, Y., Zheng, Y., Wu, F., et al., 2014a. Identification of maize long non-coding RNAs responsive to drought stress. PLoS One 9, e98958. https://doi.org/10.1371/journal.pone.0098958.

Zhang, Y.C., Liao, J.Y., Li, Z.Y., Yu, Y., Zhang, J.P., Li, Q.F., et al., 2014b. Genome-wide screening and functional analysis identify a large number of long noncoding RNAs involved in the sexual reproduction of rice. Genome Biol. 15, 512.

Zhang, X., Zhong, T., Liu, L., Ouyang, X., 2015. Impact of soil heavy metal pollution on food safety in China. PLoS One 10, e0135182. https://doi.org/10.1371/journal.pone.0135182.

Zhu, Y.G., Williams, P.N., Meharg, A.A., 2008. Exposure to inorganic arsenic from rice: a global health issue? Environ. Pollut. 154, 169–171. https://doi.org/10.1016/j.envpol.2008.03.015.

Zhu, Q.H., Stephen, S., Taylor, J., Helliwell, C.A., Wang, M.B., 2014. Long noncoding RNAs responsive to *Fusarium oxysporum* infection in *Arabidopsis thaliana*. New Phytol. 201, 574–584. https://doi.org/10.1111/nph.12537.

Chapter 9

Long noncoding RNAs and their implication in novel trait development in soybean

Kaushik Ghose[a] and Krishan Mohan Rai[a,b]

[a]*Fiber and Biopolymer Research Institute, Department of Plant and Soil Science, Texas Tech University, Lubbock, TX, United States,* [b]*Department of Plant and Microbial Biology, University of Minnesota, Saint Paul, MN, United States*

9.1 Introduction

Recent advances in next-generation sequencing technologies, including transcriptome analysis, helped to unravel the mysteries of intergenic regions in between the genic or coding regions, which for a long time has been considered as transcriptional noise by plant molecular biologist. It is already known that in case of the human genome, even though 93% is transcribable, only 2% of these transcripts represent protein-coding genes (Collins et al., 2004). Transcripts that do not code for protein-coding genes are called noncoding RNA (ncRNA). According to their origin and mode of action, noncoding RNA (ncRNA) can be classified into housekeeping ncRNAs and regulatory ncRNAs. Housekeeping ncRNAs, which are mostly involved in translational machinery and cellular functions, are comprised of ribosomal RNA (rRNA), transfer RNA (tRNA), small nuclear RNA (snRNA), and small nucleolar RNA (snoRNA) in size range of 50–200 nucleotides. On the other side, the regulatory noncoding RNAs can be broadly classified into two categories, short noncoding RNAs (small ncRNAs) which are mostly 20–35 nucleotides long and long noncoding RNA (long ncRNA, lncRNA) which are greater than 200 nucleotides long (Djebali et al., 2012; Thomas et al., 2014; Morris and Mattick, 2014).

Most of the lncRNAs have a single exon; the length of the protein coded by lncRNA is typically less than 100 amino acids. The level of expression of lncRNA is also reported to be significantly lower than that of regular protein-coding mRNA. Expression of lncRNAs is reported to be much more organelle-specific and has been observed in both the nucleus and cytoplasm and points towards their role in a diverse number of cellar activities. Based on information available on lncRNA location within the chromosome, it can be classified into

Long Noncoding RNAs in Plants: Roles in development and stress
https://doi.org/10.1016/B978-0-12-821452-7.00009-X

natural antisense transcripts, bidirectional, intronic, overlapping, and long intergenic noncoding RNAs (Mercer et al., 2009).

So far the biological functions of lncRNAs are concerned, a significant number of lncRNAs have been implicated as controller of a wide range of fundamental cellular mechanisms such as mRNA splicing, organelle-specific localization, mRNA turnover, and expression as well as protein translation (Geisler and Coller, 2013; Kashi et al., 2016). Even though most lncRNAs are transcribed by RNA Pol II, it is also reported that Pol IV and Pol V can also transcribe lncRNAs in plants (Wierzbicki et al., 2008; Li et al., 2014a,b). Even though most of the plants lncRNAs are reported to be polyadenylated, reports are also there of plant non-polyadenylated lncRNA like mammalian and yeast lncRNA (Heo and Sung, 2011; Shin and Chekanova, 2014). In a recent study, by performing poly(A)−RNA-seq for seedlings of *Arabidopsis thaliana*, Chao et al. identified differential expression of 58 lncRNAs without poly(A) tail under various stress conditions (Di et al., 2014). lncRNAs are versatile in their mode of action in plants. lncRNAs can regulate plant gene expression as cis- or trans-regulatory factors. They can also form scaffolds that could potentially employ chromatin-modifying complexes to target chromatin (Heo and Sung, 2011; Heo et al., 2013). lncRNA could act as molecular decoys by binding to the gene coding mRNA or regulatory miRNA based on sequence complementarity denying access to regulatory protein and effector complexes, leading to inhibition of translation of protein-coding mRNA and miRNA-induced cleavage of target genes (Franco-Zorrilla et al., 2007; Wu et al., 2013). Acting as decoy lncRNA could also modulate alternative splicing machinery in plants (Bardou et al., 2014; Lai et al., 2013; Pefanis et al., 2015; Shumayla et al., 2017). Even though lncRNA cannot be translated, it was reported recently that lncRNA could act as translational regulators by co-expressing and associating with ribosomal proteins (Carlevaro-Fita et al., 2016). To complement the excellent progress that has already been made in the field of Arabidopsis, rice, corn, wheat, and soybean lncRNA molecular mechanism, increasing effort should be made to develop plant lncRNAs database and functional annotation of lncRNA in agriculturally important plant species (Jin et al., 2013; Szczesniak et al., 2016; Tyagi et al., 2018).

Legume seeds, including soybean, are one of the major sources for human and animal food. Furthermore, soybean contributes to over half of global oilseed production. Considering the tremendous agricultural importance as a source of food and oils, many studies have already been done focusing on the soybean genomic and transcriptomic aspects controlling important agronomic traits. The whole-genome sequencing of soybean elucidated the complex paleopolyploid structure of the genome (Schmutz et al., 2010). In terms of species diversity, Arabidopsis and soybean were reported to be separated from their common ancestor 92 million years ago (Zhu et al., 2003) and the soybean genome has been subjected to minimum two genome duplication events, leading to homeologous relationships between chromosomes and gene loci (Shoemaker et al., 2006).

LincRNAs are characteristically a lot different than the protein-coding mRNA. The median length of LincRNAs in soybean was 320 bp with a mean length of 467.3 bp in contrast to the 365.7 bp median protein-coding mRNA transcript length (Golicz et al., 2018).

In most cases, the lincRNA contains a single exon, with the largest LincRNA having a maximum of 4 exons. In most cases, LincRNAs didn't have isoforms, and their transcript copy number or level of expression was much lower as compared to protein-coding mRNAs. The shorter length (Zhang et al., 2014; Hao et al., 2015; Khemka et al., 2016) and low expression levels of LincRNA are consistent across plant species (Liu et al., 2012; Li et al., 2014a,b; Zhang et al., 2014; Hao et al., 2015).

In the case of soybean, like other plant species, LincRNAs show highly tissue-specific expression. The tissue specificity of lncRNA expression was found to be more precise than protein-coding genes. Similar to rice and maize (Wang et al., 2015a), expression of lncRNA around the centromere was observed in soybean as well. A significant proportion of these lncRNAs were in close proximity to protein-coding genes. The median DNA length from protein-coding mRNA to the adjacent lncRNA was 1064 bp. The short distance is indicative of the fact that lncRNA could act as a trans-activating regulator by employing activators, repressors, and epigenetic modifiers (Wang and Chang, 2011; Kornienko et al., 2013). In this review, we summarize the present-day understanding of lncRNAs, review the current development of regulatory roles of lncRNAs, and shed light on the newly emerged functions of lncRNAs in regulating biotic, abiotic stress responses, and reproductive development processes in soybean.

9.2. lncRNA in plant immunity

9.2.1 Nematodes

In a recent study on Root-knot nematodes (RKNs, Meloidogyne spp.) and tobacco, Li et al. identified 5206 differentially expressed lncRNAs by RNA sequencing data analysis. The study identified several classes of lncRNAs involved in hormone signal transduction, induced systemic resistance, localized hypersensitive response, and plant cell wall modification. Among these 15 lncRNAs, expression validated by qRT-PCR was reported to be specifically involved in nematode stress response (Li et al., 2018a,b). Xing et al. (2019) also reported differential expression 1826 candidate lncRNAs between resistant and susceptible cultivars upon grapevine infection with the facultative parasite, *Lasiodiplodia theobromae*, responsible for severe trunk disease.

9.2.2 Bacterium

Plants recognize bacterial elicitor molecules by two different molecular mechanisms. The first one, known as pattern recognition receptors (PRRs), are cell surface immune receptors, whereas the second one is the cytosol-localized

NB-LRR protein type receptors (Jones and Dangl, 2006). In addition to the first line of defense like thickening of cell wall, low pH of apoplast, secretion of cell wall degrading enzymes, or other antibacterial components, plants also have developed a well-programmed innate immune system comprised of pathogen-associated molecular patterns (PAMPs)-triggered innate immunity (PTI) and effector-triggered innate immunity (ETI). The PTI and ETI are closely related and form the core molecular basis of plant innate immunity (Yan et al., 2018a,b). Through several studies, it is already well-established that LncRNAs played a crucial role in plant defense response to pathogens, but there is a lot to explore in identifying the molecular role of certain lncRNAs in plant immunity on a crop to crop basis. A large number of lncRNAs were reported to be induced upon Fusarium oxysporum infection in Arabidopsis, but without any functional information (Zhu et al., 2014). A comprehensive analysis by Seo et al. (2017) identified long noncoding RNA, ELENA1, which caused increased resistance in Arabidopsis against Pseudomonas syringe. They proposed that ELENA1 positively interacts with the transcription factor MED19a, which in turn upregulates the expression of defense gene PR1. Fifty-nine lncRNAs were reported to be involved in defense response to the bacterium in soybean (Golicz et al., 2018) and they are reported in Table 9.1.

9.2.3 lncRNA and fungal resistance

Diseases caused by fungal pathogens could be devastating and lead to significant crop yield loss. It has already been documented in multiple publications that lncRNA plays a crucial role in regulating fungal disease resistance in plants. Induction of SIGRX expression by Tomato lncRNA16397 subsequently leads to diminished free radical accumulation and reduced cell damage promoting tomato resistance against *Phytophthora infestans* (Park et al., 2013; Cui et al., 2017). In a separate study, Joshi et al. (2016) identified 931 *B. napus* lncRNAs differentially expressed upon infection with Sclerotinia. Expression of 12 of these 931 lncRNAs was further confirmed using qRT-PCR. They identified TCONS_00000966, *B. napus* antisense lncRNA, with 90% sequence similarity to a *B. napus* defensin gene and showed significant upregulation, proving its involvement in the transcriptional modulation of *B. napus* defense response against *S. sclerotiorum* infection. In addition to that, they also identified nine lncRNAs having sequence overlap with upstream-regulatory regions of the *B. napus* differentially expressed genes. In a study involving barley near-isogenic lines harboring resistance QTL associated with FHB resistance, Huang et al. (2016) identified differential expression of 12,366 long noncoding RNAs (lncRNAs), 604 of which were responsive to FHB. Furthermore, using strand-specific RNA sequencing, Kwenda et al. (2016) identified 1113 potato stem lincRNAs. Additional time course analysis identified 559 lncRNAs differentially expressed between a resistant and a susceptible potato cultivar against *Pectobacterium carotovorum* subsp. brasiliense. The co-expression analysis

TABLE 9.1 Description of the already identified lncRNAs and their functional annotation.

Functional class	Total number	lncRNAs code
Defense response to the bacterium	59	NC_GMAXPA00029253,NC_GMAXPA00063364,NC_GMAXPA00088480,NC_GMAXPA00089810NC_GMAXPA00126147,NC_GMAXPA00129699,NC_GMAXPA00138052,NC_GMAXPA00145847,NC_GMAXST00000968,NC_GMAXST00001178,NC_GMAXST00001207,NC_GMAXST00001477NC_GMAXST00004103, NC_GMAXST00005070,NC_GMAXST00006775,NC_GMAXST00006891NC_GMAXST00007012,NC_GMAXST00007496,NC_GMAXST00011980,NC_GMAXST00014659,NC_GMAXST00015906,NC_GMAXST00016776,NC_GMAXST00018336'NC_GMAXST00019614,NC_GMAXST00021640,NC_GMAXST00023886,NC_GMAXST00025107,NC_GMAXST00027706,NC_GMAXST00028806,NC_GMAXST00028808,NC_GMAXST00028810,NC_GMAXST00028965,NC_GMAXST00029028,NC_GMAXST00030086,NC_GMAXST00031742,NC_GMAXST00032485, NC_GMAXST00034158,NC_GMAXST00035294,NC_GMAXST00035400,NC_GMAXST00037450,NC_GMAXST00038787,NC_GMAXST00041627,NC_GMAXST00042492,NC_GMAXST00042795,NC_GMAXST00045137,NC_GMAXST00046497,NC_GMAXST00047586,NC_GMAXST00049855,NC_GMAXST00050271,NC_GMAXST00054458,NC_GMAXST00054863,NC_GMAXST00055048,NC_GMAXST00055049,NC_GMAXST00055058,NC_GMAXST00057692,NC_GMAXST00059792,NC_GMAXST00059857,NC_GMAXST00059858,NC_GMAXST00060570
Defense response against fungus	31	NC_GMAXPA00028821,NC_GMAXPA00031677,NC_GMAXPA00047198,NC_GMAXPA00072518,NC_GMAXPA00076151,NC_GMAXPA00086054,NC_GMAXPA00098247,NC_GMAXPA00109776,NC_GMAXPA00114665,NC_GMAXPA00120464,NC_GMAXPA00123084,NC_GMAXPA00131685,NC_GMAXPA00146987,NC_GMAXST00002590'NC_GMAXST00005070,NC_GMAXST00008136,NC_GMAXST00009362,NC_GMAXST00020234,NC_GMAXST00022492,NC_GMAXST00026008,NC_GMAXST00026302,NC_GMAXST00032484,NC_GMAXST00032485,NC_GMAXST00034856,NC_GMAXST00038973,NC_GMAXST00042932,NC_GMAXST00045137,NC_GMAXST00050049,NC_GMAXST00055454,NC_GMAXST00057778,NC_GMAXST00059949
Incompatible defense response against fungi	26	NC_GMAXPA00001666,NC_GMAXPA00009011,NC_GMAXPA00042634,NC_GMAXPA00103991,NC_GMAXPA00140360,NC_GMAXST00003033,NC_GMAXST00012471,NC_GMAXST00017268,NC_GMAXST00017396,NC_GMAXST00019176,NC_GMAXST00020226,NC_GMAXST00023567,NC_GMAXST00026603,NC_GMAXST00026604,NC_GMAXST00028965,NC_GMAXST00029028,NC_GMAXST00032939,NC_GMAXST00034582,NC_GMAXST00037241,NC_GMAXST00037450, NC_GMAXST00040093, NC_GMAXST00042390, NC_GMAXST00045283, NC_GMAXST00045503, NC_GMAXST00045798 and, NC_GMAXST00060188

Continued

TABLE 9.1 Description of the already identified lncRNAs and their functional annotation—cont'd

Functional class	Total number	lncRNAs code
Defense response against insects	17	NC_GMAXPA00001941,NC_GMAXPA00047198,NC_GMAXPA00070954,NC_GMAXPA00112837,NC_GMAXST00004095,NC_GMAXST00005636,NC_GMAXST00025750,NC_GMAXST00033505,NC_GMAXST00034856, NC_GMAXST00037702,NC_GMAXST00042409,NC_GMAXST00051347,NC_GMAXST00053826, NC_GMAXST00054885, NC_GMAXST00055032, NC_GMAXST00055454 and, NC_GMAXST00057778
Defense response to oomycetes	85	NC_GMAXPA00001367,NC_GMAXPA00004036,NC_GMAXPA00007557,NC_GMAXPA00007693,NC_GMAXPA00007767,NC_GMAXPA00009105,NC_GMAXPA00015323,NC_GMAXPA00015522, NC_GMAXPA00018249,NC_GMAXPA00018994,NC_GMAXPA00019465,NC_GMAXPA00028825,NC_GMAXPA00030803,NC_GMAXPA00034954,NC_GMAXPA00049664,NC_GMAXPA00055863,NC_GMAXPA00060879,NC_GMAXPA00064421,NC_GMAXPA00070065,NC_GMAXPA00073739, NC_GMAXPA00080548,NC_GMAXPA00085442,NC_GMAXPA00086057,NC_GMAXPA00088700,NC_GMAXPA00089149,NC_GMAXPA00092876,NC_GMAXPA00100156,NC_GMAXPA00100797,NC_GMAXPA00109944,NC_GMAXPA00113924,NC_GMAXPA00121329,NC_GMAXPA00122695, NC_GMAXPA00128505,NC_GMAXPA00129389,NC_GMAXPA00131261,NC_GMAXPA00140171,NC_GMAXPA00147771,NC_GMAXST00000324,NC_GMAXST00000325,NC_GMAXST00000667,NC_GMAXST00001916,NC_GMAXST00003405,NC_GMAXST00006891,NC_GMAXST00010091,NC_GMAXST00010563,NC_GMAXST00010635,NC_GMAXST00011794,NC_GMAXST00011796,NC_GMAXST00011797,NC_GMAXST00011800, NC_GMAXST00012839,NC_GMAXST00012928,NC_GMAXST00016617,NC_GMAXST00017142,NC_GMAXST00019954,NC_GMAXST00022590,NC_GMAXST00023213,NC_GMAXST00023775,NC_GMAXST00024154,NC_GMAXST00025909,NC_GMAXST00029446,NC_GMAXST00036329,NC_GMAXST00036621,NC_GMAXST00036622, NC_GMAXST00040044,NC_GMAXST00040556,NC_GMAXST00042753,NC_GMAXST00045391,NC_GMAXST00046210,NC_GMAXST00047452,NC_GMAXST00047576,NC_GMAXST00049705,NC_GMAXST00049706,NC_GMAXST00050260,NC_GMAXST00050783,NC_GMAXST00052478, NC_GMAXST00052638,NC_GMAXST00053622,NC_GMAXST00054922,NC_GMAXST00055147,NC_GMAXST00055842,NC_GMAXST00056536,NC_GMAXST00058229,NC_GMAXST00059656 and,NC_GMAXST00060557
Defense response to viruses		NC_GMAXPA00009105,NC_GMAXPA00023797,NC_GMAXST00001178,NC_GMAXST00001670,NC_GMAXST00015676,NC_GMAXST00016888,NC_GMAXST00023089,NC_GMAXST00035744, NC_GMAXST00038973 and, NC_GMAXST00050870

they performed identified the association of 17 of these lincRNAs with 12 potato defense-related genes. A separate study in Arabidopsis by strand-specific RNA sequencing leads to the identification of several noncoding natural antisense transcripts involved in Arabidopsis defense response against *F. oxysporum* (Zhu et al., 2014). An interesting study performed by Zhang et al. in tetraploid cotton discovered expression dominance of induced lncRNAs in the Dt subgenome compared to At sub-genome upon challenge with *Verticillium dahlia*, a fungal pathogen for cotton (Zhang et al., 2018b). In a genome-wide lncRNA expression analysis study, Wen et al. (2018) identified 4709 lncRNAs, 2518 of which upregulated and 2191 downregulated upon *D. zeae* infection in rice. They also predicted a large number of lncRNAs to be target mimic of miR396 and osa-miR156, suggesting their role in the *D. zeae* rice pathosystem (Li et al., 2018a,b). In case of soybean, genome-wide analysis identified 31 lncRNAs were reported to be involved in defense response against fungus and they are reported in Table 9.1. In addition to that, 26 soybean lncRNAs were reported to be involved particularly in incompatible defense response against fungi, and they are reported in Table 9.1.

9.2.4 Resistance against insect

Even though a large number of lncRNAs are related to insect immunity of *Apis mellifera, Heliconius melpomene, Bombyx mori*, Drosophila, *Anopheles gambiae*, *Aedes gambiae*, and *Danaus plexippus,* in-depth functional characterization of this lncRNA is scarce. Several lncRNAs of *Tribolium castaneum* were identified to be functional on the template strand (Shippy et al., 2008). In another study, Zhou et al. reported the identification and characterization of four conserved, lowly expressed organelle-specific lncRNAs in silkworm. In a recent report, Li et al. reported the differential expression of silk gland-enriched intermediate-size noncoding RNAs in silkworm *Bombyx mori*. Additionally, Satyavathi et al. (2013, 2017) reported the differential expression of silkworm lncRNA upon infection with resistant (SBNP1) and susceptible (CSR2) strains, BmNPV, at the fifth larval instar. In the case of soybean, genome-wide analysis of lncRNA by Golicz et al. identified 17 defense lncRNA involved in defense response against the insect and they are reported in Table 9.1.

9.2.5 Resistance against oomycetes

In a recent study by Jiang et al. (2019), they identified three tomato lncRNAs, lncRNA23468, lncRNA01308, and lncRNA13262, containing conserved eTM sites for miR482b. Of the three, overexpression of lncRNA23468 causes downregulation of miR482b expression which in turn upregulates the expression of NBS-LRRs, causing enhanced resistance to *P. infestans*. 85 soybean lncRNAs have been identified by Golicz et al. (2018) to be involved in defense response to oomycetes pathogen and they are reported in Table 9.1.

9.2.6 Resistance against viruses

LncRNA can play crucial role in resistance against viruses in plants. Differential expression of several lncRNA in response against tomato yellow leaf curl virus (TYLCV) has already been reported (Wang et al., 2016). In this tomato YLCV pathosystem, negative correlation was observed between lncRNA and miRNA expression. On the contrary, positive correlation was observed between the lncRNA and miRNA target genes. The authors validated their hypothesis by virus-induced gene silencing of lnc0195. They identified that tomato lnc0195 causes downregulation of miR166 by miRNA target mimicry, which in turn causes upregulation of class III HD-ZIP family of transcription factors. In addition, tomato lnc0195 and lnc1077 could act as decoy for miRNA166 and miR399 and provide viral resistance in tomato. In case of soybean, genome-wide analysis by Golicz et al. identified 10 lncRNAs involved in defense response against viruses. They are reported in Table 9.1.

9.3. lncRNA and plant response to salt tolerance

Soil salinity has huge negative impact on agricultural crop production across the world. According to a study conducted by UNESCO and FAO, 9.5438 billion hectares of land have high salinity. Similar to other environmental stress factors, salt stress can have significantly damaging impact on yield and productivity of crops (Wang et al., 2016). Soil salt concentration above certain limit causes ionic toxicity, osmotic imbalance, and free radical-mediated oxidative stress in plants, which in turn deter plant growth and development (Zhang et al., 2018a,b). High salt concentration in the soil can negatively impact several biological and physiological processes like leaf chlorophyll content and photosynthesis (Sui and Han, 2014; Liu et al., 2017; Zhang et al., 2014; Kalaji et al., 2016; Sui et al., 2015; Cheng et al., 2015; Li et al., 2012; Zhang et al., 2010), ionic balances (Ruiz et al., 2016; Han et al., 2010; Feng et al., 2015; Song et al., 2017; Shaheen et al., 2016; Dai et al., 2014; Guo et al., 2016; Zao et al., 2010), seed germination, seedling establishment and maturation (Song et al., 2016; Zhou et al., 2016a,b; Guo et al., 2012; Vu et al., 2015; Yao et al., 2010; Wang et al., 2015a,b; Xu et al., 2016; Guo et al., 2015; Guo et al., 2017; Liu et al., 2018; Zhang et al., 2015), and membrane permeability (Cui et al., 2018; Dinneny, 2015; Liu et al., 2015; Tavakkoli et al., 2011; Tao et al., 2012; Cheng et al., 2014).

Until today, not much information is available regarding the involvement of lncRNA in plant salt stress. One such well-cited work has been done by Ying-Ping et al. (2018) in mulberry, where they documented the emerging role of a novel lncRNA MuLnc1. Cleavage of MuLnc1 by mul-miR3954 generates 21 nucleotides secondary siRNA si161579, which in turn silence the calmodulin-like protein gene MuCML27. The heterologous expression of MuCML27 in Arabidopsis provides increased tolerance to drought and high salt concentration.

In a separate study in maize, Huanca-Mamani et al. (2018) performed a comprehensive differential expression profiling of lncRNA using RNAseq and qRT-PCR under salt and boron exposure, leading to the identification of 1710 lncRNAs. The only reported study in soybean root has been done by Chen et al. (2019), where they have cultured soybean plants under high salt concentration for 15 continuous days starting from seed germination and identified 3030 long intergenic noncoding RNAs (lincRNAs) and 275 natural antisense transcripts (lncNATs) in soybean roots using Strand-specific whole transcriptome sequencing and stringent bioinformatics analysis. The newly discovered lncRNAs are evenly distributed across the genome and have a lower number of exons compared to protein-coding mRNA. A large proportion of these newly identified salt stress-induced lncRNAs targeted essential catalytic proteins in the soybean stress response pathway. Even though reduction of protein-coding mRNA and lncRNA length was observed under continuous exposure to high salt concentration, similar to that reported in Arabidopsis (Srivastava et al., 2018), the authors proposed that reduction in length of wide range of RNA molecules under stress might be a part of internal mechanism, which helps the plants to save metabolic energy. In addition to that, the authors have also performed two methylome maps by bisulfite sequencing with a single-base resolution, providing valuable insights into the regulation of stress-associated lncRNAs and protein-coding genes by an epigenetic mechanism.

9.4. lncRNA in nodulation and rhizobium interaction

The soybean early nodulin gene pGmENOD40 has been identified by a group of researchers in the early 90s, which did not show structural similarity to other proteins due to the absence of methionine in it. By in situ hybridization, they identified that the gene had an intricate expression pattern at different stages of nodule development, and the expression is inducible by intracellular bacteria or infection threads in both determinate and indeterminate nodules suggesting ENOD40's role in transport (Yang et al., 1993). Later in early 2000, researchers have determined that ENNOD is not specific for legumes only, and homologs are present in nonleguminous plants as well, implicating its role in critical biological function other than nodulation in leguminous plants. By performing an in vitro experiment, Rohrig et al. (2002) showed that the ENDO40 is initially transcribed into a polycistronic eukaryotic mRNA which was then translated into two peptides, 12 and 24 aa residues (peptides A and B). By affinity purification and peptide mass fingerprinting, they have identified that both the peptides specifically bind nodulin 100 subunits of sucrose synthase, suggesting ENOD40's role in modulating sucrose availability in nitrogen-fixing nodules. In a separate study, Girard et al. (2003) reported a detailed analysis of the secondary structure of enod40 mRNA. Even though the enod40 gene is highly conserved across leguminous and nonleguminous species, they reported the presence of unique structural domains present in leguminous species only, suggesting its role in indeterminate nodules development. In a more recent study, Wang et al. (2014) determined the molecular mechanisms of soybean rhizobium infection and nodules development is controlled by miR172c. They observed induction of miR172c transcripts upon *B. japonicum* infection or treatment with lipo-oligosaccharide Nod factor, causing a significant change in nodule number and development. They determined that miR172c functions by suppressing Nodule Number Control 1, which acts as a direct regulator of the early nodulin gene, enod40. Very recently, Yan et al. reported nodulation control in soybean through Brassinosteroids. They have identified a soybean BRI1-EMS suppressor 1 (BES1)/brassinazole-resistant 1 (BZR1) family transcription factor, GmBEHL1, as a transcriptional inhibitor of soybean nodulation, which acts in association with Nodule Number Control 1. As reported earlier, Nodule Number Control 1 acts as a direct regulator of the early nodulin gene, enod40.

9.5. Role of lncRNA in flower development and reproduction

Several studies conducted in other plant species have already documented that lncRNAs play a crucial role in reproductive growth and development. The most extensively studied plant lncRNAs are associated with transcriptional and post-transcriptional regulation of flower development and reproduction machinery (Yu and Zhu, 2019).

Even though a large number of lncRNAs are differentially expressed in multiple studies, only a limited number of them are functionally characterized. In mammals, lncRNAs, namely Xist, H19, Kcnq1ot1, bxd, and HOTAIR, have already been documented to be involved in embryonic development (Kalantry et al., 2009; Bartolomei et al., 1991; Umlauf et al., 2004; Rinn et al., 2007). Similarly, several plant lncRNAs like COLDAIR, COOLAIR, LDMAR, CsM10, and Zm401 also played a crucial role in reproductive control. Two Arabidopsis lincRNAs, COOLAIR and COLDAIR lncRNAs, which function as flowering repressor from FLOWERING LOCUS C (FLC) can also operate similarly in rice, suggesting the conservation of lncRNA functionality across species (Heo and Sung, 2011; Ding et al., 2012; Cho et al., 2005; Ma et al., 2008; Zhang and Chen, 2013). In rice, photoperiod-controlled male sterility is reported to be regulated by the lncRNA, LONG-DAY-SPECIFIC MALE-FERTILITY-ASSOCIATED RNA (LDMAR) (Ding et al., 2012). According to a study done by Komiya et al. (2014), a significant number of lncRNAs could lead to the production of 21-nucleotide phasiRNAs, which bind with the Argonaute (AGO) protein MEL1 in rice germline cells, pointing towards the role of rice lncRNAs in premeiotic germ cells development.

Zhang et al. performed the strand-specific whole transcriptome analysis from several rice reproductive tissues after 5 and 14 days of germination and identified rice lncRNAs specifically involved during the process of reproduction. A significant proportion of the lncRNA identified to be functionally characterized by them was explicitly expressed in pollen. In a recent soybean genome-wide lncRNA analysis study, Golicz et al. identified the expression of the largest number of lncRNA in floral tissues. The latest genome-wide lincRNA study performed by Golicz et al. contributed to our understanding that a number of soybean genes involved in controlling growth and reproductive development exist in multiple copies and lncRNAs could play a vital role in providing the extra level of molecular control which is not identified by studies where only protein-coding genes are considered. This was followed by a somewhat smaller number of lncRNAs expressed in shoot apical meristem and leaf, pointing towards a crucial role of lincRNAs in soybean flowering and reproductive developmental processes. The study also reported that the flower expressed 1891 number of total lncRNA, out of which 51 were uniquely expressed in flower bud only.

The expression pattern of lncRNAs in shoot epical meristem was in accord with the already reported expression in chickpea and other plants (Khemka et al., 2016). In general, the tissue-specific expression pattern of lncRNAs seems to be consistent across species. Overall, they have reported a unique expression of 366 lincRNAs in the floral transition samples, 99% of which were expressed following short-day treatment. Eighty-nine of these lncRNAs were uniquely expressed in leaf, 128 in SAM, and149 lincRNAs expressed in both leaf and SAM. The study made lncRNAs a potential target for floral development studies in soybean. When the lncRNA expression pattern in different tissues was compared to the protein-coding gene expression, lncRNAs showed much more tissue specificity with an expression level similar to the transcription factors. The authors concluded that lncRNAs might play a transcription factor kind of role in soybean floral development.

9.6. Conclusion and future prospective

With the advancement of high-throughput next-generation sequencing, many lncRNAs have been implicated in playing a crucial role in controlling the operational mechanism of multiple plant biological processes. But the functional characterization of these vast numbers of plant lncRNAs is still at its infancy. Through genome-wide identification studies, we already have good knowledge of the structure and chromosomal location of soybean lncRNAs. But studies done on functional characterization of soybean lncRNAs are minimal. The advent of the latest genome editing technologies could accelerate this process of lncRNA functionality testing. Recently, work done by Li et al. (2018a,b) identified and functionally characterized tomato lncRNA1459 using (CRISPR) / Cas9-mediated genome editing technology. Precise editing of soybean lncRNAs could lead to the development of several important agronomic traits.

Acknowledgment

I would like to thank Dr. Venugopal Mendu, Associate Professor, Department of Plant and Soil Science, Texas Tech University for his constant support and encouragement.

References

Bardou, F., Ariel, F., Simpson, C.G., Romero-Barrios, N., Laporte, P., Balzergue, S., Brown, J.W.S., Crespi, M., 2014. Long noncoding RNA modulates alternative splicing regulators in Arabidopsis. Dev. Cell 30, 166–176.

Bartolomei, M.S., Zemel, S., Tilghman, S.M., 1991. Parental imprinting of the mouse H19 gene. Nature 351, 153–155. https://doi.org/10.1038/351153a0.

Carlevaro-Fita, J., Rahim, A., Guigó, R., Vardy, L.A., Johnson, R., 2016. Cytoplasmic long noncoding RNAs are frequently bound to and degraded at ribosomes in human cells. RNA 22, 867–882.

Chen, R., Li, M., Zhang, H., et al., 2019. Continuous salt stress-induced long non-coding RNAs and DNA methylation patterns in soybean roots. BMC Genomics 20, 730. https://doi.org/10.1186/s12864-019-6101-7.

Cheng, S., Yang, Z., Wang, M., et al., 2014. Salinity improves chilling resistance in Suaeda salsa. Acta Physiol. Plant. 36 (7), 1823–1830.

Cheng, T., Chen, J., Zhang, J., Shi, S., Zhou, Y., Lu, L., Wang, P., Jiang, Z., Yang, J., Zhang, S., Shi, J., 2015. Physiological and proteomic analyses of leaves from the halophyte Tangut Nitraria reveals diverse response pathways critical for high salinity tolerance. Front. Plant Sci. 6, 30.

Cho, J., Koo, D.H., Nam, Y.W., Han, C.T., Lim, H.T., Wook, J., Hur, B.Y.K., 2005. Isolation and characterization of cDNA clones expressed under male sex expression conditions in a monoecious cucumber plant (Cucumis sativusL.cv. Winter long). Euphytica 146, 271–281. https://doi.org/10.1007/s10681-005-9023-1.

Collins, F.S., Lander, E.S., Rogers, J., et al., 2004. International human genome sequencing consortium, finishing the euchromatic sequence of the human genome. J. Tanpakushitsu Kakusan Koso Protein Nucl. Acid Enzym. 431, 931–945.

Cui, J., Luan, Y., Jiang, N., Bao, H., Meng, J., 2017 Feb. Comparative transcriptome analysis between resistant and susceptible tomato allows the identification of lncRNA16397 conferring resistance to Phytophthora infestans by co-expressing glutaredoxin. Plant J. 89 (3), 577–589. https://doi.org/10.1111/tpj.13408. Epub 2017 Feb 3, 27801966.

Cui, F., Sui, N., Duan, G., et al., 2018. Identification of metabolites and transcripts involved in salt stress and recovery in peanut. Front. Plant Sci. 9, 217.

Dai, L.Y., Zhang, L.J., Jiang, S.J., et al., 2014. Saline and alkaline stress genotypic tolerance in sweet sorghum is linked to sodium distribution. J. Acta Agric. Scand. 64 (6), 471–481.

Di, C., Yuan, J., Wu, Y., Li, J., Lin, H., Hu, L., Zhang, T., Qi, Y., Gerstein, M.B., Guo, Y., Lu, Z.J., 2014. Characterization of stress-responsive lncRNAs in Arabidopsis thaliana by integrating expression, epigenetic and structural features. Plant J. 80, 848–861. https://doi.org/10.1111/tpj.12679.

Ding, J., Qing, L., Yidan, O., Hailiang, M., Pingbo, Z., Jialing, Y., Caiguo, X., Xianghua, L., Jinghua, X., Qifa, Z., 2012. A long noncoding RNA regulates photoperiod-sensitive male sterility, an essential component of hybrid rice. Proc. Natl. Acad. Sci. U. S. A. 109 (7), 2654–2659. https://doi.org/10.1073/pnas.1121374109.

Dinneny, J.R., 2015. Traversing organizational scales in plant salt-stress responses. Curr. Opin. Plant Biol. 23, 70–75.

Djebali, S., Davis, C.A., Merkel, A., Dobin, A., Lassmann, T., Mortazavi, A., Tanzer, A., Lagarde, J., Lin, W., Schlesinger, F., Xue, C., Marinov, G.K., Khatun, J., Williams, B.A., Zaleski, C., Rozowsky, J., Röder, M., Kokocinski, F., Abdelhamid, R.F., Alioto, T., Antoshechkin, I., Baer, M.T., Bar, N.S., Batut, P., Bell, K., Bell, I., Chakrabortty, S., Chen, X., Chrast, J., Curado, J., Derrien, T., Drenkow, J., Dumais, E., Dumais, J., Duttagupta, R., Falconnet, E., Fastuca, M.,

Fejes-Toth, K., Ferreira, P., Foissac, S., Fullwood, M.J., Gao, H., Gonzalez, D., Gordon, A., Gunawardena, H., Howald, C., Jha, S., Johnson, R., Kapranov, P., King, B., Kingswood, C., Luo, O.J., Park, E., Persaud, K., Preall, J.B., Ribeca, P., Risk, B., Robyr, D., Sammeth, M., Schaffer, L., See, L.H., Shahab, A., Skancke, J., Suzuki, A.M., Takahashi, H., Tilgner, H., Trout, D., Walters, N., Wang, H., Wrobel, J., Yu, Y., Ruan, X., Hayashizaki, Y., Harrow, J., Gerstein, M., Hubbard, T., Reymond, A., Antonarakis, S.E., Hannon, G., Giddings, M.C., Ruan, Y., Wold, B., Carninci, P., Guigó, R., Gingeras, T.R., 2012. Landscape of transcription in human cells. Nature 489 (7414), 101–108. https://doi.org/10.1038/nature11233.

Feng, Z.T., Deng, Y.Q., Zhang, S.C., Liang, X., Yuan, F., Hao, J.L., Zhang, J.C., Sun, S.F., Wang, B.S., 2015. K (+) accumulation in the cytoplasm and nucleus of the salt gland cells of Limonium bicolor accompanies increased rates of salt secretion under NaCl treatment using Nano SIMS. J. Plant Sci. 238, 286–296.

Franco-Zorrilla, J.M., Valli, A., Todesco, M., Mateos, I., Puga, M.I., Rubio-Somoza, I., Leyva, A., Weigel, D., García, J.A., Paz-Ares, J., 2007. Target mimicry provides a new mechanism for regulation of micro RNA activity. Nat. Genet. 39, 1033–1037.

Geisler, S., Coller, J., 2013. RNA in unexpected places: long non-coding RNA functions in diverse cellular contexts. Nat. Rev. Mol. Cell Biol. 14 (11), 699–712. https://doi.org/10.1038/nrm3679.

Girard, G., Roussis, A., Gultyaev, A.P., Pleij, C.W., Spaink, H.P., 2003. Structural motifs in the RNA encoded by the early nodulation gene enod40 of soybean. Nucleic Acids Res. 31 (17), 5003–5015.

Golicz, A.A., Singh, M.B., Bhalla, P.L., 2018. The long intergenic noncoding RNA (LincRNA) landscape of the soybean genome. Plant Physiol. 176 (3), 2133–2147.

Guo, Y., Jia, W., Song, J., et al., 2012. Thellungilla halophila is more adaptive to salinity than Arabidopsis thaliana at stages of seed germination and seedling establishment. Acta Physiol. Plant. 34 (4), 1287–1294.

Guo, J., Suo, S., Wang, B.S., 2015. Sodium chloride improves seed vigour of the euhalophyte Suaeda salsa. J. Seed Sci. Res. 25 (3), 335–344.

Guo, P., Wei, H.X., Zhang, W.J., et al., 2016. Physiological responses of alfalfa to high level salt stress: root ion flux and stomatal characteristics. Int. J. Agric. Biol. 18 (1), 125–133.

Guo, J., Li, Y., Han, G., et al., 2017. NaCl markedly improved the reproductive capacity of the euhalophyte Suaeda salsa. J. Funct. Plant Biol. 45 (3), https://doi.org/10.1071/FP17181.

Han, N., Shao, Q., Bao, H., et al., 2010. Cloning and characterization of a Ca 2þ/H þ., antiporter from halophyte Suaeda salsa, L. J. Plant Mol. Biol. Rep. 29 (2), 449–457.

Hao, Z., Fan, C., Cheng, T., Su, Y., Wei, Q., Li, G., 2015. Genome-wide identification, characterization and evolutionary analysis of long intergenic noncoding RNAs in cucumber. PLoS One. 10, e0121800.

Heo, J.B., Sung, S., 2011. Vernalization-mediated epigenetic silencing by a long intronic noncoding RNA. Science 331, 76–79. https://doi.org/10.1126/science.1197349.

Heo, J.B., Lee, Y.S., Sung, S., 2013. Epigenetic regulation by long noncoding RNAs in plants. Chromosome Res 21, 685–693.

Huanca-Mamani, W., Arias-Carrasco, R., Cárdenas-Ninasivincha, S., Rojas-Herrera, M., Sepúlveda-Hermosilla, G., Caris-Maldonado, J.C., Bastías, E., Maracaja-Coutinho, V., 2018. Long non-coding RNAs responsive to salt and boron stress in the hyper-arid Lluteño maize from Atacama Desert. Genes 9, 170.

Huang, Y., Li, L., Smith, K.P., Muehlbauer, G.J., 2016. Differential transcriptomic responses to Fusarium graminearum infection in two barley quantitative trait loci associated with Fusarium head blight resistance. BMC Genomics 17, 387. https://doi.org/10.1186/s12864-016-2716-0.

Jiang, N., Cui, J., Shi, Y., et al., 2019. Tomato lncRNA23468 functions as a competing endogenous RNA to modulate NBS-LRR genes by decoying miR482b in the tomato-Phytophthora infestans interaction. Hortic Res. 6, 28. https://doi.org/10.1038/s41438-018-0096-0.

Jin, J., Liu, J., Wang, H., Wong, L., Chua, N.H., 2013. PLncDB: plant long noncoding RNA database. Bioinformatics 29, 1068–1071.

Jones, J.D., Dangl, J.L., 2006. The plant immune system. Nature 444 (7117), 323–329.

Joshi, R.K., Megha, S., Basu, U., Rahman, M.H., Kav, N.N., 2016. Genome wide identification and functional prediction of long non-coding RNAs responsive to Sclerotinia sclerotiorum infection in Brassica napus. PLoS One 11 (7), 0158784. https://doi.org/10.1371/journal.pone.0158784.

Kalaji, H.M., Jajoo, A., Oukarroum, A., et al., 2016. Chlorophyll a fluorescence as a tool to monitor physiological status of plants under abiotic stress conditions. Acta Physiol Plant 38, 102. https://doi.org/10.1007/s11738-016-2113-y.

Kalantry, S., Purushothaman, S., Bowen, R.B., Starmer, J., Magnuson, T., 2009. Evidence of Xist RNA-independent initiation of mouse imprinted X-chromosome inactivation. Nature 460, 647–651.

Kashi, K., Henderson, L., Bonetti, A., Carninci, P., 2016. Discovery and functional analysis of lncRNAs: Methodologies to investigate an uncharacterized transcriptome. Biochim. Biophys. Acta 1859 (1), 3–15.

Khemka, N., Singh, V.K., Garg, R., Jain, M., 2016. Genome-wide analysis of long intergenic noncoding RNAs in chickpea and their potential role in flower development. Sci. Rep. 6, 33297.

Komiya, R., Ohyanagi, H., Niihama, M., Watanabe, T., Nakano, M., Kurata, N., Nonomura, K., 2014. Rice germline-specific Argonaute MEL1 protein binds to phasiRNAs generated from more than 700 lincRNAs. Plant J. 78, 385–397. https://doi.org/10.1111/tpj.12483.

Kornienko, A.E., Guenzl, P.M., Barlow, D.P., Pauler, F.M., 2013. Gene regulation by the act of long non-coding RNA transcription. BMC Biol. 11, 59.

Kwenda, S., Birch, P.R., Moleleki, L.N., 2016. Genome-wide identification of potato long intergenic noncoding RNAs responsive to Pectobacterium carotovorum subspecies brasiliense infection. BMC Genomics 17 (1), 614. https://doi.org/10.1186/s12864-016-2967-9.

Lai, F., Orom, U.A., Cesaroni, M., Beringer, M., Taatjes, D.J., Blobel, G.A., Shiekhattar, R., 2013. Activating RNAs associate with mediator to enhance chromatin architecture and transcription. Nature 494, 497–501.

Li, X., Liu, Y., Chen, M., et al., 2012. Relationships between ion and chlorophyll accumulation in seeds and adaptation to saline environments in Suaeda salsa populations. J. Giornale Botanico Italiano. 146 (Supp. 1), 142–149.

Li, L., Eichten, S.R., Shimizu, R., Petsch, K., Yeh, C.T., Wu, W., Chettoor, A.M., Givan, S.A., Cole, R.A., Fowler, J.E., et al., 2014a. Genome-wide discovery and characterization of maize long non-coding RNAs. Genome Biol. 15, R40.

Li, D.D., Liu, Z.C., Huang, L., Jiang, Q.L., Zhang, K., Qiao, H.L., Jiao, Z.J., Yao, L.G., Liu, R.Y., Kan, Y.C., 2014b. The expression analysis of silk gland-enriched intermediate-size non-coding RNAs in silkworm Bombyx mori. Insect Sci. 21 (4), 429–438. https://doi.org/10.1111/1744-7917.12063.

Li, R., Fu, D., Zhu, B., Luo, Y., Zhu, H., 2018a. CRISPR/Cas9-mediated mutagenesis of lncRNA1459 alters tomato fruit ripening. Plant J. 94, 513–524.

Li, X., Xing, X., Xu, S., Zhang, M., Wang, Y., Wu, H., et al., 2018b. Genome-wide identification and functional prediction of tobacco lncRNAs responsive to root-knot nematode stress. PLoS One 13 (11), e0204506.

Liu, J., Jung, C., Xu, J., Wang, H., Deng, S., Bernad, L., Arenas-Huertero, C., Chua, N.-H., 2012. Genome-wide analysis uncovers regulation of long intergenic noncoding RNAs in Arabidopsis. Plant Cell 24, 4333–4345.

Liu, J., Cai, H., Liu, Y., et al., 2015. A study on physiological characteristics and comparison of salt resistance of two Medicago sativa at the seedling stage. Acta Pratacul. Sin. 22 (3), 250–256.

Liu, S., Wang, W., Li, M., et al., 2017. Antioxidants and unsaturated fatty acids are involved in salt tolerance in peanut. J. Acta Physiol. Plantarum. 39 (9), 207.

Liu, Q., Liu, R., Ma, Y., et al., 2018. Physiological and molecular evidence for Naþ and Cl exclusion in the roots of two Suaeda salsa populations. J. Aquat. Bot. 146, 1–7.

Ma, J., Yan, B., Qu, Y., Qin, F., Yang, Y., Hao, X., Yu, J., Zhao, Q., Zhu, D., Ao, G., 2008. Zm401, a short-open reading-frame mRNA or noncoding RNA, is essential for tapetum and microspore development and can regulate the floret formation in maize. J. Cell. Biochem. 105, 136–146. https://doi.org/10.1002/jcb.21807.

Mercer, T.R., Dinger, M.E., Mattick, J.S., 2009. Long non-coding RNAs: insights into functions. J. Nat. Rev. Genet. 10 (3), 155–159.

Morris, K.V., Mattick, J.S., 2014. The rise of regulatory RNA. Nat. Rev. Genet. 15 (6), 423–437. https://doi.org/10.1038/nrg3722.

Park, Y., Hwang, J., Kim, K., et al., 2013. Development of the gene-based SCARs for the Ph-3 locus., which confers late blight resistance in tomato. J. Sci. Hortic. 164 (2), 9–16.

Pefanis, E., Wang, J., Rothschild, G., Lim, J., Kazadi, D., Sun, J., Federation, A., Chao, J., Elliott, O., Liu, Z.P., et al., 2015. RNA exosome-regulated long non-coding RNA transcription controls super-enhancer activity. Cell 161, 774–789.

Xing, Q., Zhang, W., Liu, M., Li, L., Li, X., Yan, J., 2019. Genome-wide identification of long non-coding RNAs responsive to infection in grapevine. Evol. Bioinform. 15, 1–10. https://doi.org/10.1177/1176934319841362.

Rinn, J.L., Kertesz, M., Wang, J.K., Squazzo, S.L., Xu, X., Brugmann, S.A., Goodnough, L.H., Helms, J.A., Farnham, P.J., Segal, E., Chang, H.Y., 2007. Functional demarcation of active and silent chromatin domains in human HOX loci by noncoding RNAs. Cell 129, 1311–1323. https://doi.org/10.1016/j.cell.2007.05.022.

Rohrig, H., Schmidt, J., Miklashevichs, E., Schell, J., John, M., 2002. Soybean ENOD40 encodes two peptides that bind to sucrose synthase. Proc. Natl. Acad. Sci. U. S. A. 99 (4), 1915–1920.

Ruiz, K.B., Biondi, S., Martinez, E.A., et al., 2016. Quinoa-a model crop for understanding salt-tolerance mechanisms in halophytes. Plant Biosyst. 150 (2), 357–371.

Satyavathi, V., Srividya, S., Ghosh, R., 2013. Identification of long noncoding RNAs (LncRNAs) involved in immune response during Baculoviral infection in Bombyx mori 2 Nd global conference on entomology. At Kuching., Sarawak., Malaysia.

Satyavathi, V., Ghosh, R., Subramanian, S., 2017. Long non-coding RNAs regulating immunity in insects. Noncoding RNA 3(1), E14. doi:https://doi.org/10.3390/ncrna3010014. Review. PubMed PMID: 29657286; PubMed Central PMCID: PMC5832008.

Schmutz, J., Cannon, S.B., Schlueter, J., Ma, J., Mitros, T., Nelson, W., Hyten, D.L., Song, Q., Thelen, J.J., Cheng, J., et al., 2010. Genome sequence of the palaeopolyploid soybean. Nature 463, 178–183.

Seo, J.S., Sun, H.X., Park, B.S., Huang, C.H., Yeh, S.D., Jung, C., Chua, N.H., 2017. ELF18-INDUCED LONG-NONCODING RNA associates with mediator to enhance expression of innate immune response genes in Arabidopsis. Plant Cell 29 (5), 1024–1038. https://doi.org/10.1105/tpc.16.00886.

Shaheen, H.L., Iqbal, M., Azeem, M., et al., 2016. K-priming positively modulates growth and nutrient status of salt-stressed cotton (*Gossypium hirsutum*) seedlings. J. Arch. Agron. Soil Sci. 62 (6), 759–768.

Shin, J.-H., Chekanova, J.A., 2014. Arabidopsis RRP6L1 and RRP6L2 function in FLOWERING LOCUS C silencing via regulation of antisense RNA synthesis. PLoS Genet. 4 (10), e1004612. https://doi.org/10.1371/journal.pgen.1004612.

Shippy, T.D., Ronshaugen, M., Cande, J., et al., 2008. Analysis of the *Tribolium* homeotic complex: Insights into mechanisms constraining insect Hox clusters. Dev. Genes Evol. 218, 127–139. https://doi.org/10.1007/s00427-008-0213-4.

Shoemaker, R.C., Schlueter, J., Doyle, J.J., 2006. Paleopolyploidy and gene duplication in soybean and other legumes. Curr. Opin. Plant Biol. 9, 104–109.

Shumayla, Sharma, S., Taneja, M., Tyagi, S., Singh, K., Upadhyay, S.K., 2017. Survey of high throughput RNA-seq data reveals potential roles for lncRNAs during development and stress response in bread wheat. Front. Plant Sci. 8, 1019.

Song, J., Zhou, J., Zhao, W., et al., 2016. Effects of salinity and nitrate on production and germination of dimorphic seeds applied both through the mother plant and exogenously during germination in Suaeda salsa. J. Plant Species Biol. 31 (1), 19–28.

Song, J., Shi, W., Liu, R., et al., 2017. The role of the seed coat in adaptation of dimorphic seeds of the euhalophyte Suaeda salsa to salinity. J. Plant Species Biol. 32 (2), 107–114.

Srivastava, A.K., Lu, Y., Zinta, G., Lang, Z., Zhu, J.K., 2018. UTR-dependent control of gene expression in plants. Trends Plant Sci. 23, 248–259.

Sui, N., Han, G., 2014. Salt-induced photoinhibition of PSII is alleviated in halophyte Thellungiella halophila, by increases of unsaturated fatty acids in membrane lipids. J. Acta Physiol. Plant. 36 (4), 983–992.

Sui, N., Yang, Z., Liu, M., Wang, B., 2015. Identification and transcriptomic profiling of genes involved in increasing sugar content during salt stress in sweet sorghum leaves. BMC Genomics 16, 534. https://doi.org/10.1186/s12864-015-1760-5.

Szczesniak, M.W., Rosikiewicz, W., Makałowska, I., 2016. CANTATAdb: a collection of plant long non-coding RNAs. Plant Cell Physiol. 57, e8.

Tao, L.I., Liu, R.J., Xin-Hua, H.E., et al., 2012. Enhancement of superoxide dismutase and catalase activities and salt tolerance of euhalophyte Suaeda salsa L. by Mycorrhizal fungus Glomus mosseae. J. Pedosphere 22 (2), 217–224.

Tavakkoli, E., Fatehi, F., Coventry, S., et al., 2011. Additive effects of Naþ and Cl- ions on barley growth under salinity stress. J. Exp. Bot. 62 (6), 2189–2203.

Thomas, R., et al., 2014. The noncoding RNA revolution—Trashing old rules to forge new ones Cech. Cell 157 (1), 77–94.

Tyagi, S., Sharma, A., Upadhyay, S.K., 2018. Role of next-generation RNA-Seq data in discovery and characterization of long non-coding RNA in plants. In: Çiftçi, Y.Ö. (Ed.), Next Generation Plant Breeding. IntechOpen, https://doi.org/10.5772/intechopen.72773.

Yu, T., Zhu, H., 2019. Long non-coding RNAs: rising regulators of plant reproductive development. Agronomy 9, 2. Article-number 53.

Umlauf, D., Goto, Y., Cao, R., Cerqueira, F., Wagschal, A., Zhang, Y., Feil, R., 2004. Imprinting along the Kcnq1 domain on mouse chromosome 7 involves repressive histone methylation and recruitment of Polycomb group complexes. Nat. Genet. 36, 1296–1300. https://doi.org/10.1038/ng1467.

Vu, T.S., Zhang, D.W., Xiao, W.H., et al., 2015. Mechanisms of combined effects of salt and alkaline stresses on seed germination and seedlings of *Melilotus Officinalis* (Fabaceae) in northeast of China. Pakistan J. Bot. 47 (5), 1603–1611.

Wang, K.C., Chang, H.Y., 2011. Molecular mechanisms of long noncoding RNAs. Mol. Cell 43, 904–914.

Wang, Y., Wang, L., Zou, Y., Chen, L., Cai, Z., Zhang, S., Zhao, F., Tian, Y., Jiang, Q., Ferguson, B.J., Gresshoff, P.M., Li, X., 2014. Soybean miR172c targets the repressive AP2 transcription factor NNC1 to activate ENOD40 expression and regulate nodule initiation. Plant Cell 26 (12), 4782–4801. https://doi.org/10.1105/tpc.114.131607.

Wang, H., Niu, Q.W., Wu, H.W., Liu, J., Ye, J., Yu, N., Chua, N.H., 2015a. Analysis of non-coding transcriptome in rice and maize uncovers roles of conserved lncRNAs associated with agriculture traits. Plant J. 84, 404–416.

Wang, F., Xu, Y.G., Wang, S., et al., 2015b. Salinity affects production and salt tolerance of dimorphic seeds of Suaeda salsa. J. Plant Physiol. Biochemist 95, 41–48.

Wang, J., Yu, W., Yang, Y., Li, X., Chen, T., Liu, T., Ma, N., Yang, X., Liu, R., Zhang, B., 2016. Genome-wide analysis of tomato long non-coding RNAs and identification as endogenous target mimic for microRNA in response to TYLCV infection. Sci. Rep. 5, 16946. https://doi.org/10.1038/srep16946. Erratum in: Sci Rep. 14 (6), 32828.

Wen, Q.L., Yu, L.J., Feng, Q.L., Fang, Q.W., Fang, J.F., Jun, W., Jin, Y.Z., Yang, X., Wei, G.Z., Jie, Y., 2018. Genome-wide identification and characterization of long non-coding RNAs responsive to Dickeya zeae in rice. RSC Adv. 8 (60), 34408–34417. https://doi.org/10.1039/C8RA04993A.

Wierzbicki, A.T., Haag, J.R., Pikaard, C.S., 2008. Noncoding transcription by RNA polymerase pol IVb/pol V mediates transcriptional silencing of overlapping and adjacent genes. Cell 135, 635–648. https://doi.org/10.1016/j.cell.2008.09.035.

Wu, H.J., Wang, Z.M., Wang, M., Wang, X.J., 2013. Widespread long noncoding RNAs as endogenous target mimics for microRNAs in plants. Plant Physiol. 161, 1875–1884.

Xu, Y., Liu, R., Sui, N., Shi, W., Wang, L., Tian, C., Song, J., 2016. Changes in endogenous hormones and seed-coat phenolics during seed storage of two Suaeda salsa populations. J. Bot. 64 (4), 325–332.

Yan, X., Niu, X., Tao, J., 2018a. Research advances on pathogen-induced plant innate immunity. Mol. Plant Breed. 16 (03), 821–831.

Yan, Q., Wang, L., Li, X., 2018b. GmBEHL1, a BES1/BZR1 family protein, negatively regulates soybean nodulation. Sci. Rep. 8, 7614. https://doi.org/10.1038/s41598-018-25910-x.

Yang, W.C., Katinakis, P., Hendriks, P., Smolders, A., de, V.F., Spee, J., van Kammen, A., Bisseling, T., Franssen, H., 1993. Characterization of GmENOD40, a gene showing novel patterns of cell-specific expression during soybean nodule development. Plant J. 3 (4), 573–585.

Yao, S., Chen, S., Zhao, J., et al., 2010. Effect of three salts on germination and seedling survival of dimorphic seeds of *Chenopodium album*. J. Bot. 88 (9), 821–828.

Gai Ying-Ping, Yuan Shuo-Shuo, Zhao Ya-Nan, Zhao Huai-Ning, Zhang Hua-Liang, Ji Xian-Ling, 2018. A novel LncRNA, MuLnc1, associated with environmental stress in mulberry (*Morus multicaulis*). Front. Plant Sci. 9, 669. doi:https://doi.org/10.3389/fpls.2018.00669, ISSN = 1664-462X

Zhang, Y.C., Chen, Y.Q., 2013. Long noncoding RNAs: new regulators in plant development. Biochem. Biophys. Res. Commun. 436, 111–114. https://doi.org/10.1016/j.bbrc.2013.05.086.

Zhang. S., Song, J., Wang, He., Feng, G., 2010. Effect of salinity on seed germination., ion content and photosynthesis of cotyledons in halophytes or xerophyte growing in Central Asia, J. Plant Ecol. 3 (4), 259–267.

Zhang, Y.C., Liao, J.Y., Li, Z.Y., et al., 2014. Genome-wide screening and functional analysis identify a large number of long noncoding RNAs involved in the sexual reproduction of rice. Genome Biol. 15 (12), 512. https://doi.org/10.1186/s13059-014-0512-1.

Zhang, T., Song, J., Fan, J., et al., 2015. Effects of saline-waterlogging and dryness/moist alternations on seed germination of halophyte and xerophyte. J. Plant Species Biol. 30 (3), 231–236.

Zhang, K., Chen, S., Yan, H., et al., 2018a. Effects of exogenous nitric oxide on physiology and antioxidant activity of tobacco seedlings under salt stress. J. Crops. 03, 123–128.

Zhang, L., Wang, M., Li, N., Wang, H., Qiu, P., Pei, L., Xu, Z., Wang, T., Gao, E., Liu, J., Li, S., Hu, Q., Miao, Y., Lindsey, K., Tu, L., Zhu, L., Zhang, X., 2018b. Long noncoding RNAs involve in resistance to Verticillium dahliae, a fungal disease in cotton. Plant Biotechnol. J. 16 (6), 1172–1185. https://doi.org/10.1111/pbi.12861.

Zhou, J.C., Fu, T.T., Sui, N., et al., 2016a. The role of salinity in seed maturation of the euhalophyte Suaeda salsa. J. Plant Biosyst. 150 (1), 83–90.

Zhou, Q.Z., Zhang, B., Yu, Q.Y., Zhang, Z., 2016b. BmncRNAdb: a comprehensive database of non-coding RNAs in the silkworm, *Bombyx mori*. BMC Bioinform. 17 (1), 370. https://doi.org/10.1186/s12859-016-1251-y.

Zhu, Y.L., Song, Q.J., Hyten, D.L., van Tassell, C.P., Matukumalli, L.K., Grimm, D.R., Hyatt, S.M., Fickus, E.W., Young, N.D., Cregan, P.B., 2003. Single nucleotide polymorphisms in soybean. Genetics 163, 1123–1134.

Zhu, Q.H., Stephen, S., Taylor, J., Helliwell, C.A., Wang, M.B., 2014. Long noncoding RNAs responsive to *Fusarium oxysporum* infection in *Arabidopsis thaliana*. New Phytol. 201, 574–584.

Chapter 10

Comprehending lncRNA-mediated gene regulation during abiotic stresses and reproductive development in legumes

Priyanka Dhakate, Heena Ambreen, Anirban Chakraborty, Baljinder Singh, Manisha Yadav, Manish Tiwari, Gourav Singh, Vimal Pandey, and Sabhyata Bhatia
National Institute of Plant Genome Research, New Delhi, India

10.1 Introduction

Legumes rank third in world crop production after grasses and oilseed family (Latef and Ahmad, 2015). They are an irreplaceable source of proteins from plant origin contributing toward human and livestock diet, especially in developing countries. In addition, the legume family is central in regulating the overall agriculture and environmental health as these crops interact symbiotically with rhizobia to allow fixation of atmospheric nitrogen and simultaneously augment realized crop potential (Graham and Vance, 2003; Young et al., 2003; Dita et al., 2006). Therefore, in present day, farming systems, especially smallholder farmers, often employ legumes in crop rotation or as an intercrop to ensure decreased weed and pest infestation and also to replenish soil nutrients (Sheaffer and Seguin, 2003; Matusso et al., 2014; Stagnari et al., 2017; Jensen et al., 2010). Logistic relevance of legumes in maintaining sustainable ecosystem globally has driven concerted efforts across plant scientists to better understand legume biology.

Regardless of the significance of legumes in agriculture systems worldwide, the present day understanding of molecular mechanisms affecting legume productivity and adaptability specifically during abiotic stresses remains severely limited (Sita et al., 2017; Nadeem et al., 2019). Biotechnological tools including advanced "omics" approaches have been directly implicated to decipher the molecular basis of stress responses and for improving the current knowledge of key molecular mechanisms regulating reproductive development and abiotic stress tolerance in legumes (Hasanuzzaman et al., 2013; Meena et al., 2017;

Long Noncoding RNAs in Plants: Roles in development and stress
https://doi.org/10.1016/B978-0-12-821452-7.00010-6

Iquebal et al., 2019). Over the past decade, genome-wide expression analysis has been carried out extensively in several legumes to identify master regulators and regulatory networks triggered in response to various endogenous and exogenous cues. Most of these studies filtered a major part of the genome transcribed into nonprotein-coding RNAs (ncRNAs) as "transcriptional noise" (Wang et al., 2017a,b; Wan et al., 2019; Ntini and Marsico, 2019). However, over the past decade, significance of these molecules such as long noncoding RNA (lncRNA) and microRNA (miRNA) has come into focus as they have been found to fine-tune gene expression.

LncRNAs are over 200 nucleotides noncoding RNA transcripts transcribed by RNA Polymerase II, with no discernible protein-coding potential (Liu et al., 2012; Jin et al., 2013; Wang et al., 2014; Wang and Chekanova, 2017). Although recent reports suggest that lncRNAs code for short peptides, their functional relevance remains invalidated (Lin et al., 2020). LncRNAs share structural similarities with mRNAs and also show varied subcellular localizations and differential cellular expression patterns, suggesting their diverse roles in plant development. These molecules regulate gene expression in *cis* or in *trans* modes via transcriptional, posttranscriptional, and epigenetic mechanisms (Bhat et al., 2016; Marchese et al., 2017). These modes of regulation have provided more diversity to lncRNA functions. The *trans*-acting lncRNAs regulate larger co-expression networks, while the *cis*-regulating lncRNAs have more localized effects. A complex interplay between transcription factors and lncRNAs that are involved in regulatory networks has been reported (Golicz et al., 2018; Yamaguchi and Abe, 2012). Promoters of lncRNAs have a distinct structure from protein-coding genes. However, many transcription factors have been reported to bind to lncRNA gene promoters and affect their expression (Li et al., 2015; Ariel et al., 2015; Golicz et al., 2018). Consequently, lncRNA loci have been found enriched in regions that are proximal to regions where master regulators of developmental processes are located (Golicz et al., 2018).

Recently, their relevance in regulating reproductive development pathways and abiotic processes has been validated in various plant species. However, our present day understanding about plant lncRNAs is still in infancy. Reports illuminating the significance of plant lncRNAs-mediated gene regulation have increased over the last decade. However, only limited studies have reported identification and characterization of lncRNAs from legumes. In the present chapter, we discuss studies that have illuminated functional relevance of lncRNAs in legumes during abiotic stress and reproductive development.

10.2 Identification and expression pattern of legume lncRNAs

Gene regulation mediated via noncoding RNAs (ncRNAs) such as miRNAs is a well-established process across both the plant and animal kingdoms. Over the last decade, pivotal role of a new class of ncRNAs, lncRNAs, has been demonstrated as regulatory elements modulating a variety of biological processes, directly and (or) indirectly in multiple species. Existence and significance of lncRNAs were first reported in primates with identification of polyadenylated brain-specific lncRNAs (Lukiw et al., 1992). Ever since their cardinal role in regulating gene expression was demonstrated, identification of lncRNAs from various organisms has been continuously reported. However, in comparison to animals wherein functional significance of lncRNAs has been extensively documented, the biogenesis and mechanism of action by plant lncRNAs still remain elusive (Bhatia et al., 2017; Wekesa et al., 2019). Nevertheless, a few studies have addressed the importance of lncRNAs across various model and crop plants.

In plants, lncRNAs show numerous structural features and are transcribed by different RNA polymerases. These elements have cardinal roles in a variety of biological processes including plant growth and adaptation (Amor et al., 2009), protein localization (Campalans et al., 2004; Zhu and Wang, 2012), phosphate homeostasis (Franco-Zorrilla et al., 2007; Wang et al., 2017a,b), and chromatin modification (Heo and Sung, 2011; He, 2013). A large number of these identified lncRNAs have displayed significant changes in their expression levels during reproductive development and under stress conditions across all studied plant species including tomato (Liu et al., 1997), rice (Wasaki et al., 2003; Ding et al., 2012a,b; Zhou et al., 2012), wheat (Shumayla et al., 2017), and *Arabidopsis* (Franco-Zorrilla et al., 2007; Swiezewski et al., 2009; Heo and Sung, 2011). Identification of novel lncRNAs has primarily employed genome-wide transcriptomics using RNA sequencing and DNA microarrays (Zhu and Wang, 2012; Tyagi et al., 2018). Advancement in techniques such as lncRNA microarray, RNA capture sequencing (RNA CaptureSeq), along with improved bioinformatics platforms and computational tools has eased the identification and functional characterization of lncRNAs. Nevertheless, only limited information is available to clearly illustrate molecular mechanisms they govern in plants, especially in legumes.

One of the first lncRNAs discovered in plants was from a legume. In *Glycine max*, *GmENOD40*, a soybean early nodulin gene was found to transcribe a lncRNA of 700 nucleotides governing nodule organogenesis (Yang et al., 1993). Shortly after, two homologs of *GmENOD40*, *MsENOD40*, and *MtENOD40*, were characterized from differentiating nodule primordia of *Medicago sativa* and *Medicago truncatula*, respectively (Crespi et al., 1994). Both of these transcripts, identified as "riboregulators" like *GmENOD40*, displayed stable secondary structures and were localized to cytoplasm of nodule primordia to regulate their nodule growth and differentiation (Crespi et al., 1994). An early nodulin gene, *MsENOD12A*, was found to be regulated by *ENOD40* to induce dedifferentiation and division of root cortical cells in *M. truncatula* (Charon et al., 1997). In legumes, existence of *ENOD40* has been identified from *G. max* (Yang et al., 1993), *Phaseolus vulgaris* (Papadopoulou et al., 1996), *Vigna radiata* (GenBank Accession No. AF061818), *Sesbania rostrata* (Corich et al., 1998), *Lotus japonicus* (Flemetakis et al., 2000), *M. truncatula* (Crespi et al., 1994),

M. sativa (Asad et al., 1994; Crespi et al., 1994), *Trifolium repens* (AJ000268), *Pisum sativum* (Matvienko et al., 1994), *Vicia sativa* (Vijn et al., 1995), and *Lupinus luteus* (GenBank Accession No. AF352375). *ENOD40* has also been reported from nonleguminous species such as rice (Kouchi et al., 1999). In rice, two homologs of *ENOD40*, *ObENOD40*, and *OsENOD40*, from *Oryza brachyantha* (wild genotype) and *Oryza sativa* (cultivated genotypes) have been reported (Kouchi et al., 1999). In addition, homologs of *ENOD40* have been reported from *Nicotiana tabacum* (Van de Sande et al., 1996), *O. sativa* (Kouchi et al., 1999), *O. brachyantha* (Kouchi et al., 1999), *Hordeum vulgare* (Larsen, 2003), and *Lolium perenne* (Larsen, 2003).

A comparative account of *ENOD40* transcripts from leguminous and nonleguminous species indicated presence of five conserved domains with an additional domain restricted to leguminous species which is involved in development of indeterminate nodules. Structure of three domains was found to be more structured in comparison to other two domains. Motifs such as U-rich internal loops and bulges were observed in all the analyzed transcripts, indicating regulatory relevance of *ENOD40* transcripts (Girard et al., 2003). In *Arachis hypogea*, a total of 6407 putative lncRNAs have been identified (Clevenger et al., 2016). However, no orthologs of *ENOD40* have been reported from both diploid and tetraploid peanut (Bertioli et al., 2016; Clevenger et al., 2016). In addition, a group of intergenic lncRNAs specifically expressed in peanut nodules were suggested to regulate nodule development (Clevenger et al., 2016).

The role of lncRNAs in axillary bud differentiation has been implicated in *P. sativum* wherein a total of 47,322 putative lncRNAs with a median length of 293 nt were identified from developing axillary buds (Kerr et al., 2017). Of these, 25% of the lncRNA candidates were found to harbor various types of repetitive sequences. Limited sets of pea lncRNAs, ~6% and ~3%, exhibited shared sequence homology with *Medicago* and *Arabidopsis*, respectively (Kerr et al., 2017). Further, differential gene expression analysis identified expression of 31 putative lncRNA candidates, specifically during axillary bud development. An improved genome assembly of *M. truncatula* A17 generated using PacBio sequencing approach discerned 1037 lncRNAs associated with nodule development, of which 35% and 38% showed upregulation in nodules or/and in the nodule differentiation zone, respectively (Pecrix et al., 2018). In addition, the genomic positions associated with identified lncRNA genes displayed variable methylation patterns and contained histone marks. These observations substantiated the dynamic involvement of epigenetic mechanisms to regulate lncRNA expression that, in turn, governs nodule development in legumes (Pecrix et al., 2018). Chao et al. (2019) employed single molecule sequencing technology to identify 17,740 lncRNA candidates from *M. sativa*. Further screening led to discernment of a total of 867 high confidence lncRNAs with longer mean length in comparison to lncRNAs identified by Illumina short-read sequencing.

Nithin et al. (2017) have reported a total of 3919 lncRNAs from *Cajanus cajan*. Of these, 87 lncRNAs were predicted as miRNA targets, suggesting existence of an extensive ncRNA-based regulatory network. Singh et al. (2017) identified and compared genomic features of 4815 and 4384 novel lncRNAs candidates from rice and chickpea, respectively. These candidates were not reported in previous studies from rice (Zhang et al., 2014; Yi et al., 2015; Wang et al., 2015; Paytuví et al., 2016; Szcześniak et al., 2016) and chickpea (Khemka et al., 2016). In chickpea, 55% of lncRNAs harbored single exon with an average of 2.13 exons per transcript. Median exon length of lncRNAs (301 bp) was found longer as compared to coding transcripts (155 bp; Singh et al., 2017). This can be attributed to the fact that less number of exons in lncRNAs results in increased exon length (Hao et al., 2015). In addition, chickpea lncRNAs and mRNAs were found to be AU- and GC-rich (Singh et al., 2017). Similar observations have been reported from other plant species (Zhang et al., 2014; Wang et al., 2015; Hao et al., 2015; Khemka et al., 2016; Lu et al., 2016; Singh et al., 2017). LncRNAs identified from various legumes, till date, are summarized in Table 10.1. Most lncRNAs identified from legumes are expected to share obvious similarities such as tissue-specific expression patterns with their other legume counterparts. However, an in-depth identification and validation of lncRNAs both from the model and crop legumes is warranted to better ascertain function and mechanism of lncRNAs in legumes.

10.3 Molecular mechanisms of lncRNA-mediated gene regulation

The direct mechanism of action of lncRNAs transcripts involves their association with chromatin modifying complexes, leading to substantial changes in chromatin that result into an altered gene expression. HOX transcript antisense RNA (HOTAIR)-mediated histone modulation is a classical example of such a regulation mechanism wherein its binding to the polycomb repressive complex 2 (PRC2) and LSD1 (in complex with CoREST/REST) results in recruitment of the entire complex to the chromatin, leading to changes in methylation at specific histone residues (Tsai et al., 2010). Additionally, several studies have indicated an indirect role of lncRNAs where these transcripts serve as precursors for biogenesis of small RNAs (Arikit et al., 2013; Ulitsky, 2018). Broadly, four models have been proposed to explain lncRNA-mediated regulation of plant development. The first model proposes that lncRNAs are transcribed by RNA polymerase II in response to exogenous cues such as drought, temperature, salinity, and light (Guttman et al., 2009). In *Arabidopsis*, constitutive expression of *DROUGHT-INDUCED LnCRNA (DRIR)* under normal environmental conditions did not affect the expression of any stress-responsive genes. However, under drought and salinity stresses, *DRIR* was found to closely regulate expression of genes which regulate abiotic stresses (Qin et al., 2017). Hence, it is suggested that *DRIR* regulates the activity of *FUT4* or NAC transcription factors to deploy adaptive strategies under abiotic stresses (Borah et al., 2018). The second mechanism employed by lncRNAs is to negatively regulate their downstream targets

TABLE 10.1 LncRNAs identified from various legume species.

S. No.	Name	Legume species	Biological function	Tissue specificity	References
1.	*GmENOD40*	*Glycine max*	Nodule organogenesis	Nodule	Yang et al. (1993)
2.	*MsENOD40*	*Medicago sativa*	Role in mitosis or protein synthesis	Nodule, leaf, and root meristematic cells	Asad et al. (1994)
3.	*MsENOD40* and *MtENOD40*	*Medicago sativa* and *Medicago truncatula*	Nodule growth and differentiation	Nodule primordia	Crespi et al. (1994)
4.	*pGmENOD40*	*Pisum sativum*	Nodule development	Early and mature nodule	Matvienko et al. (1994)
5.	*VsENOD5*, *VsENOD12*, and *VsENOD40*	*Vicia sativa*	Nodule development	Root epidermis and nodule primordial	Vijn et al. (1995)
6.	*ENOD40*	*Phaseolus vulgaris*	Nodule and lateral root development	Root pericycle and nodule primordium	Papadopoulou et al. (1996)
7.	*MsENOD12A*	*Medicago truncatula*	Dedifferentiation and division of root cortical cells	Root cortical cells	Charon et al. (1997)
8.	*SrENOD40-1*	*Sesbania rostrata*	Vascular bundle development and nodule development	Root primordia, leaf primordia, stipule primordia, nodule primordia, and vascular bundles	Corich et al. (1998)
9.	*LjENOD40-1* and *LjENOD40-2*	*Lotus japonicus*	Organogenesis	Stems, fully developed flowers, green seed pods, nodule vascular system, and hypocotyls	Flemetakis et al. (2000)

by serving as miRNA precursors (Hirsch et al., 2006). The role of lncRNAs as miRNA precursor first came into light with the identification of a *MIR162a* locus in the intron of a nonprotein-coding RNA-encoding gene (*npcRNA78*). The *npcRNA78* exhibited differential expression across various *Arabidopsis* tissues and was found to be the target of *DCL1* (Hirsch et al., 2006). A genome-wide analysis performed in *Arabidopsis* using its full-length cDNA resulted in identification of a total of 76 nonprotein-coding RNA transcripts. Two of the identified nonprotein-coding RNAs, viz., *npc83* and *npc521*, contained mature sequences of miR869a and miR160c, respectively (Amor et al., 2009). A similar study performed in *Populus tomentosa* led to identification of 126 differentially expressed lncRNAs under low-nitrogen nutrient conditions, of which at least 14 lncRNAs were predicted to process 25 different mature miRNAs (Chen et al., 2016). Studies carried out by Zhang et al. (2018) in *Brassica* revealed a large set of miRNAs that were derived from lncRNA precursors. Out of the total identified lncRNAs using whole genome re-sequencing (WGRS) data of *Brassica*, 0.74% represented precursors of 20 miRNAs from 15 different miRNA families in *Brassica napus* (Zhang et al., 2018). Similarly, in *Brassica oleracea*, 0.37% of the total identified lncRNAs were precursors of nine miRNAs, representing eight different families. Likewise, 1.15% constituted precursors of 19 miRNAs from 15 families in *Brassica rapa*. Moreover, few miRNAs with ubiquitous distribution in all higher plants, viz., miR156, miR166, miR172, miR393, were found to be derived from lncRNA precursors. Using the same cDNA database, Amor et al. (2009) reported siRNAs that were mapped to both the strands of 5′ npcRNAs, indicating that lncRNAs can also act as a precursor of small interfering RNAs (siRNA). While production of 24 nt heterochromatic small RNA (hcRNA) involves action of Dicer-like3 (*DCL3*) and RNA-dependent RNA polymerase 2 (*RDR2*) acting upon long lncRNA precursors, the generation of 21 nucleotides transacting short interfering RNA (tasiRNAs) depends upon the action of RNA-dependent RNA polymerase 6 (*RDR6*) and Dicer-like4 (*DCL4*) on small lncRNA fragments that have been processed by miRNAs (Vaucheret, 2006). Expression of a pair of genes in antisense direction results in formation of double-stranded transcripts that are further processed by *DCL2* and *DCL1* to generate natural antisense short interfering RNA (natsiRNAs, Borsani et al., 2005). Abundance of two nonprotein-coding RNA (npc34 and npc351) in *rdr2/6* mutants of *Arabidopsis* suggested that these npcRNAs are either substrates of *RDR* or are targets of RDR-dependent small RNAs (Amor et al., 2009).

In addition to serving as precursor molecules, lncRNAs utilize a "decoy" mechanism to negatively regulate gene transcription and accumulation of small RNAs (Wang and Chang, 2011). Target mimicry is a kind of mechanism identified in miRNA regulation in which a "decoy" RNA binds to its complementary target miRNA, thereby leading to accumulation of genuine targets of miRNA. These target mimics can either be endogenous (referred to as endogenous target mimics) or can be artificially engineered. The role of lncRNAs as target mimic was first studied in *Arabidopsis* where the presence of a motif in a nonprotein-encoding gene *INDUCED BY PHOSPHATE STARVATION 1* (*IPS1*) that was complementary to the phosphate starvation-induced miR399f was reported (Franco-Zorrilla et al., 2007). The overexpression of the gene led to sequestering of miRNA, thus resulting in accumulation of target *PHO2* transcripts. The decoy activity of the gene was reportedly due to the presence of a 3′ nucleotide bulge in its binding site (between the 10th and 11th positions at 3′ end) of miR399. Similar decoy mechanisms were found in *Zea mays* wherein 86 lncRNAs were predicted to function as decoys for 58 miRNAs (Fan et al., 2015). These decoy sites were found conserved across at least five monocot species and are, therefore, anticipated to spatiotemporally regulate miRNA function across these monocot species (Wu et al., 2013).

Using advanced computational strategies, Wu et al. (2013) identified noncoding genes that were target of at least 20 known miRNAs in *Arabidopsis* and rice. Transgenic *Arabidopsis* lines with overexpressing ath-eTM160-1 (an endogenous target mimic for miR160) had more serration and smaller leaves in comparison to the wild-type plants. This was apparently due to down regulation of miR160 with concomitant increase in its target mRNA transcripts, *auxin response factor16* (*ARF16*), *ARF17*, and *ARF10*. Likewise, overexpression of rice osa-eTM160-3 in *Arabidopsis* led to dwarf-sized phenotype along with severe developmental defects, serrated leaves, and an accelerated flowering time. However, constructs transformed with point mutations in the miRNA-binding site of osa-eTM160-3 did not manifest any expression (miR160 and its targets) or phenotypic changes in *Arabidopsis*, suggesting that eTM160 of *Arabidopsis* and rice are highly conserved as their miR160 counterparts (Wu et al., 2013).

The characteristics of lncRNA as target mimic can be employed as strategies for crop improvement, especially in cases where miRNAs negatively regulate traits of agronomic importance. This was evident from a genome-wide study carried out in Chinese cabbage which revealed 40 lncRNA transcripts that act as eTM for 36 different families and showed differential expression under heat stress. The targets of the eTM included miRNAs with well-documented role in plant growth and development in addition to various stress responses. For example, a heat-responsive lncRNA, TCONS_00048391 targeting bra-miR164a, showed high expression in the sensitive variety "GHA" upon heat stress for 12h, which ultimately led to up regulation of the target transcript (NAC1, Bra030820) in the variety. However, meager transcript levels of TCONS_00048391 in the tolerant variety "XK" resulted in subsequent enhancement of bra-miR164a transcripts (Wang et al., 2019).

The third molecular mechanism is to guide, bind, and position ribo-nucleoprotein complexes either in *-cis* or/and *-trans*. These mechanisms are difficult to be predicted based on lncRNA sequences. In *Arabidopsis thaliana*, ALTERNATIVE SPLICING COMPETITOR (*ASCO*)-RNA, a lncRNA, binds to Nuclear Speckles RNA-binding protein (NSRs) to alter their alternative splicing, resulting in compromised lateral root initiation (Bardou et al., 2014). Another example worked out in detail is from *Arabidopsis* wherein induction of *ELENA1* imparting immunity to the plant in response to pathogen attack was validated. The study employing overexpression and knockdown mutant *Arabidopsis* lines displayed that the *ELF18-INDUCED LONG-NONCODING RNA1 (ELENA1)*, a lncRNA, and *MED19a* subunit function independently to positively regulate expression of genes encoding *PATHOGENESIS-RELATED* (*PR1* and *PR2*), salicylic acid and β-1,3-glucanases. Therefore, this indicates that *ELENA1* influences its targets both in *-cis* and *-trans* (Seo et al., 2017).

The fourth type of lncRNA-mediated mechanism is the "scaffold" mechanism where lncRNA molecule serves as a platform to assemble required molecular players. In *A. thaliana, COLD-ASSISTED INTRONIC NONCODING RNA* (*COLDAIR*), a lncRNA, was found to recruit polycomb repressive complex 2 (*PRC2*) during the transition from vegetative to reproductive phase (Heo and Sung, 2011). These studies reflect upon the mechanism of action undertaken by lncRNAs, demonstrating their versatility as regulatory molecules during various biological processes. Fig. 10.1 provides the diagrammatic representation for various molecular mechanisms employed by lncRNAs to mediate gene regulation.

10.4 Functional relevance of lncRNAs in legumes during abiotic stress

The emerging ecological impacts of dynamic climate changes have limited crop production by up to 70% (Wallace et al., 2003; Bellard et al., 2012). It is difficult to assess the accurate yield losses incurred due to the changing environment. However, significant losses in crop production attributable to abiotic stresses exacerbated with the growing human population further constrained already limited environmental resources (Wallace et al., 2003; Pereira, 2016). Increasing competition for reducing environmental resources along with drastic climatic changes threatens global food security, and in turn, invigorates development of climate resilient crops (Wheeler and Von Braun, 2013; Pereira, 2016). Abiotic stresses that account for major yield losses in plants are drought, floods, salinity, heat, cold, and accumulation of heavy metals in soils. Plants have devised adaptive strategies to survive these stresses by generating suited molecular responses. Perception of stimulus via signal transduction pathways triggers downstream transactional machinery to modulate gene expression under various stress conditions to result in appropriate physiological responses (Pereira, 2016; Bhatia et al., 2020). Role of several transcription factors, miRNAs and lncRNAs, under various abiotic stresses is well-reported in plants. However, an extensive account of lncRNAs-regulating gene expression during abiotic stresses in legumes is not yet available. In the present section, studies from legumes wherein lncRNAs mediated strategies during abiotic stresses have been highlighted.

Water deficiency at any stage of plant development, especially during reproductive transition and grain filling, results in compromised crop productivity,

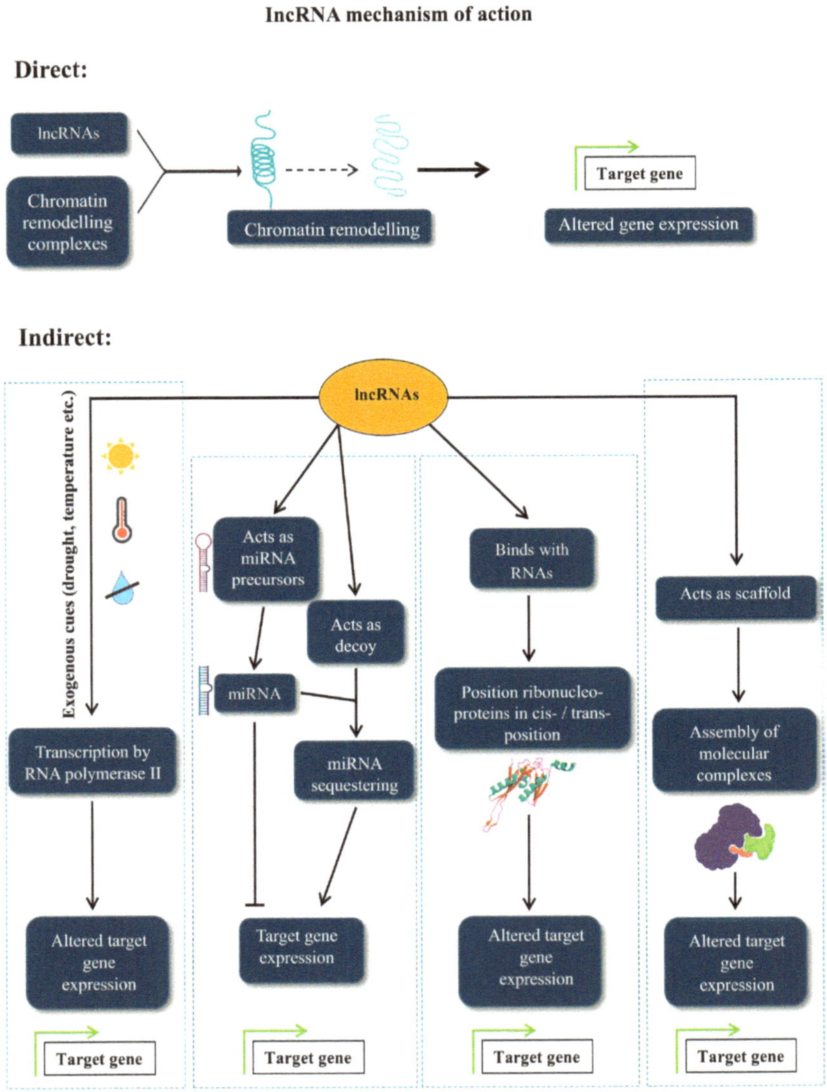

FIG. 10.1 Diagrammatic representation molecular mechanisms of lncRNA-mediated gene regulation.

plant biomass, and seed quality (Delmer, 2005; Pushpavalli et al., 2015; Nadeem et al., 2019). However, the extent of realized damage has invariably been found directly proportional to the duration and intensity of drought stress, genotypic variability, and plant developmental stage. Legumes display a range of sensitivity in response to the onset of drought, but realized yield has been found significantly decreased in all genotypes studied so far. Numerous studies assessing the effects of drought stress from the molecular to the whole plant level have attributed the reduction of crop potential to several significant factors such as decreased assimilate translocation and carbon fixation (Zlatev and Lidon, 2012), reduced germination and photosynthetic efficiency (Chowdhury et al., 2016), delayed flowering time (Samarah and Alqudah, 2011), reduced pods and seeds per pod (Croser et al., 2003; Liu et al., 2014), and sterile pollen grains (Sehgal et al., 2018). In addition, due to limited transpiration rate, nutrient deficiency often occurs concurrently with drought stress (Šurbanovski and Grant, 2014). Our understanding of the functional relevance of lncRNAs during such phases of water and nutrient deficiency in legumes still remains limited. Recently, expression of 136 lncRNAs was found elevated in response to drought stress in *P. vulgaris* (Pereira et al., 2019). Differential expression analysis of these genes revealed elevated expression of only one lncRNA in the leaf tissue and six lncRNAs in root tissues with high confidence (Pereira et al., 2019). In Lentil, a total of 8730 putative lncRNAs were predicted through genome-wide transcriptional analysis of root and leaf tissues under short- and long-term drought conditions of water deficiency (Morgil et al., 2019). Drought-responsive lncRNAs have been identified in various nonleguminous plants such as *Setaria italica* (Qi et al., 2013), *Z. mays* (Qiu et al., 2014), *Populus trichocarpa* (Shuai et al., 2014), *Manihot esculenta* (Zhang et al., 2014), and *Brassica juncea* (Bhatia et al., 2020).

Nutrient deficiency is another major abiotic stress that impacts plant growth and development (Bista et al., 2018). This stress can occur independently or can present itself as a side effect of drought or salinity stress (Atkinson and Urwin, 2012; Pandey et al., 2015; Bista et al., 2018). One of the major plant nutrients that becomes limited during phases of water deficiency is phosphorus (Wang et al., 2015). Phosphorus (P) is central to plant growth and development owing to its capacity to mediate regulation of numerous enzymatic reactions and signal transduction pathways (Chiou and Lin, 2011; Dai et al., 2016; Wang et al., 2017a,b). Although P is present in higher amount in soils, availability of inorganic phosphate (Pi) for uptake by plants is low in about 70% of the cultivated land (Plaxton and Tran, 2011; Li et al., 2016). Recently, novel lncRNAs induced in response to Pi deficiency were identified in *M. truncatula* (Wang et al., 2017a,b). In this study, roots and leaves from three-week old *M. truncatula* seedlings grown in Pi-sufficient medium (0.5 mM Pi) followed by Pi-deficient (1 μM Pi) medium displaying significant reduction in dry root biomass and P concentration in leaves and roots were employed. A total of 10,785 unique lncRNAs (4811 intergenic, 599 intronic, 161 sense, and 5214 antisense lncRNAs) distributed across 8 *Medicago* chromosomes were identified. Around 350 and 220 lncRNAs showed differential expression in leaves and roots, respectively, and were referred as *P*hosphate *D*eficiency-*I*nduced *L*ncRNAs (*PDILs*). In Pi-deficient soils, plants employ numerous genes such as *SUMO E3 ligase* (*SIZ1*), PHOSPHATE STARVATION RESPONSE 1 (*PHR1*), miR399, and *PHOSPHATE2* (*PHO2*) to regulate acquisition of available Pi from soils (Bari et al., 2006; Chiou and Lin, 2011). Briefly, in response to Pi deficiency, PHR1 is induced to positively regulate expression of miR399 that targets *PHO2* which encodes an enzyme that degrades Pi transporters. Therefore, miR399-mediated cleavage of *PHO2* results in increased Pi acquisition during Pi-deficient conditions (Bari et al., 2006; Huang et al., 2013; Park et al., 2014). Additionally, participation of lncRNAs in plants was implicated to regulate signal transduction of Pi with identification of two *Mt4*-like lncRNAs, *At4* and *IPS1*, in *Arabidopsis* (Bazin and Bailey-Serres, 2015). *At4* and *IPS1* were found to share a conserved motif that was partially complimentary to miR399. Therefore, both *At4* and *IPS2* function as target mimics and protect *PHO2* transcripts by competitively binding to miR399 (Shin et al., 2006; Franco-Zorrilla et al., 2007). While the functional characterization of *At4* and *IPS1* has been conducted by analyzing *Arabidopsis* overexpression lines and mutants (Shin et al., 2006; Franco-Zorrilla et al., 2007), the biological function of *Mt4* in Pi homeostasis still remains elusive in *M. truncatula*. However, Wang et al. (2017a,b) characterized function of a Pi deficiency-responsive lncRNA, *PDIL1*, paralog of *Mt4* and proposed a model illuminating the role of lncRNAs in regulating Pi-deficient signaling pathway. According to the suggested model, two molecular pathways mediated by lncRNAs were found to converge to constitute Pi-dependent signaling networks. In Pi-deficient soils, decreased expression of *PDIL2* and *PDIL3* results in accumulation of their target Pi transporter *Medtr1g074930*. Simultaneously, well-established *PHR1*-miR399-*PHO2* is central to regulating Pi acquisition in plants and is triggered wherein increased PHRI accumulation leads to increased expression of miR399 to mediate cleavage of *PHO2* transcripts, protecting Pi transporters to ensure optimal Pi acquisition. Thus, both these pathways work in coordination in response to Pi deficiency and enhance Pi acquisition in Pi-deficient soils. Contrastingly, the expression of *Mt4*-like lncRNAs, such as *PDIL1*, negatively regulated Pi acquisition. Furthermore, *PDIL1* can serve as a target mimic by binding competitively to miR399 as it shares a conserved motif that is identified by both PDIL1 and *miR399*, resulting in limited Pi acquisition. These prescribed molecular approaches in tandem allow plants to regulate various developmental and physiological processes at optimum Pi concentration.

Salinity stress is also one of the major abiotic stresses, especially in arid and semiarid regions, which causes severe reduction in average yield across all crops by up to 70% (Wang et al., 2003). The increased salt concentration from both disarranged agricultural practices and natural processes inhibits plant growth and development, thereby substantially affecting realized yield (Tester and Bacic, 2005). Salinity principally enforces an osmotic stress that is closely followed by ion toxicity. Therefore, high salt concentrations in the soil affect plant growth either by inhibiting water uptake by roots due to osmotic inhibition or by inducing specific toxic ion effect (Silva et al., 2008). Both osmotic and salt tolerance in plants are well-orchestrated complex traits that are regulated via numerous signaling pathways (Chen and Polle, 2010; Zhang and Shi, 2013). Our understanding of these regulatory pathways controlled by transcription factors from various gene families, such as NAC, DREB, MYB, bHLH, and WRKY, is well-documented across both the model and crop plants. However, information regarding regulation of these networks via noncoding RNAs, especially by lncRNAs, in legumes is inadequate. Wang et al. (2015) subjected leaves and roots from *M. truncatula* seedlings to 5 h of osmotic and salt stress followed by high-throughput sequencing, as after precisely 5 h of stress, onset of early stage stress-response was found to activate associated regulatory networks. Following data analysis, a total of 23,324 unique lncRNAs (10,426 intronic, 5794 intergenic, 3558 sense, and 3546 antisense) distributed evenly across 8 *Medicago* chromosomes were revealed (Wang et al., 2015). Interestingly, given the similar number of raw reads were obtained for both leaf and root samples, an increased number of both lncRNAs and mRNAs were found in roots in comparison to leaf samples. However, the number of lncRNAs with an increased expression was higher than mRNAs expression in leaves (Wang et al., 2015). These patterns of expression did not change following induction of osmotic and salt stress. Further, in leaves, accumulation of 7863 and 7361 lncRNAs was found to change in response to induced osmotic and salinity stress, respectively. However, in roots, 5561 and 7874 lncRNAs were found affected in presence of osmotic and salinity stress, respectively. Of these, 1783 and 2148 lncRNAs from leaves and roots, respectively, was found to be affected under both induced stresses. Given the possibility of fine-tuning of gene regulation mediated by lncRNA, interactive networks were analyzed to reveal complex interplay between lncRNAs and mRNAs near their vicinity (100 kb) during both stress conditions. For instance, in leaves, 3 lncRNAs were found to regulate 13 protein-coding transcripts that regulate energy synthesis, oxidation/reduction reaction, signal transduction, and transcription during salt stress. Further, in roots, lncRNA *TCONS_00046739* was found to regulate cytochrome P450 during salt stress. In leaves, lncRNAs, *TCONS_00100258* and *TCONS_00118328*, were found to target genes that putatively code transmembrane proteins under salt stress. Interestingly, another lncRNA, *TCONS_00097188*, was located upstream of coding region of a gene *Medtr6g006990* which is associated with GO terms like carbonate dehydratase activity and carbon utilization. As carbonic anhydrase is central to accumulation of CO_2 in the rubisco active site, it was, therefore, suggested that *TCONS_00097188* probably regulates photosynthesis by modulating expression of *Medtr6g006990* under abiotic stresses. In the same study, lncRNA *TCONS_00047650* was implicated to express from regulatory region of *Medtr3g069280* that encodes phosphatidylinositol-specific phospholipase C. *Medtr3g069280* belonging to GO terms GO:0004435 (Phosphatidylinositol phospholipase C activity) and GO:0007165 (Signal transduction) and was found upregulated in response to both osmotic and salt stresses. These studies implicate the role of other related lncRNAs in regulating plant growth and development under salt and osmotic stress in other legume species.

Symptoms of oxidative stress for plants are often indicated by increased accumulation of molecules known as reactive oxygen species (ROS). To defend against these molecules, plants employ numerous antioxidant enzymes. In

Medicago, a lncRNA, *TCONS_00116877*, was mapped approximately 4 Kbp upstream of the start site of *Medtr7g094600* (Wang et al., 2015). Interestingly, *Medtr7g094600* encoding for glutathione peroxidase (*POD*) was found up regulated under osmotic stress, implicating *TCONS_00116877*-mediated regulation of the gene during osmotic stress (Wang et al., 2015). Effect of salinity stress on plant growth is directed via ionic toxicity (Zhu, 2002). In *Medicago* roots, an elevated expression of *Medtr1g081900* coding for the Na^+/H^+ exchanger (*NHX*) gene was found under salt stress (Wang et al., 2015). This vacuolar antiporter protein is responsible for sequestration of Na^+ in vacuoles (Yamaguchi et al., 2005). A lncRNA *TCONS_00020253* was predicted to regulate expression of this vacuolar Na^+/H^+ antiporter, given its location upstream of the gene (Wang et al., 2015). Nevertheless, these networks only predicted the interaction between lncRNAs and protein-coding genes present in their vicinity; therefore, extensive validation of these regulatory interactions is warranted.

Among atmospheric abiotic stresses, extreme temperature stress such as both heat and cold stress limits crop productivity and incurs huge agricultural losses worldwide. Cold stress induces critical physiological changes including photosynthesis-related damages, chlorosis, unregulated apoptosis, decreased membrane fluidity, and wilting, which altogether ultimately compromise plant growth and development (Pareek et al., 2017). On the other hand, high temperature induces synthesis of detoxifying enzymes, transporters, and the genes encoding regulatory factors such as heat stress transcription factors (HSFs), which, in turn, regulate the expression of heat-responsive genes and phytohormones such as abscisic acid (ABA), salicylic acid (SA), and ethylene (Dobrá et al., 2015; Mickelbart et al., 2015). Development of crops tolerant to these extreme temperatures is, thus, necessary. Therefore, elucidating regulatory landscape of plants exposed to extreme temperature conditions is required to identify key molecular targets which can be genetically altered for development of tolerant lines. Heat stress-responsive lncRNAs have been reported in plants including *A. thaliana* (Wunderlich et al., 2014), *B. rapa* (Song et al., 2016; Wang et al., 2019), *Populus simonii* (Ci et al., 2015), *Triticum aestivum* (Shumayla et al., 2017), *Raphanus sativus* (Wang et al., 2019), *Cucumis sativus* (He et al., 2020), and *B. juncea* (Bhatia et al., 2020).

Emerging evidences regarding the dynamic expression of lncRNAs on exposure to cold conditions imply their strong role as regulatory components for cold response and acclimatization. In cold-stressed *Arabidopsis* plants, numerous lncRNAs, *COOLAIR*, *COLDAIR*, and *COLDWRAP*, work in coordination to regulate the expression of a major floral repressor gene, *FLOWERING LOCUS C* (*FLC*), thereby controlling flowering time. The antisense lncRNA *COOLAIR* on prolong exposure to cold mediates chromatin silencing of the FLC gene (Swiezewski et al., 2009; Liu et al., 2010). Further, the intron-derived *COLDAIR* lncRNA recruits the Polycomb Repressive Complex 2 (PRC2) to facilitate *FLC* silencing during vernalization (Heo and Sung, 2011). The lncRNA, *COLDWRAP*, transcribed from the repressed promoter of *FLC* reinforces *FLC* silencing (Kim and Sung, 2017). Together, these cold-triggered lncRNAs work in a coordinated environment for stable repression of the *FLC* gene. In a similar line of experiments to evaluate the role of lncRNAs during early exposure to cold in *Arabidopsis*, Kindgren et al. (2018) identified two lncRNAs *SVALKA* and its read-through lncRNA transcript, as *CBF1*, which in cooperation tightly regulate the expression of a transcription factor, C-repeat/dehydration-responsive element binding factors (CBFs), known to be involved in promoting cold tolerance in several plant species. Although constitutive expression of CBFs increases cold tolerance, it is accompanied by fitness penalties causing reduced biomass. Therefore, its expression is critically fine-tuned and repressed by the lncRNA *SVALKA-asCBF1* cascade, which triggers RNAPII (RNA polymerase II) collision at 3′ end of the *CBF1* gene.

Progressive advancement in next-generation sequencing and computational tools has paved the way for global identification and prediction of novel lncRNAs in spatiotemporal manner and/or under stress-induced conditions. RNA-sequencing data from cold and heat-exposed *B. rapa* plants discerned ~10,000 lncRNA, of which ~9687 were of novel origin. A total of 589 lncRNAs were expressed at three variant cold temperatures (−4°C, 0°C, and 4°C) and their target prediction analysis revealed 67 gene targets related to cold-responsive proteins including *CBF1*, *COLD REGULATED* (*COR6.6*), and *LATE EMBRYOGENESIS ABUNDANT* (*LEA14*; Song et al., 2016). Likewise, analysis of the ssRNA sequencing data from cassava shoot apices and young leaves led to identification of 682 high confidence lncRNAs expressed during cold, drought stress, and control conditions (Li et al., 2017). Further examination revealed ~198 cold-responsive differentially expressed cassava lncRNAs. Genome-wide analysis of transcriptome data resulted in identification of 12,462 lncRNAs in cold-stressed wild banana plants (Liu et al., 2018). A survey of lncRNA expression during cold conditions in grapevine resulted in identification of ~203 known and ~2088 novel cold-inducible lncRNA transcripts (Wang et al., 2019). Among the differentially expressed lncRNAs from grapevine, 242 lncRNAs were predicted to target 326 protein-coding genes in a *cis*-regulatory manner, while 62 lncRNAs targeted approximately hundred protein-coding genes in a *trans*-regulatory pattern. In entirety, the cold-responsive lncRNAs targets belonged to the class of stress-associated genes including late embryogenesis abundant protein genes, *CBF4* transcription factor genes, *WRKY* transcription factor genes, and peroxisome biogenesis protein genes. A few lncRNAs were characterized as miRNA precursors and their mimicking activity as miRNA targets was also studied in grapevine. Cold-stressed wheat plants at reproductive stage exhibited differential expression of ~31 lncRNAs, of which lncRNA Traes_2BS_7A04BF5D5 was predicted to target WCOR413, a cold acclimation gene that is responsible for dehydration protection and membrane cryoprotection, while lncRNA Traes_2DL_ABD08139B targeted flowering-promoting factor 1-like protein 3 via miRNA target mimic mechanism (Díaz et al., 2019). Recently, ~135 cold-responsive lncRNAs with differential expression and (or) exhibiting differential pattern of alternative splicing have been identified in *Arabidopsis*, highlighting the importance of alternative splicing in lncRNA processing (Calixto et al., 2019). The described case studies adequately reflect the importance of lncRNAs in regulation of critical plant growth and developmental processes during cold stress in various plant species. Despite continuous efforts of scientific community to understand the underlying regulatory mechanism governing cold stress-induced phenotype in legumes, to the best of our knowledge, not a single report is available wherein cold-responsive lncRNAs have been explored and studied in legume crops. Therefore, it is imperative that studies should be undertaken in order to understand the complex level of regulation imposed by lncRNAs during cold exposure in legume species.

10.5 Regulatory role of lncRNAs during legume reproductive development

The regulatory role of lncRNAs has been strongly implied during reproductive development in several plants including leguminous species. These noncoding classes of RNAs influence all the major stages of plant reproduction, including perception of initial flowering signals, floral organogenesis, seed development, and dormancy induction. An ordered and highly coordinated signaling pathway operating via well-characterized lncRNAs, *COLDAIR*, *COOLAIR*, and *COLDWRAP*, determines flowering time in *Arabidopsis*. A lncRNA, CsM10, has been reported to play putative role in male sex differentiation in cucumber (Cho et al., 2005). In maize, lncRNA *Zm401* has been identified as a key regulator for anther development. Nonfunctionality of this lncRNA influences expression profiles of critical genes of pollen development (*ZmMADS2*, *MZm3-3*, and *ZmC5*), causing aberrant microspore and tapetum development and ultimately leading to male sterility (Ma et al., 2008). An optimum expression of a long-day-specific male-fertility-associated lincRNA (*LDMAR*) was found essential for pollen development and regulation of photoperiod-sensitive male sterility (PSMS) under long-day conditions in rice (Ding et al., 2012a,b). Another rice lncRNA, osa-eTM160, in a target mimicry manner impaired expression of miRNA, osa-miR160, thus regulating seed setting and size in rice plants (Wang et al., 2017a,b). In sugar beet, three lncRNAs, *BvRAV1*-like, *BvFT1*, and *BvFT2*, were found to be associated with the process of vernalization, marking the transition from vegetative to reproductive growth (Liang et al., 2017). A RNA-seq analysis of pollen developmental stages identified 12,051 putative lncRNAs in *B. rapa*, of which 14 lncRNAs were found to be co-expressed with 10 pollen related genes (Huang et al., 2018). Recently, a genome-wide screening of rice ovule developmental stages revealed several lncRNAs with strong implication in ovule development orchestrated through their target mimicry activity or as precursors of miRNAs (Liu et al., 2019).

In legumes, limited studies have explored the potential regulatory functions of lncRNAs during reproductive development. A preliminary investigation of soybean embryos transcriptomic profiles recognized a large set of 13,652 lncRNAs, which are likely to demonstrate regulatory role in seed maturation and desiccation-related biological processes in soybean (Aghamirzaie et al., 2015). Utilizing RNA-seq data from eight developmental stages of flower and three vegetative tissues, Khemka et al. (2016) recognized 2248 putative lncRNAs in chickpea. These predicted lncRNAs displayed variable expression profiles among all the studied tissues with few lncRNAs (~29%) exhibiting tissue specificity for their expression. Intriguingly, higher lncRNA expression was observed in tissues with actively dividing cells and flower developmental stages. Around 13% of the total lncRNAs showed transcript accumulation that was specific to floral development. The lncRNAs specific to flower bud were enriched in GO terms related to developmental processes, floral whorl development, and homeostasis processes, while the lncRNAs with expression restricted to flower development stages showed an enriched representation of GO terms, including external stimuli, chemical homeostasis, developmental process, and different signaling pathways. Additionally, a total of 144 lncRNAs were distinguished as probable targets of 236 miRNAs, establishing the target mimicry nature of lncRNAs during reproductive development in chickpea (Khemka et al., 2016). Recently, a systematic survey of pigeon pea seed and pod developmental stages discerned 3019 lncRNAs through transcriptome sequencing. A large set of 3768 mRNA were predicted to be targeted by putative lncRNAs. Around 302 lncRNAs formed endogenous target mimics (eTMs) and regulated miRNAs, which, in turn, controlled expression of important genes related to development of seed and pod (Das et al., 2019). This study provided critical insights about the regulatory mechanism operating during seed and pod development in pigeon pea. The above-described studies demonstrate the crucial biological functions undertaken by lncRNAs during various aspects of reproductive development in multiple plant species. Albeit, the real potential of these regulatory molecules during plant reproduction is still elusive and requires an immediate and extensive investigation for holistic understanding of vital processes.

10.6 Future directions and conclusions

It is not surprising that majority of plant lncRNAs are expressed at low levels as compared to mRNAs as their expression is cell-specific and (or) is induced in response to specific endogenous and (or) exogenous cues (Liu et al., 2012). Given increasing relevance of lncRNAs in regulating significant signal transduction pathways that are central in determining reproductive and adaptive success of plants, an in-depth analysis and validation of these transcripts is further required. While in most plant-based studies lncRNA candidates have been identified and validated, it is rather unfortunate that only limited information about this aspect is available in legumes, especially crop legumes.

Representation of plant lncRNAs majorly comes from model species, given the ease of use of the genetic resources and biological information associated with these plants. Extrapolation of this information to other crop systems, especially legumes, is rather difficult as lncRNAs share only limited homology. Our focus should also be on characterization of lncRNAs that display cell-specific expression in absence and (or) presence of exogenous and (or) endogenous cues. In vitro fractionation to enrich specific cell types coupled with single cell transcriptomic analysis performed to suitable depth promises comprehensive identification of most cell-specific lncRNAs from legumes during reproductive biology and various stresses as well (Casson et al., 2005; Wang and Chang, 2011). Following identification of putative candidates, the information should be easily accessible to augment lncRNA research in legumes. We believe that a comprehensive identification of legume-specific lncRNAs driven by advanced technical platforms is necessary to unravel new paradigms in legume biology.

References

Aghamirzaie, D., Batra, D., Heath, L.S., Schneider, A., Grene, R., Collakova, E., 2015. Transcriptome-wide functional characterization reveals novel relationships among differentially expressed transcripts in developing soybean embryos. BMC Genomics 16 (1), 928.

Amor, B.B., Wirth, S., Merchan, F., Laporte, P., d'Aubenton-Carafa, Y., Hirsch, J., Maizel, A., Mallory, A., Lucas, A., Deragon, J.M., Vaucheret, H., 2009. Novel long non-protein coding RNAs involved in Arabidopsis differentiation and stress responses. Genome Res. 19 (1), 57–69.

Ariel, F., Romero-Barrios, N., Jégu, T., Benhamed, M., Crespi, M., 2015. Battles and hijacks: noncoding transcription in plants. Trends Plant Sci. 20 (6), 362–371.

Arikit, S., Zhai, J., Meyers, B.C., 2013. Biogenesis and function of rice small RNAs from noncoding RNA precursors. Curr. Opin. Plant Biol. 16 (2), 170–179.

Asad, S., Fang, Y., Wycoff, K.L., Hirsch, A.M., 1994. Isolation and characterization of cDNA and genomic clones of MsENOD40, transcripts are detected in meristematic cells of alfalfa. Protoplasma 183 (1–4), 10–23.

Atkinson, N.J., Urwin, P.E., 2012. The interaction of plant biotic and abiotic stresses: from genes to the field. J. Exp. Bot. 63 (10), 3523–3543.

Bardou, F., Ariel, F., Simpson, C.G., Romero-Barrios, N., Laporte, P., Balzergue, S., Brown, J.W.S., Crespi, M., 2014. Long noncoding RNA modulates alternative splicing regulators in Arabidopsis. Dev. Cell 30, 166–176.

Bari, R., Pant, B.D., Stitt, M., Scheible, W.R., 2006. PHO2, microRNA399, and PHR1 define a phosphate-signaling pathway in plants. Plant Physiol. 141 (3), 988–999.

Bazin, J., Bailey-Serres, J., 2015. Emerging roles of long non-coding RNA in root developmental plasticity and regulation of phosphate homeostasis. Front. Plant Sci. 6, 400.

Bellard, C., Bertelsmeier, C., Leadley, P., Thuiller, W., Courchamp, F., 2012. Impacts of climate change on the future of biodiversity. Ecol. Lett. 15 (4), 365–377.

Bertioli, D.J., Cannon, S.B., Froenicke, L., Huang, G., Farmer, A.D., Cannon, E.K., Liu, X., Gao, D., Clevenger, J., Dash, S., Ren, L., 2016. The genome sequences of Arachis duranensis and Arachis ipaensis, the diploid ancestors of cultivated peanut. Nat. Genet. 48 (4), 438.

Bhat, S.A., Ahmad, S.M., Mumtaz, P.T., Malik, A.A., Dar, M.A., Urwat, U., Ganai, N.A., 2016. Long non-coding RNAs: mechanism of action and functional utility. Non-coding RNA Res. 1 (1), 43–50.

Bhatia, G., Goyal, N., Sharma, S., Upadhyay, S.K., Singh, K., 2017. Present scenario of long noncoding RNAs in plants. Non-Coding RNA 3 (2), 16.

Bhatia, G., Singh, A., Verma, D., Sharma, S., Singh, K., 2020. Genome-wide investigation of regulatory roles of lncRNAs in response to heat and drought stress in *Brassica juncea* (Indian mustard). Environ. Exp. Bot. 171, 103922.

Bista, D.R., Heckathorn, S.A., Jayawardena, D.M., Mishra, S., Boldt, J.K., 2018. Effects of drought on nutrient uptake and the levels of nutrient-uptake proteins in roots of drought-sensitive and-tolerant grasses. Plan. Theory 7 (2), 28.

Borah, P., Das, A., Milner, M.J., Ali, A., Bentley, A.R., Pandey, R., 2018. Long non-coding RNAs as endogenous target mimics and exploration of their role in low nutrient stress tolerance in plants. Genes 9 (9), 459.

Borsani, O., Zhu, J., Verslues, P.E., Sunkar, R., Zhu, J.K., 2005. Endogenous siRNAs derived from a pair of natural cis-antisense transcripts regulate salt tolerance in Arabidopsis. Cell 123 (7), 1279–1291.

Calixto, C.P.G., Tzioutziou, N.A., James, A.B., Hornyik, C., Guo, W., Zhang, R., Brown, J.W.S., 2019. Cold-dependent expression and alternative splicing of Arabidopsis long non-coding RNAs. Front. Plant Sci. 10, 235.

Campalans, A., Kondorosi, A., Crespi, M., 2004. Enod40, a short open reading frame-containing mRNA, induces cytoplasmic localization of a nuclear RNA binding protein in Medicago truncatula. Plant Cell 16 (4), 1047–1059.

Casson, S., Spencer, M., Walker, K., Lindsey, K., 2005. Laser capture microdissection for the analysis of gene expression during embryogenesis of *Arabidopsis*. Plant J. 42, 111–123.

Chao, Y., Yuan, J., Guo, T., Xu, L., Mu, Z., Han, L., 2019. Analysis of transcripts and splice isoforms in *Medicago sativa* L. by single-molecule long-read sequencing. Plant Mol. Biol. 99 (3), 219–235.

Charon, C., Johansson, C., Kondorosi, E., Kondorosi, A., Crespi, M., 1997. enod40 induces dedifferentiation and division of root cortical cells in legumes. Proc. Natl. Acad. Sci 94 (16), 8901–8906.

Chen, S., Polle, A., 2010. Salinity tolerance of Populus. Plant Biol. 12 (2), 317–333.

Chen, M., Wang, C., Bao, H., Chen, H., Wang, Y., 2016. Genome-wide identification and characterization of novel lncRNAs in Populus under nitrogen deficiency. Mol. Genet. Genomics 291 (4), 1663–1680.

Chiou, T.J., Lin, S.I., 2011. Signaling network in sensing phosphate availability in plants. Annu. Rev. Plant Biol. 62, 185–206.

Cho, J., Koo, D.H., Nam, Y.W., Han, C.T., Lim, H.T., Bang, J.W., Hur, Y., 2005. Isolation and characterization of cDNA clones expressed under male sex expression conditions in a monoecious cucumber plant (*Cucumis sativus* L. cv. Winter Long). Euphytica 146 (3), 271–281.

Chowdhury, J.A., Karim, M.A., Khaliq, Q.A., Ahmed, A.U., Khan, M.S.A., 2016. Effect of drought stress on gas exchange characteristics of four soybean genotypes. Bangl. J. Agric. Res. 41 (2), 195–205.

Ci, D., Song, Y.P., Tian, M., Zhang, D.Q., 2015. Methylation of miRNA genes in the response to temperature stress in *Populus simonii*. Front. Plant Sci. 6, 921.

Clevenger, J., Chu, Y., Scheffler, B., Ozias-Akins, P., 2016. A developmental transcriptome map for allotetraploid Arachis hypogaea. Front. Plant Sci. 7, 1446.

Corich, V., Goormachtig, S., Lievens, S., Van Montagu, M., Holsters, M., 1998. Patterns of ENOD40 gene expression in stem-borne nodules of *Sesbania rostrata*. Plant Mol. Biol. 37 (1), 67–76.

Crespi, M.D., Jurkevitch, E., Poiret, M., d'Aubenton-Carafa, Y., Petrovics, G., Kondorosi, E., Kondorosi, A., 1994. *enod40*, a gene expressed during nodule organogenesis, codes for a non-translatable RNA involved in plant growth. EMBO J. 13, 5099.

Croser, J.S., Clarke, H.J., Siddique, K.H.M., Khan, T.N., 2003. Low-temperature stress: implications for chickpea (*Cicer arietinum* L.) improvement. Crit. Rev. Plant Sci. 22 (2), 185–219.

Dai, X., Wang, Y., Zhang, W.H., 2016. OsWRKY74, a WRKY transcription factor, modulates tolerance to phosphate starvation in rice. J. Exp. Bot. 67 (3), 947–960.

Das, A., Nigam, D., Junaid, A., Tribhuvan, K.U., Kumar, K., Durgesh, K., et al., 2019. Expressivity of the key genes associated with seed and pod development is highly regulated via lncRNAs and miRNAs in Pigeon pea. Sci. Rep. 9 (1), 1–14.

Delmer, D.P., 2005. Agriculture in the developing world: connecting innovations in plant research to downstream applications. Proc. Natl. Acad. Sci. 102 (44), 15739–15746.

Díaz, M.L., Soresi, D.S., Basualdo, J., Cuppari, S.J., Carrera, A., 2019. Transcriptomic response of durum wheat to cold stress at reproductive stage. Mol. Biol. Rep. 46 (2), 2427–2445.

Ding, J., Lu, Q., Ouyang, Y., Mao, H., Zhang, P., Yao, J., Xu, C., Li, X., Xiao, J., Zhang, Q., 2012a. A long noncoding RNA regulates photoperiod-sensitive male sterility, an essential component of hybrid rice. Proc. Natl. Acad. Sci. U. S. A. 109, 2654–2659.

Dita, M.A., Rispail, N., Prats, E., Rubiales, D., Singh, K.B., 2006. Biotechnology approaches to overcome biotic and abiotic stress constraints in legumes. Euphytica 147 (1–2), 1–24.

Ding, J., Shen, J., Mao, H., Xie, W., Li, X., Zhang, Q., 2012b. RNA-directed DNA methylation is involved in regulating photoperiod-sensitive male sterility in rice. Mol. Plant 5 (6), 1210–1216.

Dobrá, J., Černý, M., Štorchová, H., Dobrev, P., Skalák, J., Jedelský, P.L., Vanek, T., 2015. The impact of heat stress targeting on the hormonal and transcriptomic response in *Arabidopsis*. Plant Sci. 231, 52–61.

Fan, C., Hao, Z., Yan, J., Li, G., 2015. Genome-wide identification and functional analysis of lincRNAs acting as miRNA targets or decoys in maize. BMC Genomics 16, 793.

Flemetakis, E., Kavroulakis, N., Quaedvlieg, N.E., Spaink, H.P., Dimou, M., Roussis, A., Katinakis, P., 2000. *Lotus japonicus* contains two distinct ENOD40 genes that are expressed in symbiotic, nonsymbiotic, and embryonic tissues. Mol. Plant-Microbe Interact. 13 (9), 987–994.

Franco-Zorrilla, J.M., Valli, A., Todesco, M., Mateos, I., Puga, M.I., Rubio-Somoza, I., Leyva, A., Weigel, D., García, J.A., Paz-Ares, J., 2007. Target mimicry provides a new mechanism for regulation of microRNA activity. Nat. Genet. 39, 1033–1037.

Girard, G., Roussis, A., Gultyaev, A.P., Pleij, C.W., Spaink, H.P., 2003. Structural motifs in the RNA encoded by the early nodulation gene enod40 of soybean. Nucleic Acids Res. 31 (17), 5003–5015.

Golicz, A.A., Bhalla, P.L., Singh, M.B., 2018. LncRNAs in plant and animal sexual reproduction. Trends Plant Sci. 23 (3), 195–205.

Graham, P.H., Vance, C.P., 2003. Legumes: importance and constraints to greater use. Plant Physiol. 131 (3), 872–877.

Guttman, M., Amit, I., Garber, M., French, C., Lin, M.F., Feldser, D., Huarte, M., Zuk, O., Carey, B.W., Cassady, J.P., et al., 2009. Chromatin signature reveals over a thousand highly conserved large non-coding RNAs in mammals. Nature 458, 223–227.

Hao, Z., Fan, C., Cheng, T., Su, Y., Wei, Q., Li, G., 2015. Genome-wide identification, characterization and evolutionary analysis of long intergenic noncoding RNAs in cucumber. PLoS One 10, e0121800.

Hasanuzzaman, M., Nahar, K., Alam, M., Roychowdhury, R., Fujita, M., 2013. Physiological, biochemical, and molecular mechanisms of heat stress tolerance in plants. Int. J. Mol. Sci. 14 (5), 9643–9684.

He, Y., 2013. Noncoding RNA-mediated chromatin silencing (RmCS) in plants. Mol. Biol. 2, e106.

He, X., Guo, S., Wang, Y., Wang, L., Shu, S., Sun, J., 2020. Systematic identification and analysis of heat-stress-responsive lncRNAs, circRNAs and miRNAs with associated co-expression and ceRNA networks in cucumber (*Cucumis sativus* L.). Physiol. Plant. 168 (3), 736–754.

Heo, J.B., Sung, S., 2011. Vernalization-mediated epigenetic silencing by a long intronic noncoding RNA. Science 331 (6013), 76–79.

Hirsch, J., Lefort, V., Vankersschaver, M., Boualem, A., Lucas, A., Thermes, C., d'Aubenton-Carafa, Y., Crespi, M., 2006. Characterization of 43 non-protein-coding mRNA genes in Arabidopsis, including the MIR162a-derived transcripts. Plant Physiol. 140 (4), 1192–1204.

Huang, T.K., Han, C.L., Lin, S.I., Chen, Y.J., Tsai, Y.C., Chen, Y.R., Chen, Y.S., 2013. Identification of downstream components of ubiquitin-conjugating enzyme PHOSPHATE2 by quantitative membrane proteomics in Arabidopsis roots. Plant Cell 25 (10), 4044–4060.

Huang, L., Dong, H., Zhou, D., Li, M., Liu, Y., Zhang, F., et al., 2018. Systematic identification of long non-coding RNA s during pollen development and fertilization in *B. rapa*. Plant J. 96 (1), 203–222.

Iquebal, M.A., Sharma, P., Jasrotia, R.S., Jaiswal, S., Kaur, A., Saroha, M., 2019. RNAseq analysis reveals drought-responsive molecular pathways with candidate genes and putative molecular markers in root tissue of wheat. Sci. Rep. 9 (1), 1–18.

Jensen, E.S., Peoples, M.B., Hauggaard-Nielsen, H., 2010. Faba bean in cropping systems. Field Crops Res. 115, 203–216.

Jin, J., Liu, J., Wang, H., Wong, L., Chua, N.-H., 2013. PLncDB: plant long non-coding RNA database. Bioinformatics 29 (8), 1068–1071.

Kerr, S.C., Gaiti, F., Beveridge, C.A., Tanurdzic, M., 2017. *De novo* transcriptome assembly reveals high transcriptional complexity in *Pisum sativum* axillary buds and shows rapid changes in expression of diurnally regulated genes. BMC Genomics 18 (1), 221.

Khemka, N., Singh, V.K., Garg, R., Jain, M., 2016. Genome-wide analysis of long intergenic noncoding RNAs in chickpea and their potential role in flower development. Sci. Rep. 6, 33297.

Kim, D.H., Sung, S., 2017. Vernalization-triggered intragenic chromatin loop formation by long noncoding RNAs. Dev. Cell 40 (3), 302–312.

Kindgren, P., Ard, R., Ivanov, M., Marquardt, S., 2018. Transcriptional read-through of the long non-coding RNA SVALKA governs plant cold acclimation. Nat. Commun. 9 (1), 1–11.

Kouchi, H., Takane, K.I., So, R.B., Ladha, J.K., Reddy, P.M., 1999. Rice ENOD40: isolation and expression analysis in rice and transgenic soybean root nodules. Plant J. 18 (2), 121–129.

Larsen, K., 2003. Molecular cloning and characterization of a cDNA encoding a ryegrass (*Lolium perenne*) *ENOD40* homologue. J. Plant Physiol. 160 (6), 675–687.

Latef, A.A.H.A., Ahmad, P., 2015. Legumes and breeding under abiotic stress: an overview. In: Legumes Under Environmental Stress: Yield, Improvement and Adaptations. John Willey & Sons Ltd., USA, p. 315.

Li, Z.F., Zhang, Y.C., Chen, Y.Q., 2015. miRNAs and lncRNAs in reproductive development. Plant Sci. 238, 46–52.

Li, N., Wang, J., Song, W.Y., 2016. Arsenic uptake and translocation in plants. Plant Cell Physiol. 57 (1), 4–13.

Li, S., Yu, X., Lei, N., Cheng, Z., Zhao, P., He, Y., Peng, M., 2017. Genome-wide identification and functional prediction of cold and/or drought-responsive lncRNAs in cassava. Sci. Rep. 7, 45981.

Liang, N., Cheng, D., Cui, J., Dai, C., Luo, C., Liu, T., Li, J., 2017. Vernalisation mediated LncRNA-like gene expression in *Beta vulgaris*. Funct. Plant Biol. 44 (7), 720–726.

Lin, X., Lin, W., Ku, Y.S., Wong, F.L., Li, M.W., Lam, H.M., Ngai, S.M., Chan, T.F., 2020. Analysis of soybean long non-coding RNAs reveals a subset of small peptide-coding transcripts. Plant Physiol. https://doi.org/10.1104/pp.19.01324.

Liu, C., Muchhal, U.S., Raghothama, K.G., 1997. Differential expression of *TPS11*, a phosphate starvation-induced gene in tomato. Plant Mol. Biol. 33, 867–874.

Liu, F., Marquardt, S., Lister, C., Swiezewski, S., Dean, C., 2010. Targeted 3′ processing of antisense transcripts triggers Arabidopsis FLC chromatin silencing. Science 327 (5961), 94–97.

Liu, J., Jung, C., Xu, J., Wang, H., Deng, S., Bernad, L., Arenas-Huertero, C., Chua, N.-H., 2012. Genome-wide analysis uncovers regulation of long intergenic noncoding RNAs in Arabidopsis. Plant Cell 24 (11), 4333–4345.

Liu, F., Jensen, C.R., Andersen, M.N., 2014. Drought stress effect on carbohydrate concentration in soybean leaves and pods during early reproductive development: its implication in altering pod set. Field Crops Res. 86 (1), 1–13.

Liu, W., Cheng, C., Lin, Y., XuHan, X., Lai, Z., 2018. Genome-wide identification and characterization of mRNAs and lncRNAs involved in cold stress in the wild banana (*Musa itinerans*). PLoS ONE 13 (7), e0200002.

Liu, H., Wang, R., Mao, B., Zhao, B., Wang, J., 2019. Identification of lncRNAs involved in rice ovule development and female gametophyte abortion by genome-wide screening and functional analysis. BMC Genomics 20 (1), 90.

Lu, X., Chen, X., Mu, M., Wang, J., Wang, X., Wang, D., Yin, Z., Fan, W., Wang, S., Guo, L., et al., 2016. Genome-wide analysis of long non-coding RNAs and their responses to drought stress in cotton (*Gossypium hirsutum* L.). PLoS One 11, e0156723.

Lukiw, W.J., Handley, P., Wong, L., McLachlan, D.C., 1992. BC200 RNA in normal human neocortex, non-Alzheimer dementia (NAD), and senile dementia of the Alzheimer type (AD). Neurochem. Res. 17 (6), 591–597.

Ma, J., Yan, B., Qu, Y., Qin, F., Yang, Y., Hao, X., Yu, J., Zhao, Q., Zhu, D., Ao, G., 2008. Zm401, a short-open reading-frame mRNA or noncoding RNA, is essential for tapetum and microspore development and can regulate the floret formation in maize. J. Cell. Biochem. 105 (1), 136–146.

Marchese, F.P., Raimondi, I., Huarte, M., 2017. The multidimensional mechanisms of long noncoding RNA function. Genome Biol. 18 (1), 206.

Matusso, J.M.M., Mugwe, J.N., Mucheru-Muna, M., 2014. Potential role of cereal-legume intercropping systems in integrated soil fertility management in smallholder farming systems of Sub-Saharan Africa. Res. J. Agric. Environ. Manag. 3 (3), 162–174.

Matvienko, M., Van De Sande, K., Yang, W.C., Van Kammen, A., Bisseling, T., Franssen, H., 1994. Comparison of soybean and pea *ENOD40* cDNA clones representing genes expressed during both early and late stages of nodule development. Plant Mol. Biol. 26 (1), 487–493.

Meena, K.K., Sorty, A.M., Bitla, U.M., Choudhary, K., Gupta, P., Pareek, A., Singh, H.B., 2017. Abiotic stress responses and microbe-mediated mitigation in plants: the omics strategies. Front. Plant Sci. 8, 172.

Mickelbart, M.V., Hasegawa, P.M., Bailey-Serres, J., 2015. Genetic mechanisms of abiotic stress tolerance that translate to crop yield stability. Nat. Rev. Genet. 16 (4), 237–251.

Morgil, H., Tardu, M., Cevahir, G., Kavakli, İ.H., 2019. Comparative RNA-seq analysis of the drought-sensitive lentil (*Lens culinaris*) root and leaf under short-and long-term water deficits. Funct. Integr. Genomics, 1–13.

Nadeem, M., Li, J., Yahya, M., Wang, M., Ali, A., Cheng, A., Xiaobo, W., Ma, C., 2019. Grain legumes and fear of salt stress: focus on mechanisms and management strategies. Int. J. Mol. Sci. 20 (4), 799.

Nithin, C., Thomas, A., Basak, J., Bahadur, R.P., 2017. Genome-wide identification of miRNAs and lncRNAs in *Cajanus cajan*. BMC Genomics 18 (1), 878.

Ntini, E., Marsico, A., 2019. Functional impacts of non-coding RNA processing on enhancer activity and target gene expression. J. Mol. Cell Biol. 11, 868–879.

Pandey, P., Ramegowda, V., Senthil-Kumar, M., 2015. Shared and unique responses of plants to multiple individual stresses and stress combinations: physiological and molecular mechanisms. Front. Plant Sci. 6, 723.

Papadopoulou, K., Roussis, A., Katinakis, P., 1996. Phaseolus *ENOD40* is involved in symbiotic and non-symbiotic organogenetic processes: expression during nodule and lateral root development. Plant Mol. Biol. 30 (3), 403–417.

Pareek, A., Khurana, A., K Sharma, A., Kumar, R., 2017. An overview of signalling regulons during cold stress tolerance in plants. Curr. Genomics 18 (6), 498–511.

Park, B.S., Seo, J.S., Chua, N.H., 2014. NITROGEN LIMITATION ADAPTATION recruits PHOSPHATE2 to target the phosphate transporter PT2 for degradation during the regulation of Arabidopsis phosphate homeostasis. Plant Cell 26 (1), 454–464.

Paytuví, G.A., Hermoso, P.A., Anzar Martínez de Lagrán, I., Sanseverino, W., Aiese Cigliano, R., 2016. GREENC: a Wiki-based database of plant lncRNAs. Nucleic Acids Res. 44, D1161–D1166.

Pecrix, Y., Staton, S.E., Sallet, E., et al., 2018. Whole-genome landscape of *Medicago truncatula* symbiotic genes. Nat. Plants 4, 1017–1025. https://doi.org/10.1038/s41477-018-0286-7.

Pereira, A., 2016. Plant abiotic stress challenges from the changing environment. Front. Plant Sci. 7, 1123.

Pereira, W.J., Melo, A.T.D.O., Coelho, A.S.G., Rodrigues, F.A., Mamidi, S., Alencar, S.A.D., Lanna, A.C., Valdisser, P.A.M.R., Brondani, C., Nascimento-Júnior, I.R.D., Borba, T.C.D.O., 2019. Genome-wide analysis of the transcriptional response to drought stress in root and leaf of common bean. Genet. Mol. Biol. 43 (1), e20180259.

Plaxton, W.C., Tran, H.T., 2011. Metabolic adaptations of phosphate-starved plants. Plant Physiol. 156 (3), 1006–1015.

Pushpavalli, R., Zaman-Allah, M., Turner, N.C., Baddam, R., Rao, M.V., Vadez, V., 2015. Higher flower and seed number leads to higher yield under water stress conditions imposed during reproduction in chickpea. Funct. Plant Biol. 42 (2), 162–174.

Qi, X., Xie, S., Liu, Y., Yi, F., Yu, J., 2013. Genome-wide annotation of genes and noncoding RNAs of foxtail millet in response to simulated drought stress by deep sequencing. Plant Mol. Biol. Plant Mol. Biol. 83 (4–5), 459–473.

Qin, T., Zhao, H., Cui, P., Albesher, N., Xiong, L.A., 2017. Nucleus-localized long non-coding RNA enhances drought and salt stress tolerance. Plant Physiol. 175, 1321–1336.

Qiu, C.W., Zhao, J., Chen, Q., Wu, F., 2014. Genome-wide characterization of drought stress-responsive long non-coding RNAs in Tibetan wild barley. Environ. Exp. Bot. 164, 124–134.

Samarah, N., Alqudah, A., 2011. Effects of late-terminal drought stress on seed germination and vigor of barley *Hordeum vulgare* L.). Arch. Agron. Soil Sci. 57 (1), 27–32.

Sehgal, A., Sita, K., Siddique, K.H., Kumar, R., Bhogireddy, S., Varshney, R.K., Nayyar, H., 2018. Drought or/and heat-stress effects on seed filling in food crops: impacts on functional biochemistry, seed yields, and nutritional quality. Front. Plant Sci. 9, 1705.

Seo, J.S., Sun, H.-X., Park, B.S., Huang, C.-H., Yeh, S.-D., Jung, C., Chua, N.-H., 2017. *ELF18-INDUCED LONG-NONCODING* RNA associates with mediator to enhance expression of innate immune response genes in Arabidopsis. Plant Cell 29, 1024–1038.

Sheaffer, C.C., Seguin, P., 2003. Forage legumes for sustainable cropping systems. J. Crop. Prod. 8 (1–2), 187–216.

Shin, H., Shin, H.S., Chen, R., Harrison, M.J., 2006. Loss of At4 function impacts phosphate distribution between the roots and the shoots during phosphate starvation. Plant J. 45 (5), 712–726.

Shuai, P., Liang, D., Tang, S., Zhang, Z., Ye, C.Y., Su, Y., Yin, W., 2014. Genome-wide identification and functional prediction of novel and drought-responsive lincRNAs in *Populus trichocarpa*. J. Exp. Bot. 65 (17), 4975–4983.

Shumayla, S., Sharma, Taneja, M., Tyagi, S., Singh, K., Upadhyay, S.K., 2017. Survey of high throughput RNA-seq data reveals potential roles for lncRNAs during development and stress response in bread wheat. Front. Plant Sci. 8, 1019.

Silva, C., Martínez, V., Carvajal, M., 2008. Osmotic versus toxic effects of NaCl on pepper plants. Biol. Plant. 52 (1), 72–79.

Singh, U., Khemka, N., Rajkumar, M.S., Garg, R., Jain, M., 2017. PLncPRO for prediction of long non-coding RNAs (lncRNAs) in plants and its application for discovery of abiotic stress-responsive lncRNAs in rice and chickpea. Nucleic Acids Res. 45 (22), e183-e183.

Sita, K., Sehgal, A., Hanumantha Rao, B., Nair, R.M., Vara Prasad, P.V., Kumar, S., et al., 2017. Food legumes and rising temperatures: effects, adaptive functional mechanisms specific to reproductive growth stage and strategies to improve heat tolerance. Front. Plant Sci. 8, 1658.

Song, X., Liu, G., Huang, Z., Duan, W., Tan, H., Li, Y., Hou, X., 2016. Temperature expression patterns of genes and their coexpression with LncRNAs revealed by RNA-Seq in non-heading Chinese cabbage. BMC Genomics 17 (1), 297.

Stagnari, F., Maggio, A., Galieni, A., Pisante, M., 2017. Multiple benefits of legumes for agriculture sustainability: an overview. Chem. Biol. Technol. Agric. 4 (1), 2.

Šurbanovski, N., Grant, O.M., 2014. The emerging role of aquaporins in plant tolerance of abiotic stress. In: Emerging Technologies and Management of Crop Stress Tolerance. Academic Press, pp. 431–447.

Swiezewski, S., Liu, F., Magusin, A., Dean, C., 2009. Cold-induced silencing by long antisense transcripts of an Arabidopsis Polycomb target. Nature 462 (7274), 799–802.

Szcześniak, M.W., Rosikiewicz, W., Makałowska, I., 2016. CANTATAdb: a collection of plant long non-coding RNAs. Plant Cell Physiol. 57, e8.

Tester, M., Bacic, A., 2005. Abiotic stress tolerance in grasses. From model plants to crop plants. Plant Physiol. 137 (3), 791–793.

Tsai, M.C., Manor, O., Wan, Y., Mosammaparast, N., Wang, J.K., Lan, F., et al., 2010. Long noncoding RNA as modular scaffold of histone modification complexes. Science 329 (5992), 689–693.

Tyagi, S., Sharma, A., Upadhyay, S., 2018. Role of next-generation RNA-Seq data in discovery and characterization of long non-coding RNA in plants. In: Çiftçi, Y.Ö. (Ed.), Next Generation Plant Breeding. IntechOpen, https://doi.org/10.5772/intechopen.72773.

Ulitsky, I., 2018. Interactions between short and long noncoding RNAs. FEBS Lett. 592 (17), 2874–2883.

Van de Sande, K., Pawlowski, K., Czaja, I., Wieneke, U., Schell, J., Schmidt, J., Walden, R., Matvienko, M., Wellink, J., van Kammen, A., Franssen, H., 1996. Modification of phytohormone response by a peptide encoded by *ENOD40* of legumes and a nonlegume. Science 273 (5273), 370–373.

Vaucheret, H., 2006. Post-transcriptional small RNA pathways in plants: mechanisms and regulations. Genes Dev. 20 (7), 759–771.

Vijn, I., Yang, W.C., Pallisgård, N., Jensen, E.Ø., van Kammen, A., Bisseling, T., 1995. *VsENOD5*, *VsENOD12* and *VsENOD40* expression during Rhizobium-induced nodule formation on *Vicia sativa* roots. Plant Mol. Biol. 28 (6), 1111–1119.

Wallace, J.S., Acreman, M.C., Sullivan, C.A., 2003. The sharing of water between society and ecosystems: from conflict to catchment-based co-management. Philos. Trans. R. Soc. Lond. B Biol. Sci. 358, 2011–2026.

Wan, D., Ma, J., Bai, X., Luo, W., Shao, X., Bai, Q., et al., 2019. Genome-wide identification of long noncoding RNAs (lncRNAs) and their responses to salt stress in two closely related poplars. Front. Genet. 10, 777.

Wang, K.C., Chang, H.Y., 2011. Molecular mechanisms of long noncoding RNAs. Mol. Cell 41–47.

Wang, H.L.V., Chekanova, J.A., 2017. Long noncoding RNAs in plants. In: Long Non Coding RNA Biology. Springer, Singapore, pp. 133–154.

Wang, W., Vinocur, B., Altman, A., 2003. Plant responses to drought, salinity and extreme temperatures: towards genetic engineering for stress tolerance. Planta 218 (1), 1–14.

Wang, H., Chung, P.J., Liu, J., Jang, I.-C., Kean, M.J., Xu, J., Chua, N.-H., 2014. Genome-wide identification of long noncoding natural antisense transcripts and their responses to light in Arabidopsis. Genome Res. 24 (3), 444–453.

Wang, H., Niu, Q.-W., Wu, H.-W., Liu, J., Ye, J., Yu, N., Chua, N.-H., 2015. Analysis of non-coding transcriptome in rice and maize uncovers roles of conserved lncRNAs associated with agriculture traits. Plant J. 84, 404–416.

Wang, C.Y., Liu, S.R., Zhang, X.Y., Ma, Y.J., Hu, C.G., Zhang, J.Z., 2017a. Genome-wide screening and characterization of long non-coding RNAs involved in flowering development of trifoliate orange (*Poncirus trifoliata* L. Raf.). Sci. Rep. 7, 43226.

Wang, M., Wu, H.J., Fang, J., Chu, C., Wang, X.J., 2017b. A long noncoding RNA involved in rice reproductive development by negatively regulating osa-miR160. Sci. Bull. 62 (7), 470–475.

Wang, A., Hu, J., Gao, C., Chen, G., Wang, B., Lin, C., Zhou, G., 2019. Genome-wide analysis of long non-coding RNAs unveils the regulatory roles in the heat tolerance of Chinese cabbage (*Brassica rapa* ssp. chinensis). Sci. Rep. 9 (1), 1–14.

Wasaki, J., Yonetani, R., Shinano, T., Kai, M., Osaki, M., 2003. Expression of the *OsPI1* gene, cloned from rice roots using cDNA microarray, rapidly responds to phosphorus status. New Phytol. 158, 239–248.

Wekesa, J.S., Luan, Y., Chen, M., Meng, J., 2019. A hybrid prediction method for plant lncRNA-protein interaction. Cells 8 (6), 521.

Wheeler, T., Von Braun, J., 2013. Climate change impacts on global food security. Science 341 (6145), 508–513.

Wu, H.-J., Wang, Z.-M., Wang, M., Wang, X.J., 2013. Widespread long noncoding RNAs as endogenous target mimics for microRNAs in plants. Plant Physiol. 161, 1875–1884.

Wunderlich, M., Gross-Hardt, R., Schoffl, F., 2014. Heat shock factor HSFB2a involved in gametophyte development of *Arabidopsis thaliana* and its expression is controlled by a heat-inducible long non-coding antisense RNA. Plant Mol. Biol. 85, 541–550.

Yamaguchi, A., Abe, M., 2012. Regulation of reproductive development by non-coding RNA in *Arabidopsis*: to flower or not to flower. J. Plant Res. 125 (6), 693–704.

Yamaguchi, T., Aharon, G.S., Sottosanto, J.B., Blumwald, E., 2005. Vacuolar Na^+/H^+ antiporter cation selectivity is regulated by calmodulin from within the vacuole in a Ca^{2+}- and pH-dependent manner. Proc. Natl. Acad. Sci. 102 (44), 16107–16112.

Yang, W.C., Katinakis, P., Hendriks, P., Smolders, A., Vries, F., Spee, J., Kammen, A., Bisseling, T., Franssen, H., 1993. Characterization of *GmENOD40*, a gene showing novel patterns of cell-specific expression during soybean nodule development. Plant J. 3, 573–585.

Yi, X., Zhang, Z., Ling, Y., Xu, W., Su, Z., 2015. PNRD: a plant non-coding RNA database. Nucleic Acids Res. 43, D982–D989.

Young, N.D., Mudge, J., Ellis, T.N., 2003. Legume genomes: more than peas in a pod. Curr. Opin. Plant Biol. 6 (2), 199–204.

Zhang, J.L., Shi, H., 2013. Physiological and molecular mechanisms of plant salt tolerance. Photosynth. Res. 115 (1), 1–22.

Zhang, Y.C., Liao, J.Y., Li, Z.Y., Yu, Y., Zhang, J.P., Li, Q.F., Qu, L.H., Shu, W.S., Chen, Y.Q., 2014. Genome-wide screening and functional analysis identify a large number of long noncoding RNAs involved in the sexual reproduction of rice. Genome Biol. 15, 512.

Zhang, J., Wei, L., Jiang, J., Mason, A.S., Li, H., Cui, C., Chai, L., Zheng, B., Zhu, Y., Xia, Q., Jiang, L., 2018. Genome-wide identification, putative functionality and interactions between lncRNAs and miRNAs in Brassica species. Sci. Rep. 8 (1), 1–11.

Zhou, H., Liu, Q., Li, J., Jiang, D., Zhou, L., Wu, P., Lu, S., Li, F., Zhu, L., Liu, Z., et al., 2012. Photoperiod-and thermo-sensitive genic male sterility in rice are caused by a point mutation in a novel noncoding RNA that produces a small RNA. Cell Res. 22, 649–660.
Zhu, J.K., 2002. Salt and drought stress signal transduction in plants. Annu. Rev. Plant Biol. 53, 247–273.
Zhu, Q.H., Wang, M.B., 2012. Molecular functions of long non-coding RNAs in plants. Genes 3 (1), 176–190.
Zlatev, Z., Lidon, F.C., 2012. An overview on drought induced changes in plant growth, water relations and photosynthesis. Emir. J. Food Agric. 57–72.

Chapter 11

Long noncoding RNAs in tomato: Roles in development and stress response

Zhao Yang, Chengcheng Yang, Zhenchao Yang, and Yongjun Wu
College of Life Sciences, College of Science, College of Horticulture, Northwest A&F University, Xianyang, China

11.1 Introduction

Long noncoding RNA (lncRNA) is a type of noncoding RNA with a length of about 200 nt to 100 kb (Kim and Sung, 2012). Through cloning, sequencing, and gene mapping methods, Pachnis et al. (1988) discovered the first lncRNA H19 associated with X chromosome. Originally, lncRNA was considered as a kind of transcription noise like other noncoding RNAs. Recent research over the years has shown that lncRNA has an important role in many biological processes (Ponting et al., 2009; Shumayla et al., 2017). There is more work done on human or animal lncRNA than on plant lncRNA (Liu et al., 2015; Lu and Huang, 2017). LncRNA can regulate the expression of cis-acting genes or trans-acting genes, and the regulation of the expression level of cis-acting genes may represent the extensive functions of lncRNA (Ponting et al., 2009).

The upstream of the lncRNA gene may interact with promoters or other cis-acting elements of co-expressed genes to regulate gene expression (Chen, 2016; Signal et al., 2016). By overlapping with the transcription factor-binding site, the transcription of lncRNA can prevent the transcription factor from binding to this site (Chen, 2016; Ponting et al., 2009; Signal et al., 2016). In addition, lncRNA near the transcription factor-binding site can enhance the binding of transcription factor to this site (Signal et al., 2016). Generally, lncRNA can be classified as sense, natural antisense, bidirectional, intron, and intergenic lncRNA (Pachnis et al., 1988; Ponting et al., 2009). Different types of lncRNA have different regulatory effects. Plant lncRNAs are involved in processes such as growth, development, and stress response (Sani et al., 2018; Tyagi et al., 2018). For example, compared to wild-type plants, overexpression of lncRNA

Long Noncoding RNAs in Plants: Roles in development and stress
https://doi.org/10.1016/B978-0-12-821452-7.00011-8

npc48 in Arabidopsis increased rosette diameter and leaf sawtooth and delayed flowering time (Amor et al., 2009). In maize leaves, a total of 664 lncRNAs responding to drought stress were identified (Zhang et al., 2014). Among them, 567 lncRNAs were upregulated and 97 lncRNAs were downregulated. More lncRNAs have been reported in different plants by deep sequencing (Di et al., 2014; Hotto et al., 2011; Li et al., 2013).

As an important vegetable crop, tomato (*Solanum lycopersicum*) is rich in proteins, lipids, carbohydrates, calcium, iron, phosphorus, and other minerals. Tomato is also an important model plant for studying plant-pathogen interactions. In recent years, tomato lncRNAs have been reported. For example, lncRNAs were identified by Wang et al. (2015a) in tomato plants infected with tomato yellow leaf curl virus (TYLCV), and expression of slylnc0049 and slylnc0761 was found to inhibit TYLCV infection in tomatoes. Li et al. (2018) reported that low expression of lncRNA 1459 may inhibit tomato fruit maturation, while changes in expression of other lncRNAs would affect tomato maturity.

11.2 lncRNA related to tomato development

The tomato (*Solanum lycopersicum*) as a model plant for studying the development of fleshy fruits, especially flower development, is good and directly affects the quantity and quality of the later fruit. The evolution of the ovaries and fruits has made seed dispersal easier, a key process for the successful propagation of flowering plants (Wang et al., 2018b). In this case, transcriptional regulation is the basis of an important mechanism controlling the development of reproductive organs (Wang et al., 2018b). Functional genomics studies have shown that noncoding RNAs regulate a variety of developmental and reproductive stages, from carpel formation and ovarian development to softening of mature fruits (Xiao et al., 2020). The final mature fruit is the result of processes at various levels of regulation, including the mechanism of lncRNA regulation (Xiao et al., 2020). Although the understanding of lncRNA regulation in plants is still limited, lncRNA is related to seed morphology, flowering time, and recent control of fruit maturity (Berry and Dean, 2015; Ding et al., 2012; Li et al., 2018; Wang et al., 2014).

11.2.1 The role of lncRNA in tomato flower development

Plant productivity depends on inflorescences, flower-bearing shoots that originate from the stem cell populations of shoot meristems (Xu et al., 2016). Inflorescence architecture determines flower production, which can vary dramatically both between and within species. Sympodial growth is most well-understood in tomato, where hundreds of multiflowered inflorescences form from the coordinated activities of sequentially maturing vegetative and reproductive sympodial meristems (Lippman et al., 2008; Pnueli et al., 1998).

11.2.1.1 Classic case of lncRNA regulating flower development

LncRNA is a type of noncoding RNA that plays an important regulatory role in various developmental processes (Stav et al., 2019). LncRNA can regulate flowering, but the function of lncRNA during tomato flowering is unclear. The flowering period is an important stage of plant development and affects fruit formation and yield. COOLAIR and COLDAIR are two types of lncRNA, and they can inhibit the expression of FLOERING LOCUS C (FLC) in the lncR2Epi regulatory pathway under cold stress (Csorba et al., 2014; Heo and Sung, 2011). FLC is a MADS box transcription factor that inhibits flowering in cold environments (Michaels and Amasino, 1999). COOLAIR affects the reduction of H3K36me3 or H3K4me2 in FLC (Marquardt et al., 2014; Shafiq et al., 2016). In addition, COLDAIR can affect flowering by inhibiting FLC expression during vernalization (Kim et al., 2017; Liu et al., 2015; Shafiq et al., 2016). Therefore, the study of lncRNAs related to tomato flowering is of great significance for further understanding the regulatory mechanism of lncRNAs involved in flowering tomato development.

11.2.1.2 Study of lncRNA during flowering

To explore the profile and function of lncRNA during tomato flowering, Yang et al. (2019b) performed strand-specific paired-end RNA sequencing of tomato leaves, flowers, and roots. They identified 10,919 lncRNAs, including 248 new lncRNAs, of which 65 new lncRNAs were significantly differentially expressed (DE) in flowers, leaves, and roots. Gene Ontology (GO) and Kyoto Gene and Genome Encyclopedia (KEGG) analyses were performed to identify the cis-target genes of DE lncRNA. The results show that lncRNAs may play an important role in the growth, development, and apoptosis of flowering tomato plants by regulating the formation of the inner membrane of flower tissues, binding with various molecules, affecting metabolic pathways, and inducing apoptosis. In addition, lncRNA can regulate the mRNA expression of tomato during flowering period by forming competitive endogenous RNA and further regulate various biological metabolic pathways of tomato.

11.2.1.3 Study of alternative splicing lncRNA during flowering

Alternative splicing (AS) regulates multiple biological processes in plants, including flowering, circadian rhythm, and stress response (Xu et al., 2016). Using RNA-seq, Sun and Xiao (2015) investigated AS patterns in tomato seedlings, flowers, and young fruits and found that 59.3% of the expressed multi-exon genes in these tissues underwent AS. Although the frequency of AS events was similar in seeds, flowers, and early-maturing fruits, the fruit of each gene produced more splice variants. In addition, the study found that 2507 new transcribed regions (NTRs) were not related to the annotated chromosomal regions, from which 1690 putative lncRNAs were identified.

Yang et al. (2020) studied the AS pattern of lncRNA in tomato flower, leaf, and root tissues at the early flowering stage. Using RNA-Seq, they found that 72.55% of lncRNAs in these tissues experienced AS, resulting in a total of 16,995 AS events. Their analysis of candidate target genes for tissue-specific AS lncRNA showed that these candidate target genes of lncRNA may be involved in the regulation of circadian rhythm, plant immunity, cellulose synthesis, and phosphate-containing complex metabolic processes. This provides new insights into the complexity of the transcriptome and the regulation of AS.

11.2.2 The role of lncRNA in tomato fruit maturity

lncRNA plays a key role in several biological processes in plants (Zhang and Chen, 2013), and some lncRNAs have been identified related to tomato fruit maturity (Zhu et al., 2015). LncRNAs are widely found in plant transcriptomes, and emerging evidence suggests that they play a key role in a variety of developmental processes (Scarano et al., 2017). LncRNA may regulate gene expression through a variety of mechanisms, and this mechanism has been extensively reviewed by Chekanova (2015). Tomato development is a complex biological process that involves tightly regulated cell division and differentiation, cell expansion, and eventual maturity (Kumar et al., 2013).

11.2.2.1 *Potential functions of* lncRNA-1459 *and* lncRNA1840 *regulating fruit maturity*

RNA-seq analysis compared the transcriptomes of tomato cils *Ailsa Craig* and *rin* fruits and identified more than 3000 tomato lncRNAs, several of which were differentially expressed in *rin* (Yang et al., 2019b). In the same study, two lncRNAs (lncRNA1459 and lncRNA1840) that were strongly downregulated in *rin* were selected for VIGS-based silencing assays in fruits. Silencing of both lncRNAs produces immature portions in the fruit injection region, similar to the effect observed in VIGS-based RIN silencing. To better understand the function of lncRNA1459, intergenic lncRNA, Li et al. (2018) used genome editing techniques induced by clustered regularly spaced short palindromic repeats (CRISPR)/associated protein 9 (Cas9) to produce mutants with loss of lncRNA1459 function (Doudna and Charpentier, 2014; Feng et al., 2013). Mutated fruits show a delayed ripening phenotype, which is related to the inhibition of ethylene and carotenoid biosynthesis, and the downregulation of genes related to maturity.

11.2.2.2 *Potential functions of* lncRNA-314 *regulating fruit maturity*

Many fruit-specific high-expression genes and lncRNAs play crucial roles in fruit ripening (Vrebalov et al., 2002). *LncRNA-314* was predicted to be involved in tomato fruit ripening because it is highly expressed in fruits at the breaker ripening stage and suppressed in different ripening mutants (Wang et al., 2016). *LncRNA-314* is conserved and highly expressed in the ripening fruit of *Lycopersicon* tomato only. Nearest-neighbor analysis combined with co-expression analysis is an effective method used to predict the potential functions of lncRNAs (Pauli et al., 2012; Wang et al., 2015b). *Solyc10g024420.1*, a member of the ABC transporter C family located 20kb upstream of *lncRNA-314* in the complementary strand, was screened as the co-expressed gene of *lncRNA-314* (Wang et al., 2016). *Solyc10g024420.1* was also highly expressed in fruits at the breaker ripening stage and suppressed in different ripening mutants; this gene also showed higher expression in *Lycopersicon* tomatoes (M82) than that in *S. pennellii* fruit, which did not express *lncRNA-314* (Wang et al., 2016). Previous studies illustrated that lncRNA could also activate chromatin proteins to regulate the expression of neighboring genes (Wang et al., 2011; Xiang et al., 2014). Therefore, it will be interesting to investigate if *lncRNA-314* acts as a positive regulator of its neighboring gene (*Solyc10g024420.1*), controlling tomato fruit ripening.

11.2.2.3 The role of lncRNAs in ethylene response

LeERF1 is a tomato ethylene response factor (ERF) near the end of the ethylene signal transduction pathway. LncRNAs were identified in tomato and their role in ethylene signaling was elucidated (Wang et al., 2018c). Deep sequencing was used for the overexpression and suppression of LeERF1 transgenic and control tomato fruits. A total of 397 lncRNAs were identified, including 169 previously unidentified tomato lncRNAs. Among them, 12 were differentially expressed in transgenic tomato fruits and control tomato fruits. A number of lncRNA target genes were identified, including many genes related to ethylene signals, including auxin response factor and auxin-inducing protein, F-box protein, ERF, and MADS-box protein. In addition, they proposed a regulatory model that highlights the relationship between lncRNA and its targets involved in ethylene signal transduction, thereby laying the foundation for solving the role of LncRNA in ethylene reactions.

11.3 lncRNA involved in tomato stress response

Like other plants, tomatoes encounter a variety of biological and abiotic adversities in their lifetime. Studies showed that plant lncRNAs play important roles in plant biotic and abiotic stress responses (Qin et al., 2017; Wang et al., 2018a).

11.3.1 Biotic stress response

Many viruses can cause severe damages and yield loss to tomato, such as tobacco mosaic virus (TMV) (Takamatsu et al., 1990), tomato mosaic virus (ToMV), tomato yellow leaf curl virus (TYLCV) (Yang et al., 2019a), and so on. In the long evolutionary process, tomato has established a variety of defense mechanisms in the face of virus invasion. LncRNA may be involved in these mechanisms.

TYLCV can cause devastating diseases to tomatoes worldwide, which are transmitted by the whitefly *Bemisia tabaci* (Scholthof et al., 2011). Symptoms of TYLCV infection in young plants include stunted growth and development, curled leaf margins upward, significant reduction in leaf size, mottled and yellowing of young leaves, flower shedding, and severe yield reduction (Glick et al., 2009; Yang et al., 2019a). Wang et al. (Yang et al., 2019a) used strand-specific RNA-seq method to study the transcriptome changes in tomato in response to TYLCV infection and analyzed tomato lncRNA related to TYLCV resistance. In their study, they predicted 1565 lncRNAs and explicitly identified lnRNAs associated with TYLCV infection. In another related research, using mutants selected from neutron-irradiated tomato plants, Zhou et al. (Wang et al., 2015a) conducted RNA-sequencing and identified 1563 lncRNAs. This result is very close to the result of Wang et al. (Kim and Mutschler, 2005). In addition, by comparing TYLCV-infected plants to noninfected plants, they identified 19 upregulated and 11 downregulated lncRNAs.

As a biotrophic pathogen, *phytophthora infestans* (*P. infestans*) is the causal agent of late blight (LB) in tomato, which occurs worldwide and causes serious economic loss for tomatoes (Kim and Mutschler, 2005; de Miranda et al., 2010). Using comparative transcriptome method, Cui et al. (2017) obtained approximately 9880 unique lncRNAs. Most interestingly, they found that lncRNA16397 could act as an antisense transcript of SlGRX22; their results showed that tomato lncRNA16397 induced the expression of Glutaredoxin (SlGRX) in tomato solanum to reduce accumulation of reactive oxygen species (ROS) and reduce cell membrane damage, thereby enhancing resistance to *Phytophthora infestans*. Recently, similar work from the same team showed that lncRNA42705 and lncRNA08711 could act in tomato resistance to the pathogen by affecting MYB (proto-oncogene, transcription factor) expression (Cui et al., 2020). In their study, they found 46 miRNAs could be decoyed by 88 lncRNAs, forming 39 miRNA-lncRNA duplexes, among which 15 lncRNAs were found to suppress the expression of miR482d-5p. Their results showed that lncRNAs might function as ceRNAs to decoy miRNAs and affect their target genes 35 in tomato plants, increasing resistance to disease.

Except for large-scale identification of lncRNA at the omics level, some studies began to focus on the function of a specific lncRNA related to tomato viral infection.

Based on their previous research, Jiang et al. (2019) found tomato lncRNA23468 to induce glutathione expression to enhance resistance to *Phytophthora infestans*. To reveal its mechanism, ceRNA was first predicted using bioinformatics methods. The results showed that 89 lncRNAs were found to bind to 46 miRNAs, and miR168a-5p, miR168b-5p, and miR477-3p were mutated by lncRNA34658. In order to investigate whether miR482b's ceRNA and lncRNA23468 have an effect on the response of tomato to *Phytophthora infestans*, relative accumulation levels were detected in tomato leaves infected with *Phytophthora infestans*. Their results show that lncRNA23468 can be used as ceRNA to regulate the nucleotide-binding site leucine-rich repeat (NBS-LRR) gene by mutagenesis of miR482b. In other words, lncRNA23468 increases the expression of NBSLRRs by inhibiting miR482b level.

Transcription factors play key roles in plant defense mechanisms through activating the expression of defense-related genes (Cui et al., 2018). In their study, Cui et al. (2019) found that 148 genes were activated in WRKY1-overexpressing tomato. To investigate why WRKY1 (Tryptophan, Arginine, Lysine, Tyrosine Motif) activates the expression of lncRNA33732, they cloned the 5′ upstream fragment of lncRNA33732 and constructed mutation fragment of the promoters of lncRNA33732 using oligonucleotide-directed mutagenesis; they also used a yeast one-hybrid assay to validate the binding of WRKY1 to the promoter of lncRNA33732. Their results suggested that WRKY1 may lead to an increase in expression of lncRNA33732 through recognizing the W-box in its promoter region during tomato resistance to *P. infestans*.

11.3.2 Abiotic stress response

Except biotic stress, drought, cold, heat, and other abiotic stress could limit plant growth and productivity. Researches have shown that noncoding RNAs play key roles in abiotic stress response in plants (Khraiwesh et al., 2012). In recent years, evidence has supported that lncRNAs may involve in multiple abiotic stress responses (Wang et al., 2017). More than 1060 stress-related lncRNAs have been reported in plant (Xuan et al., 2015). Unfortunately, there are few reports on tomato lncRNA response to abiotic stress.

To reveal the regulatory roles of lncRNAs in tomato fruit under chilling treatment, Wang et al. (2018d) identified 1411 lncRNAs using deep sequencing and bioinformatics methods, among which 239 were significantly differentially expressed. Targets prediction showed that 63 target genes were found to encode enzymes related to chilling injury, including redox reaction-related enzymes (Peroxidase, Phenylalanine ammonia lyase, Catalase), heat and cold shock protein (Heat shock protein 70, Cold shock protein/3), energy metabolism-related enzymes (NADH dehydrogenase, ATPase,), salicylic acid and abscisic acid metabolism (Abscisic acid receptor PYL4, Salicylic acid-binding protein), and the low temperature-induced transcription factor WRKY. The result suggested that lncRNAs may be involved in chilling injury in tomato fruits.

11.4 Concluding remarks

Although lncRNAs of tomato have been increasingly studied during the past 5 years, there are some problems in the research of tomato lncRNAs compared with other noncoding RNAs.

Firstly, although studies have shown that expression of lncRNAs shows a tissue-specific character, studies on tomato lncRNA are mostly carried on fruit. Since 2015, more than 10 studies focused on tomato fruit, especially on fruit ripening. Secondly, tomato is confronted with various biotic and abiotic stresses during its growth periods; these factors all could influence tomato yield and quality. Studies showed that many microRNAs of tomato have been identified under kinds of stress. But, lncRNA studies of tomato are only carried out under few stresses. LncRNAs still need to be identified under different stresses. Lastly, although thousands of lncRNAs of tomato have been identified, a few functions of them were proved; functions of many tomato lncRNAs are still unknown. On the other hand, even if some are proved, all used overexpression lncRNA method (Cui et al., 2019; Jiang et al., 2019). Other more effective or convincible methods also need to be applied.

Altogether, as one of new regulators in plant development and stress response, much more research could be done for tomato lncRNA, including identification of new lncRNAs and function verification.

References

Amor, B.B., Wirth, S., Merchan, F., Laporte, P., d'Aubenton-Carafa, Y., Hirsch, J., Maizel, A., Mallory, A., Lucas, A., Deragon, J.M., 2009. Novel long non-protein coding RNAs involved in Arabidopsis differentiation and stress responses. Genome Res. 19, 57–69.

Berry, S., Dean, C., 2015. Environmental perception and epigenetic memory: mechanistic insight through FLC. Plant J. 83, 133–148.

Chekanova, J.A., 2015. Long non-coding RNAs and their functions in plants. Curr. Opin. Plant Biol. 27, 207–216.

Chen, L.-L., 2016. Linking long noncoding RNA localization and function. Trends Biochem. Sci. 41, 761–772.

Csorba, T., Questa, J.I., Sun, Q., Dean, C., 2014. Antisense COOLAIR mediates the coordinated switching of chromatin states at FLC during vernalization. Proc. Natl. Acad. Sci. 111, 16160–16165.

Cui, J., Luan, Y., Jiang, N., Bao, H., Meng, J., 2017. Comparative transcriptome analysis between resistant and susceptible tomato allows the identification of lnc RNA 16397 conferring resistance to *Phytophthora infestans* by co-expressing glutaredoxin. Plant J. 89, 577–589.

Cui, J., Xu, P., Meng, J., Li, J., Jiang, N., Luan, Y., 2018. Transcriptome signatures of tomato leaf induced by *Phytophthora infestans* and functional identification of transcription factor SpWRKY3. Theor. Appl. Genet. 131, 787–800.

Cui, J., Jiang, N., Meng, J., Yang, G., Liu, W., Zhou, X., Ma, N., Hou, X., Luan, Y., 2019. LncRNA33732-respiratory burst oxidase module associated with WRKY1 in tomato-*Phytophthora infestans* interactions. Plant J. 97, 933–946.

Cui, J., Jiang, N., Hou, X., Wu, S., Zhang, Q., Meng, J., Luan, Y., 2020. Genome-wide Identification of lncRNAs and analysis of ceRNA networks during tomato resistance to *Phytophthora infestans*. Phytopathology 110, 456–464.

de Miranda, B.E.C., Suassuna, N.D., Reis, A., 2010. Mating type, mefenoxam sensitivity, and pathotype diversity in *Phytophthora infestans* isolates from tomato in Brazil. Pesq. Agrop. Brasileira 45, 671–679.

Di, C., Yuan, J., Wu, Y., Li, J., Lin, H., Hu, L., Zhang, T., Qi, Y., Gerstein, M.B., Guo, Y., 2014. Characterization of stress-responsive lnc RNA s in A rabidopsis thaliana by integrating expression, epigenetic and structural features. Plant J. 80, 848–861.

Ding, J., Lu, Q., Ouyang, Y., Mao, H., Zhang, P., Yao, J., Xu, C., Li, X., Xiao, J., Zhang, Q., 2012. A long noncoding RNA regulates photoperiod-sensitive male sterility, an essential component of hybrid rice. Proc. Natl. Acad. Sci. 109, 2654–2659.

Doudna, J.A., Charpentier, E., 2014. The new frontier of genome engineering with CRISPR-Cas9. Science 346, 1258096.

Feng, Z., Zhang, B., Ding, W., Liu, X., Yang, D.-L., Wei, P., Cao, F., Zhu, S., Zhang, F., Mao, Y., 2013. Efficient genome editing in plants using a CRISPR/Cas system. Cell Res. 23, 1229–1232.

Glick, E., Levy, Y., Gafni, Y., 2009. The viral etiology of tomato yellow leaf curl disease—a review. Plant Prot. Sci. 45, 81–97.

Heo, J.B., Sung, S., 2011. Vernalization-mediated epigenetic silencing by a long intronic noncoding RNA. Science 331, 76–79.

Hotto, A.M., Schmitz, R.J., Fei, Z., Ecker, J.R., Stern, D.B., 2011. Unexpected diversity of chloroplast noncoding RNAs as revealed by deep sequencing of the Arabidopsis transcriptome. G3 1, 559–570.

Jiang, N., Cui, J., Shi, Y., Yang, G., Zhou, X., Hou, X., Meng, J., Luan, Y., 2019. Tomato lncRNA23468 functions as a competing endogenous RNA to modulate NBS-LRR genes by decoying miR482b in the tomato-*Phytophthora infestans* interaction. Horticult. Res. 6, 1–11.

Khraiwesh, B., Zhu, J.-K., Zhu, J., 2012. Role of miRNAs and siRNAs in biotic and abiotic stress responses of plants. Biochim. Biophys. Acta 1819, 137–148.

Kim, M.-J., Mutschler, M.A., 2005. Transfer to processing tomato and characterization of late blight resistance derived from Solanum pimpinellifolium L. L3708. J. Am. Soc. Hortic. Sci. 130, 877–884.

Kim, E.-D., Sung, S., 2012. Long noncoding RNA: unveiling hidden layer of gene regulatory networks. Trends Plant Sci. 17, 16–21.

Kim, D.-H., Xi, Y., Sung, S., 2017. Modular function of long noncoding RNA, COLDAIR, in the vernalization response. PLoS Genet. 13, e1006939.

Kumar, R., Khurana, A., Sharma, A.K., 2013. Role of plant hormones and their interplay in development and ripening of fleshy fruits. J. Exp. Bot. 65, 4561–4575.

Li, S., Liberman, L.M., Mukherjee, N., Benfey, P.N., Ohler, U., 2013. Integrated detection of natural antisense transcripts using strand-specific RNA sequencing data. Genome Res. 23, 1730–1739.

Li, R., Fu, D., Zhu, B., Luo, Y., Zhu, H., 2018. CRISPR/Cas9-mediated mutagenesis of lncRNA1459 alters tomato fruit ripening. Plant J. 94, 513–524.

Lippman, Z.B., Cohen, O., Alvarez, J.P., Abu-Abied, M., Pekker, I., Paran, I., Eshed, Y., Zamir, D., 2008. The making of a compound inflorescence in tomato and related nightshades. PLoS Biol. 6.

Liu, X., Hao, L., Li, D., Zhu, L., Hu, S., 2015. Long non-coding RNAs and their biological roles in plants. Genom. Proteom. Bioinform. 13, 137–147.

Lu, C., Huang, Y.H., 2017. Progress in long non-coding RNAs in animals. Yi Chuan 39, 1054–1065.

Marquardt, S., Raitskin, O., Wu, Z., Liu, F., Sun, Q., Dean, C., 2014. Functional consequences of splicing of the antisense transcript COOLAIR on FLC transcription. Mol. Cell 54, 156–165.

Michaels, S.D., Amasino, R.M., 1999. FLOWERING LOCUS C encodes a novel MADS domain protein that acts as a repressor of flowering. Plant Cell 11, 949–956.

Pachnis, V., Brannan, C.I., Tilghman, S.M., 1988. The structure and expression of a novel gene activated in early mouse embryogenesis. EMBO J. 7, 673–681.

Pauli, A., Valen, E., Lin, M.F., Garber, M., Vastenhouw, N.L., Levin, J.Z., Fan, L., Sandelin, A., Rinn, J.L., Regev, A., 2012. Systematic identification of long noncoding RNAs expressed during zebrafish embryogenesis. Genome Res. 22, 577–591.

Pnueli, L., Carmel-Goren, L., Hareven, D., Gutfinger, T., Alvarez, J., Ganal, M., Zamir, D., Lifschitz, E., 1998. The SELF-PRUNING gene of tomato regulates vegetative to reproductive switching of sympodial meristems and is the ortholog of CEN and TFL1. Development 125, 1979–1989.

Ponting, C.P., Oliver, P.L., Reik, W., 2009. Evolution and functions of long noncoding RNAs. Cell 136, 629–641.

Qin, T., Zhao, H., Cui, P., Albesher, N., Xiong, L., 2017. A nucleus-localized long non-coding RNA enhances drought and salt stress tolerance. Plant Physiol. 175, 1321–1336.

Sani, H.M., Hejazian, M., Khatibi, S.M.H., Ardalan, M., Vahed, S.Z., 2018. Long non-coding RNAs: an essential emerging field in kidney pathogenesis. Biomed. Pharmacother. 99, 755–765.

Scarano, D., Rao, R., Corrado, G., 2017. In silico identification and annotation of non-coding RNAs by RNA-seq and de novo assembly of the transcriptome of tomato fruits. PLoS One 12, e0171504.

Scholthof, K.-B.G., Adkins, S., Czosnek, H., Palukaitis, P., Jacquot, E., Hohn, T., Hohn, B., Saunders, K., Candresse, T., Ahlquist, P., 2011. Top 10 plant viruses in molecular plant pathology. Mol. Plant Pathol. 12, 938–954.

Shafiq, S., Li, J., Sun, Q., 2016. Functions of plants long non-coding RNAs. Biochim. Biophys. Acta 1859, 155–162.

Shumayla, Sharma, S., Taneja, M., Tyagi, S., Singh, K., Upadhyay, S.K., 2017. Survey of high throughput RNA-seq data reveals potential roles for lncRNAs during development and stress response in bread wheat. Front. Plant Sci 8, 1019.

Signal, B., Gloss, B.S., Dinger, M.E., 2016. Computational approaches for functional prediction and characterisation of long noncoding RNAs. Trends Genet. 32, 620–637.

Stav, S., Atilho, R.M., Arachchilage, G.M., Nguyen, G., Higgs, G., Breaker, R.R., 2019. Genome-wide discovery of structured noncoding RNAs in bacteria. BMC Microbiol. 19, 66.

Sun, Y., Xiao, H., 2015. Identification of alternative splicing events by RNA sequencing in early growth tomato fruits. BMC Genomics 16, 948.

Takamatsu, N., Watanabe, Y., Meshi, T., Okada, Y., 1990. Mutational analysis of the pseudoknot region in the 3'noncoding region of tobacco mosaic virus RNA. J. Virol. 64, 3686–3693.

Tyagi, S., Sharma, A., Upadhyay, S.K., 2018. Role of next-generation RNA-seq data in discovery and characterization of long non-coding RNA in plants. Next generation plant breeding, Yelda Özden Çiftçi. IntechOpen. https://doi.org/10.5772/intechopen.72773.

Vrebalov, J., Ruezinsky, D., Padmanabhan, V., White, R., Medrano, D., Drake, R., Schuch, W., Giovannoni, J., 2002. A MADS-box gene necessary for fruit ripening at the tomato ripening-inhibitor (rin) locus. Science 296, 343–346.

Wang, K.C., Yang, Y.W., Liu, B., Sanyal, A., Corces-Zimmerman, R., Chen, Y., Lajoie, B.R., Protacio, A., Flynn, R.A., Gupta, R.A., 2011. A long noncoding RNA maintains active chromatin to coordinate homeotic gene expression. Nature 472, 120–124.

Wang, Y., Fan, X., Lin, F., He, G., Terzaghi, W., Zhu, D., Deng, X.W., 2014. Arabidopsis noncoding RNA mediates control of photomorphogenesis by red light. Proc. Natl. Acad. Sci. 111, 10359–10364.

Wang, J., Yu, W., Yang, Y., Li, X., Chen, T., Liu, T., Ma, N., Yang, X., Liu, R., Zhang, B., 2015a. Genome-wide analysis of tomato long non-coding RNAs and identification as endogenous target mimic for microRNA in response to TYLCV infection. Sci. Rep. 5, 16946.

Wang, M., Yuan, D., Tu, L., Gao, W., He, Y., Hu, H., Wang, P., Liu, N., Lindsey, K., Zhang, X., 2015b. Long noncoding RNAs and their proposed functions in fibre development of cotton (Gossypium spp.). New Phytol. 207, 1181–1197.

Wang, X., Ai, G., Zhang, C., Cui, L., Wang, J., Li, H., Zhang, J., Ye, Z., 2016. Expression and diversification analysis reveals transposable elements play important roles in the origin of L ycopersicon-specific lnc RNA s in tomato. New Phytol. 209, 1442–1455.

Wang, J., Meng, X., Dobrovolskaya, O.B., Orlov, Y.L., Chen, M., 2017. Non-coding RNAs and their roles in stress response in plants. Genom. Proteom. Bioinform. 15, 301–312.

Wang, J., Yang, Y., Jin, L., Ling, X., Liu, T., Chen, T., Ji, Y., Yu, W., Zhang, B., 2018a. Re-analysis of long non-coding RNAs and prediction of circRNAs reveal their novel roles in susceptible tomato following TYLCV infection. BMC Plant Biol. 18, 104.

Wang, M., Zhao, W., Gao, L., Zhao, L., 2018b. Genome-wide profiling of long non-coding RNAs from tomato and a comparison with mRNAs associated with the regulation of fruit ripening. BMC Plant Biol. 18, 75.

Wang, Y., Gao, L., Li, J., Zhu, B., Zhu, H., Luo, Y., Wang, Q., Zuo, J., 2018c. Analysis of long-non-coding RNAs associated with ethylene in tomato. Gene 674, 151–160.

Wang, Y., Gao, L., Zhu, B., Zhu, H., Luo, Y., Wang, Q., Zuo, J., 2018d. Integrative analysis of long non-coding RNA acting as ceRNAs involved in chilling injury in tomato fruit. Gene 667, 25–33.

Xiang, J.-F., Yin, Q.-F., Chen, T., Zhang, Y., Zhang, X.-O., Wu, Z., Zhang, S., Wang, H.-B., Ge, J., Lu, X., 2014. Human colorectal cancer-specific CCAT1-L lncRNA regulates long-range chromatin interactions at the MYC locus. Cell Res. 24, 513–531.

Xiao, Y., Kang, B., Li, M., Xiao, L., Xiao, H., Shen, H., Yang, W., 2020. Transcription of lncRNA ACoS-AS1 is essential to trans-splicing between SlPsy1 and ACoS-AS1 that causes yellow fruit in tomato. RNA Biol. 17, 596–607.

Xu, C., Park, S.J., Van Eck, J., Lippman, Z.B., 2016. Control of inflorescence architecture in tomato by BTB/POZ transcriptional regulators. Genes Dev. 30, 2048–2061.

Xuan, H., Zhang, L., Liu, X., Han, G., Li, J., Li, X., Liu, A., Liao, M., Zhang, S., 2015. PLNlncRbase: a resource for experimentally identified lncRNAs in plants. Gene 573, 328–332.

Yang, Y., Liu, T., Shen, D., Wang, J., Ling, X., Hu, Z., Chen, T., Hu, J., Huang, J., Yu, W., 2019a. Tomato yellow leaf curl virus intergenic siRNAs target a host long noncoding RNA to modulate disease symptoms. PLoS Pathog. 15, e1007534.

Yang, Z., Yang, C., Wang, Z., Yang, Z., Chen, D., Wu, Y., 2019b. LncRNA expression profile and ceRNA analysis in tomato during flowering. PLoS One 14, .

Yang, Z., Yang, Z., Yang, C., Wang, Z., Chen, D., Xie, Y., Wu, Y., 2020. Identification and genetic analysis of alternative splicing of long non-coding RNAs in tomato initial flowering stage. Genomics 112, 897–907.

Zhang, Y.-C., Chen, Y.-Q., 2013. Long noncoding RNAs: new regulators in plant development. Biochem. Biophys. Res. Commun. 436, 111–114.

Zhang, W., Han, Z., Guo, Q., Liu, Y., Zheng, Y., Wu, F., Jin, W., 2014. Identification of maize long non-coding RNAs responsive to drought stress. PLoS One 9, .

Zhu, B., Yang, Y., Li, R., Fu, D., Wen, L., Luo, Y., Zhu, H., 2015. RNA sequencing and functional analysis implicate the regulatory role of long non-coding RNAs in tomato fruit ripening. J. Exp. Bot. 66, 4483–4495.

Chapter 12

Genome-wide identification and functional analysis of lincRNAs in cucumber

Zhiqiang Hao and Guanglin Li
College of Life Science, Shaanxi Normal University, Xi'an, China

12.1 Introduction

Long intergenic noncoding RNAs (lincRNAs) are a type of RNA molecule participating in gene expression regulation, which are larger than 200 nucleotides (nt) in length and transcribed from intergenic regions of genome. LincRNAs play a central role in gene expression regulation by acting as signals, decoys, guides, and scaffolds (Guttman and Rinn, 2012; Mercer et al., 2009). Many studies in the past few years showed that plant lincRNAs have many important functions in different biological processes, such as response to abiotic stress (cold or salt, etc.) and plant development (Deng et al., 2018; Wang et al., 2019).

Often, lincRNAs have properties of lacking protein-coding potential and conservation among species, which make it difficult to identify and analyze functions of lincRNAs based on sequence conservation. New strategies, such as machine learning and ceRNA hypothesis, were developed and used widely in lincRNA studies (Kong et al., 2007; Salmena et al., 2011). Along with the development of sequencing technology, more and more public data by next-generation sequencing are available, which make it possible to identify different kinds of RNAs systematically, such as lincRNAs. It is meaningful to explore the connections among different RNA types by constructing co-expression network or sequence-associated network based on ceRNAs hypothesis, which can infer the function of lincRNAs from those correlations.

Cucumber is an important vegetable in daily life. In our previous study, many characters of cucumber lincRNAs were found, such as low expression levels, fewer exons, and shorter lengths in average than that in mRNAs. Here, we used more than two hundred samples to identify comprehensive lincRNAs in cucumber and explore their functions in different ways, which can increase the knowledge of plant lincRNAs' repertoires.

Long Noncoding RNAs in Plants: Roles in development and stress
https://doi.org/10.1016/B978-0-12-821452-7.00012-X

12.2 Results and discussion

12.2.1 Computational identification of lincRNAs

About 727 Gb RNA-Seq data were collected from a total of 257 cucumber samples. We used Hisat2 and Stringtie to map reads to cucumber genome (v2) and assemble transcripts, respectively. A total of 96,832 transcripts were generated, including 87,941 multi-exon transcripts and 25,600 annotated mRNA sequences. The unannotated RNA transcripts (71,232) were subjected to our pipeline for identification of lincRNAs (Fig. 12.1). A length filter was used to discard those sequences with length less than 200 nt. Almost all sequences (71,226) satisfied this filter. And then, ORF (open reading frame) filter was used to exclude those transcripts with ORF length more than 300 nt (between start codons and stop codons); 13,223 sequences were kept. Nr homology filter was used to filter those sequences with significant hits in Nr (nonredundant protein sequences database). 5503 transcripts were kept after Nr filter. CPC (coding potential calculator) filter was used to remove transcripts with coding potential (Kong et al., 2007). 5268 transcripts were left for next analysis. The final step is known as RNA filter, which used Rfam database to filter known RNA types, such as small nucleolar RNAs (snoRNAs), small nuclear (snRNAs), transfer RNAs (tRNAs), and ribosomal RNAs (rRNAs). 5173 transcripts were considered as long noncoding RNA candidates. After all, we selected those transcripts located in intergenic regions of cucumber genome and defined those transcripts as lincRNAs. 2972 lincRNAs were identified and used for next analysis. The above steps for lincRNA identification were implemented by in-house perl scripts and essential tools such as blast and linux working environment.

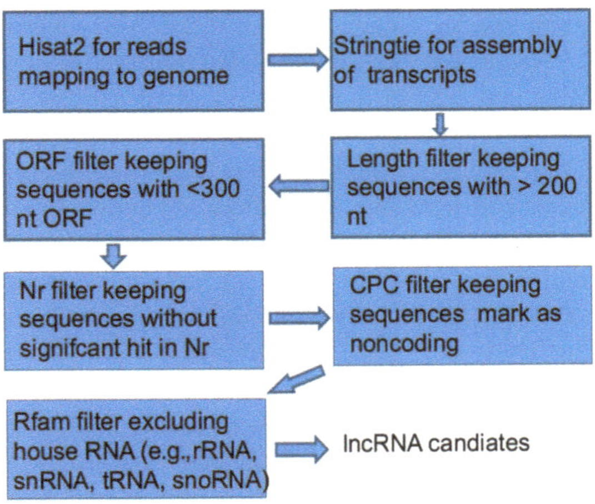

FIG. 12.1 The pipeline for identification of lncRNAs. ORF: Open reading frame. CPC: coding potential calculator. Hisat2 was a tool for reads mapping to genome, and Stringtie(v1.3.3b) was used for transcripts assemble. Nr: nonredundant protein sequence database. Rfam: RNA database.

In our pipeline, ORF filter was powerful to exclude potential protein-coding sequence and more than 80% sequences were filtered, but it may cause some false negative results; in that case, noncoding sequences with long ORF were dropped. By random simulation examination with same amount and sequence length, we found that around 20% random sequences possessed long ORFs more than 300 nt. It showed that ORF filter in our pipeline is useful but not perfect, which may cause about 20% sequence losses. Nevertheless, it was tolerable and effective for strict and confident lincRNA identification.

12.2.2 Co-expression network construction between lincRNAs and mRNA

257 RNA-seq sample data were used to calculate expression levels of cucumber lincRNAs and mRNAs. An expression matrix including mRNAs and lincRNAs was made and R scripts were used to calculate the pearson correlation in pairwise. Briefly, the main steps were as follows: (1) the mRNA and lincRNAs whose expression variances ranked in top 75% were selected for next step separately; (2) *PCC* (pearson' correlation coefficient), *p*-value of *PCC* and adjusted *p*-value were calculated for each pair gene using Fisher's asymptotic test and Bonferroni correction method; (3) the co-expression relations with *PCC* ranking in top 5% and bottom 5% and the adjusted *p*-values less than 0.05 were selected for further analysis. A total of 11,658,425 relationships existed in co-expression network, including 2207 lincRNAs and 17,584 mRNAs nodes. 78,860, 1,179,234, and 10,400,331 links were formed between lincRNAs and lincRNAs, between lincRNAs and mRNAs, and between mRNA and mRNAs, respectively. It is amazing that more than five hundred relatives per node for average show complex gene regulation within network.

12.2.3 Functional analysis of lincRNAs based on co-expression network

We used two different methods, hub-based and module-based methods, to dig lincRNAs' functions from co-expression network of lincRNAs and mRNAs. By using hub-based method, 2062 lincRNAs were recognized as hub genes (a hub gene must contain not less than ten mRNAs as pairs). For module-based method, a total of 39 modules include 2056 lincRNAs (every module must contain not less than one lincRNA and ten mRNAs) (Fig. 12.2).

1979 lincRNAs were shared in both two methods. For example, module-1 contained 446 lincRNAs and 4307 mRNAs, and many GO terms were enriched in module-1, such as nucleic acid metabolic process, chromosome organization, regulation of gene expression, regulation of biosynthetic process, etc. For module-2, 210 lincRNAs and 3675 mRNAs were contained and participated in peptide metabolic process, translated, and biosynthetic process. For module-3, 462 lincRNAs and 1914 mRNAs can be found in this module, which have

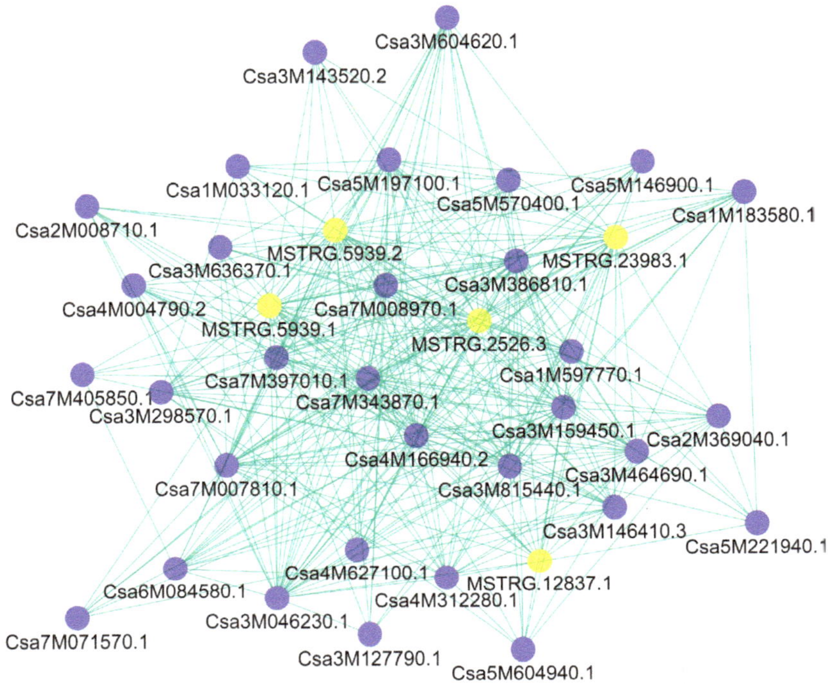

FIG. 12.2 Gene co-expression network example for module-16. Blue nodes represented mRNAs, and yellow nodes represented lincRNAs. This module contained 339 edges, 32 mRNA nodes, and 5 lincRNA nodes. 339 edges represented 118 mRNA-lincRNA relations, 9 lincRNA-lincRNA relations, and 212 mRNA-mRNA relations. The *PCC* for displayed edges satisfied >0.8, and every node was linked by at least six edges.

specific GO terms enriched, such as reactive oxygen species metabolic process, response to oxidative stress, and response to nitrate. For module-4, 138 lincRNAs and 1575 mRNAs were found in the module; the overrepresented GO terms were about purine nucleoside metabolic process and others. For module-5, 75 lincRNAs and 1064 mRNAs were presented in this module; the most significant GO term was about photosynthesis and more than half mRNAs (117/216) for the specific GO term in total gene set were presented. There were 98 lincRNAs and 683 mRNAs in module-6; some important GO terms were found: protein phosphorylation, protein modification process, regulation of transcription, etc. And, for module-7, 15 lincRNAs and 753 mRNAs were found and the associated GO terms were about regulation of biological process, regulation of macromolecular metabolic process, etc.; For module-12, 34 lincRNAs and 284 mRNAs were included; the inferred GO terms were carbohydrate metabolic process, sucrose metabolic, starch metabolic process, etc.

Hub-based method could annotate specific lincRNAs by enriching co-expressed mRNAs. For example, a lincRNA (ID:MSTRG.6240.1) was co-expressed with 303 mRNAs, which were annotated by GO terms: hydrogen peroxide catabolic process, response to oxidative stress, and root development. For lincRNA (ID:MSTRG.16804.1), 173 mRNAs were related with it, and the overrepresented GO terms were pectin catabolic, carbohydrate metabolic process, glucan metabolic process, cell wall organization, etc.

12.2.4 Construction of miRNA-lincRNAs-mRNAs network

534 unique miRNA sequences from 261 miRNA families were collected. 1321 lincRNAs were predicted as 460 miRNAs' targets and 178 lincRNAs were predicted as 109 miRNAs' decoys, separately (Fig. 12.3). There were 10,958 mRNAs and 2074 mRNAs that were predicted as 532/109 miRNAs' targets and decoys. 839 lincRNAs can be targeted by two or more miRNAs that came from different miRNA families. Amazingly, lincRNA MSTRG.10028.1 was targeted by 26 different miRNAs in the network (miR394/miR390/miR528/miR160/miR319/miR159, etc.).

12.2.5 Function of lincRNAs acting as miRNA decoys

Based on ceRNA hypothesis, lincRNAs as miRNAs decoys can regulate mRNAs levels when the mRNA was targeted by same miRNAs. Then, we can infer the

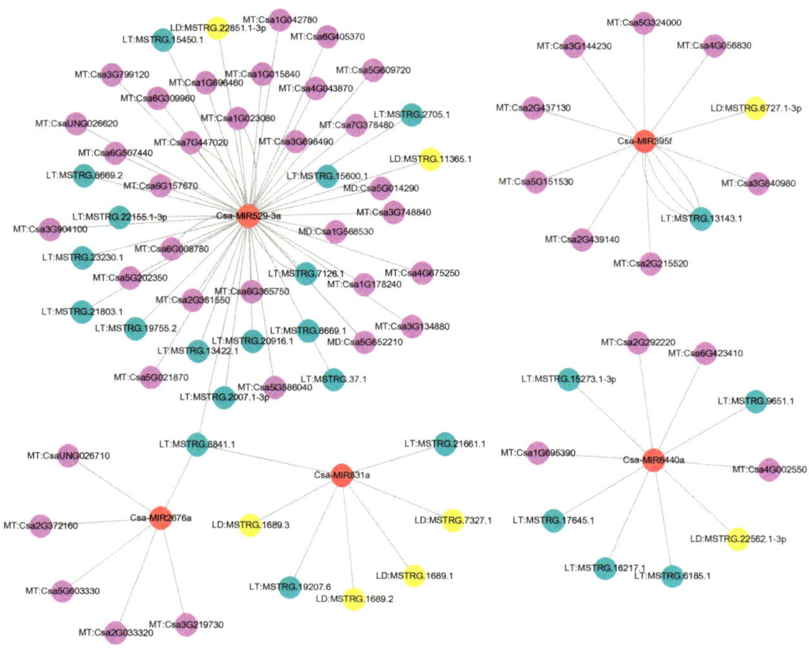

FIG. 12.3 The subnetwork for miRNA-regulated network of miRNA-lincRNA-mRNA. Red nodes: miRNAs. Blue nodes: lincRNAs acting as miRNA targets. Yellow nodes: lincRNAs acting as miRNA decoys. Purple nodes: mRNAs acting as miRNA targets or decoys. *MT/MD*, mRNA target/decoy; *LD/LT*, lincRNA decoy/target. Five miRNAs and correlated mRNAs and lincRNAs were showed in this subnetwork.

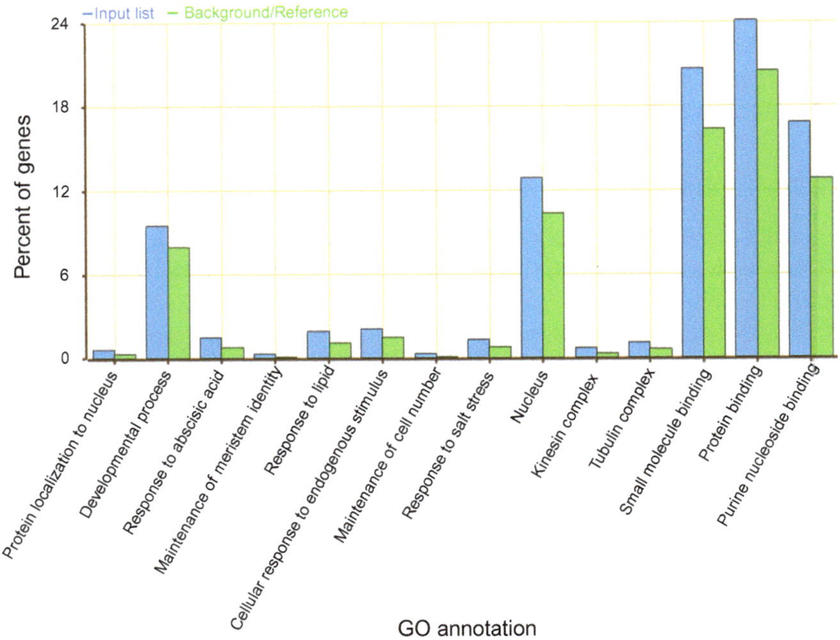

FIG. 12.4 GO term enrichment of lincRNAs acting as miRNA decoys in cucumber. X axis represented the GO annotations and Y axis showed the percentage of genes. Blue bars showed the enrichment status of mRNAs acting as miRNA targets, in which those miRNAs acted on lincRNA as decoys at same time. Green bars showed the percentages of total genes with specific GO terms.

lincRNA functions from mRNAs indirectly by miRNA-mediated manner. 109 miRNAs linked 178 lincRNAs and 4145 mRNAs in specific miRNA-lincRNA-mRNA network, in which same miRNAs can act on mRNAs as miRNA targets and lincRNAs as decoys simultaneously. By GO enrichment analysis of 4145 mRNAs, we can found that lincRNAs as miRNA decoys were involved in multiple biological processes, cell components, and molecular functions. They were mainly about developmental process, response to abscisic acid, response to salt stress, response to lipid, nucleus, tubulin complex, protein binding, and small molecule binding (Fig. 12.4).

12.2.6 Function of lincRNAs acting as miRNA targets

To speculate the functions of 1321 lincRNAs acting as miRNAs targets, we extracted a co-expression subnet that contained 16444 mRNA nodes, 980 lincRNA nodes, and 537,224 mRNA-lincRNA edges. In this network, 36,479 mRNA-lincRNA edges that contained 6710 mRNA nodes and 471 lincRNAs have negative *PCC*. By GO enrichment analysis of the mRNAs that were

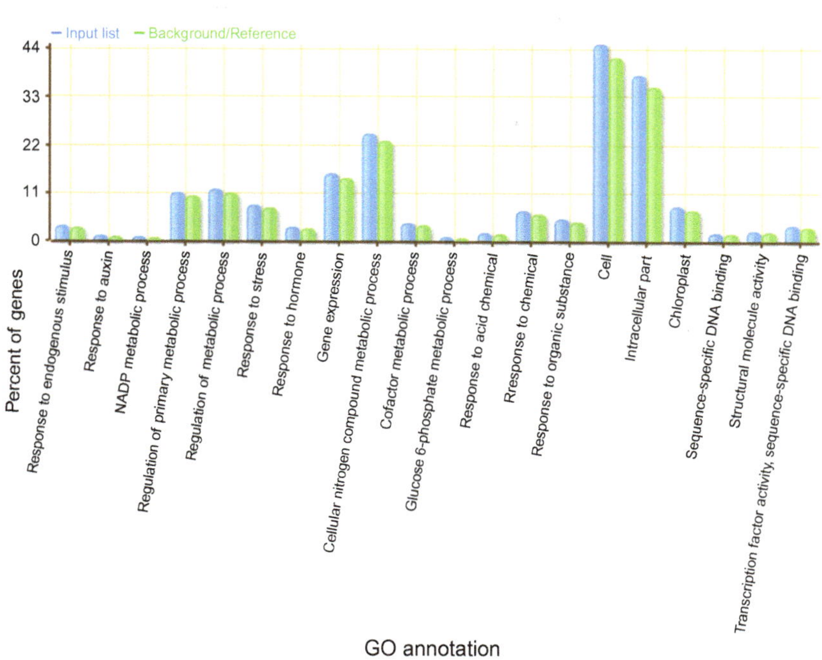

FIG. 12.5 GO term enrichment of lincRNAs acting as miRNA targets in cucumber. X axis represented the GO annotations and Y axis showed the percentage of genes. Blue bars showed the enrichment status of mRNAs, in which those mRNAs were co-expressed with lincRNA as miRNA targets at same time. Green bars showed the percentages of total genes with specific GO terms.

co-expressed with lincRNAs positively or negatively, those lincRNAs were involved in response to auxin, response to stress, gene expression, cell, chloroplast, and sequence-specific DNA binding (Fig. 12.5).

12.3 Methods

12.3.1 Data collection

1. RNA-Seq data, there were total 257 RNA-Seq data of cucumber from 22 projects (PRJNA263870,PRJNA293283,PRJNA307098,PRJNA307279,PRJNA312872,PRJNA321023,PRJNA359788,PRJNA380322,PRJNA381300,PRJNA400678,PRJNA403979,PRJNA431537,PRJNA431715,PRJNA445328,PRJNA448682,PRJNA475903,PRJNA483461,PRJNA531735,PRJNA540430,PRJNA546604,PRJNA80169).
2. Cucumber miRNA data were from studies (Ma et al., 2018; Zhang et al., 2018).
3. Cucumber genome v2 was downloaded from website (ftp://cucurbitgenomics.org/pub/cucurbit/genome/cucumber/chinese_long/v2/).

12.3.2 Softwares

1. Perl v5.28.1, basic scripts were written by perl language under linux environment Ubuntu19.04.
2. GO enrichment for hub genes and module genes were performed by online tool from (http://cucurbitgenomics.org/goenrich), Dataset was selected as Cucumber v2. And Agrigo, another web-based GO analysis toolkit, was used to perform enrichment analysis for miRNA-regulated network.
3. Method and script regarding how to identify target/decoy of miRNAs came from our previous studies (Li et al., 2017).

References

Deng, F., Zhang, X., Wang, W., Yuan, R., Shen, F., 2018. Identification of Gossypium hirsutum long non-coding RNAs (lncRNAs) under salt stress. BMC plant biology 18 (1), 23.

Guttman, M., Rinn, J.L., 2012. Modular regulatory principles of large non-coding RNAs. Nature 482 (7385), 339–346.

Kong, L., Zhang, Y., Ye, Z.Q., Liu, X.Q., Zhao, S.Q., Wei, L., Gao, G., 2007. CPC: assess the protein-coding potential of transcripts using sequence features and support vector machine. Nucleic Acids Research 35 (Web Server issue), W345–W349.

Li, G., Hao, Z., Fan, C., Wu, X., 2017. Genome-Wide Function Analysis of lincRNAs as miRNA Targets or Decoys in Plant. Springer International Publishing.

Ma, C., Yang, J., Cheng, Q., Mao, A., Zhang, J., Wang, S., Weng, Y., Wen, C., 2018. Comparative analysis of miRNA and mRNA abundance in determinate cucumber by high-throughput sequencing. PloS one 13 (1), e0190691.

Mercer, T.R., Dinger, M.E., Mattick, J.S., 2009. Long non-coding RNAs: insights into functions. Nature reviews Genetics 10 (3), 155–159.

Salmena, L., Poliseno, L., Tay, Y., Kats, L., Pandolfi, P.P., 2011. A ceRNA hypothesis: the Rosetta Stone of a hidden RNA language? Cell 146 (3), 353–358.

Wang, P., Dai, L., Ai, J., Wang, Y., Ren, F., 2019. Identification and functional prediction of cold-related long non-coding RNA (lncRNA) in grapevine. Scientific reports 9 (1), 6638.

Zhang, X., Lai, Y., Zhang, W., Ahmad, J., Qiu, Y., Zhang, X., Duan, M., Liu, T., Song, J., Wang, H., et al., 2018. MicroRNAs and their targets in cucumber shoot apices in response to temperature and photoperiod. BMC genomics 19 (1), 819.

Chapter 13

Role of long noncoding RNAs in Brassicaceae family

Ira Vashisht[a,*] and Neha Sharma[b,*]
[a]*Crop Genetics and Information Group, School of Computational and Integrative Sciences, Jawaharlal Nehru University, New Delhi, India,* [b]*Division of Crop Improvement, ICAR-Central Potato Research Institute, Shimla, India*

13.1 Introduction

Brassicaceae family consists of 435 genera and 3675 species worldwide (Ahmed et al., 2019). In addition to the member model species Arabidopsis, the family includes several species like *Brassica napus*, *B. juncea*, *B. rapa*, *B. oleracea*, *B. carinata*, *B. nigra*, and *Raphanus sativus*, which have high nutritional value and are of great economic importance worldwide. Due to the presence of phenolics, glucosinolates, carotenoids, tocopherols, and ascorbic acid in these crops, this family is well-known for health-promoting benefits. These crops face various biotic and biotic stresses including heat, light, cold, hypoxia, drought, salt, and bacterial, fungal, and viral infections under field conditions, which decrease their yield and affect their biochemical properties. Plants have various mechanisms to sense and retort to different environmental conditions, which include different gene networks, multifaceted metabolic pathways, and physiological traits.

Transcriptional or posttranslational gene regulation has been known to play an important role in attaining cellular homeostasis under abiotic and biotic stresses. The noncoding RNAs (ncRNAs) are known to play critical role in modulation of transcription, translation, and other cellular functions (Yamada, 2017; Shen et al., 2018) and are further divided into small ncRNAs (sncRNAs) and long noncoding RNAs (lncRNAs). Among these, sncRNAs of size 20–30 nucleotides include small interfering RNAs (siRNAs), microRNAs (miRNAs), and piwi-interacting RNAs (piRNAs), which are widely considered for their role in RNA interference (Brant and Budak, 2018). However, lncRNAs are transcripts of more than 200 nucleotides in length typically ranging from 1000–10,000 nucleotides, which can be grouped on the basis of their position and orientation with respect to their neighboring protein-coding genes as long noncoding natural

[*] Authors share equal contribution.

Long Noncoding RNAs in Plants: Roles in development and stress
https://doi.org/10.1016/B978-0-12-821452-7.00013-1

antisense transcripts (lncNATs), long intergenic ncRNAs (lincRNAs), intronic noncoding RNAs (incRNAs), and overlapping lncRNAs (Bazin et al., 2017). lncRNAs depict low expression levels in tissue and cell-specific manner. They regulate gene expression by *cis*-acting (both the regulatory RNA and the target gene are transcribed from the same locus) or *trans*-acting (regulatory RNA and the target gene are transcribed from the different locus) in the genome (Rai et al., 2018).

Recently, lncRNAs have emerged as crucial molecules that guide regulation of various biological processes in plants such as flowering time (Wu et al., 2019; Yang et al., 2019), reproduction (Zhang et al., 2014), root development (Bazin and Bailey-Serres, 2015), gene silencing (Wu et al., 2013), fruit development and ripening (Bai et al., 2019), and responses to biotic and abiotic stresses (Nejat and Mantri, 2017; Sun et al., 2018; Shumayla et al., 2017), along with other important developmental pathways. RNA polymerase II involving multi-subunit protein complex transcribes most of the lncRNAs with a 5′-7mGTP-cap and 3′-polyadenylated tail. However, RNA polymerase IV and V also generate some lncRNAs which influence processes like determination of heterochromatin boundaries and recognition of target loci by siRNAs (Böhmdorfer et al., 2016). Structurally, RNA polymerase IV products are 5′ monophosphorylated and non-polyadenylated at 3′ ends (Li et al., 2014). RNA polymerase V transcripts possess 5′ triphosphate and lack 3′ polyadenylated tail (Wierzbicki et al., 2008). lncRNAs regulate the gene expression by different methods, which include, transcription as miRNA precursors that aid in cleavage of gene of interest (Lauressergues et al., 2015); RNA polymerase IV and V-transcribed lncRNAs decrease the expression level of a target gene by inducing RNA-directed DNA methylation (Kim and Sung, 2012); and another approach involves lncRNAs which can enable chromatin modification by recruiting chromatin remodeling protein complex Polycomb Repressive Complex (PRC) (Hou et al., 2019).

Genome-wide comparisons indicated that sequences as well as positional conservation govern the expression of protein-coding genes across the species (Tang et al., 2008). However, noncoding lncRNAs depict a different scenario as a comparative genomics approach with Cleomaceae and Brassicaceae revealed that although long noncoding RNAs are positionally conserved and expressed, they exhibit divergence at the nucleotide level (Mohammadin et al., 2015). Thus, positional and nucleotide conservation shows a negative correlation when it comes to lncRNAs in plants as well as animals (Batista and Chang, 2013). It has been estimated that $<2\%$ of putative lncRNAs detected in *A. thaliana* (Liu et al., 2012) show conservation across plant species (Mohammadin et al., 2015). Liu and colleagues pioneered the identification of lncRNAs in Arabidopsis using multiple technical platforms like RNA-Seq, tilling arrays, and qPCR. The study showed the organ-preferential as well as stress-responsive nature of lncRNAs. Interestingly, these lncRNAs were not processed via small-RNA-dependent pathways and were coregulated by proteins, Cap Binding Proteins (CBP 20, 80) and Serrate.

Novel lncRNAs can be identified by both experimental and computational approaches using widely used genome-wide transcriptome analysis, microarrays, fluorescence in situ hybridization (FISH), and RNA sequencing (Tyagi et al., 2018; Yu and Zhu, 2019). Several lncRNAs have been identified in different plant species; however, studies related to functional characterization are in infancy as compared to those on humans and animals (Bazin and Bailey-Serres, 2015; Yu and Zhu, 2019). In order to provide transcript information more accurately, there is a strong need to optimize high-throughput sequencing technologies and update databases such as NCBI-Gene, Ensembl, NONCODE, and LNCipedia with identified functions of lncRNAs in plants. Recently, genome editing tool clustered regularly interspaced short palindromic repeats (CRISPR)/-associated protein 9 (Cas9) and has emerged as an efficient and powerful way to identify function of lncRNAs through loss-of-function approach (Mali et al., 2013; Yu and Zhu, 2019). Bhatia and colleagues have compiled various available databases such as TAIR10, PlantNATsDB, PLncDB, NONCODE v4, PNRD, lncRNAdb v2.0, PLNlncRbase, GreeNC, CANTATAdb, DsTRD, RNACentral, and PLncRNAdb, which will aid in managing, depositing, and understanding plant lncRNAs (Bhatia et al., 2017).

In this article, we have summarized the role of lncRNAs associated with regulation of different biological processes related to rooting, photomorphogenesis, vernalization, flowering, and reproductive development. Further, involvement of lncRNAs in regulation of drought, heat, phosphate, and various pathogenic stress responses has been reviewed.

13.2 Brassicaceae family

Brassicaceae or Cruciferae is one of the most economically important plant families in the world. The major genera and species of this family include following plants, bittercress (*Cardamine* sp.), bok choy (*Brassica rapa*, var. *chinensis*), brown mustard (*B. juncea*), broccoli (*B. oleracea*, var. *italica*), Brussels sprouts (*B. oleracea*, var. *gemmifera*), cabbage (*B. oleracea*, var. *capitata*), cauliflower (*B. oleracea*, var. *botrytis*), collard (*B. oleracea*, var. *acephala*), kale (*B. oleracea*, var. *acephala*), kohlrabi (*B. oleracea*, var. *gongylodes*), napa cabbage (*B. rapa*, var. *pekinensis*), rape (*B. napus*, var. *napus*), rutabaga (*B. napus*, var. *napobrassica*), turnip (*B. rapa*, var. *rapa*), candytuft (*Iberis* sp.), peppergrass (*Lepidium* sp.), radish (*Raphanus sativus*), rock cress (*Arabis* sp.), sea kale (*Crambe maritima*), sea rocket (*Cakile* sp.), sweet alyssum (*Lobularia maritima*), thale cress (*Arabidopsis thaliana*), wallflower (*Erysimum* sp.), watercress (*Nasturtium officinale*), white mustard (*Sinapis alba*), whitlow grass (*Drab* sp.), wild radish (*R. raphanistrum*), winter cress (*Barbarea* sp.), woad (*Isatis tinctoria*), and yellow cress (*Rorippa* sp.). The most commonly consumed vegetables of this family include cabbage, cauliflower, radish, Brussels sprouts and turnip, kohlrabi, and broccoli, which play an important part in the human diet worldwide (Manchali et al., 2012). They are good source of dietary fiber, calcium, provitamin A, and vitamin C and are characterized by high contents of glucosinolates (Samuolienė et al., 2019). They also accumulate high levels of antioxidant phytochemicals, such as ascorbic acid, carotenoids, phenolic compounds, or tocopherols. Therefore, this family contributes to various health-promoting effects by reducing the risk of chronic ailments like cardiovascular diseases and some types of cancers (Raiola et al., 2018). Brassica family is also known for its oilseed crops and is the third most important source of edible vegetable oils after soybean and palm oil worldwide (Cartea et al., 2019). Oilseed brassicas, generally known as rapeseed, include *B. napus*, *B. juncea*, *B. carinata*, and *B. rapa*. In *Brassica* oil, linoleic acid provides the nutrition, while oleic acid provides thermostability, which is required in cooking oil. It is also appropriate for some industrial applications such as anti-blocking agents in polyethylene films, adhesives in printing, and anticorrosive materials in the steel sheet metal industry (Cartea et al., 2019).

13.3 Role of lncRNAs in plant development

Recently, lncRNAs have emerged as regulators of plant development like their small noncoding counterparts, miRNAs and siRNAs. Role of lncRNAs has been reported in regulation of root and panicle development in Brassicaceae (Yanfei et al., 2014; Bazin and Bailey-Serres, 2015). Wang and colleagues performed a genome-wide analysis to identify photo-responsive lncRNAs in *A. thaliana* where they identified NAT-lncRNAs associated with histone modifications (Wang et al., 2014a). Simultaneously, another group characterized the role of HIDDEN TREASURE 1 (HID1) noncoding RNAs in photomorphogenesis. It was found to be a positive regulator of red light-mediated photomorphogenesis via repression of PHYTOCHROME-INTERACTING FACTOR 3 (PIF3). Further, it was also indicated that HID1 appears to have conservation across plant species (Wang et al., 2014b). It would be interesting to investigate the conservation status of the regulatory noncoding RNAs and HID1 across species.

If we talk about the Brassicaceae family, the most critical role of lncRNAs has been seen in reproductive development where it regulates multiple aspects ranging from flowering time pollen development to plant fertility. Multiple mechanisms of action have been observed in different studies which have explored role of lncRNAs among members of Brassicaceae family. lncRNA-mediated regulation has been observed to be crucial for pollen development in rice, maize, and *Brassica* species (Ma et al., 2010; Ding et al., 2012a,b; Song et al., 2013). An 828 nucleotide lncRNA, *BcMF11*, regulates reproductive development in *Brassica campestris* as a dip in its expression level affects pollen maturity (Ma et al., 2010). Another study reported effect of silencing *BcMF11*, which led to abnormal development of tapetum and microspore resulting in male sterility (Song et al., 2013).

In some species, they have been observed to act as endogenous target mimics (eTMs) for miRNAs, thereby blocking their activity (Wu et al., 2013). In the course of this mechanism, they often emerge as activators of gene expression. Studies in Arabidopsis and rice have identified several eTMs which act as sponges for multiple miRNAs including miR172, miR165, miR166, and miR159 (Franco-Zorrilla et al., 2007). Thus, lncRNAs were observed to control flower and seed development via action as sponge for miR160 and miR166. Activity of lncRNAs extends to short interfering RNAs (siRNAs) also, which has been elucidated in monocots like rice (Ding et al., 2012a,b).

lncRNAs have also been seen to act as guides for binding chromatin-modifying enzymes for elicitation of local chromatin structure modifications to influence flower development. Transcriptional regulation has been observed via control of *FLOWERING LOCUS C (FLC)* gene expression by *cold-induced long antisense intragenic RNA (COOLAIR)* (Swiezewski et al., 2009). Further, lncRNAs as scaffolds have emerged as the most common mechanism of regulatory control which guides assembly of multiple types of molecular complexes in formation of ribonucleoprotein complexes (Jing et al., 2008; Miao-Chih et al., 2010). The phenomenon has been exemplified by *cold-assisted intronic noncoding RNA, (COLDAIR)*, which via binding to *polycomb repressive complex-2 protein (PRC2) CURLY LEAF (CLF)* could attach to FLC locus. This complex results in trimethylation of histone chromatin marker (*H3K27me3*) (Heo and Sung, 2011). *COLDAIR* plays a critical role in vernalization which is a well-studied and important phenomenon in plants, especially for long-day plants like members of Brassicaceae where it directly impairs growth and productivity. Interestingly, it possesses a dual role for PRC2 protein; acting as scaffold as well as guide for targeting PRC2 to required sites (Swiezewski et al., 2009). COOLAIR and COLDAIR, though both correspond to cold stress, exhibit different mechanisms. While the former induces silencing by triggering transcription termination, the latter acts by directly recruiting silencing factors (Heo and Sung, 2011). Both these cold stress responders appear to act in a spatiotemporal specific pattern for transcriptional regulation. Taking a closer look at the role and mechanism of lncRNAs in flowering, *FLC* is the chief locus associated with vernalization in *A. thaliana* (Helliwell et al., 2006). When plants are not exposed to low temperature, *FLC* expression levels are high and high levels of trimethylation are observed for *H3K4* and *H3K6* too. On other hand, *H3K27me3* expression is low, but with onset of winters, FLC expression goes down under the influence of pathway-genes and repressors (Johanson et al., 2000; He, 2009; Richard, 2010). *COOLAIR* present at the 3′ end of *FLC* accumulates during vernalization (Carbonell and Rodriguezescudero, 2015) and has a putative role in silencing the locus. Recent findings have also identified FT, a flowering time regulatory gene which controls the expression of *FLC* locus via *COOLAIR* (Chen and Penfield, 2018). Thus, the feedback loop governed by *COOLAIR* is a major checkpoint for flowering time as well as seed dormancy.

Another lncRNA associated with vernalization is *COLDAIR*, 5′-end capped, non-polyadenylated, and transcribed from intron 1 of *FLC* locus (Csorba et al., 2014). It can stimulate accumulation of *H3K9me2* and *H3K27me3*, via recruitment of *PRC2*, resulting in silencing of *FLC* expression (Helliwell et al., 2011; Heo and Sung, 2011). Transcription of *COOLAIR* precedes *COLDAIR* during the course of vernalization (Michaels and Amasino, 1999). FLC-inhibiting promoter keeps *FLC* expression in a stable repressive state by *producing cold of winter-induced noncoding RNA from promoter (COLDWRAP)* (Kim and Sung, 2017). Although intricacies of co-regulatory relationship among these are yet to be explored, the role of lncRNAs in flowering is well-established. Further, genome-wide studies in *A. thaliana* using custom-made NAT arrays identified another lncRNA *called CDF5 LONG NONCODING RNA (FLORE)* (Ding et al., 2012a,b; Henriques et al., 2017; Liang et al., 2017; Liu et al., 2018). *CDF5* is associated with delay in flowering via binding to *FT* and promoters of *CONSTANS*, while *FLORE* promotes transcript expression levels of *FT* (Ding et al., 2012a,b; Henriques et al., 2017; Liang et al., 2017; Liu et al., 2018). Liu and colleagues analyzed the comparative transcriptomes in vernalized *Brassica rapa*, which elucidated the relationship of *BraZF-HD21* and lncRNA locus TCONS_00035129. *BraZF-HD21* is a photo-period-responsive gene which is active during vernalization too. The lncRNA locus was found to be associated with plant hormone signal transduction cascade involved in production of gibberellin A_3, thereby establishing the role of lncRNAs in hormone-mediated regulation of vernalization (Liu et al., 2018). A recent short-term cold treatment study in *Brassica rapa* indicated that cold stress-responsive lncRNAs can vary among species of the family. NAT lncRNAs were found in *BrFLC* and *BrMAF* genes, which were found to be different from studies in *A. thaliana* during vernalization (Shea et al., 2019). These findings present the necessity for investigation of species-specific lncRNAs for better understanding of lncRNA-mediated regulation of vernalization.

13.4 Role of lncRNAs in stress response

Abiotic as well as biotic stress leads to multiple physiological, biochemical, and phenotypic changes, ultimately affecting the photosynthetic capacity and subsequent plant productivity (Hirayama and Shinozaki, 2010). Understanding of the cellular processes affected under myriad stress is a prerequisite for development of stress management and crop improvement strategies. Initially, stress-inducible molecular response studies were focussed on small noncoding RNAs like siRNAs and miRNAs, which have started moving towards lncRNAs. Among the poly-adenylated and non-polyadenylated class of lncRNAs, latter have shorter transcripts and depict lower expression levels. However, this class shows substantial expression specificity in stress responses (Di et al., 2014). Additionally, drought and heat stress further enrich the expression of these lncRNAs as observed in *A. thaliana* (Di et al., 2014). Wu and colleagues observed the role of AtR8 lncRNA in hypoxic stress and shed light on its conservation status in Brassicaceae. Five possible orthologs of *AtR8* depicted significant conservation at the level of secondary structure, but less conservation at the level of nucleotide sequence. Similarly, *At4* lncRNA has been linked with phosphate starvation response and *npc48* and *npc536* have been associated with plant development and stress response in Arabidopsis (Amor et al., 2009; Shin et al., 2006; Franco-Zorrilla et al., 2007; Wirth and Crespi, 2009). Wang and colleagues used NGS to identify differentially expressed lncRNAs in *B. hexaploid*, *B. carinata*, and *B. rapa* to find lncRNAs responsive to immune responses, reproduction, pigmentation, and heat and cold treatments (Wang et al., 2018).

A natural long noncoding antisense RNA, *asHSF2α*, has been known to act as a regulator of heat stress response as it counteracts the expression of heat stress transcription factor, *HSFB2α*. Knockdown and overexpression studies of *asHSF2α* and *HSFB2α* revealed their antagonistic relationship in *A. thaliana*, thus emphasizing the role of lncRNAs in biomass production during early vegetative development (Wunderlich et al., 2014). *Induced by Phosphate Starvation 1 (IPS1)* is an endogenous lncRNA, which controls activity of miR399 by acting as an eTM (endogenous target mimic) (Franco-Zorrilla et al., 2007; Wu et al., 2013). IPS1 acts like a decoy for miR399, thereby preventing its pairing with complementary mRNA targets (Franco-Zorrilla et al., 2007). It was the first eTM lncRNA discovered in *A. thaliana*, which is involved in control of phosphate response. Two transcripts involved in phosphate homeostasis, *IPS1 (Induced by phosphate starvation 1)* and *PHO2 (Phosphate 2)*, have a similar binding site which requires competitive pairing with miR399. IPS1, however, forms a bulge on the position meant for cleavage via binding to miR399, which in turn prevents its cleavage and allows sequestration of miRNA leading to deregulation of its original target PHO2.

eTMs have recently been investigated in several plant species having completely sequenced genomes (Banks et al., 2012; Ivashuta et al., 2011; Meng et al., 2012). Summonwar and colleagues investigated the role of lncRNAs in *Plasmodiophora brassicae*-infected roots of *Brassica napus* lines. A total of 530 differentially expressed lncRNAs were identified including long intergenic RNAs and natural antisense transcripts from clubroot-susceptible and resistant lines. Several noncoding RNAs were active as target mimics of microRNAs (miRNAs) or as precursors of miRNAs. Thus, an evidence of lncRNAs in conferring resistance to diverse *P. brassicae* pathogens was observed (Summanwar et al., 2019). Similarly, Wang and colleagues discovered noncoding natural antisense transcripts associated with approximately 70% of annotated mRNAs, while Zhu and colleagues identified *Fusarium oxysporum*-induced lncRNAs in *A. thaliana* (Wang et al., 2014a,b; Zhu et al., 2014). Analysis in *B. napus* led to identification of lncRNAs responsive to *Sclerotinia sclerotiorum* infection (Joshi et al., 2016). Several lncRNAs were found to be precursors for miRNAs like miR156, miR169, and miR395, which have been known for their role in mediation of phytopathogen response in plants.

Finally, lncRNAs have also been known to affect the commercial aspect of Brassicaceae family via potential regulation of *BnaC08g11970D*, which encodes a protein present in oil bodies and is responsible for lipid accumulation in seeds (Shen et al., 2018).

13.5 Future perspectives

lncRNA research is still in its infancy in plants as most of the studies have been initiated recently (Zhu and Wang, 2012). The advances in computational biology and next-generation sequencing have permitted the speedy identification and functional characterization of novel and conserved lncRNAs in plants. Therefore, there is a necessity to improve high-throughput sequencing technologies and utilize databases to identify lncRNAs and their functions more accurately in Brassicaceae family. Functional cross talk between miRNAs and lncRNAs to regulate the development and stress response is an essential topic of research, along with elucidation of complex regulatory networks. It is, therefore, important to explore the role of lncRNAs in other important crops of Brassicaceae family for future research.

13.6 Conclusions

lncRNAs are involved in different biological processes in plants. In this review, we have discussed the roles of lncRNAs in development and stress responses in Brassicaceae. lncRNA studies have highlighted tissue-specific expression of lncRNAs, thereby indicating their crucial role in plant development. We have also underlined the significant information on lncRNAs and their involvement in different functions in abiotic and biotic stresses. Although information on regulation of abiotic and biotic stress response by lncRNAs is gradually growing, complete understanding of their mechanisms of action is still needed to be accomplished. lncRNAs as eTMs can be used in transgenics which can encourage solutions to produce superior crop cultivars. Regulation of reproductive development via lncRNAs can be utilized in developing sterile lines for producing hybrids. The advancements in lncRNAs research will bring new opportunities to explore different aspects of plant development and their response towards biotic and abiotic stresses in designing crop improvement strategies.

References

Ahmed, W., Xia, Y., Li, R., Bai, G., Siddique, K.H.M., Guo, P., 2019. Non-coding RNAs: functional roles in the regulation of stress response in Brassica crops. Genomics. https://doi.org/10.1016/j.ygeno.2019.08.011.

Amor, B., Wirth, S., Merchan, F., Laporte, P., d'Aubenton-Carafa, Y., Hirsch, J., et al., 2009. Novel long non-protein coding RNAs involved in Arabidopsis differentiation and stress responses. Genome Res. 19, 57–69.

Bai, L., Chen, Q., Jiang, L., et al., 2019. Comparative transcriptome analysis uncovers the regulatory functions of long noncoding RNAs in fruit development and color changes of Fragaria pentaphylla. Hortic. Res. 6, 42.

Banks, I.R., Zhang, Y., Wiggins, B.E., Heck, G.R., Ivashuta, S., 2012. RNA decoys: an emerging component of plant regulatory networks? Plant Signal. Behav. 7, 1188–1193.

Batista, P.J., Chang, H.Y., 2013. Long noncoding RNAs: cellular address codes in development and disease. Cell 152, 1298–1307.

Bazin, J., Bailey-Serres, J., 2015. Emerging roles of long non-coding RNA in root developmental plasticity and regulation of phosphate homeostasis. Front. Plant Sci. 9 (6), 400.

Bazin, J., Baerenfaller, K., Gosai, S.J., Gregory, B.D., Crespi, M., Bailey-Serres, J., 2017. Global analysis of ribosome-associated noncoding RNAs unveils new modes of translational regulation. Proc. Natl. Acad. Sci. U. S. A. 114, E10018–E10027.

Bhatia, G., Goyal, N., Sharma, S., Upadhyay, S.K., Singh, K., 2017. Present scenario of long non-coding RNAs in plants. Non-Coding RNA 3 (2), 16.

Böhmdorfer, G., Sethuraman, S., Rowley, M.J., Krzyszton, M., Rothi, M.H., Bouzit, L., Wierzbicki, A.T., 2016. Long non-coding RNA produced by RNA polymerase V determines boundaries of heterochromatin. elife 5, e19092.

Brant, E.J., Budak, H., 2018. Plant small non-coding RNAs and their roles in biotic stresses. Front. Plant Sci. https://doi.org/10.3389/fpls.2018.01038.

Carbonell, P., Rodriguezescudero, A.I., 2015. The Solanum commersonii genome sequence provides insights into adaptation to stress conditions and genome evolution of wild potato relatives. Plant Cell 27, 954–968.

Cartea, E., Haro-Bailón, A.D., Padilla, G., Obregón-Cano, S., Rio-Celestino, M., Ordás, A., 2019. Seed oil quality of Brassica napus and Brassica rapa germplasm from Northwestern Spain. Foods 8 (8), 292.

Chen, M., Penfield, S., 2018. Feedback regulation of COOLAIR expression controls seed dormancy and flowering time. Science 360, 1014–1017.

Csorba, T., Questa, J.I., Sun, Q., Dean, C., 2014. Antisense COOLAIR mediates the coordinated switching of chromatin states at FLC during vernalization. Proc. Natl. Acad. Sci. U. S. A. 111, 16160–16165.

Di, C., Yuan, J., Wu, Y., Li, J., Lin, H., Hu, L., Zhang, T., Qi, Y., Gerstein, M.B., Guo, Y., Lu, Z.J., 2014. Characterization of stress-responsive lncRNAs in Arabidopsis thaliana by integrating expression, epigenetic and structural features. Plant J. 80 (5), 848–861.

Ding, J., Lu, Q., Ouyang, Y., Mao, H., Zhang, P., Yao, J., Xu, C., Li, X., Xiao, J., Zhang, Q., 2012a. A long noncoding RNA regulates photoperiod-sensitive male sterility, an essential component of hybrid rice. Proc. Natl. Acad. Sci. U. S. A. 109, 2654–2659.

Ding, J., Shen, J., Mao, H., Xie, W., Li, X., Zhang, Q., 2012b. RNA-directed DNA methylation is involved in regulating photoperiod-sensitive male sterility in rice. Mol. Plant 5, 1210–1216.

Franco-Zorrilla, J.M., Valli, A., Todesco, M., Mateos, I., Puga, M.I., Rubio-Somoza, I., Leyva, A., Weigel, D., Garcia, J.A., Paz-Ares, J., 2007. Target mimicry provides a new mechanism for regulation of microRNA activity. Nat. Genet. 39, 1033–1037.

He, Y., 2009. Control of the transition to flowering by chromatin modifications. Mol. Plant 2, 554–564.

Helliwell, C.A., Wood, C.C., Robertson, M., James, P.W., Dennis, E.S., 2006. The Arabidopsis FLC protein interacts directly in vivo with SOC1 and FT chromatin and is part of a high-molecular-weight protein complex. Plant J. 46, 183–192.

Helliwell, C.A., Masumi, R., Jean, F.E., Buzas, D.M., Dennis, E.S., 2011. Vernalization-repression of Arabidopsis FLC requires promoter sequences but not antisense transcripts. PLoS One 6 (6), e21513.

Henriques, R., Wang, H., Liu, J., Boix, M., Huang, L.F., Chua, N.H., 2017. The antiphasic regulatory module comprising CDF5 and its antisense RNA FLORE links the circadian clock to photoperiodic flowering. New Phytol. 216, 854–867.

Heo, J.B., Sung, S., 2011. Vernalization-mediated epigenetic silencing by a long intronic noncoding RNA. Science 331, 76–79.

Hirayama, T., Shinozaki, K., 2010. Research on plant abiotic stress responses in the post genome era: past, present and future. Plant J. 61, 1041–1052.

Hou, J., Lu, D., Mason, A.S., Li, B., Xiao, M., An, S., Fu, D., 2019. Non-coding RNAs and transposable elements in plant genomes: emergence, regulatory mechanisms and roles in plant development and stress responses. Planta 250, 23–40.

Ivashuta, S., Banks, I.R., Wiggins, B.E., Zhang, Y., Ziegler, T.E., Roberts, J.K., et al., 2011. Regulation of gene expression in plants through miRNA inactivation. PLoS One 6, e21330.

Jing, Z., Bryan, K.S., Jennifer, A.E., Ji-Joon, S., Jeannie, T.L., 2008. Polycomb proteins targeted by a short repeat RNA to the mouse X chromosome. Science 322, 750–756.

Johanson, U., West, J., Lister, C., Michaels, S., Amasino, R., Dean, C., 2000. Molecular analysis of FRIGIDA, a major determinant of natural variation in Arabidopsis flowering time. Science 290, 344–347.

Joshi, R.K., Megha, S., Basu, U., Rahman, M.H., Kav, N.N., 2016. Genome wide identification and functional prediction of long non-coding RNAs responsive to Sclerotinia sclerotiorum infection in Brassica napus. PLoS One 11, e0158784.

Kim, E.D., Sung, S., 2012. Long noncoding RNA: unveiling hidden layer of gene regulatory networks. Trends Plant Sci. 17, 16–21.

Kim, D.H., Sung, S., 2017. Vernalization-triggered intragenic chromatin loop formation by long noncoding RNAs. Dev. Cell 40, 302–312.

Lauressergues, D., Couzigou, J.M., Clemente, H.S., Martinez, Y., Dunand, C., Becard, G., Combier, J.P., 2015. Primary transcripts of microRNAs encode regulatory peptides. Nature 520, 90–93.

Li, S., Vandivier, L.E., Tu, B., Gao, L., Won, S.Y., et al., 2014. Detection of PolIV/RDR2-dependent transcripts at the genomic scale in Arabidopsis reveals features and regulation of siRNA biogenesis. Genome Res. 25, 235–245.

Liang, N., Cheng, D., Cui, J., Dai, C., Luo, C., Liu, T., Li, J., 2017. Vernalisation mediated LncRNA-like gene expression in Beta vulgaris. Funct. Plant Biol. 44, 720–726.

Liu, J., Jung, C., Xu, J., Wang, H., Deng, S., Bernad, L., et al., 2012. Genome-wide analysis uncovers regulation of long intergenic noncoding RNAs in Arabidopsis. Plant Cell 24, 4333–4345.

Liu, T., Wu, P., Wang, Q., et al., 2018. Comparative transcriptome discovery and elucidation of the mechanism of long noncoding RNAs during vernalization in Brassica rapa. Plant Growth Regul. 85, 27–39.

Ma, J., Yan, B.Y., Qin, F., Yang, Y., Hao, X., Yu, J., Zhao, Q., Zhu, D., Ao, G., 2010. Zm401, a short-open reading frame mRNA or noncoding RNA, is essential for tapetum and microspore development and can regulate the floret formation in maize. J. Cell. Biochem. 105, 136–146.

Mali, P., Yang, L., Esvelt, K.M., Aach, J., Guell, M., DiCarlo, J.E., Norville, J.E., Church, G.M., 2013. RNA-guided human genome engineering via Cas9. Science 339, 823–826.

Manchali, S., Murthy, K.N., Patil, B.S., 2012. Crucial facts about health benefits of popular cruciferous vegetables. J. Funct. Foods 4 (1), 94–106.

Meng, Y., Shao, C., Wang, H., Jin, Y., 2012. Target mimics: an embedded layer of microRNA-involved gene regulatory networks in plants. BMC Genomics 3, 197.

Miao-Chih, T., Ohad, M., Yue, W., Nima, M., Wang, J.K., Fei, L., Yang, S., Eran, S., Chang, H.Y., 2010. Long noncoding RNA as modular scaffold of histone modification complexes. Science 329, 689–693.

Michaels, S.D., Amasino, R.M., 1999. Flowering locus C encodes a novel MADS domain protein that acts as a repressor of flowering. Plant Cell 11, 949–956.

Mohammadin, S., Edger, P.P., Pires, J.C., Schranz, M.E., 2015. Positionally-conserved but sequence-diverged: identification of long non-coding RNAs in the Brassicaceae and Cleomaceae. BMC Plant Biol. 15 (1), 217.

Nejat, N., Mantri, N., 2017. Emerging roles of long non-coding RNAs in plant response to biotic and abiotic stresses. Crit. Rev. Biotechnol. 38 (1), 93–105.

Rai, M.I., Alam, M., Lightfoot, D.A., Gurha, P., Afzal, A.J., 2018. Classification and experimental identification of plant long non-coding RNAs. Genomics. https://doi.org/10.1016/j.ygeno.2018.04.014.

Raiola, A., Errico, A., Petruk, G., Monti, D.M., Barone, A., Rigano, M.M., 2018. Bioactive compounds in Brassicaceae vegetables with a role in the prevention of chronic diseases. Molecules 23 (1), 15.

Richard, A., 2010. Seasonal and developmental timing of flowering. Plant J. 61, 1001–1013.

Samuolienė, G., Brazaitytė, A., Viršilė, A., Miliauskienė, J., Vaštakaitė-Kairienė, V., Duchovskis, P., 2019. Nutrient levels in Brassicaceae microgreens increase under tailored light-emitting diode spectra. Front. Plant Sci. https://doi.org/10.3389/fpls.2019.01475.

Shea, D.J., Nishida, N., Takada, S., Itabashi, E., Takahashi, S., Akter, A., Miyaji, N., Osabe, K., Mehraj, H., Shimizu, M., Seki, M., 2019. Long noncoding RNAs in Brassica rapa L. following vernalization. Sci. Rep. 9 (1), 9302.

Shen, E., Zhu, X., Hua, S., Chen, H., Ye, C., Zhou, L., Liu, Q., Zhu, Q.H., Fan, L., Chen, X., 2018. Genome-wide identification of oil biosynthesis-related long non-coding RNAs in allopolyploid Brassica napus. BMC Genomics 9 (1), 745.

Shin, H., Shin, H.S., Chen, R., Harrison, M.J., 2006. Loss of At4 function impacts phosphate distribution between the roots and the shoots during phosphate starvation. Plant J. 45, 712–726.

Shumayla, Sharma, S., Taneja, M., Tyagi, S., Singh, K., Upadhyay, S.K., 2017. Survey of high throughput RNA-seq data reveals potential roles for lncRNAs during development and stress response in bread wheat. Front. Plant Sci. 8, 1019.

Song, J.H., Cao, J.S., Wang, C.G., 2013. BcMF11, a novel non-coding RNA gene from Brassica campestris, is required for pollen development and male fertility. Plant Cell Rep. 32, 21–30.

Summanwar, A., Basu, U., Rahman, H., Kav, N., 2019. Identification of lncRNAs responsive to infection by Plasmodiophora brassicae in clubroot-susceptible and-resistant Brassica napus lines carrying resistance introgressed from rutabaga. Mol. Plant-Microbe Interact. 32 (10), 1360–1377.

Sun, X., Zheng, H., Sui, N., 2018. Regulation mechanism of long non-coding RNA in plant response to stress. Biochem. Biophys. Res. Commun. 503 (2), 402–407.

Swiezewski, S., Liu, F., Magusin, A., Dean, C., 2009. Cold-induced silencing by long antisense transcripts of an Arabidopsis Polycomb target. Nature 462, 799–802.

Tang, H., Bowers, J.E., Wang, X., Ming, R., Alam, M., Paterson, A.H., 2008. Synteny and collinearity in plant genomes. Science (New York, N.Y.) 320, 486–488.

Tyagi, S., Sharma, A., Upadhyay, S.K., 2018. Role of next-generation RNA-Seq data in discovery and characterization of long non-coding RNA in plants. In: Çiftçi, Y.Ö. (Ed.), Next Generation Plant Breeding. IntechOpen, https://doi.org/10.5772/intechopen.72773.

Wang, H., Chung, P.J., Liu, J., Jang, I.C., Kean, M.J., Xu, J., et al., 2014a. Genome-wide identification of long noncoding natural antisense transcripts and their responses to light in Arabidopsis. Genome Res. 24, 444–453.

Wang, Y., Fan, X., Lin, F., He, G., Terzaghi, W., Zhu, D., et al., 2014b. Arabidopsis noncoding RNA mediates control of photomorphogenesis by red light. Proc. Natl. Acad. Sci. U. S. A. 111, 10359–10364.

Wang, R., Zou, J., Meng, J., Wang, J., 2018. Integrative analysis of genome-wide lnc RNA and mRNA expression in newly synthesized brassica hexaploids. Ecol. Evol. 12, 6034–6052.

Wierzbicki, A.T., Haag, J.R., Pikaard, C.S., 2008. Noncoding transcription by RNA polymerase Pol IVb/Pol V mediates transcriptional silencing of overlapping and adjacent genes. Cell 135, 635–648.

Wirth, S., Crespi, M., 2009. Non-protein-coding RNAs, a diverse class of gene regulators, and their action in plants. RNA Biol. 6, 161–164.

Wu, H.J., Wang, Z.M., Wang, M., Wang, X.J., 2013. Widespread long noncoding RNAs as endogenous target mimics for microRNAs in plants. Plant Physiol. 161, 1875–1884.

Wu, X., Shi, T., Iqbal, S., et al., 2019. Genome-wide discovery and characterization of flower development related long non-coding RNAs in Prunus mume. BMC Plant Biol. 19, 64.

Wunderlich, M., Gross-Hardt, R., Schöffl, F., 2014. Heat shock factor HSFB2a involved in gametophyte development of Arabidopsis thaliana and its expression is controlled by a heat-inducible long non-coding antisense RNA. Plant Mol. Biol. 85, 541–550.

Yamada, M., 2017. Functions of long intergenic non-coding (linc) RNAs in plants. J. Plant Res. 130, 67–73.

Yanfei, M., Wu, F., Yu, X., Bai, J., Zhong, W., He, Y., 2014. MicroRNA319a-targeted Brassica rapa ssp. pekinensis TCP genes modulate head shape in Chinese cabbage by differential cell division arrest in leaf regions. Plant Physiol. 164, 710–720.

Yang, Z., Yang, C., Wang, Z., Yang, Z., Chen, D., Wu, Y., 2019. LncRNA expression profile and ceRNA analysis in tomato during flowering. PLoS One 14 (1), e0210650.

Yu, T., Zhu, H., 2019. Long non-coding RNAs: rising regulators of plant reproductive development. Agronomy 9 (2), 53.

Zhang, Y.C., Liao, J.Y., Li, Z.Y., Yu, Y., Zhang, J.P., Li, Q.F., Qu, L.H., Shu, W.S., Chen, Y.Q., 2014. Genome-wide screening and functional analysis identify a large number of long noncoding RNAs involved in the sexual reproduction of rice. Genome Biol. 15, 512.

Zhu, Q.H., Wang, M.B., 2012. Molecular functions of long non-coding RNAs in plants. Genes (Basel) 3, 176–190.

Zhu, Q.H., Stephen, S., Taylor, J., Helliwell, C.A., Wang, M.B., 2014. Long noncoding RNAs responsive to Fusarium oxysporum infection in Arabidopsis thaliana. New Phytol. 201, 574–584.

Chapter 14

Role of long noncoding RNAs in vegetable crops

Aradhana Lucky Hans and Sangeeta Saxena
Department of Biotechnology, Babasaheb Bhimrao Ambedkar University, Lucknow, Uttar Pradesh, India

14.1 Introduction

The researches on the regulatory mechanism of lncRNAs in humans and animals are well-studied and are in a rapid phase of progress. However, in the area of plant research, the study of lncRNAs is yet to catch up its pace. In plants, these lncRNAs go through dynamic regulation and have role in various developmental and stress regulations. These lncRNAs are involved in biological regulatory processes including gene silencing, control of flowering time, fruit development and ripening, photomorphogenesis in seedlings, and organogenesis in roots. LncRNAs can not only regulate gene transcription and epigenetics in the nucleus (Bazin et al., 2017), but are also associated with stability of mRNA and translation in the cytoplasm (Gong and Maquat, 2011). Nevertheless, there are a large number of plant lncRNAs whose regulatory function is not discovered and identified yet. Hence, there is an urgent need for a systematic identification of lncRNAs with characterization which are involved in plant adaptation, diversity, and even speciation. In vegetables, which are further an important section of plants, the research on lncRNAs is still sprouting. The importance of vegetables and their uses is myriad and beyond description.

The changes in the environment have not only affected humans, but the stress has been increasing constantly on plants too. There might be various kinds of biotic and abiotic stresses from which plant suffers during its life span. Abiotic stress mainly includes cold, drought being more severe followed by heat, salt, and light causing limitation in their development and reproductivity. Biotic stress includes bacterial, viral, and pathogen infections and infestation of various pests that cause major damage to vegetable crops productivity. To overcome these stress-based losses in crop yield, it is essential to develop stress-tolerant cultivars of vegetable crops. Plants have the capability to respond and adapt to such stress through various biochemical and physiological complex processes, leading towards acquiring resistance. Gene regulation, both transcriptional or

Long Noncoding RNAs in Plants: Roles in development and stress
https://doi.org/10.1016/B978-0-12-821452-7.00014-3

posttranslational, plays a critical role in establishing cellular homeostasis under biotic and abiotic stresses. Generally, plants react to deprivation of nutrient and salinity mainly through transcriptional regulation, but miRNA-mediated post-transcriptional regulation is also involved in such process which plays a critical role in establishing cellular homeostasis under abiotic and biotic stresses. In recent past, various studies have investigated these stresses in multiple plants, including rice, wheat, tomato, maize, Arabidopsis, cabbage etc. According to these studies, a large number of genes have been identified which are induced by various kinds of stress and are thought to function as key regulators in stress tolerance and response. Emerging technologies like next-generation sequencing, deep sequencing, and bioinformatic approaches reveal that huge number of ncRNAs, especially miRNAs and lncRNAs, have been identified to have a very important role in regulation of expression of gene in response to stress conditions in plants (Tyagi et al., 2018).

In recent times, more proactive researches are being carried out in the field of lncRNAs and their varied role in plants. There is a growing evidence that supports the view that these lncRNAs have a crucial role in regulating responses to a wide range of biotic and abiotic stresses. These lncRNAs are generally 200 nt in length and are involved in various biological and regulatory processes. In plants, mostly the lncRNAs are transcribed by RNA pol II, while there are some that are transcribed by RNAIII or RNA IV/V. LncRNAs are categorized into three types based on their genomic origins, intergenic lncRNA (lincRNAs), intronic lncRNAs (incRNAs), and antisense lncRNAs. Although the scenario of lncRNAs in the field of animal research is at quite an advanced level, it is pacing up its rhythm in plants too. Among plants, a fairly good number of lncRNAs have been reported in model plant, Arabidopsis, and some cereals like rice, maize, wheat, fruits like grapes, kiwi, and few other species (Song and Zhang, 2017; Shumayla et al., 2017; Bhatia et al., 2019). LncRNAs in vegetables are still very sparsely reported. A much more preference towards the staple food is always there, which is important, for example, the involvement of lncRNA in the increase in yield in rice has been reported by overexpressing lncRNA LAIR (Wang et al., 2018a,b). But vegetables are equally essential for nutrition security, and thus, they are getting higher recognition as a source of high mineral, fiber, and nutrient source. They are extremely vulnerable to biotic and abiotic stress and the productivity goes down severely. Thus, the need to see to the alternatives to protect or enhance tolerance in them is the need of the hour. LncRNAs that are involved in various biological and physiological functions of vegetables are reported in potato, tomato, cabbage, cassava, cucumber, etc. (Zhang et al., 2018; Zhou et al., 2019; Kwenda et al., 2016). Detailed investigations on characterization of lncRNA-based regulatory mechanisms underlying ripening, drought, and salinity stress in tomato will be very much valuable as it is an important vegetable highly consumed raw or processed for various sauces. Pepper that is another important member in the vegetable arena seems to have been widely influenced by the regulatory roles of lncRNAs (Zuo et al., 2019).

Although numerous lncRNAs are identified, more are under the process of identification in such vegetables (Table 14.1). The mechanism, more importantly, of lncRNAs regulation as well its function needs to be established for better understanding of the whole process.

14.2 Involvement of lncRNAs in vegetable

Numerous vegetables are globally cultivated for consumption and they have high nutritional and economical value. In this fast evolving world, there are various health ailments due to unhealthy lifestyle choices. Vegetables are gaining more and more popularity and are highly recommended in the diet of an individual. The agricultural land being limited and the production requirement high, the protection from various stresses to increase yield becomes all the more important. Since lncRNAs are involved in various biological and regulatory functions, few studies are springing up in identifying and establishing new roles of lncRNAs in vegetables (Fig. 14.1). Since plants growth and development are largely regulated by lncRNAs, markers based on these would be extremely helpful in breeding programs. Studies are lacking in this area which can be explored more for the development molecular markers based on lncRNAs for molecular-breeding programs. Among markers, it is well-known that functional markers or perfect markers (if they are part of gene) are more important in genetic studies and breeding programs. Many functional marker systems have been developed in plants which include gene-SSR (Simple sequence repeat) (Li et al., 2018), EST (expressed sequence tag)-SSRs (Cai et al., 2019), target region amplification polymorphism (TRAPs), and sequence-related amplification polymorphism (SRAPs) (da Silva et al., 2016). The lncRNAs are also another kind of potential functional markers which are extremely important as they play a crucial role in gene regulation. In recent study in capsicum, some lncRNAs have been identified for the development of SSR markers. Therefore, the SSRs developed in this study can be useful in mapping QTLs or genes for different traits related to leaf, flower, fruit, and architecture of the plant. Further, the possible targets of lncRNAs were found to be those genes involved in metabolism, DNA repair, genome stability, stress resistance, cytoskeleton organization, etc. (Jaiswal et al., 2020). This clearly shows the potential of SSRs associated to lncRNAs in various studies as well as breeding programs. The lncRNAs-based markers also have higher level of transferability across different species, and therefore, would be extremely helpful in genetic diversity studies and in those cases where genome sequence is not available.

Since lncRNAs regulate the growth and development of plant in several ways, they have an important role in germination and sprouting. Another very economically and nutrition-wise important vegetable is *Solanum tuberosum* or potato. Sprouting is the key factor that affects the quality of potato. Several methods have been utilized to identify various differentially expressed genes in potato from dormancy stage to sprouting (Bachem et al., 2000; Kloosterman

TABLE 14.1 List of LncRNAs reported in different vegetables.

Sl.no	Plant	lncRNAs	Biological function	Reference
1	Soya bean (*Glycine* max)	*GmENOD40*	Nodule formation	Yang et al. (1993)
2	Tomato (*Solanum lycopersicum*)	TPS11	Phosphate homeostasis	Liu et al. (1997)
3	Cucumber (*Cucumis sativus*)	Csm10	Male Sex differentiation	Cho et al. (2005)
4	Cucumber (*Cucumis sativus*)	LincRNA (CU1NC272)	Targeted by csa-miRNA396b	Hao et al. (2015)
5	Tomato (*Solanum lycopersicum*)	LncRNA 1565	Biotic stress Tomato Yellow Leaf Curl Virus	Wang et al. (2015)
6	Potato (*Solanum tuberosum*)	LincRNA1113	Biotic Stress (*Pectobacterium* subsp. *Brasielence*)	Kwenda et al. (2016)
7	Cassava (*Manihot esculenta*)	LincRNA356	Drought Stress	Li et al. (2017)
8	Cucumber (*Cucumis sativus*)	LncRNA CsaIPS1	Responsive to early Pi deficiency	Zhang et al. (2018)
9	Pepper (*Capsicum annuum* L.)	lncRNAs MSTRG.81557.3 and MSTRG.81557.4	Target 15-cis-phytoene synthase, which is a key enzyme in carotenoid synthesis	Zuo et al. (2019)
10	Tomato (*Solanum lycopersicum*)	LncRNA_tomato_025 and LncRNA_tomato_467	Drought stress	Eom et al. (2019)
11	Tomato (*Solanum lycopersicum*)	TCONS_00020980 and TCONS_00035472	Biotic stress TLCV	Zhou et al. (2019)
12	Chinese cabbage (*Brasica rapa* ssp. *chinensis*)	LncRNA (TCONS_00048391)	Heat tolerance	Wang et al. (2019)

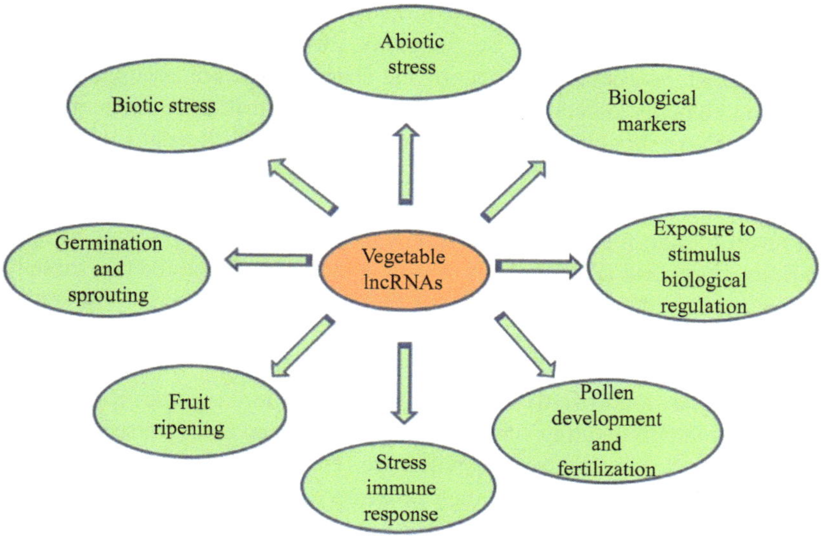

FIG. 14.1 Various biological functions of lncRNAs in vegetables.*(Adapted from Ahmed, W., Xia, Y., Li, R., Bai, G., Siddique, K.H., Guo, P., 2020. Non-coding RNAs: functional roles in the regulation of stress response in Brassica crops. Genomics 112 (2), 1419–1424, Elsevier, License number 4731820115358.)*

et al., 2008). Yet, most of the studies were from the mRNA-coding regions and very few from the noncoding regions. Therefore, lncRNAs having the role in the mechanism of regulation of sprouting in potato are very important. In one such study, a comparison between differential expression of lncRNAs present in apical meristem during dormancy release and sprouting stages was done using lncRNA sequencing. Various bioinformatic tools were involved like Kyoto Encyclopedia of Genes and Genomes (KEGG) and Gene Ontology (GO) analyses were done on the differentially expressed lncRNAs for the dentification of possible precise functions in dormancy and sprouting. A fairly good number of 386 lncRNAs was found to be differentially expressed during potato tuber sprouting and identified as putative targets of 250 miRNAs of potato (Hou et al., 2018). A few other studies on rice, Arabidopsis, also show that similar lncRNAs bind to miRNAs160 and miRNA164, thus playing the role in regulation of reproductive development (Zhang et al., 2014; Wu et al., 2013). Therefore, the differentially expressed lncRNAs can be candidate genes for genetic modification and target selection in potato-breeding programs. Hence, this has open up a way to deluge more studies on the mechanism of potato sprouting regulation done by lncRNAs.

Tomato is the second most consumed vegetables in the world. It is, therefore, a very economically important plant. One other trait that makes it more important is that fleshy fruit ripening of tomato is specific to it and not present in any other model plants. This makes it more exemplary for basic and applied research, particularly in the area of investigations-related regulation of fruit ripening (Gapper et al., 2013; Bergougnoux, 2014). The number of lncRNAs that identified their expression pattern and characteristic are largely unknown. Therefore, to better understand the process of tomato fruit ripening, more number of lncRNAs should be discovered and characterized. Recent studies show that tomato lncRNAs are shorter and have fewer exons as compared to protein-coding genes, a situation evocating lncRNAs from other model plants. The two novel tomato lncRNAs, lncRNA1459 and lncRNA1840, seems to delay the ripening process (Zhu et al., 2015). Therefore, it is suggested that these lncRNAs have an important role in fruit ripening process and need to be studied in more detail. Some lncRNAs are also studied to explore the specific role of RIPENING INHIBITOR (RIN) target lncRNAs in tomato fruit development and ripening. A total of 187 lncRNAs were identified as direct RIN targets, which showed RIN-binding sites in their promoters and had diverse expression pattern between *rin*-mutants and wild-type. Out of these, 6 target lncRNAs have direct binding with RIN in their promoters in vivo and in vitro. With the help of upcoming techniques as CRISPR/Cas9 (Clustered Regularly Interspace Short Palindromic Repeat), which is used to knock out the locus of a single lncRNA, namely lncRNA2155, it is indicated that it delayed fruit ripening in tomato (Yu et al., 2018). The role of lncRNAs in tomato development and fruit ripening becomes more prominent and clear, paving way for further detailed research.

14.3 Identification of lncRNAs in stress response in vegetables

The ever-increasing populations come with the higher food demand in coming years. To meet this, a substantial knowledge about agriculture and factors governing production needs to be understood well. The extremities of climate change caused largely by global warming have had an impact on crop yield. Vegetable crops have not been untouched in this aspect and have encountered great loss in yield. Abiotic and biotic stresses have been major players in cutting down the productivity of vegetables as well as all crops. In general term, abiotic stress includes diverse form of stress, for example, drought, heat, cold, salinity, osmotic shock, heavy metals, and frost. Recent studies related to lncRNAs indicate that more than just transcriptional noises, they can act as regulatory molecules in various biological and developmental processes. There have been emerging evidences that support the role of lncRNAs in stress responses (Eom et al., 2019). In plants, these lncRNAs execute their function in several different ways, a few of which are mentioned below (Fig. 14.2).

LncRNAs can serve as sRNA (small RNA) precursors to produce sRNAs, viz., siRNAs (Small interfering RNA) and miRNAs (micro-interfering RNA), in *Brasicca napus* and poplar (Joshi et al., 2016; Chen et al., 2016). There are various roles these lncRNAs play in response to stress conditions that are represented

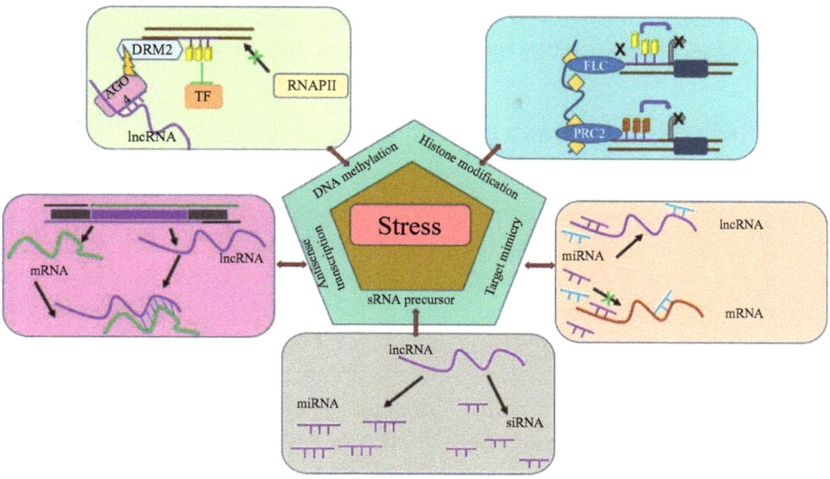

FIG. 14.2 Role of plants lncRNAs in stress conditions.

graphically in Fig. 14.2. The main mechanisms leading to the involvement of lncRNAs in stress response include target mimicry, lncRNA-mediated chromatin modification regulation pathway (lncR2Epi), sRNA precursors, natural antisense transcripts (NATs) pairs, and RNA-directed DNA methylation (RdDM) pathway. There are lncRNAs that can act as ceRNAs (competing endogenous RNA); these are targeted by specific miRNAs through target mimicry, thereby blocking the interactions between miRNAs and their targets (Franco-Zorrilla et al., 2007; Wang et al., 2015). Apparently, there are few plant lncRNAs that are involved in lncRNA-mediated chromatin modification regulation pathway (lncR2Epi). Expression of cold-induced lncRNAs, *COOLAIR* AND *COLDAIR*, is well reported to repress the expression of *FLOWERING LOCUS C (FLC)* through the regulation of lncR2Epi in cold-stressed Arabidopsis (Csorba et al., 2014; Heo and Sung, 2011). Also, a few plant antisense lncRNAs are responsive to stress through the interaction with sense mRNAs. Antisense RNAs have been reported to be involved in chromatin modification, alternative splicing, RNA stability, genomic imprinting, and transcription interference (Lindsay et al., 2013; Zhang et al., 2013). A few lncRNAs seem to play a crucial role in RNA-directed DNA methylation (RdDM) pathway towards stress response. The 24 nt siRNAs and lncRNAs are both involved in the establishment *od* de novo cytosine methylation via the RdDM pathway. The RdDM complex is recruited by siRNAs synthesized from RNAPIV-dependent transcripts. These siRNAs get attached to Argonaut 4 (AGO4) to form silencing complex. Following this, RNAPII-dependent lncRNAs recruit AGO4 through interaction with AGO-associated siRNAs. Finally, the AGO4-siRNA-lncRNA complex recruits DNA methyltransferase domains and rearrange methyl transferase 2 (DRM2), resulting in de novo methylation of cytosine in all three sequence contexts. The RdDM pathway is suggested to be very important for biotic and abiotic stress tolerance. Downregulated expressions of RdDM pathway are the key factors that significantly improved tolerance to stress (Matzke and Mosher, 2014).

14.4 Role of lncRNAs in abiotic stress in vegetables

Abiotic stress is a general term, which includes miscellaneous stresses, e.g., chilling, high temperature, osmotic shock, drought, salinity, water logging, wounding, ozone exposure, etc. These stresses are complex in their nature and controlled by networks of different factors that impede crop-breeding strategies. Several advancements have been made in the identification of stress-related genes, which have the potential to increase plant tolerance. Understanding the molecular mechanism of abiotic response is important to improve tolerance using molecular techniques. Proteins such as transcription factors (TFs) generally have been considered as main regulators of gene expression. Several researches have indicated important role of noncoding RNAs (ncRNAs) in the regulation of gene expression and cellular functions through various pathways including transcriptional gene silencing (Knowling and Morris, 2011). Recent studies have identified and established the role of ncRNAs, viz., siRNAs and miRNAs, in controlling abiotic stress tolerance in plants through sequence-dependent gene silencing at both transcriptional and posttranscriptional levels. In comparison to small ncRNAs, only a few lncRNAs have been identified and characterized to show their role in abiotic stress response in plants (Contreras-Cubas et al., 2012; Sunkar et al., 2012).

One of the most widespread and harmful among the all other abiotic stress is the drought stress. Be it soil drought or atmospheric, both lead to dehydration stress in plants. It is a recurring phenomenon and majorly impacts the agricultural productivity of all major vegetables negatively. Under the stress of drought, vegetables undergo various physiological and biochemical responses, like stomatal movement (opening and closing of stomata), change in biosynthetic pathways, morphological changes, etc. These mechanisms are important for yield protection. To deal with such stress, plants evolve complex mechanisms like deeper and thicker root systems, accumulation of osmotic adjustment compounds, and stomatal modulations. A functional novel lncRNA, namely DRIR (drought-induced lncRNA), is identified in Arabidopsis that enhances plants' sensitivity to ABA (Absicic acid) and increased tolerance to drought and salinity (Qin et al., 2017). Few studies show lncRNA-mediated gene regulation in drought response, suggesting that drought-responsive lncRNAs act as putative targets of tomato miRNA. There is some prediction of lncRNA-mRNA interaction, suggesting that drought-responsive lncRNAs act as transcriptional regulators of genes involved in stress tolerance, including response to stimulus, transporter activity, and signaling (Eom et al., 2019). Such interactions of lncRNAs with other molecules will be an interesting area to ponder upon to hone our knowledge of drought-induced molecular mechanisms. Recent study in cassava (*Manihot esculenta*) shows that two differentially expressed lncRNAs, namely LNC_001148 and LNC_000160, regulated six DEGs (Differentially expressed genes) that improved drought tolerance in cassava by modulating its stomatal density (Xiao et al., 2019). Another study in cassava 16 lncRNAs was identified as putative target mimics of known miRNAs in cassava. A good number of 318 lncRNAs are responsive to cold and drought stress that co-expressed consistently or inconsistently with their neighboring genes (Li et al., 2017). Such regulatory network analysis suggests that many lncRNAs were involved in secondary metabolites biosynthesis, sucrose metabolism pathway, and hormonal signal transduction. More such studies in future will bring new functions of lncRNAs in cassava and other vegetables too.

One of the other very important environmental stresses, viz., heat stress, is becoming extremely important due to global warming and affecting the production of many important vegetables.

One such vegetable is Non-heading Chinese cabbage (NHCC), commonly called *bok choy,* which is important in China and now globally grown. It is extremely affected by heat stress which leads to low productivity and affects edible quality. Not much is known regarding the molecular mechanism underlying tolerance for extreme temperature and temperature-regulated pathways in NHCC. Due to the limited genetic information availability, not much can be implemented in the areas of molecular breeding. Recent studies have identified large number of lncRNAs and a co-expression network of these lncRNAs and protein encoding genes has been constructed. This paves a way for explaining physiological responses to high temperature in NHCC (Song et al., 2016). Target mimicry recently has come up to be an important function of lncRNAs in plants. Another study in NHCC brings up several endogenous target mimics (eTMs) for miRNAs, out of which five were validated to be involved in heat tolerance of NH Chinese cabbage. LncRNA (TCONS_00048391) is heat-responsible eTM for bra-miR164a; this gives an indication that it could be a sponge for miRNA-binding. Also, it can be said that it is a ceRNA for the target gene NAC1 (Bra030820), which is responsible for the expression of bra-miR164a in NHCC. This finding brings new light on functions of lncRNAs in NHCC, suggesting that interaction between eTMs and miRNA may exist and play important role in heat tolerance (Wang et al., 2019).

Another important vegetable crop of Brassicaceae family is radish (*Raphanus sativus* L.); this family also includes many other important species with low temperature requirement. It is widely grown in China and worldwide too. Lack of good varieties which are suitable for summer prevents the plantation of radish during hot temperate seasons. Therefore, mechanism underlying heat stress and breeding heat-tolerant cultivars will be extremely significant in such cultivation of heat-sensitive vegetables. One such study has identified 69 differentially expressed lncRNAs and three circRNAs (Circular RNAs) through strand-specific RNA sequencing technology. Through function prediction analysis, few significant pathways related to heat stress in radish are photosynthesis antenna proteins, circadian rhythm-plant, carbon fixation in photosynthetic organisms, and plant hormone signal transduction. One lncRNA-miRNA-mRNA combination has been identified in radish related to heat stress (Yang et al., 2019). Such studies will help in further discovering the molecular regulation network in heat response in radish.

A process that requires a time period of prolonged cold for flowering is known as vernalization. *Brassica rapa* L. is one such agricultural crop that requires vernalization for flowering. Some studies have shown that lncRNAs have important roles in several cold-responsive genes involved in vernalization. Some lncRNAs (around 2088 lncRNAs) have been identified in response to cold treatment in Brassica (Shea et al., 2019). Banana (*Musa* sp.) is an economically important plant and consumed as vegetable in many parts of the world. Cold stress affects the quality and yield of banana too. Some studies have shown that some lncRNAs response to cold stress-related genes (Liu et al., 2018). This would be useful to generate cold-resistant breeding of banana for better performance.

Plants are sessile and have to face all the abiotic stress and cannot escape. Therefore, more molecular mining has to be done to identify such active lncRNAs that have myriad role in regulation of various kinds of abiotic stresses and enable the plant to cope up with the harshness of the environmental conditions.

14.5 Role of lncRNAs in biotic stress in vegetables

Plants are vulnerable to various environmental changes along with varied number of pathogen attacks. Since plants are sessile, the response to pathogen attack is complex; they have evolved a plethora of defense mechanisms to minimize the damage they would probably encounter. Vegetables are under various biotic stresses that cause the loss in yield productivity. They often suffer from bacterial, fungal, viral infections, and attacks of numerous pest and pathogens throughout their cycle of growth and development. The counter attack of plant to such pathogens attack relies on pathogen recognition at the cellular level, which is followed by a complex defense signaling network at molecular level to coordinate transcriptional reprogramming. Recent studies suggest that lncRNAs have a crucial role to play in such biological processes like plant defense mechanism, plant pathogen interaction, etc.

Tomato is one of the most economically important vegetable crops in the world constituting major portion in industrial agriculture. It encounters various attacks of pests and pathogen, causing major loss in production and yield. The Late blight (LB) disease caused by *Phytopthora infestans* is one of the most devastating diseases of tomato causing major economic losses. Several studies have come up showing the role of lncRNAs in plant-pathogen interactions. A tomato lncRNA 16,397 has shown to induce the expression of glutaredoxin genes to enhance resistance against *P. infestans* (Cui et al., 2017). Nevertheless, the mechanism by which lncRNAs in role as ceRNAs suppress the accumulation of miRNAs in tomato *P. infestans* infections was still not clear. At last, Jiang et al. found that lncRNA23468 modulated the accumulation of nucleotide-binding site leucine-rich repeats (NBS-LRRs) by suppressing miRNA482b expression in tomato plants infected with *P. infestans* (Jiang et al., 2019). These findings would help in better understanding of the regulatory mechanisms pertaining to lncRNAs and build up molecular-breeding approaches *wrt* pathogen resistance. With the advance, degradome sequencing revealed that 11 miRNAs that were decoyed by 20 lncRNAs could target 30 genes. During *P.infestans* infection, the expression levels of lncRNA42705 and lncRNA08711 showed a negative correlation with the expression level of miR159 and a positive correlation with the expression levels of MYB genes. Tomato plants in which these two lncRNAs, lncRNA42705 and lncRNA08711, were silenced displayed increased levels of miR159 and decreased levels of MYBs, respectively. From this result, we can say that lncRNAs might function as ceRNAs to decoy miRNAs and affect their target genes in tomato plants, increasing resistance to disease (Cui et al., 2019).

Another major pathogen causing serious economical loss in tomato is Tomato yellow leaf curl virus (TYLCV). A few recent studies have shown that there are some lncRNAs that have sprung up as in combating viral infections; they are involved to be associated with TYLCV defense mechanisms in tomato. Various numbers of lncRNAs have been identified and listed from resistant and susceptible tomato plants (Wang et al., 2018a,b). In another study of 30 out of 30 differentially expressed lncRNAs, 6 showed sequence similarity to tomato mRNAs; however, the functions of three mRNAs are known as of now. The lncRNA TCONS_00005642 was downregulated by TYLCV infection and seems to be associated with Basic-Helix-Loop-Helix (BHLH) Transcription Factor (TF) Solyc02g063440.2. Other two LncRNAs, TCONS_00020980 and TCONS_00035472, showed sequence similarity to pyruvate decarboxylase 2 (Solyc06082140.2) and a gene coding for Ariadne-like ubiquitine ligase (Solyc11g008590.1), respectively. Both the lncRNAs were upregulated by TYLCV infection (Zhou et al., 2019).

A plant pathogenic fungus, *Sclerotinia sclerotiorum,* causes Sclerotina, a serious disease globally of canola (*Brasica napus*), which was used as vegetable and seeds for oil worldwide. A few novel lncRNAs have been identified that are sensitive to *S. sclerotiorum* infection in *B. napus* at different time points. A total of 3181 lncRNAs were found which included 111 natural antisense transcripts, 2821 lincRNAs, 173 lncRNA isoforms, and 76 exonic overlap with reference to coding sequence. A total of 41 lncRNAs were identified as the precursors for miRNAs, namely miR156, miR169, and miR394; these were reported to have important role in mediating plant response to fungal pathogens. LncRNA TCONS_00000966 is transcribed from the antisense strand of the gene related to plant defense with about 90% overlapping regions and is induced during infection stage, indicating that it is involved in the transcriptional regulation of defense-responsive genes during *S. sclerotiorum* infection (Joshi et al., 2016).

Pectobacterium carotovorum subsp Brasiliense (Pcb1692) is an emerging soft rot bacterium causing major threat to potato tuber quality and yield. Pathogen-responsive lincRNAs have been reported in defense response against *Fusarium oxysporum* in few plants. It becomes important to identify such genes in potato for protection from such harmful pathogens. Similar to many plants, most of potato lincRNAs possess single exons. RNA sequence analysis between a tolerant and susceptible potato cultivar showed 559 lincRNAs were responsive to *Pectobacterium carotovorum subsp* Brasiliense. Later, co-expression analysis figured out that 17 from these lincRNAs were highly associated with 12 potato defense-related genes, showing the involvement of potato lincRNAs in potato defense response (Kwenda et al., 2016).

14.6 Conclusion

The short span of discovery of extensive transcription-based studies has significantly broadened the knowledge of lncRNA biology, particularly in different categories of lncRNAs. The journey has to take up a further path so as to understand the functions and mechanisms of lncRNAs in plants and particularly vegetables. Information on the regulation of synthesis and biogenesis of lncRNAs in vegetables is still in its infancy. Studies on identifying factors that control the biogenesis of lncRNAs and utilizing the information along with the components functioning with lncRNAs will enlighten us about the mechanisms of lncRNA function. Most of the plant lncRNAs are expressed at low levels which can be a result of high enrichment of lncRNAs in a specific cell type or by the induction of specific signal. Thus, the isolation and collection of specific cell types combined with transcriptome analysis of single cells might provide new insights into new lncRNA identification.

One of the important aspects of lncRNA biology is the interaction of lncRNAs with Transcription factors (TFs). They play crucial roles in cellular functions and disease response using a wide range of mechanisms. They affect the DNA-binding activity of the TFs through modifying TFs, promoting TFs phosphorylation, and controlling their nuclear localization. Some lncRNAs are also capable of interacting with mediator subunits and regulate mediator complex formation. Another type of lncRNA that needs more focus is enhancer RNAs (eRNAs). These eRNAs are transcribed from enhancer domain or transcription factor-binding sites of genes and may regulate the transcription activities of neighboring genes. In doing so, they might recruit transcription activators or repressors or control chromatin topology. Taking in context the amount of lncRNAs transcribed from flanking region of protein-coding genes, their sequence might have more number of cis-regulatory elements. These elements may play important role in *cis*-regulatory function in expression of neighboring genes and trans-regulatory for expression of potential targets through recruiting transcription activators/repressors.

Therefore, a complete understanding of the multi-facets of vegetable lncRNAs will have an additional advantage in understanding lncRNAs of other plant species. To work more in this area, it is important to take up research on stress regulatory lncRNAs to regulate stress response of various vegetable crops. Functional cross talk between lncRNAs and miRNAs and complex regulatory networks of miRNAs and lncRNAs in response to different kinds of abiotic and biotic stress need to be elucidated. Addressing stress-related problem in vegetables will enable synergistic advancement to lncRNAs, wherein many more lncRNAs are awaiting for their discovery and unleashing their functional validation. Moreover, addressing all angles of the problem will also enable synergistic advances that will allow plant lncRNAs to be better understood. Many discoveries are waiting to be unearthed for myriad plant lncRNAs.

Acknowledgment

ALH and SS would like to acknowledge the funding from DST SERB-NPDF-2019.

Conflict of interest

The authors declare that they have no conflict of interest.

References

Bachem, C., van der Hoeven, R., Lucker, J., Oomen, R., Casarini, E., Jacobsen, E., Visser, R., 2000. Functional genomic analysis of potato tuber life-cycle. Potato Res. 43 (4), 297–312.

Bazin, J., Baerenfaller, K., Gosai, S.J., Gregory, B.D., Crespi, M., Bailey-Serres, J., 2017. Global analysis of ribosome-associated noncoding RNAs unveils new modes of translational regulation. Proc. Natl. Acad. Sci. 114 (46), E10018–E10027.

Bergougnoux, V., 2014. The history of tomato: From domestication to biopharming. Biotechnol. Adv. 32 (1), 170–189.

Bhatia, G., Sharma, S., Upadhyay, S.K., Singh, K., 2019. Long non-coding RNAs coordinate developmental transitions and other key biological processes in grapevine. Sci. Rep. 9, 3552. https://doi.org/10.1038/s41598-019-38989-7.

Cai, K., Zhu, L., Zhang, K., Li, L., Zhao, Z., Zeng, W., Lin, X., 2019. Development and characterization of EST-SSR markers from RNA-Seq data in *Phyllostachys violascens*. Front. Plant Sci. 10, 50.

Chen, M., Wang, C., Bao, H., Chen, H., Wang, Y., 2016. Genome-wide identification and characterization of novel lncRNAs in Populus under nitrogen deficiency. Mol. Genet. Genomics 291 (4), 1663–1680.

Cho, J., Koo, D.H., Nam, Y.W., Han, C.T., Lim, H.T., Bang, J.W., Hur, Y., 2005. Isolation and characterization of cDNA clones expressed under male sex expression conditions in a monoecious cucumber plant (*Cucumis sativus* L. cv. Winter long). Euphytica 146 (3), 271–281.

Contreras-Cubas, C., Palomar, M., Arteaga-Vázquez, M., Reyes, J.L., Covarrubias, A.A., 2012. Non-coding RNAs in the plant response to abiotic stress. Planta 236 (4), 943–958.

Csorba, T., Questa, J.I., Sun, Q., Dean, C., 2014. Antisense COOLAIR mediates the coordinated switching of chromatin states at FLC during vernalization. Proc. Natl. Acad. Sci. 111 (45), 16160–16165.

Cui, J., Luan, Y., Jiang, N., Bao, H., Meng, J., 2017. Comparative transcriptome analysis between resistant and susceptible tomato allows the identification of lnc RNA 16397 conferring resistance to *Phytophthora infestans* by co-expressing glutaredoxin. Plant J. 89 (3), 577–589.

Cui, J., Jiang, N., Hou, X., Wu, S., Zhang, Q., Meng, J., Luan, Y., 2019. Genome-wide identification of lncRNAs and analysis of ceRNA networks during tomato resistance to *Phytophthora infestans*. Phytopathology. PHYTO-04.

da Silva, E.F., de Sousa, S.B., da Silva, G.F., Sousa, N.R., do Nascimento Filho, F.J., Hanada, R.E., 2016. TRAP and SRAP markers to find genetic variability in complex polyploid *Paullinia cupana* var. sorbilis. Plant Gene 6, 43–47.

Eom, S.H., Lee, H.J., Lee, J.H., Wi, S.H., Kim, S.K., Hyun, T.K., 2019. Identification and functional prediction of drought-responsive long non-coding RNA in tomato. Agronomy 9 (10), 629.

Franco-Zorrilla, J.M., Valli, A., Todesco, M., Mateos, I., Puga, M.I., Rubio-Somoza, I., Leyva, A., Weigel, D., García, J.A., Paz-Ares, J., 2007. Target mimicry provides a new mechanism for regulation of microRNA activity. Nat. Genet. 39 (8), 1033.

Gapper, N.E., McQuinn, R.P., Giovannoni, J.J., 2013. Molecular and genetic regulation of fruit ripening. Plant Mol. Biol. 82 (6), 575–591.

Gong, C., Maquat, L.E., 2011. lncRNAs transactivate STAU1-mediated mRNA decay by duplexing with 3′ UTRs via Alu elements. Nature 470 (7333), 284.

Hao, Z., Fan, C., Cheng, T., Su, Y., Wei, Q., Li, G., 2015. Genome-wide identification, characterization and evolutionary analysis of long intergenic noncoding RNAs in cucumber. PLoS One 10 (3). p. e0121800.

Heo, J.B., Sung, S., 2011. Vernalization-mediated epigenetic silencing by a long intronic noncoding RNA. Science 331 (6013), 76–79.

Hou, X., Du, Y., Liu, X., Zhang, H., Liu, Y., Yan, N., Zhang, Z., 2018. Genome-wide analysis of long non-coding RNAs in potato and their potential role in tuber sprouting process. Int. J. Mol. Sci. 19 (1), 101.

Jaiswal, V., Rawoof, A., Dubey, M., Chhapekar, S.S., Sharma, V., Ramchiary, N., 2020. Development and characterization of non-coding RNA based simple sequence repeat markers in capsicum species. Genomics 112 (2), 1554–1564.

Jiang, N., Cui, J., Shi, Y., Yang, G., Zhou, X., Hou, X., Meng, J., Luan, Y., 2019. Tomato lncRNA23468 functions as a competing endogenous RNA to modulate NBS-LRR genes by decoying miR482b in the tomato-*Phytophthora infestans* interaction. Horticult. Res. 6 (1), 28.

Joshi, R.K., Megha, S., Basu, U., Rahman, M.H., Kav, N.N., 2016. Genome wide identification and functional prediction of long non-coding RNAs responsive to *Sclerotinia sclerotiorum* infection in *Brassica napus*. PLoS One 11 (7). p. e0158784.

Kloosterman, B., De Koeyer, D., Griffiths, R., Flinn, B., Steuernagel, B., Scholz, U., Sonnewald, S., Sonnewald, U., Bryan, G.J., Prat, S., Bánfalvi, Z., 2008. Genes driving potato tuber initiation and growth: Identification based on transcriptional changes using the POCI array. Funct. Integr. Genomics 8 (4), 329–340.

Knowling, S., Morris, K.V., 2011. Non-coding RNA and antisense RNA. Nature's trash or treasure? Biochimie 93 (11), 1922–1927.

Kwenda, S., Birch, P.R., Moleleki, L.N., 2016. Genome-wide identification of potato long intergenic noncoding RNAs responsive to *Pectobacterium carotovorum* subspecies brasiliense infection. BMC Genomics 17 (1), 614.

Li, S., Yu, X., Lei, N., Cheng, Z., Zhao, P., He, Y., Wang, W., Peng, M., 2017. Genome-wide identification and functional prediction of cold and/or drought-responsive lncRNAs in cassava. Sci. Rep. 7, 45981.

Li, J., Zhou, R., Endo, T.R., Stein, N., 2018. High-throughput development of SSR marker candidates and their chromosomal assignment in rye (*Secale cereale* L.). Plant Breed. 137 (4), 561–572.

Lindsay, M.A., Griffiths-Jones, S., Wight, M., Werner, A., 2013. The functions of natural antisense transcripts. Essays Biochem. 54, 91–101.

Liu, C., Muchhal, U.S., Raghothama, K.G., 1997. Differential expression of TPS11, a phosphate starvation-induced gene in tomato. Plant Mol. Biol. 33 (5), 867–874.

Liu, W., Cheng, C., Lin, Y., XuHan, X., Lai, Z., 2018. Genome-wide identification and characterization of mRNAs and lncRNAs involved in cold stress in the wild banana (*Musa itinerans*). PLoS One 13 (7). p. e0200002.

Matzke, M.A., Mosher, R.A., 2014. RNA-directed DNA methylation: an epigenetic pathway of increasing complexity. Nat. Rev. Genet. 15 (6), 394–408.

Qin, T., Zhao, H., Cui, P., Albesher, N., Xiong, L., 2017. A nucleus-localized long non-coding RNA enhances drought and salt stress tolerance. Plant Physiol. 175 (3), 1321–1336.

Shea, D.J., Nishida, N., Takada, S., Itabashi, E., Takahashi, S., Akter, A., Miyaji, N., Osabe, K., Mehraj, H., Shimizu, M., Seki, M., 2019. Long noncoding RNAs in *Brassica rapa* L. following vernalization. Sci. Rep. 9 (1), 9302.

Shumayla, Sharma S., Taneja, M., Tyagi, S., Singh, K., Upadhyay, S.K., 2017. Survey of high throughput RNA-seq data reveals potential roles for lncRNAs during development and stress response in bread wheat. Front. Plant Sci. 8, 1019.

Song, Y., Zhang, D., 2017. The role of long noncoding RNAs in plant stress tolerance. In: Plant Stress Tolerance. Humana Press, New York, NY, pp. 41–68.

Song, X., Liu, G., Huang, Z., Duan, W., Tan, H., Li, Y. and Hou, X., 2016. Temperature expression patterns of genes and their coexpression with LncRNAs revealed by RNA-Seq in non-heading Chinese cabbage. BMC Genomics, 17(1), p. 297.

Sunkar, R., Li, Y.F., Jagadeeswaran, G., 2012. Functions of microRNAs in plant stress responses. Trends Plant Sci. 17 (4), 196–203.

Tyagi, S., Sharma, A., Upadhyay, S.K., 2018. Role of next-generation RNA-Seq data in discovery and characterization of long non-coding RNA in plants. In: Çiftçi, Y.Ö. (Ed.), Next Generation Plant Breeding. IntechOpen, https://doi.org/10.5772/intechopen.72773.

Wang, J., Yu, W., Yang, Y., Li, X., Chen, T., Liu, T., Ma, N., Yang, X., Liu, R., Zhang, B., 2015. Genome-wide analysis of tomato long non-coding RNAs and identification as endogenous target mimic for microRNA in response to TYLCV infection. Sci. Rep. 5, 16946.

Wang, Y., Luo, X., Sun, F., Hu, J., Zha, X., Su, W., Yang, J., 2018a. Overexpressing lncRNA LAIR increases grain yield and regulates neighbouring gene cluster expression in rice. Nat. Commun. 9 (1), 1–9.

Wang, J., Yang, Y., Jin, L., Ling, X., Liu, T., Chen, T., Ji, Y., Yu, W., Zhang, B., 2018b. Re-analysis of long non-coding RNAs and prediction of circRNAs reveal their novel roles in susceptible tomato following TYLCV infection. BMC Plant Biol. 18 (1), 104.

Wang, A., Hu, J., Gao, C., Chen, G., Wang, B., Lin, C., Song, L., Ding, Y., Zhou, G., 2019. Genome-wide analysis of long non-coding RNAs unveils the regulatory roles in the heat tolerance of Chinese cabbage (*Brassica rapa* ssp. chinensis). Sci. Rep. 9 (1), 5002.

Wu, H.J., Wang, Z.M., Wang, M., Wang, X.J., 2013. Widespread long noncoding RNAs as endogenous target mimics for microRNAs in plants. Plant Physiol. 161 (4), 1875–1884.

Xiao, L., Shang, X.H., Cao, S., Xie, X.Y., Zeng, W.D., Lu, L.Y., Chen, S.B., Yan, H.B., 2019. Comparative physiology and transcriptome analysis allows for identification of lncRNAs imparting tolerance to drought stress in autotetraploid cassava. BMC Genomics 20 (1), 514.

Yang, W.C., Katinakis, P., Hendriks, P., Smolders, A., de Vries, F., Spee, J., van Kammen, A., Bisseling, T., Franssen, H., 1993. Characterization of GmENOD40, a gene showing novel patterns of cell-specific expression during soybean nodule development. Plant J. 3 (4), 573–585.

Yang, Z., Li, W., Su, X., Ge, P., Zhou, Y., Hao, Y., Shu, H., Gao, C., Cheng, S., Zhu, G., Wang, Z., 2019. Early response of radish to heat stress by strand-specific transcriptome and miRNA analysis. Int. J. Mol. Sci. 20 (13), 3321.

Yu, T., Tzeng, D.T., Li, R., Chen, J., Zhong, S., Fu, D., Zhu, B., Luo, Y., Zhu, H., 2018. Genome-wide identification of long non-coding RNA targets of the tomato MADS box transcription factor RIN and function analysis. Ann. Bot. 123 (3), 469–482.

Zhang, X., Lii, Y., Wu, Z., Polishko, A., Zhang, H., Chinnusamy, V., Lonardi, S., Zhu, J.K., Liu, R., Jin, H., 2013. Mechanisms of small RNA generation from cis-NATs in response to environmental and developmental cues. Mol. Plant 6 (3), 704–715.

Zhang, Y.C., Liao, J.Y., Li, Z.Y., Yu, Y., Zhang, J.P., Li, Q.F., Qu, L.H., Shu, W.S., Chen, Y.Q., 2014. Genome-wide screening and functional analysis identify a large number of long noncoding RNAs involved in the sexual reproduction of rice. Genome Biol. 15 (12), 512.

Zhang, Z., Zheng, Y., Ham, B.-K., Zhang, S., Fei, Z., Lucas, W.J., 2018. Plant lncRNAs are enriched in and move systemically through the phloem in response to phosphate deficiency. J. Integr. Plant Biol. https://doi.org/10.1111/jipb.12715.

Zhou, Y., Cho, W.K., Byun, H.S., Chavan, V., Kil, E.J., Lee, S., Hong, S.W., 2019. Genome-wide identification of long non-coding RNAs in tomato plants irradiated by neutrons followed by infection with tomato yellow leaf curl virus. Peer J 7. p. e6286.

Zhu, B., Yang, Y., Li, R., Fu, D., Wen, L., Luo, Y., Zhu, H., 2015. RNA sequencing and functional analysis implicate the regulatory role of long non-coding RNAs in tomato fruit ripening. J. Exp. Bot. 66 (15), 4483–4495.

Zuo, J., Wang, Y., Zhu, B., Luo, Y., Wang, Q., Gao, L., 2019. Network analysis of noncoding RNAs in pepper provides insights into fruit ripening control. Sci. Rep. 9 (1), 8734.

Further reading

Ahmed, W., Xia, Y., Li, R., Bai, G., Siddique, K.H., Guo, P., 2020. Non-coding RNAs: functional roles in the regulation of stress response in Brassica crops. Genomics 112 (2), 1419–1424.

Chapter 15

Long noncoding RNAs in fruit crops

Vinod Kumar and Adity Majee
Molecular Biology and Biotechnology Division, CSIR-National Botanical Research Institute, Lucknow, India

15.1 Introduction

In today's world, fruit ripening is considered to be one of the most well-thoughtout research topics. The escalation in fruit crop productivity and fruit quality is directly related to the underlying mechanism of fruit ripening. Fruits contain a lot of nutritional properties such as minerals, vitamins, flavonoids, fiber, saponins, carotenoids, and phenol, and apart from increasing its crop productivity, it is mandatory to restore the rich nutritional elements within each fruit. Sugars, organic acids, and ascorbic acids are the three foremost components that play a major role in the rich taste of a fruit (Capitani et al., 2010). In recent years, the role of long noncoding RNAs is under limelight by proving their involvement in plant growth system in *Arabidopsis* (Wang et al., 2014), maize (Li et al., 2014), rice (Zhang et al., 2014), wheat (Xin et al., 2011; Shumayla et al., 2017), *Populus trichocarpa* (Shuai et al., 2014), *Medicago truncatula* (Wang et al., 2015), and tomato (Zhu et al., 2015; Wang et al., 2018a,b). The identified long noncoding RNAs in these crops have been characterized to be under the direct regulation of the genes responsible for hormonal signaling, sugar organic metabolism, flowering, sex differentiation, and other types of reproductive development (Cho and Hsieh, 2015) known to enhance the probability of modifying the agricultural agronomical traits and can be related to the crop productivity.

Fruit ripening is one of the main aspects behind the genetic modification of the fruit development stage. The fruit ripening can be characterized by certain traits such as change in color, flavor, aroma, taste, and metabolism and gene expressing occurring concurrently during this stage. In the recent years, transcriptomic analysis of several fruit crops identified the presence of direct gene targets related to sugar and carbon metabolism, anthocyanin synthesis, and so on. Apart from ripening, the flowering stage also plays an important role in determining the fruit quantity. The role of long noncoding RNAs and their direct target genes

Long Noncoding RNAs in Plants: Roles in development and stress
https://doi.org/10.1016/B978-0-12-821452-7.00015-5

responsible for flowering is earlier highlighted in Arabidopsis as well as citrus fruit crop orange (Heo and Sung, 2011; Wang et al., 2017a,b).

Till date, the contribution of the long noncoding RNAs is not much explored in the fruit crops, except a few such as apple, trifoliate orange, sea buckthorn, kiwifruit, etc. The deep transcriptomic analysis (NGS) in these fruit crops uncovered the contribution of the lncRNAs in flower patterning, fruit maturation as well as ripening. These previous analyses have become important source of information to be used for the genetic improvement of fruit crops and restoring its nutritive value. Most fruit crops are woody perennial plants such as apple, oranges, and so on, making them difficult for micro-propagation by in vitro technique; so, a solution for the ever-increasing demand for quality fruit supply is to genetically modify the crop by manipulating the fruit productivity as well as preserving the rich taste and aroma of the fruit. In this chapter, the role of long noncoding RNAs in the fruit maturation and ripening process that are identified till date is discussed in this study.

15.2 Identification and specific expression pattern of lncRNAs in fruit crops

The research on molecular functions of plant lncRNA is still at an early stage compared to those on lncRNAs in humans and animals (Zhang and Chen, 2013). To date, a large number of lncRNAs have been identified from different fruit crops (Table 15.1).

The present research to identify novel lncRNAs mainly depends on genome-wide transcriptomics, such as RNA sequencing (Tyagi et al., 2018). The majority of fruit crop lncRNAs can be divided into different types, such as Intergenic lncRNAs, antisense lncRNAs, intronic lncRNAs, and sense lncRNAs (Table 15.1). The details of lncRNAs in different fruit crops are discussed in following sections.

15.2.1 Apple (*Malus domestica*)

In the year 2010, *Malus domestica* (apple) whole genomic sequence (17 chromosomes and 57,386 putative genes) provided an opportunity for the researchers to identify novel lncRNAs present in apple and responsible for fruit development and ripening (Velasco et al., 2010). Apple is an economically important fruit crop across the world and the secret is its rich taste and aroma. Earlier studies reported transcriptomic mRNA profiles and proteomic analysis data which suggest that the direct and indirect relation of the vegetative and reproductive development of the apple plant correlates with the quality of the fruit produced (Xing et al., 2016). A comparative analysis identified novel lncRNAs in different tissues in apples; a total of 1726 high-confidence lncRNAs are found in different tissues including shoot and root tips and young fruits and stem. Among these 1726 lncRNAs, 850 are found to be under direct target genes that are associated

TABLE 15.1 List of long noncoding RNAs (lncRNAs) identified in fruit crops.

Plant species	Number of lncRNAs	Tissues	Approaches	Type(s) of lncRNAs	Reference
Apple					
Malus domestica	1726	Stem phloem, young fruits, shoot tips, and root tips	RNA-Seq	Intergenic lncRNAs Antisense lncRNAs	An et al. (2018)
	5297	Fruit	RNA-Seq	Intergenic lncRNAs Antisense lncRNAs Intronic lncRNAs Sense lncRNAs	Yang et al. (2019)
Peach					
Prunus persica	1417	Leaf, flower, and fruit	RNA-Seq	lncRNAs	Wang et al. (2013)
Trifoliate Orange					
Poncirus trifoliata	6584	Floral buds	RNA-Seq	Intergenic lncRNAs Intragenic lncRNAs Antisense lncRNAs	Wang et al. (2017a,b)
Kiwifruit					
Actinidia chinensis	7051	Fruit	RNA-Seq	Intergenic lncRNAs Overlapping lncRNAs Antisense lncRNAs	Tang et al. (2016)

Continued

TABLE 15.1 List of long noncoding RNAs (lncRNAs) identified in fruit crops—cont'd

Plant species	Number of lncRNAs	Tissues	Approaches	Type(s) of lncRNAs	Reference
Sea buckthorn					
Hippophae rhamnoides Linn	3428	Red fruit and yellow fruit	RNA-Seq	Intergenic lncRNAs Intronic lncRNAs Antisense lncRNAs	Zhang et al. (2017)
	9008	Mature green, breaker, and red-ripe fruit	RNA-Seq	Intergenic lncRNAs Antisense lncRNAs Intronic lncRNAs	Zhang et al. (2018)
Musk melon					
Cucumis melo	3857	Fruit	RNA-Seq	Intergenic lncRNAs Antisense lncRNAs	Tian et al. (2019)
Strawberry					
Fragaria vesca	5884	Flower and fruit	RNA-Seq	lncRNAs	Kang and Liu (2015)
Fragaria pentaphylla	25,699	Red fruit and white fruit	RNA-Seq	lncRNAs	Bai et al. (2019a,b)
Grapevine					
Vitis vinifera	2088	Cold-stressed shoot apices with leaves	RNA-Seq	Intergenic lncRNAs Intronic lncRNAs Antisense lncRNAs	Wang et al. (2019)
	56,441	Inflorescence, berry, and leaf	RNA-Seq	Intergenic lncRNAs Intronic lncRNAs Antisense lncRNAs	Bhatia et al. (2019)
Japanese apricot					
Prunus mume	2572	Flower buds	RNA-Seq	lncRNAs	Wu et al. (2019)

Banana					
Musa balbisiana	1319	Root, corm, Bract, lamina, midrib, petiole, peel, male flower, female flower, fruit pulp, male inflorescence, pseudo stem	RNA-Seq	lncRNAs Antisense lncRNAs	Sampangi-Ramaiah et al. (2019)
Musa acuminata	4288				
Pineapple					
Ananas comosus	2888	Leaves	RNA-Seq	Intergenic lncRNAs Antisense lncRNAs	Bai et al. (2019a,b)
	12,536	Leaves and stem	RNA-Seq	Intergenic lncRNAs Antisense lncRNAs Intronic lncRNAs Sense lncRNAs	Wang et al. (2019)
Mulberry					
Morus notabilis	1133	Root, leaf, bark, bud, and male flower	RNA-Seq	Intergenic lncRNAs Antisense lncRNAs Intronic lncRNAs	Song et al. (2016)
Pear					
Pyrus betulifolia	14,478	Six-leaf stage seedlings	RNA-Seq	Intergenic lncRNAs Antisense lncRNAs	Wang et al. (2018a,b)

with fruit ripening hormone signaling and sugar metabolism. The lncRNAs identified in young fruits are mainly related to biological processes, carbohydrate transport, metabolism, posttranslational modification, energy production and conversation, and protein turnover. The presence of enhanced expression levels of eight particular lncRNAs in the younger fruit tissues directs the limelight towards the role of lncRNAs in fruit maturation and ripening (An et al., 2018). Two lncRNAs, TCONS_00044718 and TCONS_00040914, play important role in fruit ripening and sugar metabolism by targeting alpha-amylase and hexokinase (HXK), respectively (Wang et al., 2014; Mahlow et al., 2016; An et al., 2018). Through the miRNA-lncRNA-mRNA network and endogenous target mimic (eTM) analysis, two lncRNAs, MLNC3.2 and MLNC4.6, function as endogenous target mimic for miR156a and inhibit cleavage of *SPL2- like* and *SPL33* transcription factor by miR156a during light-induced anthocyanin accumulation in apple fruit skin (Yang et al., 2019).

15.2.2 Peach (*Prunus persica*)

Peach or *Prunus persica* belongs to Rosaceae family and originated from northwest of China. Peach fruits are rich in essential nutrients as vitamins, flavanols, sugars, catechins, and so on. Apart from ranking among the top three economically important deciduous trees, it also plays a very crucial role acting as a model organism for fruit crop research due to its self-compatibility, small genomic size (~230 Mb), and shorter juvenile phase (2–3 years) (Arús et al., 2012). Over the last few decades, several ESTs genetic maps and BAC libraries were studied to focus on the molecular and the biochemical mechanism responsible for the underlying agricultural traits in peach (Joobeur et al., 2000; Yamamoto et al., 2002; Dirlewanger et al., 2004; Zhebentyayeva et al., 2008; Boudehri et al., 2009; Li and Durbin, 2009; Jiménez et al., 2010; Brandi et al., 2011). The genetic identification of functionally important genes in peach is obstructed by the absence of comprehensive transcript and physical maps. But in the year 2013, high-quality genomic sequence of double haploid peach (The International Peach Genome Initiative 2013) was released and there was a turning point towards a pro-genomic era. However, till date, mostly the molecular studies of peach depend upon the structural genomics (Shulaev et al., 2008; Martínez-Gómez et al., 2011). In a recent report, RNA sequencing was carried out in different tissues of peach and the study revealed presence of long noncoding RNAs (66.2%) in the NTR region, which further suggests that these findings can open directions in understanding the functional genomics of peach in future (Wang et al., 2013).

15.2.3 Trifoliate Orange (*Poncirus trifoliata* L. Raf)

Oranges are among one of the popular citrus fruits all over the world and their fruit quality and productivity are much dependent on their flowering pattern

(Liang et al., 1999; Zhang et al., 2009; Sun et al., 2012). Flowering in citrus fruit crops has been under focus since decades and helps in understanding the mechanism that is necessary for the genetic improvement of the citrus fruit crops as being woody plants regeneration via by micro-propagation is not a feasible and an economical option. Precocious trifoliate orange (MT) is an early flowering mutant of *Poncirus trifoliata and* revealed the presence of nine lncRNAs that act as a direct target of mi156/172 (Wang et al., 2017a,b). As reported earlier, the target of mi156 is *SPL* gene that defines a separate endogenous flowering pathway (Khan et al., 2014) and mi172 is the target of APETALA2 (AP2) gene, a negative regulator of citrus. The results concluded that there might be a crucial role of some lncRNAs in trifoliate orange pathway via mi156/172 pathway. This work laid the ground work for future studies for understanding the flowering pattern in trifoliate orange (Wang et al., 2017a,b).

15.2.4 Kiwifruit (*Actinidia chinensis*)

The flavor of kiwifruit is highly dependent on the balance between soluble sugars and nonvolatile organic acids (Nishiyama et al., 2008). Another interesting work has been carried out by a group of investigators in China and USA, in kiwifruit (*Actinidia chinensis*), "*Hongyang*" cultivar. According to the recent statistics till 2018, China ranks the topmost position among all the countries producing kiwifruit; the work reported transcriptomic analysis of nine RNA libraries from different tissues within a young to mature kiwifruit that identified a set of 7051 long noncoding RNAs. Three lncRNAs (TCONS_00016031, TCONS_00022008 and TCONS_00048293) reported overlapped with one neutral invertase (*Achn178991*) and two sugar transporter genes (*Achn017471* and *Achn319221*) involved in sugar and organic acid metabolism. The gene expression levels of a sugar transporter gene, *Achn017471,* were monitored immediately days after pollination (DAP) and it was observed that the gene expression levels increased quite significantly from immature fruit tissues (20 DAP) to mature fruit tissues (127 DAP) as determined from their FPKM values. This observation concluded the fact that long noncoding RNAs might play a crucial role in sugar content enhancement, which is a key indication towards fruit ripening. This study provided a further vision towards the probable candidate genes of the lncRNAs playing a crucial role in fruit ripening in kiwifruit (Tang et al., 2016).

15.2.5 Sea buckthorn (*Hippophae rhamnoides* Linn)

Sea buckthorn, also known as sea berry, belongs to *Elaeagnaceae* family. Sea buckthorn is mostly found in Finland, Poland, and Germany. Multiple high-throughput sequencing on three stages of fruit development in sea buckthorn fruit was analyzed in a recent report. The results revealed the presence of 9008 long noncoding RNAs, among which 118 differentially expressed lncRNAs were identified. The cis- and trans-regulated targets of the DE-lncRNAs are identified in the biosynthesis of carotenoids, ascorbic acid, and flavonoids. The report further investigates the silencing of two lncRNAs (LNC1 and LNC2) in vivo, and it revealed that acts as endogenous target mimics of mi156a and mi828a and thus reduces *SPL9* expression and further induces *MYB114*, which further leads to significant variations in anthocyanin levels during HPLC analysis (Zhang et al., 2018). According to earlier reports, *mi156-SPL* plays a vital role in anthocyanin biosynthesis in Arabidopsis (Gou et al., 2011) and R2R3 MYBs are known to bind with bHLH to form a complex that initiates the expression of late biosynthetic genes and transfer of protein genes (Gonzalez et al., 2008). This study highlights the complex regulation of anthocyanin biosynthesis in sea buckthorn fruit. Previous researcher also identified three lncRNAs (XLOC_267510, XLOC_338163, and XLOC_169881), which play cis and trans roles in regulation of carotenoid biosynthesis pathways through differentially expression of three target genes (CRTISO, ABA2, CYP707A2) between red and yellow sea buckthorn fruits (Zhang et al., 2017). For further investigations, the targets of the differentially regulated lncRNAs need to be investigated to totally understand the fruit ripening regulatory mechanism of sea buckthorn fruit.

15.2.6 Muskmelon (*Cucumis melo*)

Cucumis melo is a climacteric fruit and the mechanism behind the fruit ripening is largely unknown till date. In a recent report, the transcriptomic analysis of the different tissues at four developmental stages in a melon fruit, cv. Hetao, was analyzed, out of which 1601 differentially expressed lncRNAs were identified. Six lncRNAs out of all are found to have direct target genes (cis- or trans-regulated), LNC_000987, LNC_000693, LNC_001323, LNC_003610, LNC_001263, and LNC_003380, and are identified to be related with fruit ripening mechanism by ethylene regulation and ABA signaling pathway (Tian et al., 2019). In case of climacteric fruits such as banana, apple ethylene exposure is one of the critical criteria during fruit ripening (Osorio et al., 2013). In the analysis, *ETR1* and *ETR2* were found to be upregulated in the post-ripening stage of the fruit when LNC_000866 and LNC_001504 FPKM values are also enhanced initially, but then decreased significantly in the post-ripening stage. These results confirmed that lncRNAs might interact with *ETR1/2* and might play a crucial role in ethylene transduction during fruit ripening of *C. melo*. The report focused on the fact that the biosynthesis and rate of consumption of ethylene is the key element in *C. melo* fruits (Tian et al., 2019).

15.2.7 Strawberry (*Fragaria pentaphylla* and *Fragaria vesca*)

Strawberry, also known as Fragaria, belongs to Rosaceae family and is mostly found in northern and southern America. The economic importance of this fruit varies to a great extent based on its fruit development and ripening. Anthocyanin is one of the most crucial pigments in a strawberry fruit color formation during ripening and is derived from the phenylpropanoid pathway. In earlier studies, the role of several transcription factors like WRKY, MYB, and bHLH and their coordination in regulating few of polypropanoid pathway genes as chalcone synthase (CHS), anthocyanidin synthase (ANS), and flavanone 3-hydroxylase (F3H) in pigment formation during fruit ripening have been discussed. A research work published recently on the transcriptomic analysis of different tissues in *F. pentaphylla* is a clear direction towards decoding the role of long noncoding RNAs in fruit ripening. An increased number of long noncoding RNAs were observed in ripe fruit tissues as compared to green or immature tissues. On the other hand, photosynthetic genes were also noticed to show reduced expression after the fruit mature and ripening. Mainly, three long noncoding RNAs, TN126539_c0_g1, TN131414_c0_g9, and SN.15536, were identified to target several genes and are identified from the regulatory network of the fruit development and color formation in strawberry by analysis of the protein–protein interactions and target predictions of long noncoding RNAs and miRNAs (Bai et al., 2019a,b). In *Fragaria vesca*, 35 different tissue samples from flower and fruit were analyzed by transcriptomic and proteomic sequencing and 5884 fve-lncRNAs coded with the help of 3862 loci. Direct targets of the trans-lncRNAs (C1 and C2 clusters) were identified that were expressed during receptacle fruit-specific repression and activation. The exact mechanism of the C1 and C2 clusters is yet unknown, but it can be proposed that the synergistic and the antagonistic relations between the two clusters might serve as hubs of the network of gene regulation and the downregulation of this network might lead to positive receptacle fruit growth and initiation (Kang and Liu, 2015).

15.2.8 Grapevine (*Vitis vinifera*)

Grapevine is an economically important perennial belonging to Vitaceae family and is native from the northern hemisphere. Due to exposure to the harsh cold temperatures of the northern hemisphere, survival of the plant and also maintenance of fruit quantity and quality levels are the main environmental challenges. In a very recent report (Wang et al., 2019), a detailed transcriptomic analysis is carried out to decode the relationship between the long noncoding RNAs and cold stress tolerance genes. Transcriptomic analysis of grapevine cold-treated plants as compared to control plants showed the upregulated expression of few long noncoding RNAs such as VIT_216s0100n00030, LXLOC_027751, LXLOC_010422, and VIT_202s0025n00100. The expression correlation between cold-inducible grapevine lncRNAs and their target genes in trans-regulatory relationships was higher than in cis-regulatory relationships. Some targets of the cold stress-inducible lncRNAs may have direct targets in abiotic stress response such as VIT_202s0025g01280 (WRKY transcription factor 41) and VIT_216s0100g00380 (CBF4 transcription factor) as both CBF4 and WRKY transcription factors are known to be involved in gene expression under cold stress earlier (Zou et al., 2010). In Bhatia et al. (2019), the co-expression analysis of the grapevine long noncoding RNAs with the targets uncovered the association with biosynthetic and developmental pathways and secondary metabolite biosynthesis pathway. These highlighted the crucial role of lncRNAs to regulate extranuclear genes in plant system.

15.2.9 Japanese apricot (*Prunus mume*)

Another member of the Rosaceae family, Japanese apricot or *Prunus mume*, is another ornamental fruit crop native to China. According to a certain report, analysis between the single pistil and multiple pistil cultivars of Japanese apricot predicted the role of long noncoding RNAs in the pistil differentiation during the flower developmental stage. A total of 89 novel long noncoding RNAs were identified to be differentially expressed in single pistil cultivar as compared to multiple pistil cultivar. Further analysis of the differentially expressed known long noncoding RNAs revealed that lncRNA of XR_514690.2 downregulated its target ppe-miR172d and upregulated AP2, while another lncRNA of TCONS_00032517 induced cytokinin negative regulator gene A-ARR expression via repressing its target miRNA ppe-miR160a/b in the multiple pistil variety. The findings suggested that two long noncoding RNAs of XR_514690.2 and TCONS_00032517 play a significant role in formation of pistil differentiation in multiple pistil apricot cultivar (Wu et al., 2019).

15.2.10 Banana (*Musa* spp.)

Banana (*Musa* spp.) belongs to the Monocotyledons class, Zingiberales order, and to the Musaceae family. Banana is grown in tropical and subtropical areas and is a major staple food crop in many countries in Africa and Asia (Heslop-Harrison and Schwarzacher, 2007). The cultivated edible banana and plantains are evolved from intra- or interspecific crosses between two diploid progenitor species, *Musa acuminata* (genome designated as 'AA'), and *Musa balbisiana* (genome designated as 'BB') (Sampangi-Ramaiah and Ravishankar, 2016). In the year 2012, the whole genome sequence of *M. acuminata* ssp. *malaccensis var.* Pahang provided an opportunity for the researchers to identify novel lncRNAs and explore their function (D'hont et al., 2012). A total of 1319 and 4288 lncRNAs were identified in *M. balbisiana* and *M. acuminate*, respectively (Sampangi-Ramaiah et al., 2019). The predicted lncRNAs were classified and identified as antisense lncRNAs.

Among the identified antisense lncRNAs, 202 were common between two *Musa* species and remaining 1397 and 2024 were unique to *M. balbisiana* and *M. acuminata,* respectively. Further mapping of lncRNAs to the reference genome showed that chromosomes 8 and 3 harbor a maximum number of lncRNAs in *M. balbisiana* and *M. acuminate,* respectively. Further research on deciphering lncRNAs' association and expression pattern under different developmental cues will provide us more insight into function and their regulatory mechanism.

15.2.11 Pineapple (*Ananas comosus*)

Pineapple, an economically important tropical fruit originated from South America, belongs to *Bromiliaceae* family. Apart from being a highly nutritional fruit, pineapple has other applications in the textile industry. The role of long noncoding RNAs in molecular and the biochemical mechanism behind the Crassulacean Acid Metabolism (CAM) photosynthesis strategy remains poorly characterized till date. Whole genome sequencing of pineapple (Ming et al., 2015) opened new areas for identification of direct gene targets and long noncoding RNAs. In a recent report, transcriptomic analysis carried out on different leaf samples led to the identification of 2888 leaf lncRNAs and is expressed in highly tissue-specific manner. Two long noncoding RNAs are found to be related to two photosynthetic pathway genes, PPCK and PEPC, and revealed their diurnal expression pattern in leaves, suggesting their important roles in the CAM photosynthetic pathway. The competing endogenous RNAs (ceRNAs) of PPCK and PEPC genes also exhibited a diurnal expression pattern, suggesting that the involvement of leaf lncRNAs in carbon fixation in CAM is associated with the circadian clock (Bai et al., 2019a,b). Another transcriptomic analysis is reported regarding the ethylene-induced genes responsible during flowering in pineapple. Ethylene has long been reported to induce flowering and fruit ripening in a dose-oriented manner in several crops as well as pineapple (Bartholomew et al., 2003). In a work where illumina short reads sequencing was carried out on samples treated with high concentration (HE) and low concentration (LE) ethylene, 12,536 candidates of long noncoding RNAs were identified. The LE and HE treatments induced drastic changes in transcriptome and DNA methylome, with LE inducing the initial response to flower induction and HE inducing the subsequent response. The dose-dependent induction of FLOWERING LOCUS T-like genes (FTLs) may have contributed to dose-dependent flowering induction in pineapple by ethylene. This study revealed that there might be a candidate network that initiates flowering time in pineapple. This study can prove to be very useful in innovating new areas to understand the underlying mechanism of the role played by ethylene in regulating flower formation in pineapple (Wang et al., 2017a,b).

15.2.12 Mulberry (*Morus notabilis*)

Mulberry (*Morus notabilis*) belongs to the genus *Morus*, which comprises 10–13 species and is worldwide distributed such as in India, China, Japan, North Africa, Arabia, South Europe, etc. and is famous for its medicinal and economic importance (Berg, 2005; Venkatesh Kumar and Chauhan, 2008). Apart from forage for silkworms (*Bombyx mori* L.), Mulberry fruit is attracted worldwide for its wealthy nutrient contents and unique flavor. Mulberry fruit is a rich source of isoprenylated flavonoids, such as sanggenon-type flavanones, Diels-Alder adducts, and flavones (Nomura et al., 2003; Venkatesh Kumar and Chauhan, 2008). In the year 2013, *Morus notabilis* genome sequence (approximately 29,338 protein-coding genes identified) provided an opportunity for the researchers to identify novel lncRNAs and explore their function (He et al., 2013). A total of 1133 novel lncRNAs were identified in different tissues of mulberry and 106 of these lncRNAs exhibited a tissue-specific expression in different tissues (bark, leaf, root, male flower, and winter bud) (Song et al., 2016). The reported lncRNAs were categorized as intergenic lncRNA, intronic lncRNA, and antisense lncRNAs and accounting for 1092 (96.4%), 38 (3.4%), and 3 (0.26%), respectively. Three lncRNAs, Mn_lnc_0132, Mn_lnc_0521,and Mn_lnc_0782, were specifically expressed in male flower and located near the protein–coding genes EXB28594.1 (Protein PROLIFERA), EXB81017.1 (Serine/threonine-protein phosphatase PP), and EXC20310.1 (Phosphoenolpyruvate/phosphate translocator PPT), respectively (Song et al., 2016). One root-specific lncRNA, Mn_lnc_0714, was located adjacent to EXC06697.1 (Sulfate transporter 1.3). Hence, tissue-specific expression of lncRNAs contiguous to protein-coding genes might play crucial regulatory functions in floral organ development and root development in mulberry. Recent finding demonstrated that a novel lncRNA, *MuLnc1*, involved in response of environmental stress through a complex regulatory network of mul-miR3954-*MuLnc1*-siRNAs-*MuCML27* in mulberry (Gai et al., 2018).

15.2.13 Pear (*Pyrus betulifolia*)

Pyrus betulifolia, also known as birch leaf pear, is a widely popular deciduous tree that originates in Tibet and central China. It belongs to Rosaceae family and provides a very subtle flavor and aroma. Since pear is most commercially grown in temperate climatic regions (Wu et al., 2013), drought is one of the major abiotic stresses faced by this fruit crop. Transcriptomic analysis to identify differentially expressed drought-related genes in pear is reported earlier (Li et al., 2016). A recent attempt to identify the lncRNAs related to the drought resistance genes by high-throughput sequencing technology of RNA samples of pear leaves treated under drought stress and 251 drought-responsive lncRNAs has been done. Bioinformatical analysis of the identified long noncoding RNAs revealed putative target genes involved in organic substance metabolic processes, heterocyclic compound binding, metabolic processes, and macromolecule metabolic processes and are validated by real-time PCR. This report focused on new directions for the understanding of the mechanism behind the drought response in pear (Wang et al., 2018a,b).

15.3 Conclusions and future perspectives

Recently, the noncoding RNAs or ncRNAs have gained much importance as novel materials of genetic improvement due to their significant contribution to plant development, especially in the improvement of fruit quantity and quality traits in fruit crops. Till date, the research that had been carried out is the very initiation in this field and a lot remains to be explored. Micro-propagation of fruit crops might not be a feasible way as woody plants are difficult to grow in in vitro conditions and have longer life span (in some fruit crops), longer juvenile period, and shortage of seed production. Innovation of alternate and genetically modified ways of increasing crop productivity in fruits can be the only solution that might solve the ever-increasing demands. The transcriptomic and the proteome analyses in few fruit crops, namely, apple, orange, strawberry, kiwifruit, grapevine, muskmelon, and so on, focused on the point that the cis- or trans-acting direct gene targets of the highly expressed lncRNAs in immature to mature fruit tissues might play a very crucial role in fruit formation, maturation, and ripening. Most researches carried out till date concerning the role of long noncoding RNAs in fruit crops involve the transcriptomic analysis of different plant tissues and identification and validation of lncRNAs by real-time PCR. In very few crops, the putative target genes and their role in the biochemical and metabolic processes in a fruit crop are identified, although the area is poorly unexplored till date. A table provided below shortlists the known long noncoding RNAs and their putative gene targets and their functions (Table 15.2).

From the above summarized table, it can be assumed that mostly the long noncoding RNAs that are identified till date in fruit crops are mostly related to sugar and starch metabolism, carotenoid synthesis, flavonoid biosynthesis, anthocyanin biosynthesis, and so on and are the most crucial biosynthetic pathways related to fruit formation, maturation, and ripening. Uncovering the presence of long noncoding RNAs has helped in identifying underlying secrets of fruit ripening and focused on the research towards numerous directions. But apart from fruit ripening, the role of noncoding RNAs in fruit crops regarding flower differentiation, survival of fruit crops amidst biotic and abiotic stress, and increasing the crop productivity are the few future prospects. Digging further in this newly found expanse might also help us to understand the deep secrets in the world of long noncoding RNAs in fruit crops.

TABLE 15.2 Long noncoding RNAs and their putative gene targets and functions.

lncRNA	Putative target gene	Target gene functions	References
Apple			
TCONS_00044718 TCONS_00040914	Alpha-amylase Hexokinase	Starch metabolism Sucrose pathway	An et al. (2018)
MLNC3.2 MLNC4.6	SPL transcription factors SPL transcription factors	Anthocyanin biosynthesis Anthocyanin biosynthesis	Yang et al. (2019)
Sea buckthorn			
XLOC_267510, XLOC_338163 XLOC_169881	CRTISO ABA2 CYP707A2	Carotenoid biosynthesis Carotenoid biosynthesis Carotenoid biosynthesis	Zhang et al. (2017)
TCONS_00082246 TCONS_00085219 TCONS_01039552 TCONS_00061167 TCONS_00061354	Phytoene synthase (PSY) Chalcone synthase (CHS) Flavanone 3-hydroxylase Flavonol synthase 1 Flavonol synthase 1	Carotenoid biosynthesis Flavonoid biosynthesis Flavonoid biosynthesis Flavonol biosynthesis Flavonol biosynthesis	Zhang et al. (2018)
Japanese apricot			
TCONS_00032517 XR_514690.2	ppe-miR160a/b ppe-miR172d	Cytokinin response genes (A-ARRs) *AP2* (Floral transition)	Wu et al. (2019)

References

An, N., Fan, S., Wang, Y., Zhang, L., Gao, C., Zhang, D., Han, M., 2018. Genome-wide identification, characterization and expression analysis of long non-coding RNAs in different tissues of apple. Gene 666, 44–57.

Arús, P., Verde, I., Sosinski, B., Zhebentyayeva, T., Abbott, A.G., 2012. The peach genome. Tree Genet. Genomes 8 (3), 531–547.

Bai, L., Chen, Q., Jiang, L., Lin, Y., Ye, Y., Liu, P., Wang, X., Tang, H., 2019a. Comparative transcriptome analysis uncovers the regulatory functions of long noncoding RNAs in fruit development and color changes of Fragaria pentaphylla. Horticult. Res. 6 (1), 1–15.

Bai, Y., Dai, X., Li, Y., Wang, L., Li, W., Liu, Y., Cheng, Y., Qin, Y., 2019b. Identification and characterization of pineapple leaf lncRNAs in crassulacean acid metabolism (CAM) photosynthesis pathway. Sci. Rep. 9 (1), 1–11.

Berg, C.C., 2005. Moraceae diversity in a global perspective. Biol. Skrifter 55, 423–440.

Bartholomew, D., Paul, R., Rohrbach, K., 2003. Crop environment, plant growth and physiology. In: The Pineapple: Botany, Production and Uses. CABI Publishing Wallingford, pp. 69–108.

Bhatia, G., Sharma, S., Upadhyay, S.K., Singh, K., 2019. Long non-coding RNAs coordinate developmental transitions and other key biological processes in grapevine. Sci. Rep. 9 (1), 1–14.

Boudehri, K., Bendahmane, A., Cardinet, G., Troadec, C., Moing, A., Dirlewanger, E., 2009. Phenotypic and fine genetic characterization of the D locus controlling fruit acidity in peach. BMC Plant Biol. 9 (1), 59.

Brandi, F., Bar, E., Mourgues, F., Horváth, G., Turcsi, E., Giuliano, G., Liverani, A., Tartarini, S., Lewinsohn, E., Rosati, C., 2011. Study of 'Redhaven' peach and its white-fleshed mutant suggests a key role of CCD4 carotenoid dioxygenase in carotenoid and norisoprenoid volatile metabolism. BMC Plant Biol. 11 (1), 24.

Capitani, D., Mannina, L., Proietti, N., Sobolev, A.P., Tomassini, A., Miccheli, A., Di Cocco, M.E., Capuani, G., De Salvador, R., Delfini, M., 2010. Monitoring of metabolic profiling and water status of Hayward kiwifruits by nuclear magnetic resonance. Talanta 82 (5), 1826–1838.

Cho, K.O., Hsieh, J., 2015. The lncRNA Pnky in the brain. Cell Stem Cell 16 (4), 344–345.

D'hont, A., Denoeud, F., Aury, J.M., Baurens, F.C., Carreel, F., Garsmeur, O., Noel, B., Bocs, S., Droc, G., Rouard, M., Da Silva, C., 2012. The banana (Musa acuminata) genome and the evolution of monocotyledonous plants. Nature 488 (7410), 213–217.

Dirlewanger, E., Cosson, P., Howad, W., Capdeville, G., Bosselut, N., Claverie, M., Voisin, R., Poizat, C., Lafargue, B., Baron, O., Laigret, F., 2004. Microsatellite genetic linkage maps of myrobalan plum and an almond-peach hybrid—Location of root-knot nematode resistance genes. Theor. Appl. Genet. 109 (4), 827–838.

Gai, Y.P., Yuan, S.S., Zhao, Y.N., Zhao, H.N., Zhang, H.L., Ji, X.L., 2018. A novel LncRNA, MuLnc1, associated with environmental stress in mulberry (Morus multicaulis). Front. Plant Sci. 9, 669.

Gonzalez, A., Zhao, M., Leavitt, J.M., Lloyd, A.M., 2008. Regulation of the anthocyanin biosynthetic pathway by the TTG1/bHLH/Myb transcriptional complex in Arabidopsis seedlings. Plant J. 53 (5), 814–827.

Gou, J.Y., Felippes, F.F., Liu, C.J., Weigel, D., Wang, J.W., 2011. Negative regulation of anthocyanin biosynthesis in Arabidopsis by a miR156-targeted SPL transcription factor. Plant Cell 23 (4), 1512–1522.

He, N., Zhang, C., Qi, X., Zhao, S., Tao, Y., Yang, G., Lee, T.H., Wang, X., Cai, Q., Li, D., Lu, M., 2013. Draft genome sequence of the mulberry tree Morus notabilis. Nat. Commun. 4 (1), 1–9.

Heo, J.B., Sung, S., 2011. Vernalization-mediated epigenetic silencing by a long intronic noncoding RNA. Science 331 (6013), 76–79.

Heslop-Harrison, J.S., Schwarzacher, T., 2007. Domestication, genomics and the future for banana. Ann. Bot. 100 (5), 1073–1084.

Jiménez, S., Li, Z., Reighard, G.L., Bielenberg, D.G., 2010. Identification of genes associated with growth cessation and bud dormancy entrance using a dormancy-incapable tree mutant. BMC Plant Biol. 10 (1), 25.
Joobeur, T., Periam, N., Vicente, M.D., King, G.J., Arús, P., 2000. Development of a second generation linkage map for almond using RAPD and SSR markers. Genome 43 (4), 649–655.
Kang, C., Liu, Z., 2015. Global identification and analysis of long non-coding RNAs in diploid strawberry Fragaria vesca during flower and fruit development. BMC Genomics 16 (1), 815.
Khan, M.R.G., Ai, X.Y., Zhang, J.Z., 2014. Genetic regulation of flowering time in annual and perennial plants. Wiley Interdiscip. Rev: RNA 5 (3), 347–359.
Li, H., Durbin, R., 2009. Fast and accurate short read alignment with Burrows–Wheeler transform. Bioinformatics 25, 1754–1760.
Li, L., Eichten, S.R., Shimizu, R., Petsch, K., Yeh, C.T., Wu, W., Chettoor, A.M., Givan, S.A., Cole, R.A., Fowler, J.E., Evans, M.M., 2014. Genome-wide discovery and characterization of maize long non-coding RNAs. Genome Biol. 15 (2), R40.
Li, K.Q., Xu, X.Y., Huang, X.S., 2016. Identification of differentially expressed genes related to dehydration resistance in a highly drought-tolerant pear, Pyrus betulaefolia, as through RNA-Seq. PLoS One 11 (2).
Liang, S.Q., Zhu, W.X., Xiang, W.T., 1999. Precocious trifoliate orange (Poncirus trifoliata L. Raf.) biology characteristic and its stock experiment. ZheJiang Citrus 16, 2–4.
Mahlow, S., Orzechowski, S., Fettke, J., 2016. Starch phosphorylation: Insights and perspectives. Cell. Mol. Life Sci. 73 (14), 2753–2764.
Martínez-Gómez, P., Crisosto, C.H., Bonghi, C., Rubio, M., 2011. New approaches to Prunus transcriptome analysis. Genetica 139, 755–769.
Ming, R., VanBuren, R., Wai, C.M., Tang, H., Schatz, M.C., Bowers, J.E., Lyons, E., Wang, M L., Chen, J., Biggers, E., Zhang, J., 2015. The pineapple genome and the evolution of CAM photosynthesis. Nat. Genet. 47 (12), 1435–1442.
Nishiyama, I., Fukuda, T., Shimohashi, A., Oota, T., 2008. Sugar and organic acid composition in the fruit juice of different Actinidia varieties. Food Sci. Technol. Res. 14 (1), 67–73.
Nomura, T., Fukai, T., Hano, Y., 2003. Chemistry and biological activities of isoprenylated flavonoids from medicinal plants (moraceous plants and Glycyrrhiza species). In: Studies in natural products chemistry. 28. Elsevier, pp. 199–256.
Osorio, S., Scossa, F., Fernie, A., 2013. Molecular regulation of fruit ripening. Front. Plant Sci. 4, 198.
Sampangi-Ramaiah, M.H. and Ravishankar, K.V., 2016. Current status of Banana genome in the age of next generation sequencing. In Banana: Genomics and Transgenic Approaches for Genetic Improvement (pp. 51-59). Springer, Singapore.
Sampangi-Ramaiah, M.H., Ravishankar, K.V., Rekha, A., 2019. Long non-coding RNAs in banana: prediction, mapping and their comparative studies using Musa balbisiana and Musa acuminata transcriptome. Trees 33 (2), 359–369.
Shuai, P., Liang, D., Tang, S., Zhang, Z., Ye, C.Y., Su, Y., Xia, X., Yin, W., 2014. Genome-wide identification and functional prediction of novel and drought-responsive lincRNAs in Populus trichocarpa. J. Exp. Bot. 65 (17), 4975–4983.
Shulaev, V., Korban, S.S., Sosinski, B., Abbott, A.G., Aldwinckle, H.S., Folta, K.M., Iezzoni, A., Main, D., Arús, P., Dandekar, A.M., Lewers, K., Gardiner, S.E., Potter, D., Veilleux, E., 2008. Multiple models for Rosaceae genomics. Plant Physiol. 147, 985–1003.
Shumayla, Sharma, S., Taneja, M., Tyagi, S., Singh, K., Upadhyay, S.K., 2017. Survey of high throughput RNA-seq data reveals potential roles for lncRNAs during development and stress response in bread wheat. Front. Plant Sci. 8, 1019.

Song, X., Sun, L., Luo, H., Ma, Q., Zhao, Y., Pei, D., 2016. Genome-wide identification and characterization of long non-coding RNAs from mulberry (Morus notabilis) RNA-seq data. Genes 7 (3), 11.

Sun, L.M., Ai, X.Y., Li, W.Y., Guo, W.W., Deng, X.X., Hu, C.G., Zhang, J.Z., 2012. Identification and comparative profiling of miRNAs in an early flowering mutant of trifoliate orange and its wild type by genome-wide deep sequencing. PLoS One 7 (8), e43760.

Tang, W., Zheng, Y., Dong, J., Yu, J., Yue, J., Liu, F., Guo, X., Huang, S., Wisniewski, M., Sun, J., Niu, X., 2016. Comprehensive transcriptome profiling reveals long noncoding RNA expression and alternative splicing regulation during fruit development and ripening in kiwifruit (Actinidia chinensis). Front. Plant Sci. 7, 335.

Tian, Y., Bai, S., Dang, Z., Hao, J., Zhang, J., Hasi, A., 2019. Genome-wide identification and characterization of long non-coding RNAs involved in fruit ripening and the climacteric in Cucumis melo. BMC Plant Biol. 19 (1), 1–15.

Tyagi, S., Sharma, A., Upadhyay, S.K., 2018. Role of next-generation RNA-Seq data in discovery and characterization of long non-coding RNA in plants. In: Çiftçi, Y.Ö. (Ed.), Next Generation Plant Breeding. IntechOpen, https://doi.org/10.5772/intechopen.72773.

Velasco, R., Zharkikh, A., Affourtit, J., Dhingra, A., Cestaro, A., Kalyanaraman, A., Fontana, P., Bhatnagar, S.K., Troggio, M., Pruss, D., Salvi, S., 2010. The genome of the domesticated apple (malus × domestica Borkh.). Nat. Genet. 42 (10), 833–839.

Venkatesh Kumar, R., Chauhan, S., 2008. Mulberry: Life enhancer. J. Med. Plant Res. 2 (10), 271–278.

Wang, L., Zhao, S., Gu, C., Zhou, Y., Zhou, H., Ma, J., Cheng, J., Han, Y., 2013. Deep RNA-Seq uncovers the peach transcriptome landscape. Plant Mol. Biol. 83 (4–5), 365–377.

Wang, H., Chung, P.J., Liu, J., Jang, I.C., Kean, M.J., Xu, J., Chua, N.H., 2014. Genome-wide identification of long noncoding natural antisense transcripts and their responses to light in Arabidopsis. Genome Res. 24 (3), 444–453.

Wang, T.Z., Liu, M., Zhao, M.G., Chen, R., Zhang, W.H., 2015. Identification and characterization of long non-coding RNAs involved in osmotic and salt stress in Medicago truncatula using genome-wide high-throughput sequencing. BMC Plant Biol. 15 (1), 131.

Wang, C.Y., Liu, S.R., Zhang, X.Y., Ma, Y.J., Hu, C.G., Zhang, J.Z., 2017a. Genome-wide screening and characterization of long non-coding RNAs involved in flowering development of trifoliate orange (Poncirus trifoliata L. Raf.). Sci. Rep. 7, 43226.

Wang, J., Li, Z., Lei, M., Fu, Y., Zhao, J., Ao, M., Xu, L., 2017b. Integrated DNA methylome and transcriptome analysis reveals the ethylene-induced flowering pathway genes in pineapple. Sci. Rep. 7 (1), 1–11.

Wang, J., Lin, J., Kan, J., Wang, H., Li, X., Yang, Q., Li, H., Chang, Y., 2018a. Genome-wide identification and functional prediction of novel drought-responsive lncRNAs in pyrus betulifolia. Genes 9 (6), 311.

Wang, Y., Gao, L., Li, J., Zhu, B., Zhu, H., Luo, Y., Wang, Q., Zuo, J., 2018b. Analysis of long-non-coding RNAs associated with ethylene in tomato. Gene 674, 151–160.

Wang, P., Dai, L., Ai, J., Wang, Y., Ren, F., 2019. Identification and functional prediction of cold-related long non-coding RNA (lncRNA) in grapevine. Sci. Rep. 9 (1), 1–15.

Wu, J., Wang, Z., Shi, Z., Zhang, S., Ming, R., Zhu, S., Khan, M.A., Tao, S., Korban, S.S., Wang, H., Chen, N.J., 2013. The genome of the pear (Pyrus bretschneideri Rehd.). Genome Res. 23 (2), 396–408.

Wu, X., Shi, T., Iqbal, S., Zhang, Y., Liu, L., Gao, Z., 2019. Genome-wide discovery and characterization of flower development related long non-coding RNAs in Prunus mume. BMC Plant Biol. 19 (1), 64.

Xin, M., Wang, Y., Yao, Y., Song, N., Hu, Z., Qin, D., Xie, C., Peng, H., Ni, Z., Sun, Q., 2011. Identification and characterization of wheat long non-protein coding RNAs responsive to powdery mildew infection and heat stress by using microarray analysis and SBS sequencing. BMC Plant Biol. 11 (1), 61.

Xing, L., Zhang, D., Zhao, C., Li, Y., Ma, J., An, N., Han, M., 2016. Shoot bending promotes flower bud formation by mi RNA-mediated regulation in apple (M alus domestica B orkh.). Plant Biotechnol. J. 14 (2), 749–770.

Yamamoto, T., Mochida, K., Imai, T., Shi, I.Z., Ogiwara, I., Hayashi, T., 2002. Microsatellite markers in peach [Prunus persica (L.) Batsch] derived from an enriched genomic and cDNA libraries. Mol. Ecol. Notes 2, 298–302.

Yang, T., Ma, H., Zhang, J., Wu, T., Song, T., Tian, J., Yao, Y., 2019. Systematic identification of long noncoding RNA s expressed during light-induced anthocyanin accumulation in apple fruit. Plant J. 100 (3), 572–590.

Zhang, Y.C., Chen, Y.Q., 2013. Long noncoding RNAs: new regulators in plant development. Biochem. Biophys. Res. Commun. 436 (2), 111–114.

Zhang, J.Z., Li, Z.M., Yao, J.L., Hu, C.G., 2009. Identification of flowering-related genes between early flowering trifoliate orange mutant and wild-type trifoliate orange (Poncirus trifoliata L. Raf.) by suppression subtraction hybridization (SSH) and macroarray. Gene 430 (1–2), 95–104.

Zhang, Y.C., Liao, J.Y., Li, Z.Y., Yu, Y., Zhang, J.P., Li, Q.F., Qu, L.H., Shu, W.S., Chen, Y.Q., 2014. Genome-wide screening and functional analysis identify a large number of long noncoding RNAs involved in the sexual reproduction of rice. Genome Biol. 15 (12), 512.

Zhang, G., Duan, A., Zhang, J., He, C., 2017. Genome-wide analysis of long non-coding RNAs at the mature stage of sea buckthorn (Hippophae rhamnoides Linn) fruit. Gene 596, 130–136.

Zhang, G., Chen, D., Zhang, T., Duan, A., Zhang, J., He, C., 2018. Transcriptomic and functional analyses unveil the role of long non-coding RNAs in anthocyanin biosynthesis during sea buckthorn fruit ripening. DNA Res. 25 (5), 465–476.

Zhebentyayeva, T., Swire-Clark, G., Georgi, L., Garay, L., Jung, S., Forrest, S., Blenda, A., Blackmon, B., Mook, J., Horn, R., Howad, W., Arus, P., Main, D., Tomkins, J., Sosinski, B., Baird, W., Reighard, G., Abbott, A., 2008. A framework physical map for peach, a model Rosaceae species. Tree Genet Genomics 4, 745–756.

Zhu, B., Yang, Y., Li, R., Fu, D., Wen, L., Luo, Y., Zhu, H., 2015. RNA sequencing and functional analysis implicate the regulatory role of long non-coding RNAs in tomato fruit ripening. J. Exp. Bot. 66 (15), 4483–4495.

Zou, C., Jiang, W., Yu, D., 2010. Male gametophyte-specific WRKY34 transcription factor mediates cold sensitivity of mature pollen in Arabidopsis. J. Exp. Bot. 61 (14), 3901–3914.

Chapter 16

Long noncoding RNAs' involvement in comprehensive development of *Gossypium* species

Krishan Mohan Rai[a,b], Anshulika Rai[b], Kaushik Ghose[a], and Rahul Singh[c]
[a]*Fiber and Biopolymer Research Institute, Department of Plant and Soil Science, Texas Tech University, Lubbock, TX, United States,* [b]*Department of Plant and Microbial Biology, University of Minnesota, Saint Paul, MN, United States,* [c]*Department of Basic and Translational Sciences, School of Dental Medicine, University of Pennsylvania, Philadelphia, PA, United States*

16.1 Introduction

Cotton (*Gossypium* spp.) is one of the world's most important cash crop plants, which is widely cultivated for its natural fibers for the textile industry uses and its by-products like seed oil and seed meal further add to its commercial value. The genus *Gossypium* belongs to the Malvaceae family and it has approximately 50 diploid species distributed among 8 genomes (A–G and K genomes) and ~7 allotetraploid species (Grover et al., 2015; Gallagher et al., 2017). Out of these, only two tetraploid species, *Gossypium hirsutum* (Upland cotton, AD1 genome) and *G. barbadense* (Egyptian cotton, AD2 genome), and two of diploid species, *G. arboreum* (Tree cotton, A2) and *G. herbaceum*, are cultivated commercially (Wendel, 1989; Wendel and Cronn, 2003). Tetraploids cotton species contribute to more than 95% of the world's total cotton production and were formed by a polyploidization event occurred between two diploid species (A-genome progenitor with fibers and D-genome progenitor without fibers) 1–2 million years ago (Wendel and Cronn, 2003; Chen et al., 2007; Rai et al., 2013).

Cotton fiber is a single cell seed trichome which develops in four overlapping developmental stages: initiation, elongation, secondary cell wall deposition, and maturation stages (Haigler et al., 2012). Developmental stages with the clearly distinct role and genome complexity due to the presence of two subgenomes make cotton an excellent experimental material for the study of gene regulatory network involved in plant development and fiber development. Small RNAs, especially the miRNAs, have been shown to play important role in various

Long Noncoding RNAs in Plants: Roles in development and stress
https://doi.org/10.1016/B978-0-12-821452-7.00016-7

biological processes (Ayubov et al., 2019) such as fiber development (Guan et al., 2014; Zhao et al., 2019), anther development (Chen et al., 2018a; Zhang et al., 2018a), and biotic (Pan et al., 2019) and abiotic responses such as drought (Dong et al., 2018; Lu et al., 2019) and temperature (Wang et al., 2016a). With the development of improved next-generation sequencing techniques and novel computational tools, more and more hidden attributes of organismal genomes are unraveling. Long noncoding RNAs (lncRNAs), one of such hidden genomic features, are a new class of noncoding RNAs (ncRNAs). Although this novel class of ncRNAs is better explored in the human and mammalian systems, enthusiasm can be seen in the plant research communities based on the increased reports of genome-scale lncRNAs discoveries and their further characterization in various biological processes starting from normal plant development to stress conditions (Chen et al., 2016, 2018b; Zhang et al., 2016; Bhatia et al., 2017; Lu et al., 2017b; Nithin et al., 2017; Qin et al., 2017; Shumayla et al., 2017; An et al., 2018; De Quattro et al., 2018; Golicz et al., 2018; Shen et al., 2018; Tyagi et al., 2018; Matsui and Seki, 2019; Zhou et al., 2019; Zhu et al., 2019). In the present chapter, we are discussing the recent discoveries made in the area of lncRNAs' involvement in the various cotton-related biological processes.

16.2 Long noncoding RNAs: A brief update of their role in plants

Long noncoding RNAs (lncRNAs) are more than 200 bases in size and do not code for any proteins. LncRNAs are mainly enriched in the nucleus and can be involved in various types of biological processes. LncRNAs share several common features with mRNA, but they can have more specific expression patterns than mRNAs (Quinn and Chang, 2016). Briefly, lncRNAs can be categorized into four major classes such as long intergenic noncoding RNAs (lincRNAs), long intronic noncoding RNAs (incRNAs), long noncoding natural antisense transcripts (NATs), and overlapping lncRNAs (OT-lncRNAs).

The majority of lncRNAs were considered as transcriptional noise, but recent studies have shown the significance of lncRNAs in several biological processes. Characterization of *Arabidopsis Drought-induced lncRNA* (*DRIR*) shows its importance in salt and drought stress (Qin et al., 2017). Further, Arabidopsis lncRNAs, COOLAIR and COLDAIR, have been characterized for their involvement in the regulation of FLOWERING LOCUS C (FLC) (Liu et al., 2015). Arabidopsis *Induced by Phosphate Starvation 1* (IPS1) lncRNA binds and seizes miR399 activity, repressing the miR399-mediated cleavage of *Phosphate2* (PHO2) (Shin et al., 2006). *PHOSPHATE1;2* (*PHO1;2*) is involved in phosphate homeostasis and is regulated by antisense lncRNA in rice (Jabnoune et al., 2013; Reis and Deforges, 2017). Another rice lncRNA, long-day-specific male fertility-associated RNA (LDMAR), regulates the photoperiod-sensitive male sterility in hybrid rice (Ding et al., 2012). Thus, lncRNAs are involved in several essential biological functions and invite for further detailed identification and characterization throughout plant species.

16.3 LncRNAs: Identification and functional characterization in cotton

Researchers have shown an important role played by lncRNAs in broadly every aspect of plant development, and the information is rapidly increasing. In cotton, lncRNAs-related research has already started to give important clues, instead of an early stage. In this section, we are summarizing the recent lncRNAs-related advances made in the cotton plant.

16.3.1 Role in fiber development

Cotton fiber development, being the most crucial aspect of cotton, is also one of the most complicated biological processes. Several studies have identified lncRNAs in various stages of fiber transcriptome and efforts have also been made to find the mechanism involved in lncRNA-mediated regulation of cotton fiber development (Wang et al., 2015; Lu et al., 2016; Hu et al., 2018). Many of them have utilized the precious fiberless cotton mutants to dig deep into the fiber mechanism. In one of the pioneer lncRNAs-related studies in cotton, Wang et al. identified a total of 30,550 lincRNA loci (63% single-exonic) and 4718 lncNAT loci (77% single-exonic) belonging to the *G. barbadense*. Showing consistency with the other reports, the mean length of lncRNAs (504 bp for lincRNAs and 713 bp for lncNATs) was shorter than protein coding (PC) transcripts (1621 bp) (Table 16.1) (Wang et al., 2015). At the subgenomic level, the lncRNAs were identified in a higher number in At subgenome than the Dt subgenome. The overall expression pattern of lncRNAs was lower than that of PC genes with extensive subgenome-specific expression bias and added tissue specificity. At functional enrichment level, the At subgenome-specific lncRNAs were enriched in ribosome assembly, spermine biosynthesis process, and microtubule cytoskeleton organization, whereas Dt subgenome-specific lncRNAs were more enriched in lignin catabolic process, response to biotic stimulus, and carbon utilization. They also showed the involvement of lncRNAs and DNA methylation in the cotton fiber development using a DNA methylation inhibitor Zebularine (Wang et al., 2015). Out of the total lncRNAs, 9917 lncRNAs showed differential expression when treated with Zebularine and 94.4% of them had higher expression in the treated ovules. Using wild types (TM-1, Xuzhou-142, and YZ1), lintless/fuzzless (XZ142WX and XinWX) and linted/fuzzless (n2, GZnn, and GZNn) cotton mutants, lncRNA expression analysis was performed in initiation (−1 and 0 dpa, days postanthesis) and late initiation/early elongation stage (4 and 5 dpa). Hierarchical clustering analysis showed that most lncRNAs were preferentially expressed in WT −1, 0, 4, or 5 dpa than the lintless-fuzzless/linted-fuzzless cotton. In the same study, Wang et al. showed the role of a lncRNA pair, which is involved in miR397 biogenesis, in indirect regulation of a lignin-related gene *Laccase4* (*LAC4*). Laccases have been shown to play an important role in various stages of cotton fiber development (Balasubramanian et al., 2016).

TABLE 16.1 Details of lncRNAs' identification in cotton species.

Species	No. of lncRNAs	Distribution in subclasses			Specific conditions	Literature
		lincRNA	NATs	Intronic		
G. barbadense		30,550	4718		Fiber development	Wang et al. (2015)
G. hirsutum	10,820	9989	678	153	Drought stress	Lu et al. (2016)
G. arboreum	5996	3510	2486			Zou et al. (2016)
G. hirsutum	35,802				Fiber mutants	Hu et al. (2018)
G. hirsutum	1549 (WT) 1635 (WT) 1520 (T) 1635 (T)				Salt stress	Deng et al. (2018)
G. hirsutum *G. barbadense*		2026 1757			Biotic stress	Zhang et al. (2018b)
G. hirsutum	18,333	16,777	1556		Fiber mutants	Salih et al. (2019)

T, NaCl-treated; *WT*, Wild-type.

In a similar study in *G. arboreum* (Tree cotton), Zou et al. have identified 5996 lncRNAs from the leaves and five of the fiber developmental stages including −1, 0, 1, 10, and 15 dpa using the strand-specific transcriptome sequencing (Zou et al., 2016). These lncRNAs were expressed at lower levels than PC genes, but their expression was higher than TE-mRNAs. They further identified six expression clusters with 880 lincRNAs and 741 lncNATs as differentially expressed at the fiber initiation stage. This high number of differentially expressed lncRNAs with several clusters strongly indicates toward active involvement of lncRNAs in this fiber contributing to diploid cotton species.

In another study, Hu et al. identified a total of 35,802 lncRNAs using the 0 and 5 dpa transcriptome data generated from *G. hirsutum* Xu142, its fiberless mutant Xu142*fl*, and three lines with different lint percentages which were developed by crossing them (Hu et al., 2018). Approximately 87.8% of these lncRNAs belonged to the lincRNAs, whereas 9.7% were NATs. The expression analysis of identified lncRNAs indicated toward identified lncRNAs specificity for both, the fiber-attached and fiberless mutant. Out of eight lncRNAs (XLOC_016164, XLOC_029323, XLOC_006504, XLOC_066220, XLOC_092283, XLOC_071567, XLOC_056809, and XLOC_087972) selected for expression validation, XLOC_545639 and XLOC_039050 were highly expressed in Xu142*fl* and XLOC_079089 was highly expressed in the LL, LM, LF, and Xu142 lines. Further, the VIGS-mediated silencing of XLOC_545639 and XLOC_039050 in Xu142*fl* resulted in an increased number of fiber initials, whereas XLOC_079089 resulted in a short fiber phenotype in Xu142 (Hu et al., 2018).

To better understand the fiber developmental process, Salih et al. identified 18,333 lncRNAs with 91.5% lincRNAs and 8.5% NATs (Table 16.1) (Salih et al., 2019). Pathway analysis and Gene Ontology-based functional annotation of differentially expressed lncRNAs indicated fatty acid biosynthesis and fatty acid elongation bioprocess as the key reason behind the absence of fiber in Ligon-lintless-1 (Li) mutants. Based on the expression data, Salih et al. identified two key repressed lncRNAs, LNC_001237 and LNC_017085, which could be due to highly expressed miRNA, miR2950.

16.3.2 Role in abiotic stress response

16.3.2.1 Salt stress

The osmotic and ionic toxicity, along with oxidative stress caused by salt stress, not only disturbs the plant homeostasis, but also brings extensive changes in gene expression (Shi et al., 2015). Several transcription factors and key genes are known to be differentially expressed, but recent discoveries of salt-stress-responsive lncRNAs have opened a new area that could involve distinct mechanism. In cotton, Deng et al. have identified 1549 and 1635 lncRNAs (two replicates) unique to the salt-stress (NaCl)-treated three-leaf stage cotton seedlings in comparison to 1520 and 1635 lncRNAs unique to the two replicates of controls, respectively

(Deng et al., 2018). Overall, they identified 1117 unique lncRNAs with high reliability and all of them were classified as lincRNAs based on their genomic location (Table 16.1). Out of these 1117 unique lncRNAs, 44 were identified as differentially expressed under salt stress. The PC genes co-expressing with these lncRNAs were mainly enriched in various stress-related GO categories such as response to chemical stress, biological regulation, oxidoreductase, and transcription regulators. Seven of these lncRNAs, which were within 5 kb of their putative target protein-coding genes, might be regulating these genes via *cis*-regulation. They further identified lnc_973 and lnc_253 as target mimic for bdi-miR399a (*Brachypodium distachyon*) and ghr-156e, respectively (Deng et al., 2018). They showed lnc_973 co-expressed with miR-ghr-399 and up-regulated in salt stress. Similar reports of lncRNA IPS1 working as decoy or target mimic for the ath-mir399 and interfering into its interaction with its target gene *PHOS2* have been shown previously in Arabidopsis (Franco-Zorrilla et al., 2007).

lncRNA973 was further characterized in detail by Zhang et al. for its positive role in cotton response against salt stress (Zhang et al., 2019b). lncRNA973, which is present between genes Gh_D04G0659 and Gh_D04G0660 at 4th chromosome of Dt subgenome in *G. hirsutum*, has two transcripts X1 and X2, which are mainly localized in the nucleus. The lncRNA973 overexpression lines in Arabidopsis were more tolerant to salt stress with relatively higher root length, seed germination rate, and fresh weight (Zhang et al., 2019b). Zhang et al. also used Virus-Induced Gene Silencing (VIGS) to knockdown the lncRNA973 expression and the silenced lines showed reduced tolerance against salt stress. The ghr-miR399 expression was increased in the silenced lines which resulted in reduced *GhPHO2* expression. Several other transcription factors such as MYB5, WRKY46, NAC29, and ERF62 were also hypothesized to be regulated by lncRNA973 by Zhang et al. (2019b).

16.3.2.2 Drought stress

Drought is another type of abiotic stress that affects plant growth and has a direct influence on yield (Ranjan et al., 2012). Tolerance against drought is a polygenic trait which involves adaptive changes ranging from physiology to gene regulation. Recently, several studies have shown the involvement of lncRNAs in the response against the drought stress (Qi et al., 2013; Shuai et al., 2014; Qin et al., 2017; Sun et al., 2018; Pang et al., 2019). In cotton, Lu et al. performed the genome-wide identification for the lncRNAs in three different stress conditions (control, drought, and rewatering) using the drought-resistant upland cotton variety ZhongH177 (Lu et al., 2016). Using the stringent screening criteria, they identified 10,820 lncRNAs subclassified as 9989 lincRNAs, 153 intronic lncRNAs, and 678 NATs (Table 16.1). Surprisingly, 6470 of these lncRNAs were found to have the protein-coding potential. Further, 196 of these lncRNAs were found as a putative precursor of small RNAs including ~35% as miRNAs. The differential expression analysis showed that ~63% of intronic lncRNAs and 21% of NATs were upregulated, whereas 44% lincRNAs, 57% NATs, and 20% intronic lncRNAs were downregulated in drought stress (Lu et al., 2016). Interestingly, after removing the drought stress by rewatering, 71% of NATs again showed upregulated expression pattern indicating their specificity toward drought stress. Based on the mode of action, they found 9, 247, 20, and 28 *cis*-acting lncRNAs associated with ethylene (ETH), auxin (IAA), gibberellin (GA), cytokinin (CTK), whereas 167, 3086, 239, 334, and 2621 *trans*-acting lncRNAs were associated with ethylene (ETH), auxin (IAA), gibberellin (GA), cytokinin (CTK), and ubiquitin, respectively (Lu et al., 2016). Of the *cis*-acting lncRNAs involved in ETH, two lncRNAs, XLOC_063105 (a lincRNA) and XLOC_115463 (a NAT), probably function in regulating two adjacent coding genes, CotAD_37096 and CotAD_12502, respectively.

16.3.2.3 Drought stress and epigenetics

Drought has been shown to affect DNA methylation in several ways in various crop plants (Wang et al., 2011; Liang et al., 2014). Recently, a novel lncRNA (ecCEBPA) has been shown to regulate DNA methylation level by binding to *DNA methyltransferase1* (DNMT1) (Di Ruscio et al., 2013). In cotton, Lu et al. have shown a decrease in the expression of methyltransferase-related lncRNAs (XLOC_181940, XLOC_005742, and XLOC_289283) in the 5-azacytidine (a general inhibitor of DNA (mainly 5-cytosines) methyltransferases)-treated drought-stressed cotyledons (Lu et al., 2017a). Furthermore, an increase observed in the expression of their target genes (CotAD_50357, CotAD_14650, and CotAD_72292) also confirmed the lncRNAs' involvement. Overall, this publication proposed the role of lncRNAs in the regulation of DNA methylation via their splicing into miRNAs (RNA-directed DNA methylation, RdDM) to control the hormone content in response to drought stress.

16.3.3 Role in biotic stress

The involvement of lncRNAs associated with biotic stress in cotton is largely unsolved. Lack of disease-resistant varieties have been always a major challenge in cotton production. China has faced enormous losses in both cotton yield and quality due to Verticillium wilt (Li et al., 2017a). The causal organism for Verticillium wilt is a soil-borne fungus *Verticillium dahliae* that leads to wilting and chlorosis of leaves, browning of vascular, and ultimately plant death (Li et al., 2014). Zhang et al. initiated the discovery of disease-associated lncRNAs in cotton (Zhang et al., 2018b). Comparison of sequencing and infection data of two distinctive tetraploid cotton species, *G. hirsutum* (susceptible to Verticillium wilt) and *G. barbadense* (resistant to Verticillium wilt), facilitated the identification of 2026 and 1757 lincRNA pairs in *G. hirsutum* and *G. barbadense*, respectively, showing a pattern in expression after infection (Table 16.1). The GO-annotation shows the enrichment of kinase activator activity, fructose-bisphosphate aldolase activity, structure-specific DNA binding signal transducer activity, MAP kinase activity, and superoxide dismutase copper chaperone activity lincRNAs (Zhang et al., 2018b). One of the core differentially regulated lncRNAs, GhlncNAT-ANX2, identified in the same study, was characterized for its involvement in combating the pathogen stress. VIGS-mediated silencing of GhlncNAT-ANX2 showed enhanced resistance to *V. dahliae*, with reduced leaf defoliation and wilting (Zhang et al., 2018b).

Fusarium oxysporum is another plant fungal pathogen that causes Fusarium wilt, characterized by yellowing leaves, wilting, damage to vascular tissue, and death of an infected plant (Li et al., 2017b). The differentially expressed lncRNAs identified from transcriptome and *F. oxysporum* infection data of four sensitive and resistant varieties of *G. barbadense* showed that different varieties respond to *F. oxysporum* infection in different ways. Also, *F. oxysporum* susceptibility was positively correlated with the expression of *F. oxysporum*-responsive lncRNAs. The *F. oxysporum*-responsive lncRNAs were majorly categorized into plant hormone signal transduction, glutathione metabolism, SNARE interactions in vesicular transport, butanoate metabolism, and glycolysis/gluconeogenesis pathways (Yao et al., 2019).

16.3.4 Evolutionary significance of lncRNAs in cotton

Whole-genome duplication events and interspecific hybridizations are the known factors behind the speciation. These genome-level events sometimes trigger a cascade of new interactions and expressional reprogramming, which brings dramatic changes with enhanced traits in the resulting genomes. These forces are the primary reason behind the low level of converseness observed among the mammalian (~35% between human and mouse) (Washietl et al., 2014) as well as plant lncRNAs (Wang et al., 2016b). In a comparative study, Zhao et al. have shown that cotton lncRNAs show *Gossypium* lineage specificity (Zhao et al., 2018). Approximately 86% of the cotton lncRNAs ($n = 13{,}282$) were present in diploid (*G. arboreum* and *G. raimondii*) as well as tetraploid (*G. hirsutum*) species with only 12.43% and 3.83% homologous lncRNAs found in *Theobroma cacao* and Arabidopsis and rice, respectively (Zhao et al., 2018). Further, to show that transposable elements are one of the major driving forces behind the novel lncRNAs origin, Zhao et al. also performed an in silico comparison of lncRNAs expression between a cotton F1 hybrid and their parents. Results showed that 34.75% (leaves) and 50% (ovules) of the lncRNA loci were differentially expressed in the parent's vs F1's against the 1.71% differentially expressed lncRNAs observed between two cotton near-isogenic lines. Similar reports of transposable elements playing important roles in the origin of Lycopersicon-specific lncRNAs have been shown in tomato (Wang et al., 2016b). Further, Zhao et al. have also demonstrated the negative correlation between the lncRNAs expression and DNA CG ($r = -0.63$), CHG ($r = -0.74$), and CHH ($r = -0.71$) methylation in F1 hybrids (Zhao et al., 2018). They concluded that the genomic shock of interspecies hybridization during the course of evolution is the key factor that brings tremendous changes in the lncRNAs expression. By affecting the genes they regulate, this transcription reprogramming of the lncRNAs can influence the lineage-specific trait's absence or presence.

16.3.5 Role in other important biological processes

In a recent report, Zhang et al. have identified lncRNAs involved in the anther development in three-line hybrid cotton (CMS-D2 line A, maintainer line B, restorer-of-fertility line R) (Zhang et al., 2019a). A total of 80,695 lncRNAs were identified in the anther development, out of which 43,347, 44,739, and 46,431 lncRNAs were differentially expressed in A-B, A-R, and B-R comparisons, respectively, with 1713 of these differentially expressed lncRNAs common among all these conditions. They have also identified 18,755, 20,837, and 21,346 unique lncRNAs specifically expressed in the A, B, and R lines, respectively, and 15,692 lncRNAs were expressed in common among the A, B, and R lines. In agreement with previous similar studies, Zhang et al. also identified several lncRNAs (for example TCONS_00342368 and TCONS_00148576) serving as putative miRNA (miR171 and miR399) precursors.

16.4 Conclusions and future perspective

Cotton fiber is an important commercial commodity and fiber yield primarily depends on growth conditions and the ability to cope with the environmental stress condition. With the increasing number of studies, lncRNAs could become a novel tool to regulate the gene expression and get a stature equivalent to miRNAs. However, the research community needs to focus more on the characterization of individual candidate lncRNAs for their in-depth biological role. The development of a global classification system and a platform equivalent to miRBase could be the next important and much-needed step. Developing such a platform can help the research community in the synchronization of the individual efforts and also ensure better utilization of this information. Finally, considering the role lncRNAs play in every aspect of plant development, lncRNAs can play a vital role in crop improvement programs and could give better solutions against the recent global climatic challenges.

References

An, N., Fan, S., Wang, Y., Zhang, L., Gao, C., Zhang, D., Han, M., 2018. Genome-wide identification, characterization and expression analysis of long non-coding RNAs in different tissues of apple. Gene 666, 44–57.

Ayubov, M.S., Mirzakhmedov, M.H., Sripathi, V.R., Buriev, Z.T., Ubaydullaeva, K.A., Usmonov, D.E., Norboboyeva, R.B., Emani, C., Kumpatla, S.P., Abdurakhmonov, I.Y., 2019. Role of MicroRNAs and small RNAs in regulation of developmental processes and agronomic traits in Gossypium species. Genomics 111, 1018–1025.

Balasubramanian, V.K., Rai, K.M., Thu, S.W., Hii, M.M., Mendu, V., 2016. Genome-wide identification of multifunctional laccase gene family in cotton (Gossypium spp.); expression and biochemical analysis during fiber development. Sci. Rep. 6, 34309.

Bhatia, G., Goyal, N., Sharma, S., Upadhyay, S.K., Singh, K., 2017. Present scenario of long non-coding RNAs in plants. Non-Coding RNA 3 (2), 16.

Chen, Z.J., Scheffler, B.E., Dennis, E., Triplett, B.A., Zhang, T., Guo, W., Chen, X., Stelly, D.M., Rabinowicz, P.D., Town, C.D., Arioli, T., Brubaker, C., Cantrell, R.G., Lacape, J.-M., Ulloa, M., Chee, P., Gingle, A.R., Haigler, C.H., Percy, R., Saha, S., Wilkins, T., Wright, R.J., Van Deynze, A., Zhu, Y., Yu, S., Abdurakhmonov, I., Katageri, I., Kumar, P.A., Mehboob Ur, R., Zafar, Y., Yu, J.Z., Kohel, R.J., Wendel, J.F., Paterson, A.H., 2007. Toward sequencing cotton (Gossypium) genomes. Plant Physiol. 145, 1303–1310.

Chen, M., Wang, C., Bao, H., Chen, H., Wang, Y., 2016. Genome-wide identification and characterization of novel lncRNAs in Populus under nitrogen deficiency. Mol. Gen. Genomics. 291, 1663–1680.

Chen, J., Su, P., Chen, P., Li, Q., Yuan, X., Liu, Z., 2018a. Insights into the cotton anther development through association analysis of transcriptomic and small RNA sequencing. BMC Plant Biol. 18, 154.

Chen, L., Shi, S., Jiang, N., Khanzada, H., Wassan, G.M., Zhu, C., Peng, X., Xu, J., Chen, Y., Yu, Q., He, X., Fu, J., Chen, X., Hu, L., Ouyang, L., Sun, X., He, H., Bian, J., 2018b. Genome-wide analysis of long non-coding RNAs affecting roots development at an early stage in the rice response to cadmium stress. BMC Genomics 19, 460.

De Quattro, C., Mica, E., Pe, M.E., Bertolini, E., 2018. *Brachypodium distachyon* long noncoding RNAs: genome-wide identification and expression analysis. Methods Mol. Biol. 1667, 31–42.

Deng, F., Zhang, X., Wang, W., Yuan, R., Shen, F., 2018. Identification of *Gossypium hirsutum* long non-coding RNAs (lncRNAs) under salt stress. BMC Plant Biol. 18, 23.

Di Ruscio, A., Ebralidze, A.K., Benoukraf, T., Amabile, G., Goff, L.A., Terragni, J., Figueroa, M.E., De Figueiredo Pontes, L.L., Alberich-Jorda, M., Zhang, P., Wu, M., D'alò, F., Melnick, A., Leone, G., Ebralidze, K.K., Pradhan, S., Rinn, J.L., Tenen, D.G., 2013. DNMT1-interacting RNAs block gene-specific DNA methylation. Nature 503, 371–376.

Ding, J., Lu, Q., Ouyang, Y., Mao, H., Zhang, P., Yao, J., Xu, C., Li, X., Xiao, J., Zhang, Q., 2012. A long noncoding RNA regulates photoperiod-sensitive male sterility, an essential component of hybrid rice. Proc. Natl. Acad. Sci. U. S. A. 109, 2654–2659.

Dong, Z., Zhang, J., Zhu, Q., Zhao, L., Sui, S., Li, Z., Zhang, Y., Wang, H., Tian, D., Zhao, Y., 2018. Identification of microRNAs involved in drought stress responses in early-maturing cotton by high-throughput sequencing. Genes Genomics 40, 305–314.

Franco-Zorrilla, J.M., Valli, A., Todesco, M., Mateos, I., Puga, M.I., Rubio-Somoza, I., Leyva, A., Weigel, D., Garcia, J.A., Paz-Ares, J., 2007. Target mimicry provides a new mechanism for regulation of microRNA activity. Nat. Genet. 39, 1033–1037.

Gallagher, J.P., Grover, C.E., Rex, K., Moran, M., Wendel, J.F., 2017. A new species of cotton from Wake Atoll, *Gossypium stephensii* (Malvaceae). Syst. Bot. 42, 115–123.

Golicz, A.A., Singh, M.B., Bhalla, P.L., 2018. The long intergenic noncoding RNA (LincRNA) landscape of the soybean genome. Plant Physiol. 176, 2133–2147.

Grover, C.E., Gallagher, J.P., Jareczek, J.J., Page, J.T., Udall, J.A., Gore, M.A., Wendel, J.F., 2015. Re-evaluating the phylogeny of allopolyploid Gossypium L. Mol. Phylogenet. Evol. 92, 45–52.

Guan, X., Song, Q., Chen, Z.J., 2014. Polyploidy and small RNA regulation of cotton fiber development. Trends Plant Sci. 19, 516–528.

Haigler, C., Betancur, L., Stiff, M., Tuttle, J., 2012. Cotton fiber: a powerful single-cell model for cell wall and cellulose research. Front. Plant Sci. 3.

Hu, H., Wang, M., Ding, Y., Zhu, S., Zhao, G., Tu, L., Zhang, X., 2018. Transcriptomic repertoires depict the initiation of lint and fuzz fibres in cotton (*Gossypium hirsutum* L.). Plant Biotechnol. J. 16, 1002–1012.

Jabnoune, M., Secco, D., Lecampion, C., Robaglia, C., Shu, Q., Poirier, Y., 2013. A rice cis-natural antisense RNA acts as a translational enhancer for its cognate mRNA and contributes to phosphate homeostasis and plant fitness. Plant Cell 25, 4166–4182.

Li, C., He, X., Luo, X., Xu, L., Liu, L., Min, L., Jin, L., Zhu, L., Zhang, X., 2014. Cotton WRKY1 mediates the plant defense-to-development transition during infection of cotton by *Verticillium dahliae* by activating JASMONATE ZIM-DOMAIN1 expression. Plant Physiol. 166, 2179–2194.

Li, T., Ma, X., Li, N., Zhou, L., Liu, Z., Han, H., Gui, Y., Bao, Y., Chen, J., Dai, X., 2017a. Genome-wide association study discovered candidate genes of Verticillium wilt resistance in upland cotton (*Gossypium hirsutum* L.). Plant Biotechnol. J. 15, 1520–1532.

Li, X., Zhang, Y.N., Ding, C., Xu, W., Wang, X., 2017b. Temporal patterns of cotton Fusarium and Verticillium wilt in Jiangsu coastal areas of China. Sci. Rep. 7, 12581.

Liang, D., Zhang, Z., Wu, H., Huang, C., Shuai, P., Ye, C.-Y., Tang, S., Wang, Y., Yang, L., Wang, J., Yin, W., Xia, X., 2014. Single-base-resolution methylomes of *Populus trichocarpa* reveal the association between DNA methylation and drought stress. BMC Genet. 15 (Suppl. 1), S9.

Liu, J., Wang, H., Chua, N.H., 2015. Long noncoding RNA transcriptome of plants. Plant Biotechnol. J. 13, 319–328.

Lu, X., Chen, X., Mu, M., Wang, J., Wang, X., Wang, D., Yin, Z., Fan, W., Wang, S., Guo, L., Ye, W., 2016. Genome-wide analysis of long noncoding RNAs and their responses to drought stress in cotton (*Gossypium hirsutum* L.). PLoS One 11, e0156723.

Lu, X., Wang, X., Chen, X., Shu, N., Wang, J., Wang, D., Wang, S., Fan, W., Guo, L., Guo, X., Ye, W., 2017a. Single-base resolution methylomes of upland cotton (*Gossypium hirsutum* L.) reveal epigenome modifications in response to drought stress. BMC Genomics 18, 297.

Lu, Z., Xia, X., Jiang, B., Ma, K., Zhu, L., Wang, L., Jin, B., 2017b. Identification and characterization of novel lncRNAs in *Arabidopsis thaliana*. Biochem. Biophys. Res. Commun. 488, 348–354.

Lu, X., Yin, Z., Wang, J., Chen, X., Wang, D., Wang, S., Guo, L., Fan, W., Chen, C., Wang, X., Cui, R., Zhang, B., Han, M., Yang, X., Ye, W., 2019. Identification and function analysis of drought-specific small RNAs in *Gossypium hirsutum* L. Plant Sci. 280, 187–196.

Matsui, A., Seki, M., 2019. The involvement of long noncoding RNAs in response to plant stress. Methods Mol. Biol. 1933, 151–171.

Nithin, C., Thomas, A., Basak, J., Bahadur, R.P., 2017. Genome-wide identification of miRNAs and lncRNAs in *Cajanus cajan*. BMC Genomics 18, 878.

Pan, X., Nichols, R.L., Li, C., Zhang, B., 2019. MicroRNA-target gene responses to root knot nematode (*Meloidogyne incognita*) infection in cotton (*Gossypium hirsutum* L.). Genomics 111, 383–390.

Pang, J., Zhang, X., Ma, X., Zhao, J., 2019. Spatio-temporal transcriptional dynamics of maize long non-coding RNAs responsive to drought stress. Genes (Basel) 10.

Qi, X., Xie, S., Liu, Y., Yi, F., Yu, J., 2013. Genome-wide annotation of genes and noncoding RNAs of foxtail millet in response to simulated drought stress by deep sequencing. Plant Mol. Biol. 83, 459–473.

Qin, T., Zhao, H., Cui, P., Albesher, N., Xiong, L., 2017. A nucleus-localized long non-coding RNA enhances drought and salt stress tolerance. Plant Physiol. 175, 1321–1336.

Quinn, J.J., Chang, H.Y., 2016. Unique features of long non-coding RNA biogenesis and function. Nat. Rev. Genet. 17, 47–62.

Rai, K.M., Singh, S.K., Bhardwaj, A., Kumar, V., Lakhwani, D., Srivastava, A., Jena, S.N., Yadav, H.K., Bag, S.K., Sawant, S.V., 2013. Large-scale resource development in *Gossypium hirsutum* L. by 454 sequencing of genic-enriched libraries from six diverse genotypes. Plant Biotechnol. J. 11, 953–963.

Ranjan, A., Nigam, D., Asif, M.H., Singh, R., Ranjan, S., Mantri, S., Pandey, N., Trivedi, I., Rai, K.M., Jena, S.N., Koul, B., Tuli, R., Pathre, U.V., Sawant, S.V., 2012. Genome wide expression profiling of two accession of *G. herbaceum* L. in response to drought. BMC Genomics 13, 94.

Reis, R.S., Deforges, J., 2017. Plant non-coding RNAs and the new paradigms. In: Rajewsky, N., Jurga, S., Barciszewski, J. (Eds.), Plant Epigenetics. Springer International Publishing, Cham, pp. 163–182.

Salih, H., Gong, W., He, S., Xia, W., Odongo, M.R., Du, X., 2019. Long non-coding RNAs and their potential functions in Ligon-lintless-1 mutant cotton during fiber development. BMC Genomics 20, 661.

Shen, E., Zhu, X., Hua, S., Chen, H., Ye, C., Zhou, L., Liu, Q., Zhu, Q.H., Fan, L., Chen, X., 2018. Genome-wide identification of oil biosynthesis-related long non-coding RNAs in allopolyploid *Brassica napus*. BMC Genomics 19, 745.

Shi, G., Guo, X., Guo, J., Liu, L., Hua, J., 2015. Analyzing serial cDNA libraries revealed reactive oxygen species and gibberellins signaling pathways in the salt response of Upland cotton (*Gossypium hirsutum* L.). Plant Cell Rep. 34, 1005–1023.

Shin, H., Shin, H.S., Chen, R., Harrison, M.J., 2006. Loss of At4 function impacts phosphate distribution between the roots and the shoots during phosphate starvation. Plant J. 45, 712–726.

Shuai, P., Liang, D., Tang, S., Zhang, Z., Ye, C.Y., Su, Y., Xia, X., Yin, W., 2014. Genome-wide identification and functional prediction of novel and drought-responsive lincRNAs in *Populus trichocarpa*. J. Exp. Bot. 65, 4975–4983.

Shumayla, Sharma, S., Taneja, M., Tyagi, S., Singh, K., Upadhyay, S.K., 2017. Survey of high throughput RNA-seq data reveals potential roles for lncRNAs during development and stress response in bread wheat. Front. Plant Sci. 8, 1019.

Sun, X., Zheng, H., Sui, N., 2018. Regulation mechanism of long non-coding RNA in plant response to stress. Biochem. Biophys. Res. Commun. 503, 402–407.

Tyagi, S., Sharma, A., Upadhyay, S.K., 2018. Role of next-generation RNA-Seq data in discovery and characterization of long non-coding RNA in plants. In: Çiftçi, Y.Ö. (Ed.), *Next Generation Plant Breeding*. IntechOpen, https://doi.org/10.5772/intechopen.72773. 2018.

Wang, W.-S., Pan, Y.-J., Zhao, X.-Q., Dwivedi, D., Zhu, L.-H., Ali, J., Fu, B.-Y., Li, Z.-K., 2011. Drought-induced site-specific DNA methylation and its association with drought tolerance in rice (*Oryza sativa* L.). J. Exp. Bot. 62, 1951–1960.

Wang, M., Yuan, D., Tu, L., Gao, W., He, Y., Hu, H., Wang, P., Liu, N., Lindsey, K., Zhang, X., 2015. Long noncoding RNAs and their proposed functions in fibre development of cotton (Gossypium spp.). New Phytol. 207, 1181–1197.

Wang, Q., Liu, N., Yang, X., Tu, L., Zhang, X., 2016a. Small RNA-mediated responses to low- and high-temperature stresses in cotton. Sci. Rep. 6, 35558.

Wang, X., Ai, G., Zhang, C., Cui, L., Wang, J., Li, H., Zhang, J., Ye, Z., 2016b. Expression and diversification analysis reveals transposable elements play important roles in the origin of Lycopersicon-specific lncRNAs in tomato. New Phytol. 209, 1442–1455.

Washietl, S., Kellis, M., Garber, M., 2014. Evolutionary dynamics and tissue specificity of human long noncoding RNAs in six mammals. Genome Res. 24, 616–628.

Wendel, J.F., 1989. New World tetraploid cottons contain Old World cytoplasm. Proc. Natl. Acad. Sci. 86, 4132–4136.

Wendel, J.F., Cronn, R.C., 2003. Polyploidy and the evolutionary history of cotton. Adv. Agron. 78, 139.

Yao, Z., Chen, Q., Chen, D., Zhan, L., Zeng, K., Gu, A., Zhou, J., Zhang, Y., Zhu, Y., Gao, W., Wang, L., Zhang, Y., Qu, Y., 2019. The susceptibility of sea-island cotton recombinant inbred lines to *Fusarium oxysporum* f. sp. vasinfectum infection is characterized by altered expression of long noncoding RNAs. Sci. Rep. 9, 2894.

Zhang, H., Hu, W., Hao, J., Lv, S., Wang, C., Tong, W., Wang, Y., Wang, Y., Liu, X., Ji, W., 2016. Genome-wide identification and functional prediction of novel and fungi-responsive lincRNAs in *Triticum aestivum*. BMC Genomics 17, 238.

Zhang, B., Zhang, X., Liu, G., Guo, L., Qi, T., Zhang, M., Li, X., Wang, H., Tang, H., Qiao, X., Pei, W., Shahzad, K., Xing, C., Zhang, J., Wu, J., 2018a. A combined small RNA and transcriptome sequencing analysis reveal regulatory roles of miRNAs during anther development of Upland cotton carrying cytoplasmic male sterile *Gossypium harknessii* (D2) cytoplasm. BMC Plant Biol. 18, 242.

Zhang, L., Wang, M., Li, N., Wang, H., Qiu, P., Pei, L., Xu, Z., Wang, T., Gao, E., Liu, J., Liu, S., Hu, Q., Miao, Y., Lindsey, K., Tu, L., Zhu, L., Zhang, X., 2018b. Long noncoding RNAs involve in resistance to *Verticillium dahliae*, a fungal disease in cotton. Plant Biotechnol. J. 16, 1172–1185.

Zhang, B., Zhang, X., Zhang, M., Guo, L., Qi, T., Wang, H., Tang, H., Qiao, X., Shahzad, K., Xing, C., Wu, J., 2019a. Transcriptome analysis implicates involvement of long noncoding RNAs in cytoplasmic male sterility and fertility restoration in cotton. Int. J. Mol. Sci. 20.

Zhang, X., Dong, J., Deng, F., Wang, W., Cheng, Y., Song, L., Hu, M., Shen, J., Xu, Q., Shen, F., 2019b. The long non-coding RNA lncRNA973 is involved in cotton response to salt stress. BMC Plant Biol. 19, 459.

Zhao, T., Tao, X., Feng, S., Wang, L., Hong, H., Ma, W., Shang, G., Guo, S., He, Y., Zhou, B., Guan, X., 2018. LncRNAs in polyploid cotton interspecific hybrids are derived from transposon neofunctionalization. Genome Biol. 19, 195.

Zhao, T., Xu, X., Wang, M., Li, C., Li, C., Zhao, R., Zhu, S., He, Q., Chen, J., 2019. Identification and profiling of upland cotton microRNAs at fiber initiation stage under exogenous IAA application. BMC Genomics 20, 421.

Zhou, Y., Cho, W.K., Byun, H.S., Chavan, V., Kil, E.J., Lee, S., Hong, S.W., 2019. Genome-wide identification of long non-coding RNAs in tomato plants irradiated by neutrons followed by infection with tomato yellow leaf curl virus. PeerJ 7, e6286.

Zhu, X., Tai, X., Ren, Y., Chen, J., Bo, T., 2019. Genome-wide analysis of coding and long non-coding RNAs involved in cuticular wax biosynthesis in cabbage (*Brassica oleracea* L. var. capitata). Int. J. Mol. Sci. 20 (11), 2820.

Zou, C., Wang, Q., Lu, C., Yang, W., Zhang, Y., Cheng, H., Feng, X., Prosper, M.A., Song, G., 2016. Transcriptome analysis reveals long noncoding RNAs involved in fiber development in cotton (*Gossypium arboreum*). Sci. China Life Sci. 59, 164–171.

Chapter 17

The role of lncRNA in reproductive tissue development and flowering in plants

Kalpesh Yajnik and Purva Bhalothia
Dr. B. Lal Institute of Biotechnology, Jaipur, India

17.1 Introduction

Plants have the ability to interpret endogenous and environmental signals which guides them about reproductive growth and flowering time (Yamaguchi and Abe, 2012). We classify these environmental signals as: primary factors—photoperiodism and vernalization; secondary factors—ambient temperature and irradiance; tertiary factors on the basis of quality of light received and availability of minerals (Bernier and Périlleux, 2005).

With the advancement in the field of bioinformatics, the data of genomic sequence can now be retrieved through databases, thereby enabling the fine study of regulatory networks and pathways involved during flowering. Work presented by Yamada et al. (2003) in plants revealed that only $<50\%$ of the genome can code for proteins and it was figured out that regulatory noncoding RNAs (ncRNAs) play an important part in modulating transcription (Ørom et al., 2010). Past studies also show that certain long noncoding RNAs (lncRNAs), class of ncRNAs, are involved in reproductive development in plants (Heo and Sung, 2011).

lncRNAs belong to class of molecules having length of >200 bp and occupy an important position for regulation of certain biological processes such as gene silencing, flowering time, vernalization, photoperiodism, fruit ripening, and development of pollens/ovaries in plants. Approaches involving Next-Generation Sequencing (NGS) and strand-specific RNA sequencing (ssRNA-seq) were applied for identification of lncRNAs in *Arabidopsis*, *Brassica*, rice, and other plants too (Zhang et al., 2014; De Lucia and Dean, 2011; Shumayla et al., 2017; Tyagi et al., 2018; Huang et al., 2018).

The current book chapter includes the role of lncRNAs in epigenetic modification, its mechanism and regulation of flowering time, vernalization, and photomorphogenesis. The chapter also provides the involvement of lncRNAs in reproductive tissues including pollen, embryos, and fruits.

Long Noncoding RNAs in Plants: Roles in development and stress
https://doi.org/10.1016/B978-0-12-821452-7.00017-9

17.2 Regulatory mechanism of lncRNAs during reproductive development of plants

In plants, initial approach to analyze the regulatory mechanism of lncRNAs involved in reproductive development was done by Wu et al. (2013). They proposed that certain endogenous lncRNAs (named as endogenous target mimicries) interfere with the binding sites of miRNAs by competing them and regulate target gene aggregation. Genomic analysis of *Arabidopsis thaliana* and *Oryza sativa* revealed the presence of endogenous target mimicries (eTMs) derived from lincRNAs mimicking around 20 miRNAs. Also, certain lncRNAs such as *XLOC_003639* and *XLOC_007072*, when obtained, showed sponging of miR160 and miR166, thereby regulating seed and floral development.

Long day-specific male sterility-associated RNA (*LDMAR*), a lncRNA known to make *Psi-LDMAR* by interacting with siRNA, has the ability to methylate the DNA, thereby repressing *LDMAR* (Ding et al., 2012a,b). The study was done in *O. sativa*, where, upon mapping of lncRNA p/tms12-1, it showed the presence of 21-nucleotide counts of siRNA: *osa-smR5864w*. Thus, photoperiod/thermo-sensitive male sterility was initiated (Zhou et al., 2012).

The regulation of flowering by lncRNAs is among the well-known studied mechanisms in plants. Work done in *Arabidopsis* by Yamaguchi and Abe in 2012 presents that flowering is regulated by a complicated network of genes mentioned: *FLOWERING LOCUS C* (*FLC*), *FLOWERING LOCUS T* (*FT*), *SUPRESSOR OF OVEREXPRESSION OF CO 1* (*SOC1*), and *LEAFY* (*LFY*). *FLC* gene is believed to be a major modulator for flowering time, by inhibiting flowering under cold conditions (Michaels and Amasino, 1999). It transcribes *MADS*-box which represses flowering by regulating *FT* and *SOC1*. *FT* encodes a protein, florigen, a key component to initiate flowering. Therefore, upon environmental stimulus on leaves, *FT* protein mobilizes and reaches apical meristem, initiating flowering (Abe et al., 2005; Lin et al., 2007).

Lucia and dean in 2011 reviewed that chromatin moderation had equivalent functions for lncRNAs as observed in the *FLC* expression, providing a major step in monitoring and controlling the flowering time in *Arabidopsis* (Kim and Sung, 2012). Epigenetic modification system regulates the expression of gene by altering their state as "ON"/"OFF" (Dodd et al., 2007) and such similar expressions were observed by Angel and his colleagues in 2011 in *FLC* during colder conditions. Here, cells upon exposure to cold switch to OFF state and the expression can then be analyzed through digital analogue system (Berry and Dean, 2015).

Techniques involving genetic screening are employed to identify certain protein factors which bind to *FLC* locus and regulate the alterations in chromatin to enhance or suppress transcription of *FLC* (Crevillén and Dean, 2011). Also, the activation or inactivation of expression in *FLC* is believed to be regulated by certain posttranslational modifications of histones (Berry and Dean, 2015).

The proof of the mentioned theory is supported by certain evidences: correlation among tri-methylation at lysine 4 and lysine 36 of histone H3 (H3K4me3/H3K36me3), histone 2B ubiquitination (H2Bub1), and histone acetylation to be associated with genes when they are functionally operating (Li et al., 2007). Similar marks of histones were also observed on operating locus of *FLC* by Yang et al. (2014). These spots were replaced by tri-methylation of histone H3 at lysine 27 (H3K27me3) under repression of *FLC* condition, thus indicating the switching of chromatin toward inactivated state (Bastow et al., 2004; Angel et al., 2011). Second evidence includes positioning of histones (H3K4me3) necessary for enhancing the expression of *FLC* by activating *FLC* chromatin (Kim et al., 2005; Yang et al., 2014).

The process can further be elaborated through lncRNA *COLD-ASSISTED INTRONIC NONCODING RNA (COLDAIR)*. It binds with Polycomb repressive complex 2 (PRC2) and accumulates at transcription initiation site of *FLC*. The network formed leads to gain in H3K27me3 and loss of H3K4me3/H3K36me3 at that region (Heo and Sung, 2011). Thus, *COLDAIR* performs dual functions by serving as a scaffold in formation of PRC2 and directing the complex toward binding site, presenting the roles of lncRNAs in biological processes (Swiezewski et al., 2009).

17.3 Role of lncRNAs in plant development

Reports presenting the role of lncRNAs in plant reproductive development have been published in the past few years (Zhang et al., 2014; Wang et al., 2017b; Ding et al., 2012a,b). Here, we summarize some of the important functions of lncRNAs related to plant development such as ovules development, pollen production, flowering, and ripening of fruits.

Work done on *Arabidopsis* by Heo and Sung in 2011 demonstrates the association of *COLDAIR* with PRC2 and recruitment of PRC2 toward *FLC* locus, which downregulate the expression of *FLC*. The underlaid process eases the working of vernalization pathway and initiates flowering as cold condition deprives, as shown in Fig. 17.1. The work sets as a base for understanding the flowering mechanism and reproductive tissue development in other plants too.

To identify antisense transcripts treated by abiotic stress, ssRNA-seq is applied in *O. sativa*. Of 3819 genes of *cis*-NATs, 503 are known to be in association under specific conditions. Also, the sequenced data obtained from rice seedling revealed 54.0% of *cis*-NATs to be expressed in homogenized cells and about 17%–34% of certain *cis*-NAT groups developed as nat-siRNAs. These results confer the role of *cis*-NATs in seedling development of rice (Lu et al., 2012).

Another experiment was done in *Arabidopsis*, where expression of antisense RNA was observed during heat stress which was dependent on heat stress transcription factors—*HSFB2a* and *HSFA1a*. Knockdown line developed for analysis of gametophytic and vegetative growth, known to be influenced by *HSFB2a*, showed an arrest in growth of female gametophytes by 50% and, prior to mitosis, revealed 45% sterile ovules (Wunderlich et al., 2014).

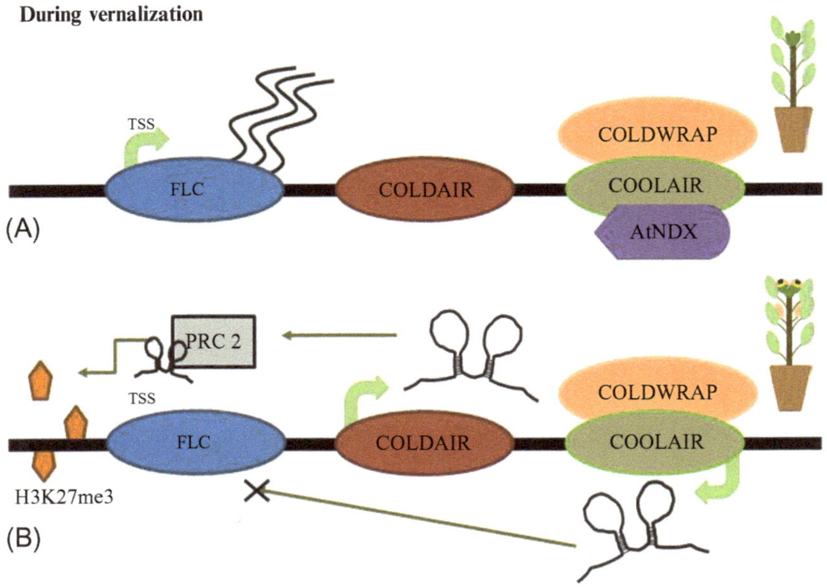

FIG. 17.1 Regulation of lncRNA in reproductive development of plants. (A) Suppression of *COOLAIR* and *COLDWRAP* by binding of *AtNDX*, leading to expression of *FLC*. (B) Suppression of *FLC* for flowering via *COOLAIR*, *COLDWRAP*, PHD-PRC2 complex, and H3K27me3. *(The figure has been adapted and modified from Zhang, Y.C., Chen, Y.Q., 2013. Long noncoding RNAs: new regulators in plant development. Biochem. Biophys. Res. Commun. 436 (2), 111–114 with the permission from Elsevier with license number-4713500366170).*

Reports provided by Zhang and their colleagues in 2014 on genomic analysis of *O. sativa* involved screening of lncRNAs. ssRNA-seq was performed from different tissues—anthers, seeds, and pistils. The results revealed that lncRNAs, identified as competing endogenous RNAs (ceRNAs), are involved in reproductive development of *O. sativa*, when high level of expression was observed in the mentioned stages. Also, they observed lncRNAs, *XLOC_057324*, implicated in development of panicle and fertility.

In order to exploit the role and pattern of inheritance of lncRNAs in *Zea mays*, genome-wide analysis and RNA-sequenced data were extracted. Of 20,163 identified lncRNAs, 1704 were found to be of high confidence. Highest expression pattern of lncRNAs was found in reproductive tissues of male and embryo sac, when compared with 13 other tissues. The expression pattern of lncRNAs in these specified tissues was known to be biologically controlled, instead of demonstrating a transcriptional noise. Also, pattern of inheritance of lncRNAs in 105 recombinant strains showed transgressive segregation, demarcating the effect of *trans*-genes in *Z. mays* (Li et al., 2014).

Cucumis sativus was selected as a model plant to study the importance of lincRNAs in sex differentiation and development of fruit using data obtained by RNA-seq from numerous tissues including male and female flowers and expanded fertilized and unfertilized ovaries. Co-expression of lincRNAs with mRNAs revealed annotation of 94 lincRNAs. It was assumed that these lincRNAs might show involvement in responses via stimuli, reproduction, growth, and certain reproductive processes. The result, hence, shows a positive selection of lincRNAs acting as a base for further research in cucumber (Hao et al., 2015).

Meiosis belongs to the stage of cell cycle where progression of reduction in the chromosomal number begins, thereby increasing the chances of reassortment (Hamant et al., 2006). Numerous disk flowers in inflorescence of *Helianthus annuus* L. permits separation of male meiocytes from defined meiotic phases (Axtell, 2015). Flórez-Zapata and their colleagues in 2016 conducted a study by sequencing the prophase I meiocytes transcriptome from various sunflower varieties having varied recombination rates. The results revealed that 70.98% of differential expressed genes which passed filters were categorized as lncRNAs and were meiosis-specific. They suggested that lncRNAs act as a protagonist in modulating meiotic expression of genes and might became the reason of such recombination rates, concluding the importance of lncRNAs in altering the expression of plant meiotic genes or in their chromatin modifications.

Wu et al. (2013) revealed the presence of binding sites of microRNA (miRNA) in lncRNAs, acting as endogenous target mimics (eTMs). One such lncRNA, osa-eTM160, acts as a repressor of osa-miR160. In *A. thaliana*, osa-miR160 is known to target ARF gene families: *osa-ARF18* and *osa-ARF22* (Liu et al., 2007). Therefore, the study was conducted by Wang et al. (2017b) to identify the roles of osa-eTM160 in *O. sativa* reproductive development. Microarray data demonstrated the highest expression in inflorescence phase. The expression between *osa-ARF18* and osa-eTM160 showed identical expression, where increased expression was observed during early phase of anther development and a decline was seen in the later stage. Thus, the function of osa-eTM160 in rice was presented by negatively modulating osa-miR160 to increase the expression of *osa-ARF18*.

Expression profiling of lncRNAs was performed in *Z. mays* through high-throughput sequencing to identify its roles in seed development. The results discovered that 753 lncRNAs were derived from six samples of maize and demonstrated differential, stage-specific expression of developing tissues. Interference between RNAs showed a new pathway for regulation of genes, thereby justifying their role in development of plants. However, functions and mechanism of lncRNAs as endogenous targets in seed development still need to be worked on. The investigation of persistent change in 840 mRNAs and 17 lncRNAs revealed presence of 7 novel lncRNAs promising to be ceRNAs. Functional analysis confirmed certain positive effects on developmental and metabolic routes in seed growth of *Z. mays*, when associated with different ceRNAs. The result, hence, provides a platform to study novel lncRNAs for further characterization in seed development (Zhu et al., 2017).

As mentioned earlier, lncRNAs are the major modulators in formation of Polycomb complex (Zhao et al., 2010). Heo and Sung, in 2011, mentioned the highest expression of *COLDAIR* to be observed in about 20 days of exposure to cold. *COLDAIR*, in association with Polycomb complex, silences *FLC* during vernalization. The study conducted by Kim and their colleagues in 2017 emphasized on another Polycomb-binding lncRNAs, *COLDWRAP*, obtained from repressed promoter site of *FLC* and plays a key role in maintaining its downregulation. At the time of vernalization, *COLDAIR* and *COLDWRAP* together form an interactive condensed loop of chromatin for silencing of *FLC* (Fig. 17.1); thus, concluding the involvement of lncRNAs in vernalization-mediated silencing of chromatin.

The first introductory report on lincRNAs as repression targets through PRC2 was conducted to identify the role of lncRNAs in *O. sativa* by transcriptome analysis. Expression of 6309 lincRNAs was observed in reproductive tissues and organs, including enhanced expression in male gametes too. Conrad et al. (2014) reasoned lower levels of H3K27me3 during panicle growth causing *emf2b* mutants to increase the expression of PC genes in *O. sativa*. Thus, on comparing the data of lincRNAs from *emf2b* mutants with wild types, derepression of lincRNAs was observed in *emf2b* mutants. Overlapping of lincRNAs in sperm cells with *emf2b* mutant lincRNAs resulted in partial relieve of epigenetic repression in male gametes. Expression of lincRNAs associated with reproductive tissues was observed in response to abiotic stress. Comprehensive analysis with different genomes revealed low conservation levels of lincRNAs at structural and sequence levels with homology of ~4% and 10% in *Z. mays* and *Brachypodium distachyon*, respectively (Johnson et al., 2018).

The technique of alternative splicing (AS) was reviewed by Ulitsky in 2016 for the production of certain lncRNAs. In plants, reports have been published involving the role of AS in determining flowering and circadian rhythms (Xin et al., 2017; Simpson et al., 2016). Therefore, a comprehensive study was conducted between endosperm and embryo of mature seeds with immature ones to analyze the effects of AS of lncRNAs in development of seed in *O. sativa*. Upon AS, processing of lncRNAs during progression from seed to endosperm was highly precise when compared with the predicted ones. Further, retention of exons of lncRNAs was increased in embryos. Also, similarities based on conserved regions, coding sequences and genome-wide analysis of lncRNAs, reveals about role of multiexonic lncRNAs which can be broadly arranged for seed development (Kiegle et al., 2018).

Characterization and in silico identification of differentially expressed lncRNAs in *Vitis vinifera* was done by Bhatia and their colleagues in 2019 through computational analysis by utilizing RNA-sequences obtained from different developmental tissues at specific stages. On determining the specific threshold values obtained by FPKM analysis, 7697 differentially expressed (DE) lncRNAs were obtained from 56,441 lncRNAs. Also, a maximum expression of 7497 DE lncRNAs were identified in inflorescence development, presenting the role of lncRNAs in reproductive tissue development. Further, lncRNAs of 115 miRNA precursors, 133 eTMs, and 560 targets were investigated and assumed to be involved in the crosstalk of regulatory RNAs. These transcripts interact with noncoding and coding RNAs along with other transcription factors and regulate certain biological functions as oxidative phosphorylation and photosynthesis in *V. vinifera*; hence, unveiling the molecular mechanisms related to growth of the plant.

In order to classify changes in the expression of lncRNAs during development of ovule and gametophytic abortion in female plants, complete transcriptome analysis via ss-RNA seq was done in majority of female-sterile lines (*fsv1*) and normal rice line (Guj99). Comprehensive analysis of these lines revealed DE lncRNAs as 1552, 233, and 197 in 3 developmental stages—megaspore mother cell, functional stage of megaspore mitosis, and matured female gametophytes. Functional analysis of the genes revealed involvement of lncRNAs in signal transduction and cellular metabolism. Here, lncRNAs may act as a bait and binds with miRNA in order to maintain proper functioning of ovule by altering genomic expression (Liu et al., 2019).

17.3.1 Vernalization

Flowering is considered as an important part of plant life cycle and is regulated majorly by receiving signals from seasonal variations in the environment. Certain plants flower after being exposed to long cold conditions (Amasino, 2005). This pattern of flowering involving a long cold exposure prior to flower is termed as vernalization (Chouard, 1960).

As described by Michaels and Amasino in 1999, *FLC* is required for suppressing flowering under cold conditions during vernalization. Prior to vernalization, higher level of *FLC* expression is observed due to increased expression of H3K4me3 and H3K36me3, termed as activation of gene. However, under colder environment, the state of *FLC* chromatin switches and get repressed as methylation of H3K27 and H3K9 increases (He, 2009). Certain lncRNAs, *COOLAIR* and *COLDAIR*, were reported to be present at *FLC* locus. Thus, the expression of *FLC* was regulated (Heo and Sung, 2011; Sun et al., 2013).

COOLAIR is transcribed from 3′ region of *FLC* in antisense direction and regulates the expression of *FLC* by its polyadenylation property. The adjacent 3′ region of polyadenylation reduces the expression of *FLC*, while distant one initiates the processing of *FLC* gene (Liu et al., 2010). Recent study reveals the involvement of *FT* gene in regulation of flowering. As mentioned by Chen and Penfield in 2018, *FLC* and *FT* are used to regulate dormancy in seeds of progenies, when changes in temperature occurred. They both act as antagonist, i.e., *FT*, when activated, lowers *FLC* expression by operating *COOLAIR*.

Intronic lncRNA-*COLDAIR* forms a complex with PRC2 and methylates the DNA. Histone methylation and deacetylation by *COLDAIR* blocks the expression of *FLC* (Crevillén et al., 2014). Another lncRNA named *COLDWRAP* is known to be expressed from upstream sequences of promoter region of *FLC*. Polycomb interplay by *COLDWRAP* seems to be fixed as binding of CLF at sense strand of *COLDWRAP* appears remarkable. It then works with *COLDAIR* cooperatively and performs inhibition of *FLC* via vernalization (Kim and Sung, 2017).

Transcriptome analysis of tissues of flower has been summarized in Table 17.1, revealing the expression patterns and effect of lncRNA in their development.

TABLE 17.1 List of plants involved in the floral development by lncRNAs.

Plants	lncRNAs and their biotypes	Approach	Remarks	References
Arabidopsis thaliana	Antisense transcript *COOLAIR*	ChIRP-seq	Role of lncRNAs in polycomb silencing, therefore regulating *FLC*.	Csorba et al. (2014)
	lncRNA *COLDAIR*	ChIRP-seq	Repression of *FLC* by epigenetic modification at the time of vernalization and interrelation of lncRNA *COLDAIR* with its chromatin-modifying complex.	Kim et al. (2017)
	Novel lncRNAs-372	ssRNA-seq	DNA motif characterization of novel lncRNAs through GO reveals its roles in flower development too.	Lu et al. (2017)
	NAT-lncRNA	ssRNA-seq; ChIRP-seq	Characterization of *MAS* responsible for suppressing flowering by enhancing H3K4me3.	Zhao et al. (2018)
	lncRNAs-*FLINC*	Illumina-seq	Differential expression was observed in lncRNAs upon increasing ambient temperature. The lncRNAs, *FLINC*, govern flowering by getting downregulated at high ambient temperature.	Severing et al. (2018)
Beta vulgaris	lncRNAs-*AGL15X1*; *AGL15X2*; *CAULIFLOWER A*	RNA-seq data	Examination of lncRNAs for their interrelation with vernalization was done along with *BvRAV1*, *BvFT1*, and *BvFT2* genes. Highest expression of *BvRAV1* was observed during vernalization. Coordinated expression of *AG15X1*, *BvRAV1*, and *BvFT2* resulted in reproductive growth and proposed a new model for flowering.	Liang et al. (2017)
Brachypodium distachyon	Antisense transcript *COOLAIR*	RNA-seq data; CLC-seq viewer	*FLC/COOLAIR* sense-antisense similarity was observed among eudicots and monocots when *BdTR3C* and *BdCOOLAIR2* were induced by cold.	Jiao et al. (2019)

Catalpa bungei	lncRNAs-12,532	Illumina-seq	Differential lncRNAs expression was performed by real-time PCR revealing involvement of 817 lncRNAs in floral transition.	Wang et al. (2018b)
Cicer arietinum	lncRNAs-2248	RNA-seq data; Computational pipeline	Differential expression and specificity in tissues of lncRNAs upon expression profiling was observed in various stages of developing flower along with assigning functions of 79% of lncRNAs.	Khemka et al. (2016)
Fragaria vesca	lncRNAs-5884	RNA-seq data; Illumina-seq	To analyze interplay between protein-coding gene targets and lncRNAs along with positive-negative expression trends; marking the inherent contribution of lncRNAs in flower development.	Kang and Liu (2015)
Morus notabilis	Intergenic-1092; Intronic-38; Antisense-3	RNA-seq data; Computational pipeline	To identify and analyze the expression of lncRNAs from different tissues; identification of role of lncRNAs alongside of protein-coding genes in growth of floral organs.	Song et al. (2016)
Poncirus trifoliata	Intergenic-2732; Intragenic-3398; Antisense-454	ssRNA-seq	Significant upregulation of 555 lncRNAs and downregulation of 276 lncRNAs proposed their role in flowering.	Wang et al. (2017a)
Prunus mume	lncRNAs-186; Novel lncRNAs-89	High-throughput RNA-seq	Prediction of DEKLs genes (421) and their analysis showed upregulation of AP2 and downregulation of ppe-miR172d through lncRNAs for pistil development.	Wu et al. (2019)
Solanum lycopersicum	Novel lncRNAs-248	Illumina-seq	Out of 248 novel lncRNAs, 65 were differentially expressed (DE). Association of DE lncRNAs with metabolic and enzymatic system proposed its role during flowering.	Yang et al. (2019b)

17.3.2 Photomorphogenesis

Light is considered as a major environmental factor for the production of energy in plants. It is the ability of plants to perceive altering light signals and adapt accordingly. Receptors as phototropins, phytochromes, and cryptochromes apprehend light of varying wavelengths and begin light-dependent signaling (Chen and Chory, 2011). Thus, we define photomorphogenesis as a light-sensitive developmental process, involving differentiation of reproductive and vegetative phase (Datta and Paul, 2019).

The regulatory process of photomorphogenesis reveals the presence of proteins for its functioning. Transcription profiling of lncNATs in *Arabidopsis* done by Wang and their colleagues in 2014 identified lncRNAs in the presence of light having concordant-discordant NAT pairs, which are known to be regulated by modifying histones. A novel lncRNA *HIDDEN TREASURE 1 (HID1)* of 236 nucleotide long has a role in photomorphogenesis as suggested by Wang et al. (2014a,b). On screening of T-DNA mutants, genes having knockdown effects revealed hypophotomorphogenic traits in the presence of continuous red light.

It was observed that *HID1* works through *phytochrome-interacting factor (PIF)*, which codes bHLH transcription factor and represses photomorphogenesis in response to light. Deletions in *HID1* showed a notable increase in *PIF3* expression, resulting in generation of long hypocotyls when exposed to continuous red light. Thus, the role of *HID1* in controlling the expression of seedling photomorphogenesis was presented.

17.3.3 Fruit ripening

Transcriptome analysis of lncRNAs was performed in plants as *O. sativa* (Zhang et al., 2014), *A. thaliana* (Liu et al., 2012), and *Z. mays* (Boerner and McGinnis, 2012) to identify their role and functions in reproductive development. But, role of lncRNAs in ripening of fruit remains unjustified because of absence of fruit ripening in these base plants. Here, we summarize the role of lncRNAs in ripening of fruit from some of the mentioned plants.

Due to higher values of tomato, it has become a base plant to elucidate the mechanism of its ripening (Gapper et al., 2013). The *ripening inhibitor (rin)* transcription factor, a mutant found in tomato, is known to abolish the normal process of ripening (Vrebalov et al., 2002). This gene, thus, acts as an important tool for further investigation in fruit ripening (Qin et al., 2012). Study conducted by Zhu et al. (2015) on tomato involves comparative analysis of lncRNAs between *rin* mutant and wild-type using ssRNA-Seq. The results showed distribution of 3679 lncRNAs in every chromosome of tomato with 85.1% of intergenic lncRNAs. Also, upregulation of 490 lncRNAs and downregulation of 187 lncRNAs in mutants were observed. Dampening of lncRNAs,1459 and 1840, in wild ones revealed a delay in ripening process; hence, indicating the role of lncRNAs in fruit ripening.

Transcription profiling of kiwifruit provided by Li and their colleagues in 2015 provided specific details of secondary metabolites, phytohormones and anthocyanin. The study sets as a base for complete analysis of lncRNAs and alternative splicing involved in development of kiwifruit. Data retrieved from numerous RNA-seq revealed 7051 lncRNAs, 29,327 transcripts by alternative splicing, and 2980 protein-coding genes. Also, 5931 differentially expressed genes were analyzed, which include genes involved in metabolism of certain amino acids and organic acids. The results provide a basis for further investigation in development and ripening of kiwifruit genes associated with regulations, thereby enhancing its agronomic value (Tang et al., 2016).

Zhang et al. (2017) performed transcriptome analyzing by ssRNA-seq in *Hippophae rhamnoides* Linn of two varieties varying in color, for identification of lncRNAs. The data obtained show the presence of 3428 lncRNAs in ripened fruit, which include 337 intronic lncRNAs, 2498 intergenic lncRNAs, and 593 antisense RNA. Certain differences in lncRNAs classified in terms of gene expression and architecture were observed. Analysis of targeted genes presents DE-lncRNAs, where the sea buckthorn was compared as 22 DE lncRNAs in yellow ones and 23 DE lncRNAs in the red ones. It was noted that DE-lncRNAs have a major role regulating biosynthesis of carotenoids, fatty acid, ascorbate, and aldarate metabolism, providing a standard for better understanding about fruit coloring.

A significant expression of capsaicinoids and ascorbic acid was observed in *Capsicum annum*, when transcriptome study was carried out via RNA-seq during ripening (Liu et al., 2013). Therefore, ssRNA-seq tool was used for identification of 2505 lncRNAs during ripening of fruit. Transcription of lncRNAs was observed from all the chromosomes, where ~95% was derived from intergenic sites and few from DE-lncRNAs. Certain PC genes transcribed by *cis/trans* positioning of lncRNA were also recognized. Here, the data reveal the association of lncRNAs in carotenoid biosynthesis and plant hormone production for fruit development (Ou et al., 2017).

Wang et al. (2018a) justified that expression of lncRNA changes along with the growth of tomato. Therefore, they conducted complete sequence analysis of lncRNAs in genome of *Solanum lycopersicum* by nonstrand-specific and strand-specific RNA seq. From different stages of fruit. The result also describes expression of novel lincRNAs in different stages of the fruit: mature green (MG), breaker (BR), and breaker plus 7 days (BR + 7). Presence of 79,332 lncRNAs, having lincRNAs, ancRNAs, and slncRNAs, was also observed. lncRNAs showed remarkable distinct feature from mRNAs, which includes shorter exons, tissue-specificity, and variation in methylation of DNA. This data, therefore, help in dissecting the mechanism related to fruit production through lncRNAs.

Zhang et al., in 2018, again worked on sea buckthorn, disclosing the role of lncRNAs in biosynthesis of anthocyanin. Complete transcripts of fruit at three different stages as MG, BR, and RR were examined through high-throughput seq. Upon assembling, they defined 9008 lncRNAs, 32 DE-miRNA, and 118

DR-lncRNAs during fruit ripening. Also, they identified 1061 *cis*- and 782 *trans*-regulating genes of DE-lncRNAs and their involvement in biosynthesis of certain flavonoids, ascorbic acid and carotenoids. Reports mentioned by Gou et al. (2011) and Gonzalez et al. (2008) reveal the presence of miR156-targeted *SPL* and *MYB-114*/bHLH/WDR which are involved independently in synthesis of anthocyanin. TRV-mediated silencing of lncRNAs, LNC1 and LNC 2, caused repression of *SPL9* and promotion in the expression of *MYB114*, respectively, to regulate the synthesis and concentration of anthocyanin.

Comprehensive analysis of transcriptome data of two variants in *Fragaria pentaphylla* at three different developing stages of fruit, mature green, turning, and ripe fruit, for identification of lncRNAs was done. The data showed 25,699 lncRNAs having distinct genotype- and stage-specific pattern of expression with 365 DE lncRNAs and 1416 protein-coding genes. Also, it was suggested that lncRNAs negatively regulate fruit ripening by suppressing photosynthesis and organization of cell wall. Casañal et al. (2013) published that family PR-10 *Fra a* gene is directly associated with anthocyanin production in strawberry by forming complex with the intermediates of flavonoid biosynthesis. Here, functional information of desired gene revealed the roles of lncRNAs in color production of fruit by altering the expression of regulatory factors. With the construction of these factors, decrement in the expression of *CHS* and *Fra a* occurred, leading to the development of colorless fruits (Bai et al., 2019).

Cucumis melo belongs to the family of Cucurbitaceae and shows significant variation in the process of fruit ripening. Climacteric stage appears to be the final one, depicting initiation of climacteric-dependent ripening of fruit (Giovannoni, 2007). Interestingly, sudden increase in the production of ethylene in fruit without any external stress is a striking property of the plant (Osorio et al., 2013), allowing scientists to focus on the molecular mechanism involved in the process. ssRNA-seq was performed to identify the role of lncRNAs during fruit ripening. Of total lncRNAs obtained, 1601 was found to be expressed differentially in different developing stages of fruit. *cis*/*trans* genes and targeted genes were analyzed in lncRNAs, where targeted genes were embellished in certain process as a response to hormone biosynthesis. An increase in the expression of LNC_000154 and LNC_002345 was observed upon co-expression and is known to co-regulate with genes in association with auxin signaling. lncRNAs corresponding with climacteric and fruit ripening were annotated along with their involvement in biosynthesis of ethylene and ABA signaling; thus, predicting the role of lncRNAs in ripening of melon and its participation in regulating plant hormones (Tian et al., 2019).

Experiment on *Malus domestica* was done by Yang et al. (2019a) to identify the role of lncRNAs in biosynthesis of anthocyanin from peels of apple cultivar. Analysis of RNA-seq revealed the presence of 5297 lncRNAs in light-driven anthocyanin production. Differential expression patterns further elaborated the involvement of lncRNAs in photosynthesis. Also, analysis of eTM anticipated that DE-lncRNAs, MLNC3.2 and MLNC4.6, are promising eTMs and enhance the expression of *SPL33* and *SPL2*-like transcribing factors. Ephemeral expression of fruit and steady transformation of callus concluded that an enhanced accumulation of anthocyanin with overexpression of *SPLs* and eTM, while inhibiting miR156a, altered the expression of the two. This justified that MLNC3.2 and MLNC4.6 mimic miR156a and hamper the breakdown of *SPL33* and *SPL2*-like genes during light-driven biosynthesis of anthocyanin.

Capsicum annuum L. is one of the important peppers in South European countries and belongs to nonclimacteric fruit developing species (Palma et al., 2014). It, thus, became a model plant for research as the processing of its ripening involves sudden ocular and biochemical alterations in fruit aroma, texture, and flavors (Klie et al., 2014). To investigate the molecular mechanism and its regulation in fruits of pepper, ssRNA-seq was adopted to examine noncoding RNAs. Of all ncRNAs examined, 11,999 lncRNAs were obtained, which included 11,085 novel and 194 known lncRNAs. KEGG and GO analysis revealed the target sites of 3266 *cis*/*trans*-acting DE lncRNAs and demonstrated the presence of certain transcription factors, enzymes involved in accumulation of flavor and color of fruit, and production of plant hormones as ethylene, gibberellic acid, and auxins. Expression profiles, thus, provide a complex structure of ncRNAs with mRNAs and its role in fruit development which gains the insight mechanism of ripening (Zuo et al., 2019).

17.4 lncRNA in cytoplasmic male sterility and pollen development

Pollen grain development is an important step to perform fertilization and proceed toward seed development in plants. During initial stages of pollination, tapetum protects the pollen grains and later provides nutrition to the matured ones. Abnormality in the pollen grains results in sterility in male plants (Datta and Paul, 2019). We define cytoplasmic male sterility (CMS) as sterility caused by nuclear or cytoplasmic elements (Hu et al., 2014). CMS is inherited maternally, where male parents unveil certain reproductive deformities. Events leading to CMS include stamens disruption and improper pollen growth (Chase and Gabay-Laughnan, 2004). Therefore, TGMS and PGMS regulated by nuclear genes and CMS have helped in enhancing the yield of crops (Mishra and Bohra, 2018).

Ding et al. (2012a,b) presented the role of lncRNAs in regulating *long day-specific male-fertility-associated RNA (LDMAR)*. Transcriptome analysis via RNA-seq revealed 1236 bps of lncRNAs modulates photoperiod-sensitive male sterility (PSMS) in *O. sativa*. Mutation leading to SNPs showed an alteration in secondary structure of *LDMAR*. Enhanced methylation in the promoter site of *LDMAR* begins reduction in their transcription level, resulting in PSMS of anthers due to early apoptosis (Fig. 17.2A).

To acknowledge the mechanism of PSMS, above-mentioned work was laid as a base. They then analyze siRNA as *Psi-LDMAR* at the promoter site of *LDMAR* (Fig. 17.2B). It was also observed that *Psi-LDMAR* was a derivative

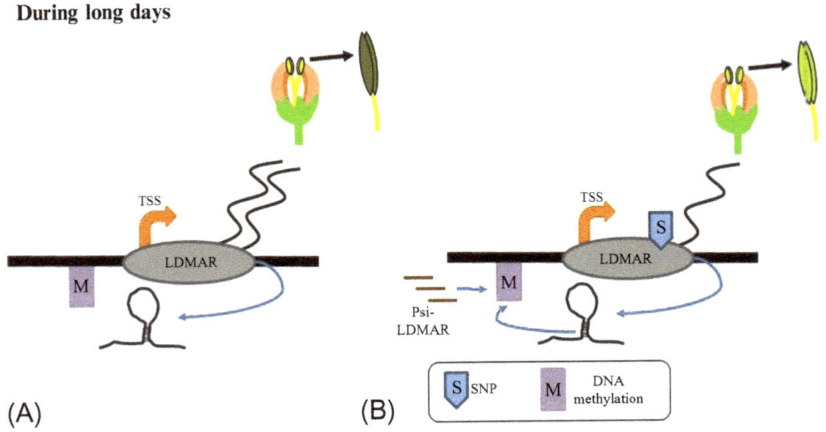

FIG. 17.2 Regulation of PSMS through LDMAR in rice. (A) Increased methylation of DNA and reduced transcription of LDMAR by its altered secondary structure caused premature apoptosis in anthers. (B) Repression in the expression of LDMAR via RNA-dependent DNA methylation (RdDM) pathway through Psi-LDMAR (siRNA) obtained from the promoter site of LDMAR ensured recovery of fertility among anthers. *(The figure has been adapted and modified from Zhang, Y.C., Chen, Y.Q., 2013. Long noncoding RNAs: new regulators in plant development. Biochem. Biophys. Res. Commun. 436 (2), 111–114 with the permission from Elsevier with license number-4713500366170).*

of *AK111270*. Therefore, upon expressing *AK111270* in PSMS rice, RNA-mediated DNA methylation was detected using bisulfite sequencing in the promoter region of *LDMAR*, causing its downregulation leading to altered day length required for recovery in fertility (Ding et al., 2012a,b). Further, Zhou et al. in 2012 worked on the identification of a unique RNA: *osa-smR5864w*, derived from *LDMAR*. He proposed that the RNA restored the fertility and a substitution mutation in *osa-smR5864w* results in development of sterile male plants. Experiments conducted on *Brassica campestris* revealed the presence of 828 nt long lncRNA-*BcMR11*, during pollen development. On developing transgenic lines, reduction in the level of *BcMF11* was observed which in turn degrades tapetum, thus inhibiting development of pollen grain (Song et al., 2013).

We define phased small-interfering RNAs (phasiRNAs) as a group of small RNAs, having 21-24 nt long sequences generated from transcripts of predecessor RNAs. Fan along with colleagues in 2016 revealed that locus of *photoperiod-sensitive male sterility (Pms1)* transcribes phasiRNAs in rice, inducing photoperiod-sensitive male sterility (PSMS). *Pms1* does so by encoding lncRNA *PMS1T*. miR2118 is known to digest *PMS1T* and produces phasiRNAs, which then further aggregate in hybrid lines of PSMS. The statement was confirmed by inducing a point mutation involving SNPs in PMS1T, resulting certain changes in fertility which lead to sharp accumulation of phasiRNAs. These findings, hence, proposed the functions of phasiRNAs in reproductive growth of *O. sativa* using illumina sequencing, demarcating its value as important modulators in developmental process.

A time series experiment involving five stages of pollen development was conducted in *Brassica rapa* for the characterization and functioning of lncRNAs. The results obtained by RNA-seq showed 12,051 lncRNAs, where co-expression of 47 *cis*-acting and 451 *trans*-acting lncRNAs with protein-coding (PC) genes was observed. Further, 14 lncRNAs were tightly co-expressed with pollen-associated PC genes; 15 lncRNAs were analyzed as eTMs for 13 miRNAs and 2 lncRNAs were imitating for miR160. These findings laid the basis for further genomic studies of lncRNAs in fertilization and pollen development of mustard (Huang et al., 2018).

Report mentioned by Zhang et al. (2019) provides the role of lncRNAs for development of anther in hybrid cotton lines as: CMS-D2 line, maintainer line, and restorer of fertility line. Transcriptome analysis for investigation of lncRNAs was performed, revealing 43,347 and 44,739 to be DE-lncRNAs. Study shows the participation of mentioned lncRNAs in CMS and restoration of fertility. Also, they identified and quantified certain lncRNAs to be predecessors of miRNAs. Further, construction of regulatory lncRNA networks reveals the presence of miRNA-lncRNA-mRNA complex and its association with fertility restoration and CMS, thus providing a base for further investigation on identification of regulatory mechanism.

17.5 Conclusions

We can conclude from the above summarized work that lncRNAs, members of regulatory ncRNAs, occupy a major role in maintaining the gene expression level and its functional annotation presents certain biological roles in plants. Data elaborated in the table represent the role of differential genes in flowering development. Also, photomorphogenesis, transcription regulation through epigenetic modification, and growth of reproductive tissues including pollens, seeds, ovules, and fruits are monitored by lncRNAs (Fig. 17.3).

With the advancement in the techniques involving the use of single molecule RNA fluorescent in situ hybridization (smFISH), one can evaluate the transcription level on cellular basis. The work performed in *Arabidopsis* enables us to assess the transcription of FLC and monitor the roles of lncRNAs in the vicinity of FLC at cellular and tissue levels. It acknowledges the molecular mechanism involved in the vernalization leading to flowering, thus enhancing our knowledge in the field of plant genomics.

A need to interpret the roles of lncRNAs in fruit ripening still remains one of the challenges to be faced. With the use of CRISPR/Cas9 technique, mutants of lncRNAs can be developed. A delay in the ripening of tomato and repression of genes involved in the process of ripening was observed when mutant of lncRNA was constructed via CRISPR/Cas9. These findings laid the base for analyzing the behavior of lncRNAs in fruit ripening and increase the outlook of research to figure out the possible molecular mechanisms involved in the process.

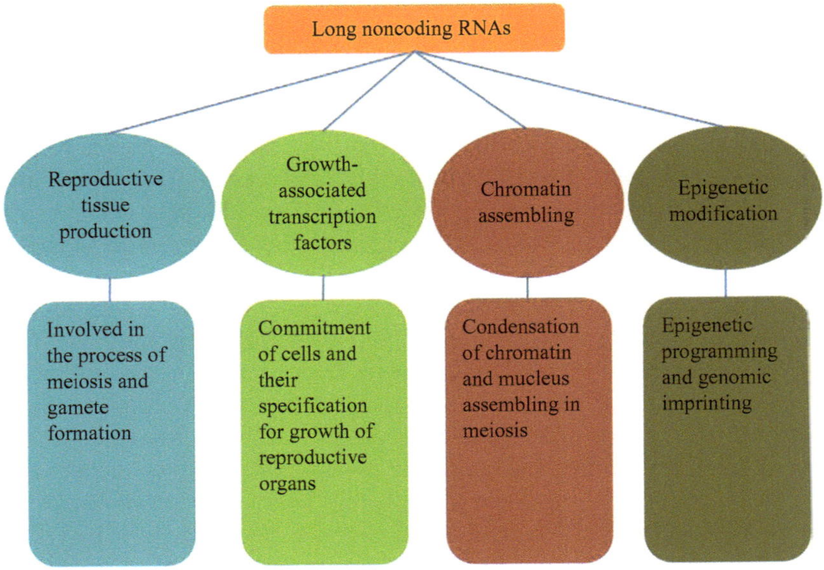

FIG. 17.3 Featured roles played by lncRNAs in procedure related to sexual reproduction, involving meiosis and gametogenesis among plants. *(The figure has been adapted and modified from Golicz, A.A., Bhalla, P.L., Singh, M.B., 2018. lncRNAs in plant and animal sexual reproduction. Trends Plant Sci. 23 (3), 195–205 with the permission from Elsevier with license number-4713491413184).*

References

Abe, M., Kobayashi, Y., Yamamoto, S., Daimon, Y., Yamaguchi, A., Ikeda, Y., et al., 2005. *FD*, a bZIP protein mediating signals from the floral pathway integrator *FT* at the shoot apex. Science 309 (5737), 1052–1056.

Amasino, R.M., 2005. Vernalization and flowering time. Curr. Opin. Biotechnol. 16 (2), 154–158.

Angel, A., Song, J., Dean, C., Howard, M., 2011. A Polycomb-based switch underlying quantitative epigenetic memory. Nature 476 (7358), 105.

Axtell, M.J., 2015. Non-coding RNAs: the small mysteries of males. Nat. Plants 1 (5), 15055.

Boerner, S., McGinnis, K.M., 2012. Computational identification and functional predictions of long noncoding RNA in *Zea mays*. Plos One 7 (8).

Bai, L., Chen, Q., Jiang, L., Lin, Y., Ye, Y., Liu, P., et al., 2019. Comparative transcriptome analysis uncovers the regulatory functions of long noncoding RNAs in fruit development and color changes of *Fragaria pentaphylla*. Hortic. Res. 6 (1), 42.

Bastow, R., Mylne, J.S., Lister, C., Lippman, Z., Martienssen, R.A., Dean, C., 2004. Vernalization requires epigenetic silencing of *FLC* by histone methylation. Nature 427 (6970), 164.

Bernier, G., Périlleux, C., 2005. A physiological overview of the genetics of flowering time control. Plant Biotechnol. J. 3 (1), 3–16.

Berry, S., Dean, C., 2015. Environmental perception and epigenetic memory: mechanistic insight through FLC. Plant J. 83 (1), 133–148.

Bhatia, G., Sharma, S., Upadhyay, S.K., Singh, K., 2019. Long non-coding RNAs coordinate developmental transitions and other key biological processes in grapevine. Sci. Rep. 9 (1), 3552.

Casañal, A., Zander, U., Muñoz, C., Dupeux, F., Luque, I., Botella, M.A., et al., 2013. The strawberry pathogenesis-related 10 (PR-10) *Fra a* proteins control flavonoid biosynthesis by binding to metabolic intermediates. J. Biol. Chem. 288 (49), 35322–35332.

Chase, C.D., Gabay-Laughnan, S., 2004. Cytoplasmic male sterility and fertility restoration by nuclear genes. In: Molecular Biology and Biotechnology of Plant Organelles. Springer, Dordrecht, pp. 593–621.

Chen, M., Chory, J., 2011. Phytochrome signaling mechanisms and the control of plant development. Trends Cell Biol. 21 (11), 664–671.

Chen, M., Penfield, S., 2018. Feedback regulation of *COOLAIR* expression controls seed dormancy and flowering time. Science 360 (6392), 1014–1017.

Chouard, P., 1960. Vernalization and its relations to dormancy. Annu. Rev. Plant Physiol. 11 (1), 191–238.

Conrad, L.J., Khanday, I., Johnson, C., Guiderdoni, E., An, G., Vijayraghavan, U., Sundaresan, V., 2014. The polycomb group gene EMF 2B is essential for maintenance of floral meristem determinacy in rice. Plant J. 80 (5), 883–894.

Crevillén, P., Dean, C., 2011. Regulation of the floral repressor gene FLC: the complexity of transcription in a chromatin context. Curr. Opin. Plant Biol. 14 (1), 38–44.

Crevillén, P., Yang, H., Cui, X., Greeff, C., Trick, M., Qiu, Q., et al., 2014. Epigenetic reprogramming that prevents transgenerational inheritance of the vernalized state. Nature 515 (7528), 587.

Csorba, T., Questa, J.I., Sun, Q., Dean, C., 2014. Antisense *COOLAIR* mediates the coordinated switching of chromatin states at *FLC* during vernalization. Proc. Natl. Acad. Sci. U. S. A. 111 (45), 16160–16165.

Datta, R., Paul, S., 2019. Long non-coding RNAs: fine-tuning the developmental responses in plants. J. Biosci. 44 (4), 77.

De Lucia, F., Dean, C., 2011. Long non-coding RNAs and chromatin regulation. Curr. Opin. Plant Biol. 14 (2), 168–173.

Ding, J., Lu, Q., Ouyang, Y., Mao, H., Zhang, P., Yao, J., et al., 2012a. A long noncoding RNA regulates photoperiod-sensitive male sterility, an essential component of hybrid rice. Proc. Natl. Acad. Sci. 109 (7), 2654–2659.

Ding, J., Shen, J., Mao, H., Xie, W., Li, X., Zhang, Q., 2012b. RNA-directed DNA methylation is involved in regulating photoperiod-sensitive male sterility in rice. Mol. Plant 5 (6), 1210–1216.

Dodd, I.B., Micheelsen, M.A., Sneppen, K., Thon, G., 2007. Theoretical analysis of epigenetic cell memory by nucleosome modification. Cell 129 (4), 813–822.

Fan, Y., Yang, J., Mathioni, S.M., Yu, J., Shen, J., Yang, X., et al., 2016. *PMS1T*, producing phased small-interfering RNAs, regulates photoperiod-sensitive male sterility in rice. Proc. Natl. Acad. Sci. 113 (52), 15144–15149.

Flórez-Zapata, N.M., Reyes-Valdés, M.H., Martínez, O., 2016. Long non-coding RNAs are major contributors to transcriptome changes in sunflower meiocytes with different recombination rates. BMC Genomics 17 (1), 490.

Gapper, N.E., McQuinn, R.P., Giovannoni, J.J., 2013. Molecular and genetic regulation of fruit ripening. Plant Mol. Biol. 82 (6), 575–591.

Giovannoni, J.J., 2007. Fruit ripening mutants yield insights into ripening control. Curr. Opin. Plant Biol. 10 (3), 283–289.

Gonzalez, A., Zhao, M., Leavitt, J.M., Lloyd, A.M., 2008. Regulation of the anthocyanin biosynthetic pathway by the TTG1/bHLH/Myb transcriptional complex in *Arabidopsis* seedlings. Plant J. 53 (5), 814–827.

Gou, J.Y., Felippes, F.F., Liu, C.J., Weigel, D., Wang, J.W., 2011. Negative regulation of anthocyanin biosynthesis in Arabidopsis by a miR156-targeted *SPL* transcription factor. Plant Cell 23 (4), 1512–1522.

Hamant, O., Ma, H., Cande, W.Z., 2006. Genetics of meiotic prophase I in plants. Annu. Rev. Plant Biol. 57, 267–302.

Hao, Z., Fan, C., Cheng, T., Su, Y., Wei, Q., Li, G., 2015. Genome-wide identification, characterization and evolutionary analysis of long intergenic noncoding RNAs in cucumber. PLoS One 10 (3), e0121800.

He, Y., 2009. Control of the transition to flowering by chromatin modifications. Mol. Plant 2 (4), 554–564.

Heo, J.B., Sung, S., 2011. Vernalization-mediated epigenetic silencing by a long intronic noncoding RNA. Science 331 (6013), 76–79.

Hu, J., Huang, W., Huang, Q., Qin, X., Yu, C., Wang, L., et al., 2014. Mitochondria and cytoplasmic male sterility in plants. Mitochondrion 19, 282–288.

Huang, L., Dong, H., Zhou, D., Li, M., Liu, Y., Zhang, F., et al., 2018. Systematic identification of long non-coding RNA s during pollen development and fertilization in *Brassica rapa*. Plant J. 96 (1), 203–222.

Jiao, F., Pahwa, K., Manning, M., Dochy, N., Geuten, K., 2019. Cold induced antisense transcription of *FLOWERING LOCUS C* in distant grasses. Front. Plant Sci. 10, 72.

Johnson, C., Conrad, L.J., Patel, R., Anderson, S., Li, C., Pereira, A., Sundaresan, V., 2018. Reproductive long intergenic noncoding RNAs exhibit male gamete specificity and polycomb repressive complex 2-mediated repression. Plant Physiol. 177 (3), 1198–1217.

Kang, C., Liu, Z., 2015. Global identification and analysis of long non-coding RNAs in diploid strawberry *Fragaria vesca* during flower and fruit development. BMC Genomics 16 (1), 815.

Khemka, N., Singh, V.K., Garg, R., Jain, M., 2016. Genome-wide analysis of long intergenic non-coding RNAs in chickpea and their potential role in flower development. Sci. Rep. 6, 33297.

Kiegle, E.A., Garden, A., Lacchini, E., Kater, M.M., 2018. A genomic view of alternative splicing of long non-coding RNAs during rice seed development reveals extensive splicing and lncRNA gene families. Front. Plant Sci. 9, 115.

Kim, D.H., Sung, S., 2017. Vernalization-triggered intragenic chromatin loop formation by long noncoding RNAs. Dev. Cell 40 (3), 302–312.

Kim, E.D., Sung, S., 2012. Long noncoding RNA: unveiling hidden layer of gene regulatory networks. Trends Plant Sci. 17 (1), 16–21.

Kim, S.Y., He, Y., Jacob, Y., Noh, Y.S., Michaels, S., Amasino, R., 2005. Establishment of the vernalization-responsive, winter-annual habit in *Arabidopsis* requires a putative histone H3 methyl transferase. Plant Cell 17 (12), 3301–3310.

Kim, D.H., Xi, Y., Sung, S., 2017. Modular function of long noncoding RNA, *COLDAIR*, in the vernalization response. PLoS Genet. 13 (7), e1006939.

Klie, S., Osorio, S., Tohge, T., Drincovich, M.F., Fait, A., Giovannoni, J.J., et al., 2014. Conserved changes in the dynamics of metabolic processes during fruit development and ripening across species. Plant Physiol. 164 (1), 55–68.

Li, B., Carey, M., Workman, J.L., 2007. The role of chromatin during transcription. Cell 128 (4), 707–719.

Li, L., Eichten, S.R., Shimizu, R., Petsch, K., Yeh, C.T., Wu, W., et al., 2014. Genome-wide discovery and characterization of maize long non-coding RNAs. Genome Biol. 15 (2), R40.

Li, W., Liu, Y., Zeng, S., Xiao, G., Wang, G., Wang, Y., et al., 2015. Gene expression profiling of development and anthocyanin accumulation in kiwifruit (*Actinidia chinensis*) based on transcriptome sequencing. PLoS One 10 (8), e0136439.

Liang, N., Cheng, D., Cui, J., Dai, C., Luo, C., Liu, T., Li, J., 2017. Vernalisation mediated LncRNA-like gene expression in *Beta vulgaris*. Funct. Plant Biol. 44 (7), 720–726.

Lin, M.K., Belanger, H., Lee, Y.J., Varkonyi-Gasic, E., Taoka, K.I., Miura, E., et al., 2007. *FLOWERING LOCUS T* protein may act as the long-distance florigenic signal in the cucurbits. Plant Cell 19 (5), 1488–1506.

Liu, P.P., Montgomery, T.A., Fahlgren, N., Kasschau, K.D., Nonogaki, H., Carrington, J.C., 2007. Repression of *AUXIN RESPONSE FACTOR10* by microRNA160 is critical for seed germination and post-germination stages. Plant J. 52 (1), 133–146.

Liu, F., Marquardt, S., Lister, C., Swiezewski, S., Dean, C., 2010. Targeted 3′ processing of antisense transcripts triggers *Arabidopsis FLC* chromatin silencing. Science 327 (5961), 94–97.

Liu, J., Jung, C., Xu, J., Wang, H., Dens, S., Bernad, L., et al., 2012. Genome-wide analysis uncovers regulation of long intergenic noncoding RNAs in *Arabidopsis*. Plant Cell 24 (11), 4333–4345.

Liu, S., Li, W., Wu, Y., Chen, C., Lei, J., 2013. De novo transcriptome assembly in chili pepper (*Capsicum frutescens*) to identify genes involved in the biosynthesis of capsaicinoids. PLoS One 8 (1), e48156.

Liu, H., Wang, R., Mao, B., Zhao, B., Wang, J., 2019. Identification of lncRNAs involved in rice ovule development and female gametophyte abortion by genome-wide screening and functional analysis. BMC Genomics 20 (1), 90.

Lu, T., Zhu, C., Lu, G., Guo, Y., Zhou, Y., Zhang, Z., et al., 2012. Strand-specific RNA-seq reveals widespread occurrence of novel *cis*-natural antisense transcripts in rice. BMC Genomics 13 (1), 721.

Lu, Z., Xia, X., Jiang, B., Ma, K., Zhu, L., Wang, L., Jin, B., 2017. Identification and characterization of novel lncRNAs in *Arabidopsis thaliana*. Biochem. Biophys. Res. Commun. 488 (2), 348–354.

Michaels, S.D., Amasino, R.M., 1999. *FLOWERING LOCUS C* encodes a novel *MADS* domain protein that acts as a repressor of flowering. Plant Cell 11 (5), 949–956.

Mishra, A., Bohra, A., 2018. Non-coding RNAs and plant male sterility: current knowledge and future prospects. Plant Cell Rep. 37 (2), 177–191.

Ørom, U.A., Derrien, T., Beringer, M., Gumireddy, K., Gardini, A., Bussotti, G., et al., 2010. Long noncoding RNAs with enhancer-like function in human cells. Cell 143 (1), 46–58.

Osorio, S., Scossa, F., Fernie, A., 2013. Molecular regulation of fruit ripening. Front. Plant Sci. 4, 198.

Ou, L., Liu, Z., Zhang, Z., Wei, G., Zhang, Y., Kang, L., et al., 2017. Noncoding and coding transcriptome analysis reveals the regulation roles of long noncoding RNAs in fruit development of hot pepper (*Capsicum annuum* L.). Plant Growth Regul. 83 (1), 141–156.

Palma, J.M., Corpas, F.J., del Río, L.A., López-Huertas, E., 2014. Pepper fruit as a model to study the metabolism of antioxidants, ROS and RNS. Free Radic. Biol. Med. 75, S39.

Qin, G., Wang, Y., Cao, B., Wang, W., Tian, S., 2012. Unraveling the regulatory network of the *MADS* box transcription factor *RIN* in fruit ripening. Plant J. 70 (2), 243–255.

Severing, E., Faino, L., Jamge, S., Busscher, M., Kuijer-Zhang, Y., Bellinazzo, F., et al., 2018. *Arabidopsis thaliana* ambient temperature responsive lncRNAs. BMC Plant Biol. 18 (1), 145.

Simpson, C.G., Fuller, J., Calixto, C.P., McNicol, J., Booth, C., Brown, J.W., Staiger, D., 2016. Monitoring alternative splicing changes in Arabidopsis circadian clock genes. In: Environmental Responses in Plants. Humana Press, New York, NY, pp. 119–132.

Shumayla, S., Sharma, Taneja, M., Tyagi, S., Singh, K., Upadhyay, S.K., 2017. Survey of high throughput RNA-seq data reveals potential roles for lncRNAs during development and stress response in bread wheat. Front. Plant Sci. 8, 1019.

Song, J.H., Cao, J.S., Wang, C.G., 2013. BcMF11, a novel non-coding RNA gene from *Brassica campestris*, is required for pollen development and male fertility. Plant Cell Rep. 32 (1), 21–30.

Song, X., Sun, L., Luo, H., Ma, Q., Zhao, Y., Pei, D., 2016. Genome-wide identification and characterization of long non-coding RNAs from mulberry (*Morus notabilis*) RNA-seq data. Genes 7 (3), 11.

Sun, Q., Csorba, T., Skourti-Stathaki, K., Proudfoot, N.J., Dean, C., 2013. R-loop stabilization represses antisense transcription at the *Arabidopsis FLC* locus. Science 340 (6132), 619–621.

Swiezewski, S., Liu, F., Magusin, A., Dean, C., 2009. Cold-induced silencing by long antisense transcripts of an *Arabidopsis* Polycomb target. Nature 462 (7274), 799.

Tang, W., Zheng, Y., Dong, J., Yu, J., Yue, J., Liu, F., et al., 2016. Comprehensive transcriptome profiling reveals long noncoding RNA expression and alternative splicing regulation during fruit development and ripening in kiwifruit (*Actinidia chinensis*). Front. Plant Sci. 7, 335.

Tian, Y., Bai, S., Dang, Z., Hao, J., Zhang, J., Hasi, A., 2019. Genome-wide identification and characterization of long non-coding RNAs involved in fruit ripening and the climacteric in *Cucumis melo*. BMC Plant Biol. 19 (1), 1–15.

Tyagi, S., Sharma, A., Upadhyay, S.K., 2018. Role of next-generation RNA-Seq data in discovery and characterization of long non-coding RNA in plants. In: Çiftçi, Y.Ö. (Ed.), Next Generation Plant Breeding. IntechOpen, p. 2018. https://doi.org/10.5772/intechopen.72773.

Ulitsky, I., 2016. Evolution to the rescue: using comparative genomics to understand long non-coding RNAs. Nat. Rev. Genet. 17 (10), 601.

Vrebalov, J., Ruezinsky, D., Padmanabhan, V., White, R., Medrano, D., Drake, R., et al., 2002. A *MADS*-box gene necessary for fruit ripening at the tomato *ripening-inhibitor (rin)* locus. Science 296 (5566), 343–346.

Wang, Y., Fan, X., Lin, F., He, G., Terzaghi, W., Zhu, D., Deng, X.W., 2014a. *Arabidopsis* noncoding RNA mediates control of photomorphogenesis by red light. Proc. Natl. Acad. Sci. 111 (28), 10359–10364.

Wang, Y., Wang, X., Deng, W., Fan, X., Liu, T.T., He, G., et al., 2014b. Genomic features and regulatory roles of intermediate-sized non-coding RNAs in *Arabidopsis*. Mol. Plant 7 (3), 514–527.

Wang, C.Y., Liu, S.R., Zhang, X.Y., Ma, Y.J., Hu, C.G., Zhang, J.Z., 2017a. Genome-wide screening and characterization of long non-coding RNAs involved in flowering development of trifoliate orange (*Poncirus trifoliata* L. Raf.). Sci. Rep. 7, 43226.

Wang, M., Wu, H.J., Fang, J., Chu, C., Wang, X.J., 2017b. A long noncoding RNA involved in rice reproductive development by negatively regulating osa-miR160. Sci. Bull. 62 (7), 470–475.

Wang, M., Zhao, W., Gao, L., Zhao, L., 2018a. Genome-wide profiling of long non-coding RNAs from tomato and a comparison with mRNAs associated with the regulation of fruit ripening. BMC Plant Biol. 18 (1), 75.

Wang, Z., Zhu, T., Ma, W., Wang, N., Qu, G., Zhang, S., Wang, J., 2018b. Genome-wide analysis of long non-coding RNAs in *Catalpa bungei* and their potential function in floral transition using high-throughput sequencing. BMC Genet. 19 (1), 86.

Wu, H.J., Wang, Z.M., Wang, M., Wang, X.J., 2013. Widespread long noncoding RNAs as endogenous target mimics for microRNAs in plants. Plant Physiol. 161 (4), 1875–1884.

Wu, X., Shi, T., Iqbal, S., Zhang, Y., Liu, L., Gao, Z., 2019. Genome-wide discovery and characterization of flower development related long non-coding RNAs in *Prunus mume*. BMC Plant Biol. 19 (1), 64.

Wunderlich, M., Groß-Hardt, R., Schöffl, F., 2014. Heat shock factor *HSFB2a* involved in gametophyte development of *Arabidopsis thaliana* and its expression is controlled by a heat-inducible long non-coding antisense RNA. Plant Mol. Biol. 85 (6), 541–550.

Xin, R., Zhu, L., Salomé, P.A., Mancini, E., Marshall, C.M., Harmon, F.G., et al., 2017. *SPF45*-related splicing factor for phytochrome signaling promotes photomorphogenesis by regulating pre-mRNA splicing in Arabidopsis. Proc. Natl. Acad. Sci. 114 (33), E7018–E7027.

Yamada, K., Lim, J., Dale, J.M., Chen, H., Shinn, P., Palm, C.J., et al., 2003. Empirical analysis of transcriptional activity in the *Arabidopsis* genome. Science 302 (5646), 842–846.

Yamaguchi, A., Abe, M., 2012. Regulation of reproductive development by non-coding RNA in *Arabidopsis*: to flower or not to flower. J. Plant Res. 125 (6), 693–704.

Yang, H., Howard, M., Dean, C., 2014. Antagonistic roles for H3K36me3 and H3K27me3 in the cold-induced epigenetic switch at *Arabidopsis* FLC. Curr. Biol. 24 (15), 1793–1797.

Yang, T., Ma, H., Zhang, J., Wu, T., Song, T., Tian, J., Yao, Y., 2019a. Systematic identification of long noncoding RNA s expressed during light-induced anthocyanin accumulation in apple fruit. Plant J. 1000 (3), 572–590.

Yang, Z., Yang, C., Wang, Z., Yang, Z., Chen, D., Wu, Y., 2019b. LncRNA expression profile and ceRNA analysis in tomato during flowering. PLoS One 14 (1), e0210650.

Zhang, Y.C., Liao, J.Y., Li, Z.Y., Yu, Y., Zhang, J.P., Li, Q.F., et al., 2014. Genome-wide screening and functional analysis identify a large number of long noncoding RNAs involved in the sexual reproduction of rice. Genome Biol. 15 (12), 512.

Zhang, G., Duan, A., Zhang, J., He, C., 2017. Genome-wide analysis of long non-coding RNAs at the mature stage of sea buckthorn (*Hippophae rhamnoides* Linn) fruit. Gene 596, 130–136.

Zhang, G., Chen, D., Zhang, T., Duan, A., Zhang, J., He, C., 2018. Transcriptomic and functional analyses unveil the role of long non-coding RNAs in anthocyanin biosynthesis during sea buckthorn fruit ripening. DNA Res. 25 (5), 465–476.

Zhang, B., Zhang, X., Zhang, M., Guo, L., Qi, T., Wang, H., et al., 2019. Transcriptome Analysis Implicates Involvement of Long Noncoding RNAs in Cytoplasmic Male Sterility and Fertility Restoration in Cotton. Int. J. Mol. Sci. 20 (22), 5530.

Zhao, J., Ohsumi, T.K., Kung, J.T., Ogawa, Y., Grau, D.J., Sarma, K., et al., 2010. Genome-wide identification of polycomb-associated RNAs by RIP-seq. Mol. Cell 40 (6), 939–953.

Zhao, X., Li, J., Lian, B., Gu, H., Li, Y., Qi, Y., 2018. Global identification of *Arabidopsis* lncRNAs reveals the regulation of *MAF4* by a natural antisense RNA. Nat. Commun. 9 (1), 5056.

Zhou, H., Liu, Q., Li, J., Jiang, D., Zhou, L., Wu, P., Chen, L., 2012. Photoperiod-and thermo-sensitive genic male sterility in rice are caused by a point mutation in a novel noncoding RNA that produces a small RNA. Cell Res. 22 (4), 649.

Zhu, B., Yang, Y., Li, R., Fu, D., Wen, L., Luo, Y., Zhu, H., 2015. RNA sequencing and functional analysis implicate the regulatory role of long non-coding RNAs in tomato fruit ripening. J. Exp. Bot. 66 (15), 4483–4495.

Zhu, M., Zhang, M., Xing, L., Li, W., Jiang, H., Wang, L., Xu, M., 2017. Transcriptomic analysis of long non-coding RNAs and coding genes uncovers a complex regulatory network that is involved in maize seed development. Genes 8 (10), 274.

Zuo, J., Wang, Y., Zhu, B., Luo, Y., Wang, Q., Gao, L., 2019. Network analysis of noncoding RNAs in pepper provides insights into fruit ripening control. Sci. Rep. 9 (1), 8734.

Chapter 18

Biotic stress response of lncRNAs in plants

Madiha Zaynab[a], Mahpara Fatima[b], Yasir Sharif[c], Muhammad Qasim[d], Mehtab Muhammad Aslam[e], Muhammad Zohaib Afzal[b], and Nelam Sajjad[e]

[a]College of Life Sciences and Oceanography, Shenzhen University, Shenzhen, PR China, [b]College of Agriculture, Fujian Agriculture and Forestry University, Fuzhou, Fujian, China, [c]College of Plant Protection, Fujian Agriculture and Forestry University, Fuzhou, Fujian, China, [d]Key Lab of Molecular Biology of Crop Pathogens and Insects, Institute of Insect Sciences, College of Agriculture & Biotechnology, Zhejiang University, Hangzhou, PR China, [e]College of Life Sciences, Fujian Agriculture and Forestry University, Fuzhou, Fujian, China

18.1 Introduction

Some RNAs transcribed from DNA, but not translated into proteins, directly play an important role in the regulation of gene expression. The RNAs that are not translated to proteins are known as noncoding RNAs (ncRNAs). The ncRNAs are categorized into small noncoding RNAs (sncRNAs) and long noncoding RNAs (lncRNAs). The length of sncRNAs usually is less than 200 nt, and these snRNAs are further classified into siRNAs (small interfering RNAs), miRNAs (microRNAs), transacting siRNAs, piwi-interacting RNAs, and natural antisense transcript siRNAs (Rinn and Chang, 2012). Likewise, usually, lncRNAs have a length of more than 200 nt and are divided into three groups depending on protein-coding genes positioned nearby, i.e., long intergenic noncoding RNAs (lincRNAs), long intronic ncRNAs, and natural antisense transcripts (NATs) (Ma et al., 2013). In plants, identification of lncRNAs started in recent years and is not widespread in comparison to other eukaryotes (Zhang et al., 2014; Tyagi et al., 2018). The lncRNAs are likely to be involved in the regulation of the expression of target genes by different molecular mechanisms (Zhu and Wang, 2012). They have also been involved in the posttranscriptional modifications of mRNA. The lncRNAs have a low potential to encode proteins, but they have a role in regulating the expression of target genes during the process of transcription and translation. It is well-documented that lncRNAs in plants play a role in the regulation of complex gene regulatory networks, which are involved in plant developmental processes and stress tolerance (Heo and Sung, 2011).

Long Noncoding RNAs in Plants: Roles in development and stress
https://doi.org/10.1016/B978-0-12-821452-7.00018-0

Thousands of lncRNAs have been discovered through in silico predictions and applied tilling arrays at the whole genome and RNA-sequencing (RNA-Seq) levels in different plant species, including *Arabidopsis thaliana* (Liu et al., 2012), *Zea mays* (Li et al., 2014), *Triticum aestivum* (Xin et al., 2011; Shumayla et al., 2017), *Oryza sativa* (Zhang et al., 2014), *Cucumis sativus* (Hao et al., 2015), and *Medicago truncatula* (Wen et al., 2007). Most of the reported lncRNAs are stress-responsive (Shafiq et al., 2016), e.g., 1212 unique candidate lncRNAs with 309 differentially expressed lncRNAs were predicted under control as well as under Pi starvation in *Arabidopsis* (Yuan et al., 2016). The 1113 lincRNAs and 17 linTARs (involved in defense) discovered through strand-specific RNA-sequencing in potato were reactive to *Pectobacterium carotovorum* subsp. *brasiliense* damage (Kwenda et al., 2016). Similarly, 1565 lncRNAs have been identified in tomato, responsive to TYLCV (tomato yellow leaf curl virus) infestation (Wang et al., 2015). However, the limited knowledge is available about the regulation of lncRNAs under stresses at the molecular level, except a few have been functionally studied. Two different lncRNAs in *Arabidopsis*, i.e., COLDAIR and COOLAIR, have been documented to be involved in the silencing of FLC (FLOWERING LOCUS C) at transcription in vernalization process via chromatin modifications (Swiezewski et al., 2009). During the recent past, an antisense and unpolyadenylated lncRNA (ASL) was identified that plays various roles in the silencing of FLC (Shin and Chekanova, 2014). Moreover, IPA1 lncRNA is induced by Pi starvation, which plays a role as miR399 target mimic and leads to reduced cleavage of PHO mRNA mediated by miR399 (Franco-Zorrilla et al., 2007).

18.2 Long noncoding RNAs and plant defense mechanism

The antiviral defense in plants involves two key mechanisms. Firstly, RNA silencing is mediated by double-stranded viral RNA that is processed by RNA silencing system of a host into siRNAs of 21–24 nt, which in turn is involved in RNA degradation through homology-dependent manner (Szittya and Burgyán, 2013). Viruses overcome this defense response through the expression of suppressors of RNA silencing that affects various steps of this defense system (Csorba et al., 2015). A few also produce ncRNAs which generate a high amount of viral siRNAs that act as a sponge for host components of RNA silencing system, and hence, inhibit the degradation of other RNA species of virus, e.g., CaMV35S, and 8S RNA (Hohn, 2015). Secondly, phytohormones, e.g., salicylic acid (SA) and 2-hydroxybenzoic acid, are involved in signaling pathways. For instance, SA is a multifaceted hormone and involved in various plant developmental processes and response to stresses (Miura and Tada, 2014). It has demonstrated that SA participates in signal transduction activating plant defense system, resulting in the suppression of viral movement and amplification (Lee et al., 2016). The perception of pathogen induces SA, which leads to the activation of responsive genes involved in the defense system, including those encoding the PR (pathogenesis-related) proteins (Alazem and Lin, 2015). Some viruses produce different proteins that hinder SA signaling (Love et al., 2012). However, so far, no studies described the role of lncRNAs produced by the virus, mitigating the SA-regulated pathways. The phytoplasma-infected plants can make effective strategies to protect themselves from invasion (Hegenauer et al., 2016). First, phytoplasma's presence is perceived by plants via PAMPs (pathogen-associated molecular patterns), which produce resistance. At this stage, plants generate a high quantity of ROS (reactive oxygen species) as well as antitoxin, which activate a hypersensitive response. Besides, few effectors, e.g., TENGU and SAP, are secreted by phytoplasmas via the general secretory system (Sec secretion system), which influences the host PAMP-activated defense signal transduction and increases colonization effectively as well as assist in their multiplication in the host plant cells. Plants recognize effectors by utilization of cellular receptor proteins, which results in activation of the immune response and finally involves activation of MAPK cascades as well as induction of disease-resistant protein.

The immune responses of the plant to phytoplasma infestation were activated and responsive genes that encode enzymes for signaling pathways, e.g., PERK1 (proline-rich receptor-like protein kinase), and important marker genes convoluted in the related processes (e.g., disease-resistant protein, SRC protein, and glucan endo-1,3-beta-glucosidase) were identified. The concentration of glucan endo-1,3-beta-glucosidase enhanced distinctly in the tobacco infected with necrotizing viruses (Beffa et al., 1993). Likewise, the gene-encoding SRC2 was highly expressed in infected peppers with *Xanthomonas axonopodis* pv. *glycines* (Kim et al., 2008). Remarkably, the upregulation of PERK1, glucan endo-1,3-beta-glucosidase, disease-resistant protein, and SRC2 protein was noticed in infested cuttings with phytoplasma. Also, the upregulation of genes encoding-resistant proteins against diseases has been reported, when infected with phytoplasma in an earlier study in *Paulownia*, while three lncRNAs, including TCONS_00021207, TCONS_00026765, and TCONS_00034613, were expected to control the expression levels of vital genes which encode disease-resistant protein, SRC2 protein, and glucan endo 1, 3-betaglucosidase, respectively. Therefore, these lncRNAs mentioned above possibly have important roles in phytoplasma-infected cuttings. Additionally, the expression level of gene-encoding cytochrome P450 also increased in the PTI cuttings. As an oxidant, cytochrome P450 can remove excess ROS and prevent cell damage. In the earlier study, *Paulownia* plants infected with phytoplasma showed an elevated level of the cytochrome. The infected cuttings with phytoplasma displayed downregulation of TCONS_00031692 and upregulation of TCONS_00007939; both of these regulate the expression of cytochrome P450-encoding gene (LCONS_00023050). Therefore, lncRNAs might be significantly involved in hypersensitive response against ROS and effector-activated immunity inside phytoplasma-infected cuttings.

It has described that co-expressed and co-localized transcripts can anticipate the possible lncRNA functions (Wilusz et al., 2009). The *cis* examination indicated high consequence with GO terms linked to morphological variations such as structural cell wall constituents, the metabolic process of chitin, the catabolic process of chitin, as well as macromolecule metabolic process of cell wall, among others. In plants, the cell wall acts as the first hurdle against pathogen attack and is mainly composed of proteins, polysaccharides, and aromatic polymers (Underwood and Somerville, 2008). The perception of fungal chitin via host plants is crucial for activating PAMP- initiated immunity (PTI) against fungal attack (Kombrink et al., 2011).

18.3 Identification of lncRNAs related to biotic stress

One study conducted on *Brassica napus* identified a total of 3181 lncRNA comprising intergenic (2821), antisense (111), exonic overlap (76), and lncRNA isoforms (173). Moreover, 41 lncRNAs are recognized as the originators of miRNA constituting miRNA156, miRNA169, and miRNA394. These miRNAs showed a prominent role in regulating plant responses against fungal pathogens (Joshi et al., 2016). A total of 931 lncRNAs have been recognized in response to *Sclerotinia sclerotiorum*, and 12 lncRNAs have been used to validate their expression using qRT-PCR that revealed different expressions (Joshi et al., 2016). Two genotypes of grapevine-susceptible (Cabernet Franc.) and -resistant (Merlot) infected with *Lasiodiplodia theobromae* were used to perform high-throughput RNA-Seq. Inclusively, it was anticipated that a total of 1826 lncRNAs were present along with intergenic lncRNA and antisense RNA transcripts (Xing et al., 2019). *Hevea brasiliensis* leaves infected with *Colletotrichum gloeosporioides* at different stages showed that dominant lncRNAs are involved in response to infection. The outcome showed that the function of lncRNA 11,205 recruiting TF (transcription factor) and other lncRNA11041 and incRNA11205 probably triggers the assembly of several disease-responsive miRNAs (Yin et al., 2019). Susceptible and tolerant banana genotypes were sequenced through the Illumina-sequencing tool for recognition of differential expression under both control and stressed environments. The resulting sequence was assembled into 201,256 and 172,434 transcriptional units for *Pratylenchus coffeae* and *Mycosphaerella eumusae*, respectively (Muthusamy et al., 2019). A total of 5142 lncRNAs comprising 3031 and 1672 intergenic and 440 antisense were acknowledged in both resistant and susceptible banana genotypes. Furthermore, lncRNA 1250 and 1284 were contrastingly expressed in response to *P. coffeae*, and *M. eumusae*, respectively (Muthusamy et al., 2019). The interaction between lncRNA and mRNA regulates the suppression of Horcolin protein (antifungal protein) accountable for ELSD sensitivity in banana genotypes (Muthusamy et al., 2019). LncRNA 1331 exhibited differential expression, GO (gene ontology), and KEGG analysis which demonstrated that various protein-encoding lncRNAs were reinforced under stressed conditions. Additionally, 55 lncRNAs out of

8414 were expected as a mimic of cotton miRNA 285. Quantitative real-time PCR (qPCR) proved that all lncRNAs were related to aphid infection (Zhang et al., 2019). Recently, one study characterized the lncRNAs of two main lignin-degrading basidiomycetes, *Coniophora puteana* and *Serpula lacrymans*, with 2712 and 2242 lncRNAs, respectively, majorly originating from intergenic regions in respective species (Borgognone et al., 2019). Previous studies specified *Phytophthora infestans* disease-resistant tomato varieties produced using WRKY1 (transcription factor). However, the mechanism underlying WRKY1 arbitrating resistance remains poorly understood. Wild-type and WRKY1 overexpressing tomato lines are used to identify contrastingly expressed genes and lncRNAs and also investigated gene networks of long noncoding RNA. The outcomes of promoter-GUS fusion and yeast hybrid assay exhibited the upregulation of lncRNA33732 by WRKY1 (Cui et al., 2019). A set of lncRNAs recognized and characterized in animals, fungi, and plants played an essential role in biological mechanisms. But, minimal information is available on its role in oomycete phytopathogens, which leads to catastrophic loss to the ecosystems (Wang et al., 2018a). *Phytophthora sojae* oomycete phytopathogen is used to identify specific RNA transcript sequencing; a total of 940 lncRNA along with 1010 isoforms were found to have shorter nucleotide length, larger exon length, the smaller number of exons, fewer GC%, and higher minimal energy contrasting with protein-encoding genes. RNA-Seq data disclose an interaction among transcript of lncRNAs and adjacent genes coding effector proteins (Wang et al., 2018c). Strand-specific RNA-Seq of susceptible and resistant banana cultivars, infected with *Fusarium oxysporum*, specified 5294 lncRNAs with high confidence provided with strict filtration along with intergenic and antisense lncRNA (Li et al., 2017). *F. oxysporum* and mock-inoculated-susceptible and resistant cultivars also revealed some differentially expressed lncRNAs (Li et al., 2017). Kang et al. (2019) focused on the discovery of functional identification and characterization of plant important lincRNAs after the viral infection. This discovery was promoted through transcriptome analysis by using RNA-Seq analysis (Kang et al., 2019). Inoculated neutron irradiated tomato seeds with tomato yellow leaf curl virus (TYLCV) and performed RNA-Seq to find out the lncRNAs. With the help of bioinformatics analysis, they found a total of 1563 lncRNAs, half of which were originated from intergenic regions and 35% were antisense lncRNAs. Almost 794 lncRNAs were functionally associated with binding function, showed attraction with other molecules, and localized in membrane and cytosol of the cell (Zhou et al., 2019). Another study conducted on potato (*Solanum tuberosum*) stem tissues identified 1113 long intergenic ncRNAs through strand-specific RNA-Seq. The size of expressed lincRNAs in all 12 chromosomes was smaller than protein-coding genes, and only a single exon was present similar to other studied plants. Further time-course RNA-Seq analysis revealed that 559 lincRNAs were responsive to *P. carotovorum* subsp. *brasiliense*, and 17 lincRNAs out of these 559 were linked with 12 defense genes (Kwenda et al., 2016). Whole-transcriptome RNA-Seq of

6-h aphid damaged cotton plants identified 8414 lncRNAs, mainly belonging to intergenic regions, i.e., 77%, where 10% was sense lncRNAs, 7% antisense, and 4% intronic lncRNAs. Further, 1331 lncRNAs showed differential expression, while KEGG pathway and GO analysis showed that many target genes of these lncRNAs were stress-responsive (Zhang et al., 2019).

18.4 Long noncoding RNAs as front-line player against pathogens

Canola is an important crop, and its seed contains almost 45% oil content. Canola production is mainly affected globally due to Sclerotinia stem rot disease caused by a pathogenic fungus (*S. sclerotiorum*). When plants encounter any environmental stress (biotic/abiotic), they evolve a different mechanism to cope with that particular stress. LncRNA plays a vital role in driving gene expression. Limited information is available on their identification and function in some plant species. Antisense lncRNA (TCONS-00000966) from *B. napus* presents 90% overlapping with defensin genes, which suggests its interaction in regulating defense-related genes upon *S. sclerotiorum* infection. These findings elaborate on the prospective involvement of lncRNAs in driving expression of those genes expressed under biotic stress (Joshi et al., 2016).

18.4.1 Role against fungi

Recently, several lncRNAs appeared as important regulators of biological mechanisms in both plants and animals, involving response against biotic and abiotic stress. Upon infection of *L. theobromae*, various lncRNAs were expected to be originators for grapevine miRNA. These findings open new insights into grapevine lncRNAs, which are involved in response to *L. theobromae* infection (Xing et al., 2019). Developing disease-resistant crops is a difficult challenge for all plant species, while the probable function of lncRNA against fungal infection remains unknown. In another study, *Gossypium barbadense*'s highly susceptible, susceptible, highly tolerant, and super highly tolerant recombinant inbred lines (RILs) along with their resistant and susceptible parents were infected with *F. oxysporum*, and their responses were compared. It was observed that aside from fungal susceptibility, the expression of infection-induced genes was remarkably increased in disease-resistant plant pathways. However, the expression of lncRNAs induced by *F. oxysporum* infection was significantly related to plant susceptibility. Bioinformatics analysis also evidenced that *F. oxysporum*-regulated lncRNAs were significantly enriched in disease-resistant-related plant pathway involving glycolysis, phytohormonal signaling pathway, glutathione metabolism, butanoate metabolism, and anthocyanin biosynthesis. Collectively, these findings propose that lncRNA would play a vital role in response to pathogenic fungi as well as in the development of disease-resistant recombinant lines (Yao et al., 2019). *H. brasiliensis* (Rubber tree) is an important economic plant, and anthracnose is a common disease predominantly caused by fungus *C. gloeosporioides*, which decrease rubber production. Whether this disease comprises noncoding RNAs or lncRNAs remains unclear. However, the development of disease-resistant varieties would be a positive and significant step toward the green revolution globally. In the last few years, considerable progress was made on identifying genes involved in disease-resistant/tolerance. Experimental and computational findings cleared that dominant lncRNAs are involved in response to pathogenic fungi (Yin et al., 2019). A study reported the disease-responding mechanism of lncRNA against the attack of *Verticillium dahliae* in cotton. It showed that induced lncRNAs indicate a biased expression pattern in allotetraploid cotton. Bioinformatics and other comparative analyses proposed distinct response mechanisms of *G. barbadense* cv. 7124 (resistant) and *Gossypium hirsutum* cv. YZ1 (susceptible). Moreover, functional examination indicated that *GhlncNAT-ANX2* and *GhlncNAT-RLP7* (2-core lncRNAs)-silenced seedling exhibited an improved tolerance against *Botrytis cinerea* and *V. dahliae* probably linked with enhanced expression of LOXA and LOX2 (Zhang et al., 2018). Aside from the role of lncRNAs in various plant developmental processes, the mechanism underlying their response to disease remains unanswered. For assessing the defense role of long nonprotein-coding RNAs (npcRNAs), two wheat genotypes (susceptible JD8, resistant JD8-pm30) were infected with powdery mildew disease caused by *Blumeria graminis*. Sequencing analysis revealed that a set of long npcRNAs were expressed differentially. Interestingly, expression pattern was the same in both susceptible and resistant genotypes. For example, TalnRNA5, TalnRNA9, TapmlnRNA19, and TapmlnRNA30 were upregulated, while TalnRNA21 downregulated. Furthermore, the expression of all long npcRNAs was tissue-specific, i.e., TapmlnRNA30 was induced in seed, TalnRNA5 in all tissues, and TalnRNA9 expressed in seed, root, and leaf. These results support the contribution of long npcRNAs in defense mechanism, even though the molecular basis is still unidentified (Xin et al., 2011).

18.4.2 Role against nematode

Researchers attempted to recognize novel lncRNAs, responsive to *M. eumusae*, which cause eumusae leaf spot disease and nematode *P. coffeae*, causing root-lesions. *P. coffeae* infection identified 5615 lncRNAs from susceptible and resistant cultivars of banana. The study also revealed 1284 and 1250 differentially expressed lncRNAs against *P. coffeae* and *M. eumusae*. Moreover, 100 lncRNA played a vital role in numerous developmental processes and drought stress responses in banana (Muthusamy et al., 2019).

18.4.3 Role against viruses

Citrus tristeza virus (CTV) belongs to *the Closterovirus* genus, which potentially caused economic lose in citrus plants and produces subgenomic noncoding

RNAs called LMT1 (low-molecular-weight tristeza 1), previously considered to be CTV replication by-product (Kang et al., 2019). The investigations of a non-LMT1 producing variant of CTV to study the possible role of LMT1 in the viral infection cycle revealed that absence of LMT1 did not affect the viral replication and virions formation.

On the other hand, the study demonstrated that mutant of viruses was significantly reduced, spreading tendency in *Nicotiana benthamiana* and establishing inability in citrus. The herbaceous host introduced with CTV-LMT1d showed that SA and SA-related pathogenic genes were upregulated as compared to herbaceous host inoculated with wild-type virus (CTV-WT). Furthermore, the study revealed that LMT1 RNA produced by ectopic expression or CTV-WT upregulated alternative oxidative genes and suppressed SA accumulation to moderate ROS accumulation in *N. benthamiana* leaves (Kang et al., 2019).

LncRNAs' function during DNA virus infection in tomatoes was assessed through RACE technique by using three replicates. Generally, predicted data presented that 1565 lncRNAs, including both antisense transcripts and long intergenic RNAs, were involved in TYLCV infection through VIGS (virus-induced gene silencing). The study trial verified the lncRNA-associated function, differentially observed between 0 and 7 dpi (days postinoculation). LncRNAs also act as CeRNAs (competing endogenous RNAs), which regulated the RNA transcripts by competing for endogenous target mimics (eTMs) TYLCV infection-causing microRNAs in tomatoes. These findings gave a new direction about lncRNAs' involvement in response to virus infection, including TYLCV in tomatoes (Wang et al., 2015).

Two major viruses with wide geographical distribution affected the tomatoes' economy, and worldwide disease spread is begomoviruses and TYLCV. Research reports present that noncoding RNAs' interaction after TYLCV infection in tomatoes helped to elucidate the mechanism of TYLCV development. TYLCV contains ssDNA circular genome and noncoding intergenic region (IR), which mediated viral DNA replication in host cells, but still no research work is available related to the direct involvement of TYLCV in host disease development. During infection, some phenotypic abnormalities can be observed, which demonstrated intergenic region-oriented protein-coding regions to promote the viral genome replication. IR sequence contains 25 nt segments, which are complementary to lncRNA designated as *SILNR1* in TYLCV-susceptible tomatoes, and 14 nt deletion in 25 nt intergenic region, but do not show any resistance. For the silencing induction derived viral small interfering RNAs (shRNAs) with 25 nt intergenic region (*SILNR1* downregulated) were associated with TYLCV evocative symptoms such as curled and stunted leaves. Obtained results suggested that lncRNAs interact with shRNAs to control infection. On the other hand, consistency promotes virus development; if overexpressed *SILNR1*, then it reduced TYLCV accumulation. Additionally, *SILNR1* silencing without virus infection induces TYLCV leaf-like phenotype. Experimental results described the host lncRNAs and viral sRNAs' relation that is unknown and help to provide a conceptual model to understand the disease induction and host immunity relation and effective strategies to control pathogenicity caused by TYLCV virus (Yang et al., 2019). Plant initiated immune response on the recognition followed by pathogen-associated recognition, and molecular-related patterns include elf18, in the past identified in *Arabidopsis* with ELF18-induced lncRNA1 (ELENA1), which act as a positive immune gene transcriptional regulation and, when associated with MED19a (Mediator subunit 19a), help to enhance the PR1 (Pathogenesis-related genes) complex promoter. In vitro and in vivo experiments presented that ELENA1 can interact with FIB2 (Fibrillarin 2). Bimolecular and co-immunoprecipitation fluorescence assay showed that FIB2 directly interacts with MED19a at specific locations such as nucleolus and nucleoplasm. *Fib2* mutant analysis presented that FIB2 (PRI) acts as a negative regulator for genes responsible for immunity. The biochemical and genetic analysis proposed that ELENA1 has the ability to dissociate the FIB2+MED19a complex and move out FIB2 from promoter PR1 and enhance its expression, which was increased by evicting FIB2 repressor from the MED19a activator. Study findings were exposed to transcriptional complexities to regulate the immune-related genes with the relation of long noncoding RNAs in plants (Seo et al., 2019).

18.4.4 Role against bacteria

Long noncoding RNAs, a class of noncoding RNAs, were implicated for gene regulation and expression both in plants and mammals, while most studies in mammals determine the biological processes, but the lncRNAs' functions in plants were still under study. Exclusively, long intergenic noncoding RNAs play defense roles in plants. These results proposed that lincRNAs have potential defense roles and this work also provides first lincRNAs library and novel lncRNAs of potato defense in response to *P. carotovorum* subsp. *brasiliense*, which was a member of soft rot phytopathogens known as *Enterobacteriaceae* (Kwenda et al., 2016). Another fatal disease PaWB (Paulownia witches broom) caused substantial economic losses. Long noncoding RNAs have played a substantial regulatory role in transcriptional and posttranscriptional regulatory pathways. Against this phytoplasma disease (PaWB), the identification of significant lncRNAs and vital roles of lncRNAs were poorly characterized. For that purpose, it was needed to perform RNA-Seq analysis on healthy *Paulownia tomentosa* and PaWB-infected *P. tomentosa*, treated with 100mg/L rifampicin. After analysis, total 28,614 distinctive mRNAs and 3693 potential long noncoding RNAs were obtained. However, when compared to the coding genes and lncRNAs, comparison designated that lncRNAs have shorter transcripts and fewer exons and demonstrated substantial expression specificity.

18.4.5 Role against insects

According to the comparison scheme, identified PaWB-associated 1063 mRNAs and 110 lncRNAs, the data identification characterized that 9

PaWB-related lncRNAs regulated 12 PaWB-related target genes (Wang et al., 2018b). For significant lncRNAs, functions were studied in different plants, while the functions of lncRNAs on cotton crops were rarely unknown under aphids (*Aphis gossypii* Glover) attack. For this purpose, whole RNAs transcriptome sequence analysis was performed on leaves at four-leaf stage (*G. hirsutum*, ZM-03199) and damaged at different stages (including 0, 6, 12, 24, 48, and 72 h) affected by cotton aphids. Molecular analysis by RT-qPCR revealed that all selected lncRNAs were comparative to aphid damage. These results provide a solid foundation for cotton lncRNAs' functions against aphids damage (Zhang et al., 2019). Recently, long noncoding RNAs become a scientist's interest due to their significant biological roles. The genome-wide characterization and identification of reputed lncRNAs in global insect (pest) called *Plutella xylostella*. During the development of diamondback moth (DBM), expression profiling presented 114 lncRNAs' expression in which the majority were temporally specified. However, the biological function was uncharacterized in lncRNAs in which many are ceRNAs and mi-RNAs precursors directly involved in miRNAs regulatory pathways. This study has importance for valuable information on further molecular-based analysis for BDM development and provides the foundations for lncRNAs' discovery function in *P. xylostella* (Wang et al., 2018c).

18.5 Conclusions

Association of long noncoding RNAs with plant immunity and dealing with pathogen virulence attract researchers to exploit RNA silencing machinery in terms of augmented plant immunity against diseases. Although considerable efforts are under progress to recognize the protection mechanism of plant lncRNAs, the precise antipathogenic defense part of their function is still vague. Similarly, lncRNAs are not getting large-scale attention for their utility in antimicrobial immunity. As it is known that pathogens have enhanced their virulence by developing lncRNAs, further studies aiming pathogen lncRNAs and the virulence development can help researchers in decoding plant-pathogen interactions. Finding missing gaps in infection-mediated lncRNAs production can be beneficial. We expect that further research on lncRNAs-mediated plant defense against different diseases, by achieving disease resistance, will help us to improve global food security.

References

Alazem, M., Lin, N.S., 2015. Roles of plant hormones in the regulation of host–virus interactions. Mol. Plant Pathol. 16, 529–540.

Beffa, R.S., Neuhaus, J.-M., Meins, F., 1993. Physiological compensation in antisense transformants: specific induction of an "ersatz" glucan endo-1, 3-beta-glucosidase in plants infected with necrotizing viruses. PNAS 90, 8792–8796.

Borgognone, A., Sanseverino, W., Aiese Cigliano, R., Castanera, R., 2019. Distribution, characteristics, and regulatory potential of long noncoding RNAs in brown-rot fungi. Int. J. Genomics 2019.

Csorba, T., Kontra, L., Burgyán, J., 2015. Viral silencing suppressors: tools forged to fine-tune host-pathogen coexistence. Virology 479, 85–103.

Cui, J., Jiang, N., Meng, J., Yang, G., Liu, W., Zhou, X., et al., 2019. LncRNA33732-respiratory burst oxidase module associated with WRKY1 in tomato-*Phytophthora infestans* interactions. Plant J. 97, 933–946.

Franco-Zorrilla, J.M., Valli, A., Todesco, M., Mateos, I., Puga, M.I., Rubio-Somoza, I., et al., 2007. Target mimicry provides a new mechanism for regulation of microRNA activity. Nat. Genet. 39, 1033.

Hao, Z., Fan, C., Cheng, T., Su, Y., Wei, Q., Li, G., 2015. Genome-wide identification, characterization and evolutionary analysis of long intergenic noncoding RNAs in cucumber. PLoS One 10, e0121800.

Hegenauer, V., Fürst, U., Kaiser, B., Smoker, M., Zipfel, C., Felix, G., et al., 2016. Detection of the plant parasite *Cuscuta reflexa* by a tomato cell surface receptor. Science 353, 478–481.

Heo, J.B., Sung, S., 2011. Vernalization-mediated epigenetic silencing by a long intronic noncoding RNA. Science 331, 76–79.

Hohn, T., 2015. RNA based viral silencing suppression in plant pararetroviruses. Front. Plant Sci. 6, 398.

Joshi, R.K., Megha, S., Basu, U., Rahman, M.H., Kav, N.N., 2016. Genome wide identification and functional prediction of long non-coding RNAs responsive to *Sclerotinia sclerotiorum* infection in *Brassica napus*. PLoS One 11, e0158784.

Kang, S.H., Sun, Y.D., Atallah, O.O., Huguet-Tapia, J.C., Noble, J.D., Folimonova, S.Y., 2019. A long non-coding RNA of *Citrus tristeza virus*: role in the virus interplay with the host immunity. Viruses 11 (5), 436.

Kim, M.I., Kim, H.S., Jung, J., Rhee, S., 2008. Crystal structures and mutagenesis of sucrose hydrolase from *Xanthomonas axonopodis* pv. glycines: insight into the exclusively hydrolytic amylosucrase fold. J. Mol. Biol. 380, 636–647.

Kombrink, A., Sanchez-Vallet, A., Thomma, B.P., 2011. The role of chitin detection in plant–pathogen interactions. Microbes Infect. 13, 1168–1176.

Kwenda, S., Birch, P.R., Moleleki, L.N., 2016. Genome-wide identification of potato long intergenic noncoding RNAs responsive to *Pectobacterium carotovorum* subspecies brasiliense infection. BMC Genomics 17, 614.

Lee, W.S., Fu, S.F., Li, Z., Murphy, A.M., Dobson, E.A., Garland, L., et al., 2016. Salicylic acid treatment and expression of an RNA-dependent RNA polymerase 1 transgene inhibit lethal symptoms and meristem invasion during tobacco mosaic virus infection in *Nicotiana benthamiana*. BMC Plant Biol. 16, 15.

Li, L., Eichten, S.R., Shimizu, R., Petsch, K., Yeh, C.T., Wu, W., et al., 2014. Genome-wide discovery and characterization of maize long non-coding RNAs. Genome Biol. 15, R40.

Li, W., Li, C., Li, S., Peng, M., 2017. Long noncoding RNAs that respond to *Fusarium oxysporum* infection in 'Cavendish' banana (*Musa acuminata*). Sci. Rep. 7, 16939.

Liu, J., Jung, C., Xu, J., Wang, H., Deng, S., Bernad, L., et al., 2012. Genome-wide analysis uncovers regulation of long intergenic noncoding RNAs in Arabidopsis. Plant Cell 24, 4333–4345.

Love, A.J., Geri, C., Laird, J., Carr, C., Yun, B.W., Loake, G.J., et al., 2012. Cauliflower mosaic virus protein P6 inhibits signaling responses to salicylic acid and regulates innate immunity. PLoS One 7, e47535.

Ma, L., Bajic, V.B., Zhang, Z., 2013. On the classification of long non-coding RNAs. RNA Biol. 10, 924–933.

Miura, K., Tada, Y., 2014. Regulation of water, salinity, and cold stress responses by salicylic acid. Front. Plant Sci. 5, 4.

Muthusamy, M., Uma, S., Suthanthiram, B., Saraswathi, M.S., Chandrasekar, A., 2019. Genome-wide identification of novel, long non-coding RNAs responsive to *Mycosphaerella eumusae* and *Pratylenchus coffeae* infections and their differential expression patterns in disease-resistant and sensitive banana cultivars. Plant Biotechnol. Rep. 13, 73–83.

Rinn, J.L., Chang, H.Y., 2012. Genome regulation by long noncoding RNAs. Annu. Rev. Biochem. 81, 145–166.

Seo, J.S., Diloknawarit, P., Park, B.S., Chua, N.H., 2019. ELF18-INDUCED LONG NONCODING RNA 1 evicts fibrillarin from mediator subunit to enhance PATHOGENESIS-RELATED GENE 1 (PR1) expression. New Phytol. 221, 2067–2079.

Shafiq, S., Li, J., Sun, Q., 2016. Functions of plants long non-coding RNAs. Biochim. Biophys. Acta (BBA)-Gene Regul. Mech. 1859, 155–162.

Shin, J.-H., Chekanova, J.A., 2014. Arabidopsis RRP6L1 and RRP6L2 function in FLOWERING LOCUS C silencing via regulation of antisense RNA synthesis. PLoS Genet. 10, e1004612.

Shumayla, Sharma, S., Taneja, M., Tyagi, S., Singh, K., Upadhyay, S.K., 2017. Survey of high throughput RNA-seq data reveals potential roles for lncRNAs during development and stress response in bread wheat. Front. Plant Sci. 8, 1019.

Swiezewski, S., Liu, F., Magusin, A., Dean, C., 2009. Cold-induced silencing by long antisense transcripts of an Arabidopsis Polycomb target. Nature 462, 799.

Szittya, G., Burgyán, J., 2013. RNA interference-mediated intrinsic antiviral immunity in plants. In: Intrinsic Immunity. Springer, pp. 153–181.

Tyagi, S., Sharma, A., Upadhyay, S.K., 2018. Role of next-generation RNA-Seq data in discovery and characterization of long non-coding RNA in plants. In: Çiftçi, Y.Ö. (Ed.), Next Generation Plant Breeding. IntechOpen. https://doi.org/10.5772/intechopen.72773.

Underwood, W., Somerville, S.C., 2008. Focal accumulation of defences at sites of fungal pathogen attack. J. Exp. Bot. 59, 3501–3508.

Wang, J., Yu, W., Yang, Y., Li, X., Chen, T., Liu, T., et al., 2015. Genome-wide analysis of tomato long non-coding RNAs and identification as endogenous target mimic for microRNA in response to TYLCV infection. Sci. Rep. 5, 16946.

Wang, Y., Xu, T., He, W., Shen, X., Zhao, Q., Bai, J., et al., 2018a. Genome-wide identification and characterization of putative lncRNAs in the diamondback moth, *Plutella xylostella* (L.). Genomics 110, 35–42.

Wang, Y., Ye, W., Wang, Y., 2018b. Genome-wide identification of long non-coding RNAs suggests a potential association with effector gene transcription in *Phytophthora sojae*. Mol. Plant Pathol. 19, 2177–2186.

Wang, Z., Li, B., Li, Y., Zhai, X., Dong, Y., Deng, M., et al., 2018c. Identification and characterization of long noncoding RNA in *Paulownia tomentosa* treated with methyl methane sulfonate. Physiol. Mol. Biol. Plants 24, 325–334.

Wen, J., Parker, B.J., Weiller, G.F., 2007. In silico identification and characterization of mRNA-like noncoding transcripts in *Medicago truncatula*. In Silico Biol. 7, 485–505.

Wilusz, J.E., Sunwoo, H., Spector, D.L., 2009. Long noncoding RNAs: functional surprises from the RNA world. Genes Dev. 23, 1494–1504.

Xin, M., Wang, Y., Yao, Y., Song, N., Hu, Z., Qin, D., et al., 2011. Identification and characterization of wheat long non-protein coding RNAs responsive to powdery mildew infection and heat stress by using microarray analysis and SBS sequencing. BMC Plant Biol. 11, 61.

Xing, Q., Zhang, W., Liu, M., Li, L., Li, X., Yan, J., 2019. Genome-wide identification of long non-coding RNAs responsive to *Lasiodiplodia theobromae* infection in grapevine. Evol. Bioinforma. 15. 1176934319841362.

Yang, Y., Liu, T., Shen, D., Wang, J., Ling, X., Hu, Z., et al., 2019. Tomato yellow leaf curl virus intergenic siRNAs target a host long noncoding RNA to modulate disease symptoms. PLoS Pathog. 15, e1007534.

Yao, Z., Chen, Q., Chen, D., Zhan, L., Zeng, K., Gu, A., et al., 2019. The susceptibility of sea-island cotton recombinant inbred lines to *Fusarium oxysporum* f. sp. vasinfectum infection is characterized by altered expression of long noncoding RNAs. Sci. Rep. 9, 2894.

Yin, H., Zhang, X., Zhang, B., Luo, H., He, C., 2019. Revealing the dominant long noncoding RNAs responding to the infection with *Colletotrichum gloeosporioides* in *Hevea brasiliensis*. Biol. Direct 14, 7.

Yuan, J., Zhang, Y., Dong, J., Sun, Y., Lim, B.L., Liu, D., et al., 2016. Systematic characterization of novel lncRNAs responding to phosphate starvation in *Arabidopsis thaliana*. BMC Genomics 17, 655.

Zhang, Y.C., Liao, J.Y., Li, Z.Y., Yu, Y., Zhang, J.P., Li, Q.F., et al., 2014. Genome-wide screening and functional analysis identify a large number of long noncoding RNAs involved in the sexual reproduction of rice. Genome Biol. 15, 512.

Zhang, L., Wang, M., Li, N., Wang, H., Qiu, P., Pei, L., et al., 2018. Long noncoding RNA s involve in resistance to *Verticillium dahliae*, a fungal disease in cotton. Plant Biotechnol. J. 16, 1172–1185.

Zhang, J., Yang, Z., Feng, P., Zhong, X., Ma, Q., Su, Q., et al., 2019. Identification and the potential roles of long non-coding RNAs in cotton leaves damaged by *Aphis gossypii*. Plant Growth Regul. 88, 215–225.

Zhou, Y., Cho, W.K., Byun, H.S., Chavan, V., Kil, E.J., Lee, S., et al., 2019. Genome-wide identification of long non-coding RNAs in tomato plants irradiated by neutrons followed by infection with Tomato yellow leaf curl virus. PeerJ 7, e6286.

Zhu, Q.H., Wang, M.B., 2012. Molecular functions of long non-coding RNAs in plants. Genes 3, 176–190.

Chapter 19

Revisiting plant response to fungal stress in view of long noncoding RNAs

Garima Bhatia[a], Neetu Goyal[a], Santosh Kumar Upadhyay[b], and Kashmir Singh[a]

[a]*Department of Biotechnology, Panjab University, Chandigarh, India,* [b]*Department of Botany, Panjab University, Chandigarh, India*

19.1 Introduction

Long noncoding RNAs or lncRNAs (transcripts longer than 200 nt, but with unknown or limited coding potential) have been associated with regulatory roles in plants across different developmental stages, during reproduction and in response to stress (Bhatia et al., 2017; Nejat and Mantri, 2018). Of the different stress conditions that a plant as a sessile organism encounters, biotic stress is the most challenging owing to the plethora of biological agents that can interact with it and cause harm. Nearly, 20%–40% of the global agricultural productivity is compromised because of biotic stress that hits the crops before or after harvest (Jones et al., 2016; Savary et al., 2012). Further, compared to bacteria or viruses, fungi have been observed to cause significant plant damage. In fact, more fungal species infect plants compared to animals; consequently, more plant-fungi systems have been studied in contrast to animal-fungi systems (Sexton and Howlett, 2006). Pathogenic fungi have a broad range of plants as hosts encompassing cereals to fruit crops and even wild grasses to woody trees (Möller and Stukenbrock, 2017). Therefore, significant research has been conducted till date aiming to alleviate the economic burden associated with fungal diseases, especially in commercially and nutritionally important plants. With the advent of the "omics" era, transcript-level profiling is now feasible and efforts have been made to comprehend plant antifungal response networks. In this direction, regulation of plant gene expression in response to fungal stress has been studied with respect to transcription factors (TFs), phytohormones, short noncoding RNAs such as miRNAs, and pathogenesis-related (PR) proteins (Amorim et al., 2016; Finkina et al., 2017; Gupta et al., 2014; Verma et al., 2016). However, the role of lncRNAs in response to fungal stress is yet to be explored in depth.

Long Noncoding RNAs in Plants: Roles in development and stress
https://doi.org/10.1016/B978-0-12-821452-7.00019-2

Given the application of lncRNAs (even as biomarkers) in understanding the underlying mechanisms of human diseases, it can be promising to focus on their role in fungal diseases in plants as well.

The present chapter includes an overview of plant-fungal interactions and the potential of lncRNAs in regulating plant defense response against the pathogenic fungi.

19.2 How fungi impact plants?

As an ecologically wide-ranging set of microbes, fungi exhibit diverse virulence strategies in order to subvert their plant hosts. For their survival (nourishment), reproduction, and dispersal, fungi exploit plant tissues in their living or dead forms. Therefore, lifestyles and interaction modes of fungi as phytopathogens vary. While some pathogenic fungi might exclusively interact in order to kill their plant host cells and derive nutrition in the form of organic substances, others refrain from toxin production so as to survive on nutrients derived from the living host. The former and latter are referred to as necrotrophic and biotrophic fungal pathogens, respectively (Glazebrook, 2005; Horbach et al., 2011). A third type, referred to as the hemibiotrophic fungal pathogens, exploits a combination of both strategies by undergoing a switch from initial biotrophy to eventual necrotrophy in their lifestyle.

The invasion tactics adopted by fungal phytopathogens to establish successful interactions with their hosts can broadly be categorized into three: first, host penetration, which involves breaching physical barriers or the plant cell wall using force or enzymes; second, evasion from plant recognition/host defense that is mediated by special molecules called fungal effectors; and third, inhibition of host defense response through altering plant pH, interfering with phytohormonal signaling, inhibiting host proteases, etc. (Horbach et al., 2011; Rodriguez-Moreno et al., 2018).

19.3 How plants respond to the fungal stress?

In order to defend themselves against pathogens, plants have developed a two-pronged innate immune system acting at basal and advanced levels. Primarily, a plant can detect a pathogen based on recognition of its invasion patterns referred to as pathogen/microbe-associated molecular patterns (PAMPs/MAMPs). These pathogen-derived PAMPs/MAMPs are usually carbohydrate- or protein-based signals that have been found to be conserved across bacteria, oomycetes, and fungi (Boller and Felix, 2009). The presence of cell-surface/transmembrane pattern recognition receptors (PRRs) in plants enables the recognition of nonself molecules. Plant PRRs include membrane-associated receptor-like kinases (RLK) and receptor-like proteins (RLP) (which have been extensively reviewed: (Boutrot and Zipfel, 2017; Couto and

Zipfel, 2016; Saijo et al., 2018)). Particularly, for fungal stress, MAMPs such as fungal cell wall components, that is, chitin and its deacetylated derivative chitosan, are recognized by plant PRRs such as lysin motif (LysM)-RLKs and -RLPs (Felix et al., 1993; Saijo et al., 2018; Shibuya et al., 1993). This triggers a quick response, that is, PAMP-triggered immunity (PTI) in the form of ion fluxes, increased Ca^{2+} and reactive oxygen species (ROS) levels, and rapid activation of mitogen-activated protein kinase (MAPKs) cascades that together cause the plant to undergo an elaborate transcriptional reprogramming (Altenbach and Robatzek, 2007; Boller and Felix, 2009; Rodriguez-Moreno et al., 2018). Subsequently, plant cell walls are reinforced by callose deposition, phytohormonal and antimicrobial compounds are produced, and stomatal closure is observed (Macho and Zipfel, 2014; Rodriguez-Moreno et al., 2018; Saijo et al., 2018; Yu et al., 2017).

When the invading fungal pathogen is not restricted by basal defense, and instead, releases effector molecules (or virulence factors) into the plant cell, the second layer of plant immunity, that is, effecter-triggered immunity (ETI), is initiated. This acts primarily inside the cell and relies on recognition by resistance (*R*) genes-encoded proteins containing nucleotide binding (NB) site–leucine-rich repeat (LRR) domains in plants (Jones and Dangl, 2006). The recognition activates oxidative burst or intense ROS generation (Boller and Felix, 2009; Glazebrook, 2005) culminating into disease resistance and, often, hypersensitive cell death response (HR) at the infection site. Therefore, it is highly effective against the invading biotrophic fungi.

Interestingly, NB-LRR activation also subjects the plant to undergo extensive transcriptional reprogramming, and immense cross-talk is involved among the concerned response pathways enabling the plant to respond differentially and effectively (Glazebrook, 2005; Jones and Dangl, 2006). This is achieved chiefly by the interplay of phytohormones, salicylic acid (SA; in the case of biotrophs), and a combination of jasmonic acid (JA) and ethylene (ET) (in the case of necrotrophs), while the other hormones aid to the maintenance of the SA-JA-ET signaling balance. Moreover, SA-dependent signaling pathways lead to activation of genes encoding pathogenesis-related (PR) proteins; some of which have been reported to possess antifungal activity. For instance, PR-10 proteins can degrade pathogenic RNAs (reviewed by Finkina et al., 2017).

Overall, during both PTI and ETI, plants involve rigorous expression of genes to mediate defense/resistance against the invading pathogens, while taking into account the different nutritional lifestyles of the latter (Plett and Martin, 2018). This highlights the importance of regulatory RNAs and TFs involved in the background, enabling the plant to undertake drastic transcriptional and metabolic reprogramming (Nejat and Mantri, 2018; Shukla et al., 2008). Therefore, examination of lncRNAs in response to the invading pathogen could deepen our understanding of plant response and enable us to comprehend the underlying regulatory mechanisms of plant defense.

19.4 Genome-wide identification of fungal-responsive lncRNAs in plants

Owing to the advancement in technologies, it is now feasible to conduct genome-wide surveys of noncoding RNAs (ncRNAs) in plants. Particularly for lncRNAs, these technologies primarily include next-generation sequencing (NGS) and ribosome profiling (Liu et al., 2017). The latter is an excellent technique to omit false positives out of the predicted lncRNAs repertoire. It has been successfully applied to fish out the false positives coding for small proteins from the predicted lncRNAs for zebrafish and humans (Bazzini et al., 2014; Pauli et al., 2014). Further, in plants, it has been used to exclude 237 protein-coding transcripts from the previously reported pool of ncRNAs in *Arabidopsis thaliana* (Hsu et al., 2016; Juntawong et al., 2014). It would be interesting to extend this technique to identify fungal-responsive plant lncRNAs with minimal false positives.

As of now, NGS persists to be a powerful tool to identify lncRNAs (Tyagi et al., 2018). Although the initial identification of fungal-responsive lncRNAs in plants can be attributed to direct cloning approach and hybridization-based microarray technology (Xin et al., 2011), it was NGS that catalyzed their genome-wide identification. Unlike the microarray-based analysis, which was limited by known gene sequences (thus, restricting the dynamic range), the RNA-sequencing approach is not limited by compulsion of preexisting knowledge of the expressed genes (Ozsolak and Milos, 2011). Therefore, it enables one to examine the regulation of transcriptional responses to stress conditions, such as fungal stress, in plants.

Table 19.1 provides an overview of genome-wide investigations conducted to identify lncRNAs in response to fungal stress in plants. Many studies did not include thorough experimental functional analysis; however, they do provide a promising starting point to gain insight into the underlying regulatory aspects of plant defense against the invading fungi. It is important to conduct such studies prior to investment of time, labor, and resources on extended experiments attempting to unravel molecular mechanisms.

19.5 Regulation of plant response to fungal stress: The lncRNA perspective

As mentioned earlier in this chapter, plant defense mechanisms to combat the invading fungal phytopathogens are mediated by two-tier immunity, that is, basal level PTI and advanced ETI. Briefly, while the former is initiated by the perception of PAMPs/MAMPs by plant PRRs, the latter involves the recognition of fungal effector molecules by specific disease resistance (*R*) genes-encoded proteins such as NB-LRR proteins (Jones and Dangl, 2006). The ultimate plant defense responses are aimed at limiting the fungal phytopathogen growth. Therefore, they vary with the lifestyle type of the infecting fungal pathogen.

TABLE 19.1 Plant lncRNAs identified in response to fungal pathogens.

S. No.	Year	Fungal phytopathogen (biotrophic, B; hemibiotrophic, H; necrotrophic, N)	Plant species	Strategy adopted	No. of lncRNAs identified	Association of lncRNAs observed with:			References
						miRNAs	Transcription factors	Phytohormones	
1.	2011	*Blumeria graminis* f. sp. *tritici* (isolate E09) (**B**)	*Triticum aestivum* (powdery mildew-susceptible and -resistant wheat cultivar)	Microarray analysis and by using computational analysis	71 lncRNAs responsive to powdery mildew infection	✓			Xin et al. (2011)
2.	2014	*Fusarium oxysporum* (strain Fo5176) (**H**)	*Arabidopsis thaliana* (ecotype Col-0)	Strand-specific RNA-sequencing	20 *F. oxysporum*-responsive lncTARs (transcriptionally active regions)		✓		Zhu et al. (2014)
3.	2016	*Puccinia striiformis* f. sp. *tritici*; *Pst* and *Blumeria graminis* f. sp. *tritici*; *Bgt* (**B**) (Chinese *Pst* race CYR31 and *Bgt* race E09)	*Triticum aestivum* (hexaploid wheat line N9134)	RNA-sequencing and computational pipeline	283 differentially expressed lincRNAs (254 DE-LincRNAs responded to the *Bgt* stress, and 52 lincRNAs in *Pst*)	✓			Zhang et al. (2016)

Continued

TABLE 19.1 Plant lncRNAs identified in response to fungal pathogens—cont'd

S. No.	Year	Fungal phytopathogen (biotrophic, B; hemibiotrophic, H; necrotrophic, N)	Plant species	Strategy adopted	No. of lncRNAs identified	Association of lncRNAs observed with:			References
						miRNAs	Transcription factors	Phytohormones	
4.	2016	*Sclerotinia sclerotiorum* (**N**)	*Brassica napus* cv. DH12075 (canola)	Stranded RNA-sequencing and computational pipeline	931 differentially expressed lncRNAs in response to *S. sclerotiorum* infection	✓	✓		Joshi et al. (2016)
5.	2016	*Fusarium graminearum* (**H**)	*Hordeum vulgare* (Barley) Near-isogenic lines (NILs) carrying Chevron-derived resistant alleles for the two QTL	RNA-sequencing	604 Fusarium head blight (FHB)-responsive lncRNAs				Huang et al. (2016)
6.	2017	*Phytophthora infestans* (**H**; oomycete)	*Solanum pimpinellifolium* L3708 (*P. infestans*-resistant line) and *S. lycopersicum* Zaofen No.2 (*P. infestans*-susceptible line)	RNA-sequencing and computational analysis	688 differentially expressed lncRNAs identified between *P. infestans*-resistant and -susceptible tomatoes				Cui et al. (2017)

7.	2017	*Fusarium oxysporum* f. sp. *cubense* (*Foc*) strain *Foc* TR4 VCG01213/16 (**H**)	*Musa acuminata* (resistant Nongke No 1, "NK" and susceptible Brazil "BX" banana cultivars)	Strand-specific paired-end RNA-sequencing and computational pipeline	5294 banana lncRNAs involved in response to *Foc*		✓	✓	Li et al. (2017)
8.	2017	*Magnaporthe oryzae* (ZB13 and Zhong strain) (**H**)	*Oryza sativa*-resistant rice lines harboring *Pi9* blast resistance gene	RNA-seq data and computational pipeline	1429, 1927, and 1981 lncRNAs in response to mock and *M. oryzae* (ZB13 and Zhong)				Jain et al. (2017)
9.	2017	*Alternaria alternata* (**N**)	*Pyrus calleryana* (Callery pear)	Single-strand RNA-sequencing and computational pipeline	144 lncRNAs in response to *A. alternata*				Kan et al. (2017)
10.	2018	*Verticillium dahliae* (**H**)	*Gossypium barbadense* cv. "7124" (resistant cotton) and *Gossypium hirsutum* cv. "YZ1" (susceptible cotton)	RNA-sequencing and computational pipeline	1236 and 1907 differentially expressed lincRNAs in *G. barbadense* and *G. hirsutum*, respectively			✓	Zhang et al. (2018)

Continued

TABLE 19.1 Plant lncRNAs identified in response to fungal pathogens—cont'd

S. No.	Year	Fungal phytopathogen (biotrophic, B; hemibiotrophic, H; necrotrophic, N)	Plant species	Strategy adopted	No. of lncRNAs identified	Association of lncRNAs observed with:			References
						miRNAs	Transcription factors	Phytohormones	
11.	2019	*Phytophthora infestans* (**H**; oomycete)	*Solanum lycopersicum* (Tomato) (A transgenic, resistant plant overexpressing WRKY1)	RNA-sequencing and computational pipeline	22 differentially expressed lncRNAs containing W-box element, the target sequence of WRKY1 TF.	✓	✓		Cui et al. (2019)
12.	2019	*Colletotrichum gloeosporioides* (**H**)	*Hevea brasiliensis* (Rubber tree)	RNA-sequencing and computational analysis	5 dominant-responsive lncRNAs	✓	✓		Yin et al. (2019)

13.	2019	*Fusarium oxysporum* f. sp. *vasinfectum* (**H**)	Four different sea-island cotton (*Gossypium barbadense*) recombinant inbred lines (RILs) with susceptible (S), highly susceptible (HS), highly resistant (HR), or super highly resistant (SHR) phenotypes	Strand-specific RNA-sequencing and computational pipeline	• 354 and 444 lncRNAs in response to *Fov* in resistant HR and SHR RILs, respectively. • 777 and 1768 lncRNAs in response to *Fov* in susceptible S and HS RILs, respectively.			✓	Yao et al. (2019)
14.	2019	*Lasiodiplodia theobromae* Strain CSS-01s (**H**)	*Vitis vinifera* (grapevine) resistant cultivar Merlot and susceptible cultivar Cabernet Franc	RNA-sequencing and computational pipeline	782 grape lncRNA candidates differentially expressed across resistant and susceptible cultivars.	✓	✓		Xing et al. (2019)

This also highlights the importance of "regulation of response" in order to enable the plant to defend itself against the invading fungal pathogen.

To better comprehend the varying plant responses subject to the invading fungi, this section has been subdivided into three major categories of fungal phytopathogens based on their nutritional strategy, that is, biotrophic, necrotrophic, and hemibiotrophic nature. Essentially, for each category, the involvement of fungal-responsive lncRNAs has been discussed in conjunction with other regulatory components of plant response, such as, TFs, miRNAs, and phytohormones as well as key genes involved in defense response.

19.5.1 Plant lncRNAs in response to biotrophic fungi

Since obligate biotrophic fungi require living plant cells to sustain nourishment (Mendgen and Hahn, 2002), plants mediate ETI-driven programmed cell death in the host as an effective defense response. It is also accompanied by the activation of salicylic acid (SA)-dependent signaling pathway that further leads to expression of defense genes such as those encoding pathogenesis-related (PR) proteins, thereby contributing to resistance (Glazebrook, 2005; Jones and Dangl, 2006). Clearly, these defense responses involve meticulous gene expression regulation in the background to facilitate extensive transcriptional reprogramming in the infected plant.

Not many studies have been conducted till date to comprehensively analyze the role of lncRNAs in regulating plant defense mechanisms against obligate biotrophs. Nevertheless, a majority of studies do revolve around the hemibiotrophic fungal pathogen-responsive lncRNAs in plants (as discussed in Section 19.5.3).

Interestingly, one of the earliest attempts to study lncRNAs in response to biotic stress in plants was conducted for an obligate fungal biotroph, *Blumeria graminis* f. sp. *tritici* (*Bgt*), which causes powdery mildew (PM) disease in wheat (Xin et al., 2011). This study revealed 71 PM-responsive lncRNAs in wheat. Of these, three lncRNAs were found to act as precursors for two miRNAs, miR2004 and miR2066, which were previously identified in wheat in response to PM (Xin et al., 2010). Particularly, higher expression levels were observed for precursor lncRNAs, TapmlnRNA19 and TalnRNA5, and for their corresponding miRNA, miR2004, upon PM infection in both PM-susceptible and -resistant wheat cultivars. The observation suggested successful processing of the two lncRNAs into their corresponding miRNAs upon PM infection. However, since this was observed for both the PM-susceptible and -resistant wheat cultivars, it was additionally suggested by Xin et al. that the two lncRNAs were involved in regulation of "basal defense response" by the plant to the biotrophic phytopathogen.

Inspired by these findings, Zhang et al. conducted genome-wide investigation of long intergenic ncRNAs (lincRNAs) in a wheat line with high resistance to *Bgt* and another biotrophic fungal pathogen, *Puccinia striiformis* f. sp. *tritici*

(*Pst*), causing wheat stripe rust (Zhang et al., 2016; Shumayla et al., 2017). In all, 283 differentially expressed lincRNAs were found in response to the two biotrophs. Particularly, 254 and 52 lincRNAs were identified in response to *Bgt* and *Pst* stress, respectively. To understand their regulatory roles, miRNA target sites were predicted and 101 and 5 lincRNAs were found as putative targets and endogenous target mimics (eTMs) of miRNAs. As eTMs, lncRNAs can rescue the actual targets of miRNAs (that is, mRNAs). Based on these findings, Zhang et al. suggested that lincRNAs could play more intricate roles with respect to miRNAs, besides merely acting as their precursors.

Based on this information, the interactions of lncRNAs, miRNAs, and corresponding mRNAs potentially regulate plant responses to obligate biotrophs such as *Bgt* and *Pst*. It would be interesting to explore how plant lncRNAs are involved in response to other biotrophs as well.

19.5.2 Plant lncRNAs in response to necrotrophic fungi

Since necrotrophic pathogens favor host cell death, the NB-LRR-mediated plant defense response for biotrophic fungi does not hamper the necrotrophs. Therefore, instead of programmed cell death and salicylic acid-dependent defense mechanisms, plant responses are mediated by jasmonic acid- and ethylene-dependent signaling to combat necrotroph fungal pathogens (Glazebrook, 2005; Jones and Dangl, 2006).

Akin to the status of biotrophic fungi, plant lncRNAs research in response to necrotrophic fungal pathogens is still at its infancy. However, hemibiotrophic fungal pathogens and lncRNAs researches have caught pace (discussed in Section 19.5.3).

In one of the first studies exploring the potential of lncRNAs in response to *Sclerotinia sclerotiorum* in *Brassica napus* (canola), 931 differentially expressed lncRNAs were reported based on RNA-seq analysis (Joshi et al., 2016). The necrotrophic fungal pathogen, which is known to affect broad host range, has been associated with Sclerotinia stem rot in the oilseed crop canola. Further, the study revealed the involvement of lncRNAs in regulation of other small regulatory RNAs such as miRNAs. In total, 41 lncRNAs were estimated to act as precursors for miRNAs, which included miR156, miR169, and miR394. Interestingly, these miRNAs have been associated with other necrotrophic fungal pathogens such as *Botrytis cinerea* (Jin and Wu, 2015).

Previously, efforts were made to understand the *B. napus-S. sclerotiorum* pathosystem by taking into account differential gene expression profiling and proteomic-level changes (Liang et al., 2008; Zhao et al., 2009). The elucidation of *S. sclerotiorum*-responsive lncRNAs has added another layer of regulation for antifungal defense networks in canola. Similar analysis can be conducted for some other commercially important hosts of this necrotrophic fungus, and based on comparative analysis, conserved lncRNAs could further be examined for the common underlying plant defense mechanisms.

19.5.3 Plant lncRNAs in response to hemibiotrophic fungi

Unlike the obligate biotrophic and necrotrophic fungal phytopathogens that secure their nourishment through living and dead host cells, respectively, hemibiotrophic fungi rely on a combination of both the strategies (Glazebrook, 2005). These pathogens exhibit effectors-mediated "switch" from biotrophy to necrotrophy, which invokes corresponding plant defense responses subject to the resistant or susceptible traits of the plant (Lee and Rose, 2010).

Interestingly, owing to their transient biotrophic early stage involving the infection process, many phytopathogenic fungi were mistakenly considered necrotrophs (Glazebrook, 2005). However, studies identifying the "switch" in the lifestyle of these pathogens, under specific conditions/host encounters, have highlighted hemibiotrophy (Chowdhury et al., 2017; Ding et al., 2011; Gan et al., 2013; Lee and Rose, 2010; Zuluaga et al., 2016).

Till date, the underlying mechanisms of lifestyle-switch by hemibiotrophic fungi and the consequent regulation of transcriptional reprogramming in plants as defense response are not well-understood (Chowdhury et al., 2017). Nevertheless, it is known that plant defense (subject to resistant or susceptible plant genotype) against hemibiotrophic fungi is effectively mediated by systematic activation of phytohormones. In particular, the biotrophic phase is marked by SA signaling, which drifts toward JA activation with the necrotrophic phase (Chowdhury et al., 2017; Zuluaga et al., 2016). This highlights the importance of coordinated gene expression regulation in the host in order to defend against the invading hemibiotrophic pathogen.

Over the decade, lncRNAs, in addition to other regulatory RNAs (such as miRNAs), TFs and phytohormones, have emerged as potent players in regulation of the plant antifungal defense response network. Below discussed are some of the hemibiotrophic fungi-responsive lncRNA studies that provide fresh insight into regulation of plant response:

19.5.3.1 lncRNAs and miRNAs

In their RNA-seq-based approach followed by computational and experimental analyses, Yin et al. revealed dominant lncRNAs in response to hemibiotrophic fungal pathogen, *Colletotrichum gloeosporioides*. The pathogen causes *Colletotrichum* Leaf Disease (CLD) in an economically important tree crop species, that is, *Hevea brasiliensis* or rubber tree, often leading to anthracnose and subsequent loss in rubber production (Yin et al., 2019). Upon infection, the expression patterns of two lncRNAs, lncRNA11205 and lncRNA11051, were observed to be in sync with those for biotic-stress-associated miRNAs, miR395 and miR397, respectively. Subsequent experiments based on overexpression of the two lncRNAs revealed increased expression levels of the two corresponding miRNAs. However, the actual targets of the miRNAs were found to be downregulated during infection by the pathogen. Interestingly, downregulation of miR397 target, *Laccase*, has been associated with increased tolerance against another hemibiotrophic fungal phytopathogen, *Verticillium dahliae*, in *Gossypium hirsutum* (cotton). This observation was attributed to the rechanneling of the phenylpropanoid pathway and the subsequent upregulation of JA signaling pathway genes upon downregulation of *Laccase* (Hu et al., 2018). Therefore, lncRNAs can potentially stimulate the accumulation of certain specific miRNAs and regulate the expression of underlying genes in defense response to hemibiotrophic fungi.

Likewise, in a study on another woody, perennial crop of economic importance, that is, *Vitis vinifera* (grapevine), lncRNAs were identified in response to *Lasiodiplodia theobromae* (Xing et al., 2019). This hemibiotrophic fungal pathogen has been observed as one of the most aggressive species causing grapevine trunk disease, specifically, Botryosphaeria dieback, which leads to severe loss in the fruit crop production (Xing et al., 2019; Yan et al., 2013). Xing et al. identified 782 differentially expressed lncRNAs in response to this hemibiotroph across susceptible and resistant grapevine cultivars, that is, Cabernet Franc and Merlot, respectively. They also identified lncRNAs as putative precursors of miRNAs such as miR156 (LNC_000402 and LNC_000403), miR169 (LNC_000032 and LNC_001084), and miR398 (LNC_001235 and LNC_001272). These miRNAs have been previously reported to be upregulated in response to *Dothiorella gregaria*, a hemibiotrophic fungal pathogen, causing infection in woody plant species *Populus beijingensis* (Chen et al., 2012; Gupta et al., 2014). Moreover, the actual targets of these miRNAs include mRNAs coding for enzymes such as copper superoxide dismutase and cytochrome *c* oxidase (as in the case of miR398) or for TFs such as squamosa-promoter binding protein-like (SPL) (as in the case of miR156). In fact, miR398 has been associated with negative regulation of PAMP-triggered immunity (PTI)-associated callose deposition in *A. thaliana*, increasing plant susceptibility to bacterial infection, particularly *Pseudomonas syringae* pv. *tomato* DC3000. This highlights the complexity of ncRNA-mediated regulation of plant defense responses (Li et al., 2010).

Moreover, lncRNAs coordinate regulation of miRNAs that, in turn, target TF, the master regulators of transcription of several genes. This lncRNA-miRNA-TF interplay is subject to the role of lncRNAs with miRNAs. For instance, *V. vinifera*-responsive lncRNAs can act both as precursors and endogenous target mimics (eTMs) of miR156 targeting *SPL/SBP* TFs, which affect floral development and plant transitions to adult stage (Bhatia et al., 2019; Xing et al., 2019).

19.5.3.2 lncRNAs and transcription factors

Besides the miRNA-mediated interplay, TFs can directly interact with lncRNAs and regulate the expression of the associated genes. In their studies on tomato-*Phytophthora infestans* interaction, Cui et al. observed the involvement of lncRNAs in imparting resistance against the hemibiotrophic oomycete infection in the plant (Cui et al., 2017, 2019). An lncRNA, lncRNA33732, was found to be activated by a TF, WRKY1, owing to the presence of W-box element (target sequence for WRKY1) in its promoter region. Since WRKY1 in tomato had previously been reported to promote resistance against *P. infestans* (Li et al., 2015), lncRNA33732 was considered as a positive regulator of plant defense against the hemibiotroph. Further experimental analysis revealed the lncRNA33732-associated induction of respiratory burst oxidase (*RBOH*), which aids in an early response to the pathogen in the form of reactive oxygen species (ROS) such as H_2O_2.

Therefore, it can be suggested that TF-mediated activation of lncRNAs can positively regulate early defense response against hemibiotrophic pathogens in plants. Using a similar experimental strategy, it would be interesting and beneficial to explore the predicted lncRNA-TF associations in other plants, for instance, *H. brasiliensis* (rubber tree). As discussed previously, lncRNAs are potentially involved in regulation of plant defense response along with miRNAs in response to hemibiotrophic fungal pathogen infection in rubber tree. Additionally, in the same study, lncRNA11254 was predicted to interact with NTL8, which is a membrane-tethered transcription factor belonging to plant-specific NAC TF family (NAM (no apical meristem), ATAF1/2, CUC2 (cup-shaped cotyledons 2)) and has been associated with development and stress response in plants (Chen et al., 2008; Yin et al., 2019). However, the underlying mechanism of the lncRNA-TF interaction-mediated plant defense response is yet to be explored.

19.5.3.3 lncRNAs and phytohormones

Phytohormones have been described as the "watchdogs" of plant stress response and their cross-talk has particularly been observed in plant defense (Verma et al., 2016). The hormonal cross-talk is inevitable upon infection by a hemibiotrophic phytopathogen, especially as it switches from biotrophy to necrotrophy. In a study on *Fusarium oxysporum* f. sp. *cubense* (*Foc*; strain Tropical Race 4) infecting the commercial cultivar "Cavendish" of *Musa acuminata* (banana), lncRNAs were identified across the resistant "NK" and susceptible "BX" cultivars of the plant (Li et al., 2017). The authors examined the expression profiles of the identified lncRNAs which are potentially involved in phytohormone signal transduction and biosynthesis. It was observed that lncRNAs were associated with mRNAs for auxin-responsive protein (*AUX/IAAs*) and TF-TGAs in the early infection stages, indicating the involvement of auxin and SA initially. Nevertheless, some lncRNAs were predominantly correlated with mRNAs for *Jasmonate-JIM-Domain* (*JAZ*), a repressor protein that negatively regulates JA signal transduction. Interestingly, this observation was made in the infected "BX" sample, indicating the susceptibility of the plant to the hemibiotrophic phytopathogen *Foc*. Further examination of the lncRNA-mediated regulation of phytohormones during infection would provide better understanding of the defense-related antifungal networks in plants.

19.6 Concluding remarks and future perspectives

Till date, the identified fungal stress-responsive lncRNAs (particularly in response to hemibiotrophic fungi) in plants have primarily provided a starting point to gain fresh insight into the regulatory mechanisms of defense response. More genome-wide level investigations should be conducted for understanding plant response, especially with respect to obligate biotrophs and necrotrophs. Future experiments (in continuation of the lncRNA identification and transcriptional profiling studies) can now be conducted to understand functional roles and subcellular localization of the responsive lncRNAs. Particularly, to understand the molecular mechanisms of the reported lncRNAs, genome editing tools, such as, transcription activator-like effector nucleases (TALENs) and clustered regulatory interspaced short palindromic repeats (CRISPR)/CRISPR-associated protein 9 (Cas9) could be used.

Many fungal-responsive lncRNAs in plants have been reported to be associated with miRNAs. Furthermore, RNA-mediated interkingdom communication has been reported with respect to miRNAs. Therefore, it would be interesting to check the potential of lncRNAs in targeting fungal transcripts, besides regulating defense responses within the plant.

Lastly, in order to attain the ultimate goal of developing fungal-resistant plants, one should consider exploiting lncRNAs as candidates since they have the potential to regulate other regulators of gene expression such as TFs and miRNAs in plant defense response. However, prior to that, comparative transcriptomic analyses should be conducted with respect to the wild and cultivated plants for understanding the holistic scenario of lncRNA-mediated regulation in response to fungal stress.

Acknowledgment

GB is thankful to ICMR, India, for awarding junior and senior research fellowship. NG is thankful to DBT India for awarding fellowship.

References

Altenbach, D., Robatzek, S., 2007. Pattern recognition receptors: from the cell surface to intracellular dynamics. Mol. Plant-Microbe Interact. https://doi.org/10.1094/MPMI-20-9-1031.

Amorim, L., Santos, R., Neto, J., Guida-Santos, M., Crovella, S., Benko-Iseppon, A., 2016. Transcription factors involved in plant resistance to pathogens. Curr. Protein Pept. Sci. 18, 335–351. https://doi.org/10.2174/1389203717666160619185308.

Bazzini, A.A., Johnstone, T.G., Christiano, R., MacKowiak, S.D., Obermayer, B., Fleming, E.S., et al., 2014. Identification of small ORFs in vertebrates using ribosome footprinting and evolutionary conservation. EMBO J. 33, 981–993. https://doi.org/10.1002/embj.201488411.

Bhatia, G., Goyal, N., Sharma, S., Upadhyay, S., Singh, K., 2017. Present scenario of long noncoding RNAs in plants. Non-Coding RNA. https://doi.org/10.3390/ncrna3020016.

Bhatia, G., Sharma, S., Upadhyay, S.K., Singh, K., 2019. Long non-coding RNAs coordinate developmental transitions and other key biological processes in grapevine. Sci. Rep. 9, 3552. https://doi.org/10.1038/s41598-019-38989-7.

Boller, T., Felix, G., 2009. A renaissance of elicitors: perception of microbe-associated molecular patterns and danger signals by pattern-recognition receptors. Annu. Rev. Plant Biol. 60, 379–406. https://doi.org/10.1146/annurev.arplant.57.032905.105346.

Boutrot, F., Zipfel, C., 2017. Function, discovery, and exploitation of plant pattern recognition receptors for broad-spectrum disease resistance. Annu. Rev. Phytopathol. 55, 257–286. https://doi.org/10.1146/annurev-phyto-080614-120106.

Chen, Y.N., Slabaugh, E., Brandizzi, F., 2008. Membrane-tethered transcription factors in *Arabidopsis thaliana*: novel regulators in stress response and development. Curr. Opin. Plant Biol. 11, 695–701. https://doi.org/10.1016/j.pbi.2008.10.005.

Chen, L., Ren, Y., Zhang, Y., Xu, J., Zhang, Z., Wang, Y., 2012. Genome-wide profiling of novel and conserved *Populus* microRNAs involved in pathogen stress response by deep sequencing. Planta 235, 873–883. https://doi.org/10.1007/s00425-011-1548-z.

Chowdhury, S., Basu, A., Kundu, S., 2017. Biotrophy-necrotrophy switch in pathogen evoke differential response in resistant and susceptible sesame involving multiple signaling pathways at different phases. Sci. Rep. 7, 1–17. https://doi.org/10.1038/s41598-017-17248-7.

Couto, D., Zipfel, C., 2016. Regulation of pattern recognition receptor signalling in plants. Nat. Rev. Immunol. https://doi.org/10.1038/nri.2016.77.

Cui, J., Luan, Y., Jiang, N., Bao, H., Meng, J., 2017. Comparative transcriptome analysis between resistant and susceptible tomato allows the identification of lncRNA16397 conferring resistance to *Phytophthora infestans* by co-expressing glutaredoxin. Plant J. 89, 577–589. https://doi.org/10.1111/tpj.13408.

Cui, J., Jiang, N., Meng, J., Yang, G., Liu, W., Zhou, X., et al., 2019. LncRNA33732-respiratory burst oxidase module associated with WRKY1 in tomato-*Phytophthora infestans* interactions. Plant J. 97, 933–946. https://doi.org/10.1111/tpj.14173.

Ding, L., Xu, H., Yi, H., Yang, L., Kong, Z., Zhang, L., et al., 2011. Resistance to hemi-biotrophic *F. graminearum* infection is associated with coordinated and ordered expression of diverse defense signaling pathways. PLoS One 6, e19008. https://doi.org/10.1371/journal.pone.0019008.

Felix, G., Regenass, M., Boller, T., 1993. Specific perception of subnanomolar concentrations of chitin fragments by tomato cells: induction of extracellular alkalinization, changes in protein phosphorylation, and establishment of a refractory state. Plant J. 4, 307–316. https://doi.org/10.1046/j.1365-313X.1993.04020307.x.

Finkina, E.I., Melnikova, D.N., Bogdanov, I.V., Ovchinnikova, T.V., Shemyakin, M.M., Ovchinnikov, Y.A., 2017. PR10 protein review. Curr. Med. Chem. 24, 1772–1787. https://doi.org/10.2174/0929867323666161610261.

Gan, P., Ikeda, K., Irieda, H., Narusaka, M., O'Connell, R.J., Narusaka, Y., et al., 2013. Comparative genomic and transcriptomic analyses reveal the hemibiotrophic stage shift of *Colletotrichum* fungi. New Phytol. 197, 1236–1249. https://doi.org/10.1111/nph.12085.

Glazebrook, J., 2005. Contrasting mechanisms of defense against biotrophic and necrotrophic pathogens. Annu. Rev. Phytopathol. 43, 205–227. https://doi.org/10.1146/annurev.phyto.43.040204.135923.

Gupta, O.P., Sharma, P., Gupta, R.K., Sharma, I., 2014. Current status on role of miRNAs during plant-fungus interaction. Physiol. Mol. Plant Pathol. 85, 1–7. https://doi.org/10.1016/j.pmpp.2013.10.002.

Horbach, R., Navarro-Quesada, A.R., Knogge, W., Deising, H.B., 2011. When and how to kill a plant cell: infection strategies of plant pathogenic fungi. J. Plant Physiol. 168, 51–62. https://doi.org/10.1016/j.jplph.2010.06.014.

Hsu, P.Y., Calviello, L., Wu, H.Y.L., Li, F.W., Rothfels, C.J., Ohler, U., et al., 2016. Super-resolution ribosome profiling reveals unannotated translation events in *Arabidopsis*. Proc. Natl. Acad. Sci. U. S. A. 113, E7126–E7135. https://doi.org/10.1073/pnas.1614788113.

Hu, Q., Min, L., Yang, X., Jin, S., Zhang, L., Li, Y., et al., 2018. Laccase GhLac1 modulates broad-spectrum biotic stress tolerance via manipulating phenylpropanoid pathway and jasmonic acid synthesis. Plant Physiol. 176, 1808–1823. https://doi.org/10.1104/pp.17.01628.

Huang, Y., Li, L., Smith, K.P., Muehlbauer, G.J., 2016. Differential transcriptomic responses to *Fusarium graminearum* infection in two barley quantitative trait loci associated with Fusarium head blight resistance. BMC Genomics 17, 1–16. https://doi.org/10.1186/s12864-016-2716-0.

Jain, P., Sharma, V., Dubey, H., Singh, P.K., Kapoor, R., Kumari, M., et al., 2017. Identification of long non-coding RNA in rice lines resistant to Rice blast pathogen *Magnaporthe oryzae*. Bioinformation 13, 249–255. https://doi.org/10.6026/97320630013249.

Jin, W., Wu, F., 2015. Characterization of miRNAs associated with *Botrytis cinerea* infection of tomato leaves. BMC Plant Biol. 15, 1–14. https://doi.org/10.1186/s12870-014-0410-4.

Jones, J.D.G., Dangl, J.L., 2006. The plant immune system. Nature. https://doi.org/10.1038/nature05286.

Jones, J.D.G., Vance, R.E., Dangl, J.L., 2016. Intracellular innate immune surveillance devices in plants and animals. Science (80-.) 354, aaf6395. https://doi.org/10.1126/science.aaf6395.

Joshi, R.K., Megha, S., Basu, U., Rahman, M.H., Kav, N.N.V., 2016. Genome wide identification and functional prediction of long non-coding RNAs responsive to *Sclerotinia sclerotiorum* infection in *Brassica napus*. PLoS One 11, 1–19. https://doi.org/10.1371/journal.pone.0158784.

Juntawong, P., Girke, T., Bazin, J., Bailey-Serres, J., 2014. Translational dynamics revealed by genome-wide profiling of ribosome footprints in *Arabidopsis*. Proc. Natl. Acad. Sci. U. S. A. 111, https://doi.org/10.1073/pnas.1317811111.

Kan, J., Liu, T., Ma, N., Li, H., Li, X., Wang, J., et al., 2017. Transcriptome analysis of Callery pear (*Pyrus calleryana*) reveals a comprehensive signalling network in response to *Alternaria alternata*. PLoS One 12, 1–17. https://doi.org/10.1371/journal.pone.0184988.

Lee, S.J., Rose, J.K.C., 2010. Mediation of the transition from biotrophy to necrotrophy in hemibiotrophic plant pathogens by secreted effector proteins. Plant Signal. Behav. 5, 769–772. https://doi.org/10.4161/psb.5.6.11778.

Li, Y., Zhang, Q.Q., Zhang, J., Wu, L., Qi, Y., Zhou, J.M., 2010. Identification of microRNAs involved in pathogen-associated molecular pattern-triggered plant innate immunity. Plant Physiol. 152, 2222–2231. https://doi.org/10.1104/pp.109.151803.

Li, J.b., Luan, Y.s., Liu, Z., 2015. SpWRKY1 mediates resistance to *Phytophthora infestans* and tolerance to salt and drought stress by modulating reactive oxygen species homeostasis and expression of defense-related genes in tomato. Plant Cell Tissue Organ Cult. 123, 67–81. https://doi.org/10.1007/s11240-015-0815-2.

Li, W., Li, C., Li, S., Peng, M., 2017. Long noncoding RNAs that respond to *Fusarium oxysporum* infection in "Cavendish" banana (*Musa acuminata*). Sci. Rep. 7, 1–13. https://doi.org/10.1038/s41598-017-17179-3.

Liang, Y., Srivastava, S., Rahman, M.H., Strelkov, S.E., Kav, N.N.V., 2008. Proteome changes in leaves of *Brassica napus* L. as a result of *Sclerotinia sclerotiorum* challenge. J. Agric. Food Chem. 56, 1963–1976. https://doi.org/10.1021/jf073012d.

Liu, D., Mewalal, R., Hu, R., Tuskan, G.A., Yang, X., 2017. New technologies accelerate the exploration of non-coding RNAs in horticultural plants. Hortic. Res. https://doi.org/10.1038/hortres.2017.31.

Macho, A.P., Zipfel, C., 2014. Plant PRRs and the activation of innate immune signaling. Mol. Cell https://doi.org/10.1016/j.molcel.2014.03.028.

Mendgen, K., Hahn, M., 2002. Plant infection and the establishment of fungal biotrophy. Trends Plant Sci. 7, 352–356. https://doi.org/10.1016/S1360-1385(02)02297-5.

Möller, M., Stukenbrock, E.H., 2017. Evolution and genome architecture in fungal plant pathogens. Nat. Rev. Microbiol. https://doi.org/10.1038/nrmicro.2017.76.

Nejat, N., Mantri, N., 2018. Emerging roles of long non-coding RNAs in plant response to biotic and abiotic stresses. Crit. Rev. Biotechnol. 38, 93–105. https://doi.org/10.1080/07388551.2017.1312270.

Ozsolak, F., Milos, P.M., 2011. RNA sequencing: advances, challenges and opportunities. Nat. Rev. Genet. https://doi.org/10.1038/nrg2934.

Pauli, A., Norris, M.L., Valen, E., Chew, G.L., Gagnon, J.A., Zimmerman, S., et al., 2014. Toddler: an embryonic signal that promotes cell movement via apelin receptors. Science (80-.) 343, https://doi.org/10.1126/science.1248636.

Plett, J.M., Martin, F.M., 2018. Know your enemy, embrace your friend: using omics to understand how plants respond differently to pathogenic and mutualistic microorganisms. Plant J. 93, 729–746. https://doi.org/10.1111/tpj.13802.

Rodriguez-Moreno, L., Ebert, M.K., Bolton, M.D., Thomma, B.P.H.J., 2018. Tools of the crook-infection strategies of fungal plant pathogens. Plant J. 93, 664–674. https://doi.org/10.1111/tpj.13810.

Saijo, Y., Loo, E.P.i., Yasuda, S., 2018. Pattern recognition receptors and signaling in plant–microbe interactions. Plant J. 93, 592–613. https://doi.org/10.1111/tpj.13808.

Savary, S., Ficke, A., Aubertot, J.N., Hollier, C., 2012. Crop losses due to diseases and their implications for global food production losses and food security. Food Secur. https://doi.org/10.1007/s12571-012-0200-5.

Sexton, A.C., Howlett, B.J., 2006. Parallels in fungal pathogenesis on plant and animal hosts. Eukaryot. Cell https://doi.org/10.1128/EC.00277-06.

Shibuya, N., Kaku, H., Kuchitsu, K., Maliarik, M.J., 1993. Identification of a novel high-affinity binding site for N-acetylchitooligosaccharide elicitor in the membrane fraction from suspension-cultured rice cells. FEBS Lett. 329, 75–78. https://doi.org/10.1016/0014-5793(93)80197-3.

Shukla, L.I., Chinnusamy, V., Sunkar, R., 2008. The role of microRNAs and other endogenous small RNAs in plant stress responses. Biochim. Biophys. Acta - Gene Regul. Mech. https://doi.org/10.1016/j.bbagrm.2008.04.004.

Shumayla, Sharma S., Taneja, M., Tyagi, S., Singh, K., Upadhyay, S.K., 2017. Survey of high throughput RNA-seq data reveals potential roles for lncRNAs during development and stress response in bread wheat. Front. Plant Sci. 8, 1019.

Tyagi, S., Sharma, A., Upadhyay, S.K., 2018. Role of next-generation RNA-Seq data in discovery and characterization of long non-coding RNA in plants. In: Çiftçi, Y.Ö. (Ed.), Next Generation Plant Breeding. IntechOpen, https://doi.org/10.5772/intechopen.72773.

Verma, V., Ravindran, P., Kumar, P.P., 2016. Plant hormone-mediated regulation of stress responses. BMC Plant Biol. 16, 86. https://doi.org/10.1186/s12870-016-0771-y.

Xin, M., Wang, Y., Yao, Y., Xie, C., Peng, H., Ni, Z., et al., 2010. Diverse set of microRNAs are responsive to powdery mildew infection and heat stress in wheat (*Triticum aestivum* L.). BMC Plant Biol. 10, https://doi.org/10.1186/1471-2229-10-123.

Xin, M., Wang, Y., Yao, Y., Song, N., Hu, Z., Qin, D., et al., 2011. Identification and characterization of wheat long non-protein coding RNAs responsive to powdery mildew infection and heat stress by using microarray analysis and SBS sequencing. BMC Plant Biol. 11, https://doi.org/10.1186/1471-2229-11-61.

Xing, Q., Zhang, W., Liu, M., Li, L., Li, X., Yan, J., 2019. Genome-wide identification of long non-coding RNAs responsive to *Lasiodiplodia theobromae* infection in grapevine. Evol. Bioinforma. 15, https://doi.org/10.1177/1176934319841362.

Yan, J.Y., Xie, Y., Zhang, W., Wang, Y., Liu, J.K., Hyde, K.D., et al., 2013. Species of *Botryosphaeriaceae* involved in grapevine dieback in China. Fungal Divers. 61, 221–236. https://doi.org/10.1007/s13225-013-0251-8.

Yao, Z., Chen, Q., Chen, D., Zhan, L., Zeng, K., Gu, A., et al., 2019. The susceptibility of sea-island cotton recombinant inbred lines to *Fusarium oxysporum* f. sp. *vasinfectum* infection is characterized by altered expression of long noncoding RNAs. Sci. Rep. 9, 1–13. https://doi.org/10.1038/s41598-019-39051-2.

Yin, H., Zhang, X., Zhang, B., Luo, H., He, C., 2019. Revealing the dominant long noncoding RNAs responding to the infection with *Colletotrichum gloeosporioides* in *Hevea brasiliensis*. Biol. Direct 14, 1–12. https://doi.org/10.1186/s13062-019-0235-z.

Yu, X., Feng, B., He, P., Shan, L., 2017. From chaos to harmony: responses and signaling upon microbial pattern recognition. Annu. Rev. Phytopathol. 55, 109–137. https://doi.org/10.1146/annurev-phyto-080516-035649.

Zhang, H., Hu, W., Hao, J., Lv, S., Wang, C., Tong, W., et al., 2016. Genome-wide identification and functional prediction of novel and fungi-responsive lincRNAs in *Triticum aestivum*. BMC Genomics 17, 1–11. https://doi.org/10.1186/s12864-016-2570-0.

Zhang, L., Wang, M., Li, N., Wang, H., Qiu, P., Pei, L., et al., 2018. Long noncoding RNAs involve in resistance to *Verticillium dahliae*, a fungal disease in cotton. Plant Biotechnol. J. 16, 1172–1185. https://doi.org/10.1111/pbi.12861.

Zhao, J., Buchwaldt, L., Rimmer, S.R., Sharpe, A., McGregor, L., Bekkaoui, D., et al., 2009. Patterns of differential gene expression in *Brassica napus* cultivars infected with *Sclerotinia sclerotiorum*. Mol. Plant Pathol. 10, 635–649. https://doi.org/10.1111/j.1364-3703.2009.00558.x.

Zhu, Q.H., Stephen, S., Taylor, J., Helliwell, C.A., Wang, M.B., 2014. Long noncoding RNAs responsive to *Fusarium oxysporum* infection in *Arabidopsis thaliana*. New Phytol. 201, 574–584. https://doi.org/10.1111/nph.12537.

Zuluaga, A.P., Vega-Arreguín, J.C., Fei, Z., Ponnala, L., Lee, S.J., Matas, A.J., et al., 2016. Transcriptional dynamics of *Phytophthora infestans* during sequential stages of hemibiotrophic infection of tomato. Mol. Plant Pathol. 17, 29–41. https://doi.org/10.1111/mpp.12263.

Chapter 20

Role of plant long noncoding RNAs in the regulation of plant metabolism

Girija Kaushal[a,*], Jyoti Singh Jadaun[b,*], Lokesh Kumar Narnoliya[a], and Sudhir P. Singh[a]
[a]*Center of Innovative and Applied Bioprocessing (CIAB), Mohali, Punjab, India,* [b]*Botany Department, Dayanand Girls Postgraduate College, Kanpur, Uttar Pradesh, India*

20.1 Introduction

The eukaryotic genome is comprised of both coding and noncoding regions. In fact, about 90% part of the genome is noncoding (Chekanova, 2015). The noncoding region is described as that part of the genome, which does not translate into any specific functional or structural protein and lies as noncoding RNA (ncRNA). The transcription of intronic regions and complementary DNA-strands results in the generation of diverse noncoding RNAs (Narnoliya et al., 2019). These ncRNAs can be further classified on the basis of their size, as small or large ncRNAs. The ncRNAs having length <50 nt are known as small ncRNAs, which include Piwi-interacting RNAs (piRNAs), small interfering RNAs (siRNAs), and microRNAs (miRNAs); whereas ncRNAs having length ≥ 200 nt are known as long ncRNAs (Bhatia et al., 2017). Noncoding RNAs are also coined as "dark matter" as they do not carry any information regarding protein translation, but can act as a key player in the regulation of gene expression. From historical point of view, the lncRNAs were first reported in humans in 1992, and later on, they are also explored in other domains of the life such as plant, fungi, yeast, prokaryotes, and even in viruses (Rai et al., 2019; Wang et al., 2017b, 2019). LncRNAs and mRNAs share similar mechanisms of biogenesis. Most of the eukaryotic lncRNAs undergo transcription via RNA polymerase II, followed by various posttranscriptional modifications, including polyadenylation, splicing, and 5′ capping (Mongelli et al., 2019). In the case of plants, in addition to RNA polymerase II, RNA polymerases IV and V are also observed to transcribe lncRNAs. The lncRNAs transcribed by RNA polymerase V are reported to be involved in RNA-directed DNA methylation (Mongelli

[*] Equal contribution.

Long Noncoding RNAs in Plants: Roles in development and stress
https://doi.org/10.1016/B978-0-12-821452-7.00020-9

et al., 2019). LncRNAs are majorly located in the nucleus and chromatin of the subnuclear compartment, and some are exclusive to the cytosol or subcellular compartments. According to a few reports, lncRNAs can have short open reading frames (ORFs) that may get converted into small peptides with unknown or less explored functions (Wang et al., 2017b). Being less abundant and conserved, lncRNAs may be specific to a particular cell or tissue type (Bhatia et al., 2017; Held et al., 2008).

In reference to humans, a significant amount of data related to lncRNAs has been generated and analyzed by using bioinformatics and computational approaches (Bhatia et al., 2017; Tyagi et al., 2018). Though the exact functional mechanism of lncRNAs is poorly understood, a mammoth exercise is undergoing throughout the globe to fully understand their mechanism to utilize them in various sectors such as genomics and personalized medicine. The research practices regarding lncRNAs are mostly oriented toward the animal system; in contrast, the other biological systems like plants are far lagging. Fortunately, with the advancements in the modern computational approaches and high-throughput sequencing technologies, viz., lncRNA microarray technique, RNA sequencing (RNA-seq), genome sequencing, RNA capture sequencing (RNA CaptureSeq), etc., a massive amount of data are being generated and analyzed (Shafiq et al., 2016). Trends in lncRNAs research also indicate the increment in the data generation, and tools developed in subsequent years.

LncRNAs play a crucial role in the gene regulation by controlling the expression level of mRNA and also influence the functioning of miRNAs and splicing machinery of the hosts (Narnoliya et al., 2017, 2019). They affect the functioning of the miRNA by mimicking the target mRNA so that the miRNA interacts with the lncRNA, instead of the original target. Thus, lncRNAs may compete with the target mRNA for the miRNA machinery and ultimately regulate the miRNA-mediated repression of the gene. The functioning of the lncRNAs can also be regulated by the miRNA by executing RNA cleavage. However, these lncRNAs can act as the precursors of the miRNAs (Golicz et al., 2018). Recent studies indicated that these noncoding RNAs could also regulate the peptide synthesis by influencing the translation machinery of the hosts. The detailed molecular functioning of lncRNAs is still in oblivion, therefore, lncRNAs are considered as the mystery molecules of the biological system. But it is hitherto truth that they are essential for the proper functioning of the living cells (Bardou et al., 2014; Franco-Zorrilla et al., 2007; Henriques et al., 2017; Heo and Sung, 2011; Jabnoune et al., 2013; Qin et al., 2017; Wunderlich et al., 2014).

20.2 Classification of lncRNAs

LncRNAs can be majorly classified on the basis of their function and genomic location (Fig. 20.1). On the basis of function, they can be classified as cis- and trans-acting lncRNAs, whereas, on the basis of their genomic locations, they are classified as sense or antisense, and intronic or intergenic lncRNA.

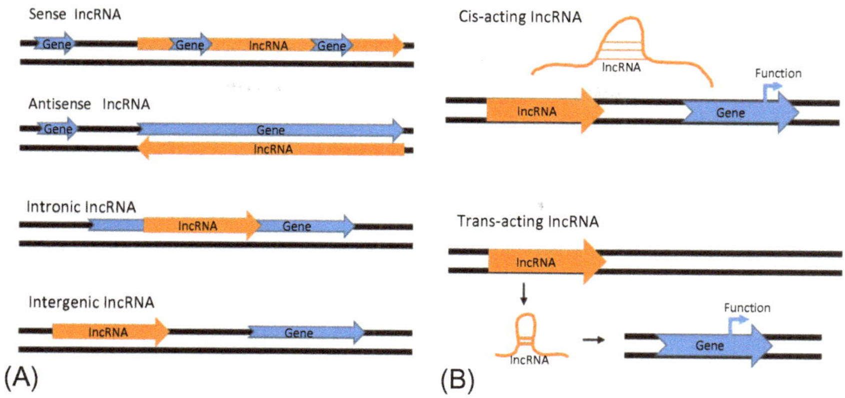

FIG. 20.1 Types of lncRNAs on the basis of their (A) genomic location and (B) functions.

20.2.1 On the basis of genomic location

On the basis of genomic location, lncRNAs are classified in two categories: one is sense or antisense, and the second is intronic or intergenic lncRNAs.

20.2.1.1 Sense and antisense lncRNAs

The transcription of ncRNA from a similar region as that of the protein-encoding gene (PEG), with the same promoter site, results in the formation of sense lncRNA (Ma et al., 2013). They may share some part with PEGs or overlap with the entire PEG. COLDAIR (COLD-Assisted Intronic noncoding RNA) is a well-known example of this class of lncRNAs (Heo and Sung, 2011). The transcription of ncRNA from an antisense strand of the protein-encoding gene (PEG) may share a similar region with that of an exon in the sense gene or may overlap the complete sequence of sense gene via intron (Shafiq et al., 2016). Rice-TWISTED LEAD (TL) is an example of antisense lncRNA (Liu et al., 2015).

Both the sense and antisense lncRNAs are known to be genuine transcripts, as proved by qRT-PCR validation, sequencing of full-length cDNA, and strand-specific sequencing. Both the types of lncRNAs can be multiexonic and can be seen using oligo-dT-guided reverse transcription and cap-analysis gene expression (Ma et al., 2013). About 87% of coding genes are reported to have their antisense transcripts, but sense lncRNAs are least explored (Zubko and Meyer, 2007). The sense transcripts may function as PEG as well as lncRNAs. Early nodulin 40 (*ENOD40*) is an example of sense lncRNAs (Crespi, 2017; Kouchi et al., 1999; Sousa et al., 2001; Yang et al., 1993).

20.2.1.2 Intronic and intergenic lncRNAs

Based on the location of transcription, lncRNAs are classified into intronic and intergenic. LncRNAs that are transcribed from introns are simply known as intronic lncRNAs, whereas those derived from intergenic regions are called intergenic lncRNAs (Ma et al., 2013). Also, both intronic and intergenic lncRNAs have different poly(A) modifications, transcriptional mechanisms, and different cellular locations. Intronic lncRNAs have not been much explored as compared to intergenic lncRNAs (Ma et al., 2013).

Long intergenic noncoding RNAs (lincRNAs) are reported to have regulatory functions in various mechanisms like splicing regulation, transcriptional regulation, posttranscriptional regulation, translational control, etc. (Shafiq et al., 2016). These lincRNAs are reported to be more conserved, tissue-specific, and more stable as compared to intronic lncRNAs. The transcript of lincRNA is low as compared to the mRNA associated with it (Ma et al., 2013). With the increasing interest in lncRNAs, about 2000 lincRNAs have been identified from various plants, such as rice, maize, wheat, potato, and sorghum (Shafiq et al., 2016; Wang et al., 2015a; Wasaki et al., 2003). ELF18-Induced Long Noncoding RNA 1 (ELENA1) is an example of lincRNAs which play a regulatory role in plant immunity (Mach, 2017; Seo et al., 2017).

20.2.2 On the basis of function

On the basis of function, lncRNAs are classified in two categories: cis-acting and trans-acting lncRNAs.

20.2.2.1 Cis-acting lncRNAs

LncRNAs that regulate the expression of a gene located on the same chromosome are known as cis-acting lncRNAs (Shafiq et al., 2016). These lncRNAs may also affect the expression of closely located genes by inducing a change in chromatin organization or transcriptional interference. These types of lncRNAs usually bind with the promoter to block the formation of the preinitiation complex or interfere with transcription factors in order to induce transcriptional disturbance (Crespi, 2017). Delay of Germination 1 (DOG1) is an example of cis-acting lncRNA, which is involved in the regulation of plant development. Thus, having a polyadenylation tail and 5′ cap enables it to suppress the seed dormancy in plants (Fedak et al., 2016).

20.2.2.2 Trans-acting lncRNAs

The trans-acting lncRNAs are more specific toward targeting the genes that are in closer vicinity of the lncRNA transcription site, majorly on the basis of complementarity (Shafiq et al., 2016). When lncRNA interacts with a miRNA and blocks its action on a specific target, it is known as trans-acting lncRNAs. This mode of operation of such lncRNAs is also termed as eTM, i.e., endogenous target mimicry (Shafiq et al., 2016). In this process, the trans-acting lncRNA acts like decoys and binds to the miRNAs, in order to hinder the interaction between miRNA and its actual target. To remain undetected by miRNA dicer machinery, these target mimics have a mismatch of minimum three nucleotides at miRNA cleavage site, thus enabling them to have more stable sequestration with the miRNA (Franco-Zorrilla et al., 2007). Unlike *cis*-lncRNAs, these can bind to transcription elongation factors to regulate the transcription (Ma et al., 2013). There are several reports on the involvement of such lncRNAs in plant immunity, like Slylnc0195 acts as target mimic of miR-166. It is an example of trans-acting lncRNAs (Nejat and Mantri, 2018).

20.3 Identification of plant lncRNAs

With the advancement in the next-generation sequencing technologies and computational tool, in-silico methods are the most common approaches to identify plant lncRNAs (Li et al., 2014; Liu et al., 2018), which need to be functionally characterized to validate their importance in regulatory machinery (Table 20.1).

In-silico methods are based on the sequencing technologies, which provide complete transcriptomic/genomic data that need to be mined, specifically the unannotated part. These unannotated transcripts are firstly confirmed with no ORFs or coding potential by using various tools like Coding Potential Calculator (CPC) (http://cpc.gao-lab.org/programs/run_cpc.jsp), POTRAIT (http://bioinformatics.cenargen.embrapa.br/portrait/), and Phylo Codon Substitution Frequency (CSF) (http://compbio.mit.edu/PhyloCSF) (Yotsukura et al., 2016). There are a few signatures that can be used to identify specific lncRNAs. For example, in the case of mammals, K4 and K36 methylation marks were used to identify lncRNAs. A similar approach was applied in the model plant organism, *Arabidopsis thaliana* (Liu et al., 2015). Further, those transcripts, which are greater than 200nt in length and have the coding potential of ≤ 0, can be used as lncRNAs for further processing. These noncoding transcripts are processed by aligning against various plant long ncRNA databases such as Lncrnadb (http://www.lncrnadb.org), PLNlncRbase http://bioinformatics.ahau.edu.cn/PLNlncRbase), and PNRD (structuralbiology.cau.edu.cn/PNRD), TAIR10 (http://www.arabidopsis.org), PlantNATsDB (http://bis.zju.edu.cn/pnatdb), PLncDB (http://chualab.rockefeller.edu/gbrowse2/homepage.html), NONCODE (http://noncode.bioinfo.org.cn), GreeNC (http://greenc.sciencedesigners.com/), CANTATAdb (http://yeti.amu.edu.pl/CANTATA), DsTRD (http://bi.sky.zstu.edu.cn/DsTRD/home.php), and RNA Central (http://rnacentral.org/). Among all these databases, NONCODE database contains lncRNA reported from multiple organisms. PLNlncRbase is having the information of functionally validated lncRNAs belonging to 43 different plant species. CANTATdb is constituted with predicted lncRNAs sequences from 37 different plant species. PlncDB and TAIR are specific to *Arabidopsis* species. LncRNAdb contains data of validated lncRNAs from 9 plant species. PlantNATsDB and GreeNc include information of nc-transcripts from 70 and 10 different plant species, respectively. After aligning with these databases (Table 20.2), the promising hits can be taken for further analysis. These predicted lncRNAs can

TABLE 20.1 Different techniques employed for identification of lncRNAs in plants.

Technique	Plant species	References (doi)
In silico	Geranium	https://doi.org/10.1002/1873-3468.13493
	Grapevine	https://doi.org/10.1038/s41598-019-43269-5
	Wild banana	https://doi.org/10.1371/journal.pone.0200002
	Maize	https://doi.org/10.1186/gb-2014-15-2-r40
	Medicago	https://doi.org/10.1186/s12870-015-0530-5
Microarray	Arabidopsis	https://doi.org/10.1105/tpc.112.102855
	Wheat	https://doi.org/10.1186/1471-2229-11-61
RNA Seq	Arabidopsis	https://doi.org/10.1016/j.bbrc.2017.05.051
	Chinese cabbage	https://doi.org/10.1186/1471-2229-13-208
	Cotton	https://doi.org/10.1038/srep16946
	Kiwi fruit	https://doi.org/10.1038/s41598-017-05377-y
	Maize	https://doi.org/10.1038/s41598-017-03878-4
	Millet	https://doi.org/10.1007/s11103-013-0104-6
	Orange	https://doi.org/10.1038/srep43226
	Peach	https://doi.org/10.1007/s11103-013-0093-5
	Populus	https://doi.org/10.1093/jxb/eru256
	Rice	https://doi.org/10.1186/s13059-014-0512-1
	Sea buckthorn	https://doi.org/10.1016/j.gene.2016.10.017
	Spikemoss	https://doi.org/10.1186/s12864-016-3266-1
	Soybean	https://doi.org/10.3389/fpls.2014.00743
	Sunflower	https://doi.org/10.1186/s12864-016-2776-1
	Tomato	https://doi.org/10.1093/jxb/erv203
	Wheat	https://doi.org/10.1038/s41598-019-43088-8

be fed into various bioinformatics tools such as LNCTAR to find its interaction with mRNAs which can help in predicting their functional profiles (Narnoliya et al., 2019).

At one point of time when NGS technology was not in practice, microarray was widely used for the identification of lncRNAs. Being cost-effective and

TABLE 20.2 Web resources available for identification of lncRNA.

Web resources	Website
AraportII	https://www.araport.org/
CANTATAdb	http://cantata.amu.edu.pl/
Green Noncoding database (GREENC)	http://greenc.sciencedesigners.com/wiki/Main_Page
NONCODE v4.0	http://www.noncode.org/index.php, http://www.noncode.org/index.php
Plant long noncoding RNA database (PLncDB)	http://chualab.rockefeller.edu/gbrowse2/homepage.html
Plant Natural Antisense transcripts database (PlantNATsDB)	http://bis.zju.edu.cn/pnatdb/
Plant ncRNA database (PNRD)	http://structuralbiology.cau.edu.cn/PNRD/
The Arabidopsis Information Resource (TAIR)	https://www.arabidopsis.org/
EVLncRNAs	http://biophy.dzu.edu.cn/EVLncRNAs
PLNlncRBase	http://bioinformatics.ahau.edu.cn/PLNlncRbase
lncRNA-screen	https://github.com/NYU-BFX/lncRNA-screen
Sebnif	http://sunlab.lihs.cuhk.edu.hk/sebnif/
UClncR	http://bioinformaticstools.mayo.edu/research/uclncr-pipeline/

simple, as compared to the RNA seq method, microarray was applied to find the gene expression profiles of transcripts (Lee and Kikyo, 2012; Shafiq et al., 2016). On the basis of probe design, this technique was further subdivided into tiling and nontiling array. Non-tiling array was used to determine the expression levels of only known genes, whereas tiling arrays can be used for both known and unknown transcribed regions. Therefore, tiling arrays were used to find ncRNA transcripts (Mockler and Ecker, 2005). Noncoding antisense transcript for *Arabidopsis* FLC locus is an example of ncRNA detection using tiling array method. In spite of being cheaper and simple in interpretation, it is not in much use because high signal saturation leads to reduce sensitivity. Further, there is a possibility of cross-hybridization among closely related species (Shafiq et al., 2016).

RNA sequencing is a well-known method that came into existence with the advancements in high-throughput sequencing methods (Wang et al., 2009). This provides a set of raw reads that are further processed to get the expression of the transcript. This method has been further modified to gain more information; for example, strand-specific RNA-seq (Levin et al., 2010) can provide the information about genomic location of a particular transcript, RNA immune precipitation sequencing (Shafiq et al., 2016) can help to discover the biological role of lncRNAs, and single-cell RNA-seq (Efroni and Birnbaum, 2016) analyzes gene expression patterns with respect to different cell types. A number of lncRNAs have been reported from various plants involved in several biological functions mentioned in Table 20.3 (Bai et al., 2019; Flórez-Zapata et al., 2016; Golicz et al., 2018; Kim et al., 2017b; Lu et al., 2017; Qi et al., 2013; Shuai et al., 2014; Wang et al., 2013, 2015b, 2017a,d; Yu et al., 2013; Zhang et al., 2014b, 2017; Shumayla et al., 2017; Zhu et al., 2015, 2017).

20.4 Role of lncRNAs in plant metabolism

20.4.1 LncRNAs as a regulator in flowering

Flowering is an essential process for the continuation of the plant's life cycle. It involves the synthesis of reproductive organs, leading to fertilization, and consequently, seed formation takes place for propagation. The vegetative phase turned into the reproductive phase to initiate flower formation, and this transition requires several signals, which may be exogenous or endogenous (Srikanth and Schmid, 2011). Mainly, five pathways are designated for flowering: (i) vernalization pathway; (ii) photoperiod pathway; (iii) gibberellic acid (GA) pathway; (iv) autonomous pathway; and (v) aging pathway (Ietswaart et al., 2012; Swiezewski et al., 2009). Vernalization and photoperiod pathways involve exogenous signaling, while GA pathway and autonomous pathway function on the endogenous status of the plant (Yamaguchi and Abe, 2012). LncRNAs are found to play a regulatory role in flowering, directly or indirectly (Ding et al., 2012; Ietswaart et al., 2012). The reproductive phase of flowering is the result of the complex networking of these pathways as they cross-talk each other to conclude the final outbreaks. In the genome, different regions are noticed to be a hotspot for flowering time regulation such as FLC (Flowering Locus C), FLT (Flowering Locus T), SOC1 (Suppressor of Overexpression of CO1), and LFY (LEAFY) (Michaels and Amasino, 1999).

FLC is a MADS-box protein family that includes five other members, MADS AFFECTING FLOWERING 1–5 (MAF1-MAF5). It is a well-studied repressor of flowering-inducing factors (Bloomer and Dean, 2017; Michaels and Amasino, 1999). Flowering depends on the transcript level of FLC, which inhibits the flowering promoting factors. The expression of FLC is controlled by vernalization, autonomous, and FRIGIDA pathways, and a number of RNA molecules are integrated into the regulation of these pathways, such as intronic

TABLE 20.3 A list of lncRNAs reported in plants with their biological functions.

Name	Species	Biological function	References (doi)
APOLO	*Arabidopsis thaliana*	Regulated plant development through auxin signaling	https://doi.org/10.1038/ng2079
ASCO	*Medicago truncatula*	Regulates the development of lateral roots	https://doi.org/10.1016/j.bbagrm.2015.06.009
ASCO-lncRNA	*Arabidopsis thaliana*	Regulates the development of lateral roots	https://doi.org/10.1016/j.devcel.2014.06.017
asHSFB2a	*Arabidopsis thaliana*	Regulates the gametophytic and vegetative development	https://doi.org/10.1007/s11103-014-0202-0
CDF5 LONG NONCODING RNA (FLORE)	*Arabidopsis thaliana*	Photomorphogenesis	https://doi.org/10.1111/nph.14703
Cis-NATPHO1;2	*Oryza sativa* (rice)	Regulates phosphate homeostasis	https://doi.org/10.1105/tpc.113.116251
COLDAIR	*Arabidopsis thaliana*	Regulates flowering	https://doi.org/10.1126/science.1197349
COOLAIR	*Arabidopsis thaliana*	Regulates flowering	https://doi.org/10.1038/nature08618
Drought induced RNA (DRIR)	*Arabidopsis thaliana*	Upregulated in drought and salt stress	https://doi.org/10.1104/pp.17.00574
ELENA1	*Arabidopsis thaliana*	Upregulated in disease caused by *Pseudomonas syringae*	https://doi.org/10.1105/tpc.17.00343
ENOD40	*Oryza sativa* (rice)	Organogenesis of root nodules	https://doi.org/10.1046/j.1365-313x.1999.00432.x
ENOD40	*Zea mays* subsp. *mays* (maize)	Organogenesis of root nodules	https://doi.org/10.1016/j.bbagrm.2015.06.009

Continued

TABLE 20.3 A list of lncRNAs reported in plants with their biological functions—cont'd

Name	Species	Biological function	References (doi)
ENOD40	*Medicago trucatula*	Organogenesis of root nodules	https://doi.org/10.1016/j.bbagrm.2015.06.009
GmENOD40	*Glycine max* (soybean)	Regulates the re-localization of proteins from nucleus to cytoplasm	https://doi.org/10.1046/j.1365-313x.1993.03040573.x
HID1	*Arabidopsis thaliana*	Regulates photomorphogenesis	https://doi.org/10.1073/pnas.1409457111
HvCesA6 lnc-NAT	*Hordeum vulgare* (barley)	Involved in the synthesis of cell wall	https://doi.org/10.1073/pnas.0809408105
incRNA20 lincRNA2752	*Populus trichocarpa*	Upregulated in abiotic stress (drought)	https://doi.org/10.1093/jxb/eru256
IPS1	*Arabidopsis thaliana*	Regulates phosphate homeostasis	https://doi.org/10.1038/ng2079
LDMAR (P/TMS12-1)	*Oryza sativa* (rice)	Male fertility regulation	https://doi.org/10.1073/pnas.1121374109
lncRNAs: 2224 lincRNAs: 1624 lncNATs: 600	*Oryza sativa* (rice)	Regulates sexual reproduction	https://doi.org/10.1186/s13059-014-0512-1
MtENOD40	*Medicago truncatula*	Regulates the re-localization of proteins	https://doi.org/10.1128/MCB.21.1.354-366.2001
OsPI1	*Oryza sativa* (rice)	Regulates phosphate homeostasis	https://doi.org/10.1046/j.1469-8137.2003.00748.x

SHO lnc-NAT	Petunia hybrid (petunia)	Regulates the synthesis of local cytokines	https://doi.org/10.1111/j.1365-313X.2007.03309.x
Slylnc0195 slylnc1077	*Solanum lycopersicum* (tomato)	Upregulated in infections caused by tomato yellow leaf curl virus	https://doi.org/10.1111/tpj.13018
TalnRNA5	Triticum (wheat)	Defense against *Blumeria graminis* f. sp. *tritici* (*Bgt*) fungus	https://doi.org/10.1186/1471-2229-11-61
TalnRNA9	*Triticum* (wheat)	Defense against *Blumeria graminis* f. sp. *tritici* (*Bgt*) fungus	https://doi.org/10.1186/1471-2229-11-61
TapmlnRNA30	*Triticum* (wheat)	Defense against *Blumeria graminis* f. sp. *tritici* (*Bgt*) fungus	https://doi.org/10.1186/1471-2229-11-62
TPS11	*Solanum lycopersicum* (tomato)	Regulates phosphate homeostasis	https://doi.org/10.1023/A:1005729309569
Zm401	*Zea mays* subsp. *mays* (maize)	Expression of pollen developmental genes	https://doi.org/10.1016/j.bbagrm.2015.06.009

lncRNA, antisense lncRNA, and small RNA (sRNA) (Sheldon et al., 2000). Three flowering-associated lncRNAs (COOLAIR, COLDAIR, and COLDWRAP) transcribed from FLC locus are presented here in detail (Fig. 20.2).

20.4.1.1 Role of lncRNAs COOLAIR in FLC regulation

COOLAIR is the antisense transcript of lncRNA, first reported in *A. thaliana* by Swiezewski and coworkers in 2007. Researches have revealed that these are originated from sRNAs in the 3′ end of FLC heterochromatic region. They covered the complete FLC gene locus and are terminated prior to the promoter region (Swiezewski et al., 2009). Due to analogy with HOTAIR (HOX Transcript Antisense RNA) lncRNAs, these are termed COOLAIR (cold-induced long antisense intragenic RNAs) (Rinn et al., 2007). Suppression of FLC in spring leads to flowering, and its expression is regulated by vernalization and autonomous pathways. Interestingly, COOLAIR lncRNAs participated in both vernalization and autonomous pathways. The autonomous pathway is important for controlling the flowering time in *Arabidopsis* and also involves in epigenetic gene silencing of FLC repressor through histone modification, transcription elongation, and RNA processing. COOLAIR has two polyadenylation sites, distal and proximal, although the result of both polyadenylation is opposite to each other. Polyadenylation of COLAIR at distal 3′ end leads to high expression, and proximal 3′ end leads to low expression of FLC (Hornyik et al., 2010; Liu et al., 2010).

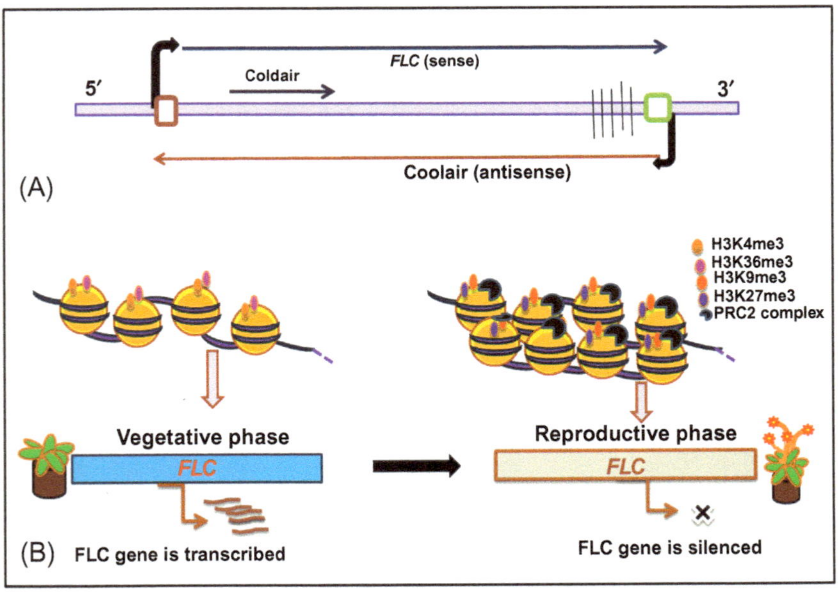

FIG. 20.2 Role of lncRNAs in flowering: (A) Generation of lncRNAs from FLC, arrows indicate the direction of transcription, (B) involvement of histone modifications during transition from vegetative phase to reproductive phase.

Components of autonomous pathway are Flowering Control Locus A (FCA)—a RNA-binding protein, FLD (Flowering Locus D), suppressor of overexpressed FCA (SOF), Cleavage stimulatory factor64 (CstF64), and CstF77—canonical 3′-processing factors, PRP8—a core spliceosome component, P-TEFb—transcription elongation factor b having homology to CDKC2-like protein, and LUMINIDEPENDENS (LD). All of these components are involved in the proximal polyadenylation of COOLAIR, controlling its expression. FLC repression is performed by H3K4me2 demethylation of the FLC gene by histone demethylase FLD. Similar to FCA, FPA, another RNA-binding protein, is also involved in proximal polyadenylation of COOLAIR and leads to repression of FLC (Hornyik et al., 2010).

Exposure to prolonged cold is needed by winter annual plants for flowering in spring, and this procedure is known as vernalization (Song et al., 2012). During vernalization, epigenetic gene silencing of FLC occurs, but the mechanism differs from the autonomous pathway (Heo and Sung, 2011; Kim et al., 2009). Chromatin modification through trimethylation of histone H3K27 (H3K27me3) of FLC is operated by complex machinery involving the polycomb repressive complex 2 (PRC2) and Plant Homeo-Domain (PHD) proteins, VERNALIZATION INSENSITIVE 3 (VIN3), VERNALIZATION1 (VRN1), VERNALIZATION 2 (VRN2), and VIN3-LIKE 1 (VIL1) VERNALIZATION 5 (VRN5) (Baulcombe and Dean, 2014; Greb et al., 2007; Tian et al., 2019). VIN3 represents a class of plant homeodomain (PHD) protein family, and PHD proteins are studied as a component of the chromatin modification system in eukaryotes (Sung et al., 2006). Similarly, VIN3 and VRN2 are needed for the methylation of histone (H3K27) at chromatin FLC (Musselman and Kutateladze, 2011). Alteration in the expression of FLC and VIN3 during vernalization suggests a strong correlation between them (Fedorenko et al., 2019).

20.4.1.2 Role of lncRNAs COLDAIR in flowering

COLDAIR (COLD-ASSISTED INTRONIC NONCODING RNA) is a polycomb-binding lncRNA, which plays a role in flowering by regulating the epigenetic gene silencing of FLC (Heo and Sung, 2011). COLDAIR lncRNA is transcribed from FLC intron in sense direction. It is 5′ capped, but without polyadenylation site, it has no alternative isoforms (Kim et al., 2017a). COLDAIR lncRNAs, associated with PRC2 component, repress the FLC expression through histone (H3K27me3) modifications. There is a modular motif in COLDAIR, which interacts with PRC2 complex, and further studies revealed that mutation in the motif leads to vernalization insensitivity in plants. This insensitivity can be overcome by ectopic expression of the wild-type COLDAIR (Kim and Sung, 2017). It was observed that the expression of COLDAIR was upregulated after COOLAIR, but before the expression of VIN3 (Heo and Sung, 2011).

20.4.1.3 COLDWRAP lncRNAs in flowering

COLDWRAP (cold of winter-induced noncoding RNA from the promoter) is another polycomb-binding lncRNA (316 nt) required for vernalization-mediated flowering. It is transcribed from the repressed promoter in sense direction, and it is also 5′capped without any polyadenylation site. The expression of COLDWRAP is upregulated during cold exposure. During vernalization, COLDWRAP lncRNA, along with COLDAIR, forms an intragenic chromatin loop in FLC locus, and it leads to stable repression of FLC, inducing flowering (Kim and Sung, 2017). COLDWRAP mutant showed less stable suppression of FLC with a lower abundance of PRC2 and H3K27, concluding that COLDWRAP operates FLC repression through PRC2-mediated complex (Kim and Sung, 2017).

20.4.2 LncRNAs in root organogenesis

With the advances in bioinformatics analysis, the role of lncRNAs has been explored in many aspects of developmental biology of plants. Being an important organ, the root has been explored for the presence and function of lncRNAs in root development. The bioinformatics analysis of the model plant (*Arabidopsis*) led to the discovery of a total 42 highly abundant lncRNAs in root involved in withstanding the environmental stress (Bazin and Bailey-Serres, 2015). One such lncRNA is named as APOLO lncRNA (AUXIN-REGULATED PROMOTER LOOP RNA), which is located to the upstream of regulatory kinase-encoding mRNA (i.e., PID). This PID mRNA deciphers the local mobilization of auxin transporter (i.e., PIN2) in the root cells. The expression of both APLOLO and PID are induced by exogenous auxin. It was observed that the plants with reduced transcripts of APOLO showed a modified root growth pattern, which is similar to *pid* gene mutated plants.

ENOD40 (early nodulin gene) is found to be involved in the development of nodes in leguminous plants (Crespi, 2017). Interaction of nuclear speckle RNA-binding proteins (MtRBP1/MtNSR1) of *Medicago* with ENOD40 showed the role of nuclear speckle RNA-binding proteins (NSRs) in lateral root development (Crespi, 2017). *Re*-localization of nuclear-located MtRBP1 into the cytoplasm was observed after overexpression of the *enod40* gene. In *A. thaliana*, paralogs of *MtNSR1*, *AtNSRa*, and *AtNSRb* were also traced, and their involvement in lateral root development was confirmed through generating the *nsransrb* double mutant, which had fewer and shorter lateral root (Bardou et al., 2014). In *A. thaliana*, an ASCO lncRNA (Alternative Splicing Competitor lncRNA) was detected as a negative regulator of NSR proteins. ASCO RNA prohibited the binding of alternative splicing machinery to the real target by mimicking as the original target. Several other lncRNAs have been found during root development, but their functional role is yet to be resolved.

20.4.3 Role of lncRNAs HID1 photomorphogenesis in seedlings

Light plays a crucial role in the growth and development of any organism, especially in plants; several physiological functions depend on the quantity and quality of light (Jiao et al., 2007). Photomorphogenesis in seedling is one of the best examples of light-regulated developmental processes. The expression of several genes depends on the light (Arsovski et al., 2012). Photomorphogenesis is carried out by a number of photoreceptors; they operate in the different light spectrums such as phytochromes in far-red and red light, cryptochromes, and phototropins in blue/UV-A light, and UV-B photoreceptor in UV-B light (Chen et al., 2004). Molecular studies revealed a number of transcription factors, which control photomorphogenesis, such as phytochrome-interacting factor 3 (PIF3). PIF3 is a representative member of the basic helix loop helix (bHLH) family, and it is known as a repressor of seedling photo morphogenesis in the dark. This study divulged that the expression of PIF3 is regulated by 236 nucleotide ncRNA, HID1 (HIDDEN TREASURE 1) (Wang et al., 2014). HID1 is a positive regulator of photomorphogenesis, and it functions under red light (Wang et al., 2014).

20.4.4 Role of lncRNAs in regulation of phosphate homeostasis

Plants evolved several adaptations to cope up with adverse conditions, such as phosphate (Pi) starvation. Recently, the role of lncRNA in Pi homeostasis is also reported by researchers in different plants such as *Arabidopsis*, *Medicago*, and tomato (Burleigh and Harrison, 1999; Liu et al., 1997). In *Arabidopsis*, two lncRNAs, At4 and IPS1 (INDUCED BY PHOSPHATESTARVATION1), were reported to be upregulated in Pi starvation condition. Three Phosphate Deficiency-Induced LncRNAs (PDILs) were also reported in *Medicago truncatula*. It was reported that PDIL1 works through the inhibition of PHOSPHATE 2 (MtPHO2), while PDIL2 and PDIL3 act at the transcriptional level. PHO2 encodes ubiquitin-conjugating E2 enzyme, and its expression is controlled by miRNA399 (Wang et al., 2017c). The mechanism behind the action of nutrition-related lncRNAs is that they may regulate gene expression through controlling the activity of miRNAs by executing endogenous target mimicry (eTM) (Borah et al., 2018).

20.4.5 Role of lncRNAs in biotic stress responses

Interaction with various bacterial, fungal, and viral organisms may lead to biotic stress in plants. Plants are known to have a well-developed immune response system against various pathogenic attacks in order to minimize the damaging effects upon pathogen encounters. This response system mostly relies upon the recognition of conserved microbial patterns known as microbe-associated molecular patterns (MAMPs) at the cellular level (Ramirez-Prado et al., 2018). These MAMPs are recognized by pattern receptors at the cell surface, also known as pattern recognition receptors (PRRs), and the defense response generated by this is known as PAMP-triggered immunity (PTI). However, there are certain pathogens that had evolved in the process of suppressing these PTI responses by various effector proteins that can affect protein activity by translocating in plant cells and enhancing disease conditions. In response to such attacks, certain resistance (R) proteins, also known as intracellular or transmembrane receptors, are evolved in plants for detecting such effector proteins released by pathogens and subsequently activating the effector-triggered immunity (ETI) response actions. Therefore, pathogen attack in plants can also trigger the signaling network of complex defense systems, which run the transcriptional reprogramming. Thus, the mechanism of transcriptional reprogramming (Nejat and Mantri, 2018) downstream the action of PTI and ETI signal system, which is performed by the action of various small RNAs (sRNAs), chromatin remodelers, transcription factors, long noncoding RNAs (lncRNAs), etc. (Fig. 20.3).

In plants, limited reports are available on the characterization of lncRNAs, related to biotic stress or plant immunity. A recent report on ELENA1 lncRNA discussed its predominant role in plant immune system (Mach, 2017). In response to microbe-associated molecular patterns, flg22 and elf18 lead to the induction of *ELENA1* expression. Further, the knockdown and overexpressing studies with the same lncRNA indicated that the resistance against *Pst* DC3000 pathogen could be governed by *ELENA1*, whereas its knockout showed susceptibility toward the pathogenic infection. The regulation of PTI-induced genes could be possible by trans-action of ELENA1; however, the detailed molecular mechanism of targeting these specific loci by this lncRNA remains unclear.

A report in 2011 about the identification of lncRNAs in wheat via computational approach stated the significance of lncRNAs against fungal infection in plants. Two wheat varieties, named PM-susceptible JD8 and PM-resistant JD8-pm30, were assessed for their defense against *Blumeria graminis* f. sp. *tritici* (*Bgt*) fungus that causes powdery mildew (PM) infection (Xin et al., 2011). Numerous lncRNAs play a crucial role in defense mechanism in a tissue-specific manner, e.g., TalnRNA9, TapmlnRNA30, and TalnRNA5 lncRNAs have been assigned to various tissue-specific regulations. TalnRNA9 has been found significantly upregulated in leaf, root, and seed. TapmlnRNA30 was found predominant in seeds. The expression of TalnRNA5 has been detected in many plant organs, including leaf, root, and internode, but most prominently in seeds than other tissues (Xin et al., 2011).

Plants infected with viral infection exhibited the changed pattern of expression in their lncRNAs due to their plant-viral interactions. During the viral invasion, lncRNAs constituted by host cells could belong to the host organism, viral organism, or due to the fusion of sequences from the host and viral cells (Prasad et al., 2019). These lncRNAs are evolved to be one of the key players in protecting plants against viral infections. Recently, a report explained the role of lncRNAs in a tomato variety named TYLCV-resistant tomato line CLN2777A, infected with tomato yellow leaf curl virus (TYLCV). In this study, a lncRNA named sly-lnc0957 was silenced via virus-induced gene silencing (VIGS), which resulted into an enhanced resistance against TYLCV in susceptible tomato line, concluding a negative regulation of lncRNA in this case; whereas another study related to TCV infection (Harvey et al., 2011) stated that the induction of viral infection leads to the accumulation of lncRNA (*LINC-AP2*) that downregulates AP2 domain-containing transcription factor (target of TCV infection). These are only the initial steps toward understanding the strategy of lncRNAs in plant defense against various microbial, fungal, and viral infections. There is a broad scope for digging the mechanism of action concerning the lncRNAs activity in plant immune response.

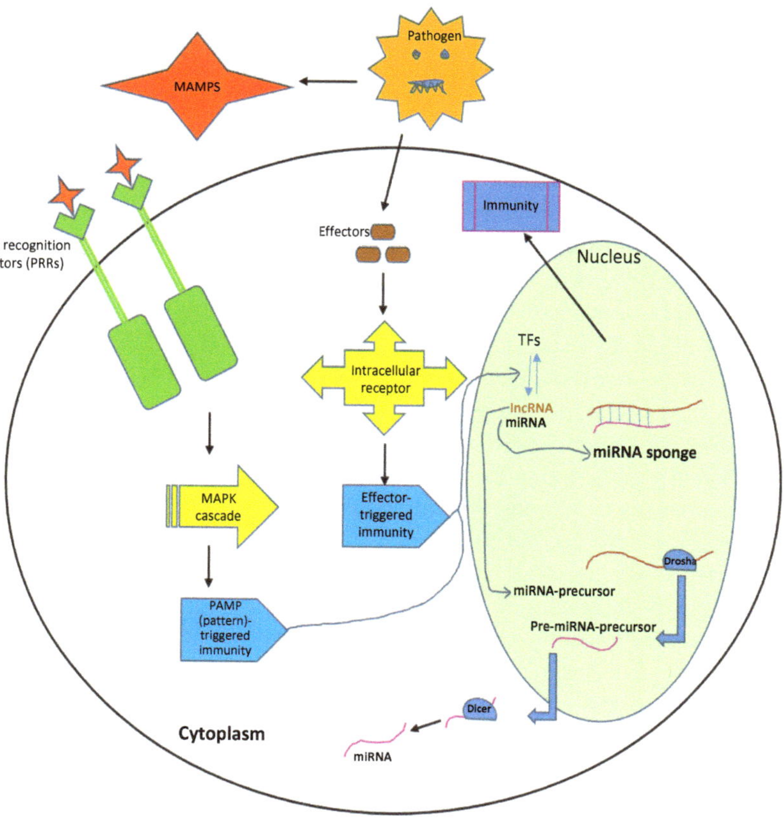

FIG. 20.3 Role of lncRNAs in plant immune response.

20.4.6 Role of lncRNAs in abiotic stress responses

The crop plants have to deal with various types of abiotic stresses like salinity, osmotic, drought, and heat stress that profoundly affect their productivity and may damage them in severe conditions. There are many factors reported in plants that can help to withstand these stress conditions and minimize the damage. New technologies had unveiled various lncRNAs that play leading vital roles along with miRNAs and protein-coding genes in plants to fight against such stress conditions. Therefore, the role of a few lncRNAs in abiotic stress conditions has been briefed in the following discussion.

A study on the model plant, *A. thaliana* ecotype Columbia (Col-0), emphasized the importance of lncRNAs in salt tolerance (Amor et al., 2009). The lncRNA named npc536 is reported to be involved in salt stress. Its overexpression helps in achieving salt tolerance with increased primary and secondary root growth. Thus, it clearly indicates that lncRNAs play a fair role in regulating salt stress by enhancing root growth. However, the detailed molecular mechanism involving hormone signaling pathway interactions is yet to be explored with respect to lncRNAs networking (Amor et al., 2009). Another study on *M. truncatula* identified the role of lncRNAs in osmotic stress (Wang et al., 2015c). The study detected a higher level of lncRNAs accumulation in leaves as compared to roots, unlike in salt stress.

Another important abiotic stress is drought. Besides various mRNAs and miRNAs, many lncRNAs are now being explored with relevant regulation in drought resistance in plants. LncRNAs are involved in histone modifications, which is an important mechanism opted by plants to fight against various stress conditions (Luo et al., 2013). Many reports have provided pieces of evidence about the involvement of lncRNAs in drought resistance (Liu et al., 2012). A significant number of drought stress-responsive regulatory lncRNAs have been identified in wheat (125) and maize (664) (Yao et al., 2010; Zhang et al., 2014a). In the case of maize, out of these 664 drought-responsive lncRNAs, 567 and 97 were found up- and downregulated, respectively (Zhang et al., 2014a). Moreover, 10 of these lncRNAs were identified to be antisense, which relates to previous reports stating that antisense lncRNA regulates chromatin remodeling under stress conditions. A study on rice detected 98 lncRNAs in response to drought conditions, out of which 31 and 67 were up- and downregulated, respectively (Chung et al., 2016). In cotton, lncRNAs involved in the regulation of photosynthesis were identified (Lu et al., 2016). It was found that 191 lncRNAs were targeting subunits of Rubisco enzyme. In this study, lncRNAs-mediated degradation of Rubisco enzyme in the cotton crop could be one of the mechanisms to resist drought effect (Lu et al., 2016).

Temperature plays a vital role in the development of the plant, and sometimes climatic changes induce heat stress, which not only retards the growth, but also leads to several diseases with a considerable risk of death in plants.

Earlier, heat stress was related to the heat-responsive genes regulated by heat stress transcription factors (HSFs) that lead to the acclimatization of heat shock proteins (HSPs) to deal with thermo-stress in plants (Zhao et al., 2016). With the increase in understanding about the role of noncoding RNAs in plant stress, several types of ncRNAs like circular RNAs (circRNAs), microRNAs (miRNAs), long noncoding RNAs (lncRNAs), and small interfering RNAs (siRNAs) are reported. In wheat, a total of 125 lncRNAs related to heat stress and powdery mildew infection were identified (Yao et al., 2010). The four lncRNAs, TalnRNA8, TalnRNA5, TalnRNA19, and TahlnRNA27, were known to be the miRNA-precursors. TalnRNA27 and TalnRNA5 have been found excessively upregulated under heat stress conditions.

20.4.7 Role of lncRNAs in secondary metabolism

Recently, our group identified potential lncRNAs from a medicinal and aromatic plant, rose-scented geranium (Narnoliya et al., 2019). In this analysis, a total of 26,784 transcripts have been identified as lncRNA with almost zero coding potential. Out of these, 466 were annotated against the publically available plant lncRNA databases. Further, 372 lncRNAs were found to have the potential to regulate terpene and tartaric acid biosynthesis. A regulatory interactome was also displayed, comprising lncRNA-miRNA-mRNA that represents a noncoding RNA (lncRNA and miRNA) regulatory network of the target mRNAs. This is the first report on lncRNAs and their possible regulatory roles in rose-scented geranium and related plant species. This study opens the door of knowledge about lncRNAs' functioning in plant metabolic pathway regulations, especially terpenes and tartrate metabolism.

20.5 Conclusion, challenges, and future remarks

Advancements in the sequencing technology have led to the discovery of diverse lncRNAs, playing crucial roles in the plant's metabolism. Many lncRNA types concerning the genomic location (sense and antisense lncRNAs), as well as functional contributions (cis- and trans-lncRNAs), are being detected from economically important plants. With an increasing interest in the scientific community toward lncRNA discovery, various bioinformatics tools and databases are being developed, e.g., CPC tool, Lncrnadb, PLNlncRbase, PNRD, TAIR10, PlantNATsDB, PLncDB, etc. The involvement of lncRNAs has been identified in various primary and secondary metabolic pathways, including regulation of flowering, organ development, and stress resistance. However, more focus is required to be given on tissue-specific, cell-type-specific, and temporal-specific lncRNAs. This area has much potential in exploring this junk part of the genome for making value out of it toward the societal interest.

Acknowledgments

Authors acknowledge Department of Biotechnology (DBT), Government of India, for financial support. JSJ and LKN acknowledge Science and Engineering Research Board (SERB) for providing N-PDF (PDF/2016/000445) and (PDF/2015/000662), respectively, and GK acknowledges DST-Inspire (DST/INSPIRE/03/2015/001777) fellowships.

References

Amor, B.B., Wirth, S., Merchan, F., Laporte, P., d'Aubenton-Carafa, Y., Hirsch, J., Maizel, A., et al., 2009. Novel long non-protein coding RNAs involved in Arabidopsis differentiation and stress responses. Genome Res. 19, 57–69.

Arsovski, A.A., Galstyan, A., Guseman, J.M., Nemhauser, J.L., 2012. Photomorphogenesis. Arabidopsis Book 10, e0147.

Bai, Y., Dai, X., Li, Y., Wang, L., Li, W., Liu, Y., Cheng, Y., et al., 2019. Identification and characterization of pineapple leaf lncRNAs in crassulacean acid metabolism (CAM) photosynthesis pathway. Sci. Rep. 9, 6658.

Bardou, F., Ariel, F., Simpson, C.G., Romero-Barrios, N., Laporte, P., Balzergue, S., Brown, J.W., et al., 2014. Long noncoding RNA modulates alternative splicing regulators in Arabidopsis. Dev. Cell 30, 166–176.

Baulcombe, D.C., Dean, C., 2014. Epigenetic regulation in plant responses to the environment. Cold Spring Harb. Perspect. Biol. 6, a019471.

Bazin, J., Bailey-Serres, J., 2015. Emerging roles of long non-coding RNA in root developmental plasticity and regulation of phosphate homeostasis. Front. Plant Sci. 6, 400.

Bhatia, G., Goyal, N., Sharma, S., Upadhyay, S., Singh, K., 2017. Present scenario of long non-coding RNAs in plants. Non-Coding RNA 3, 16.

Bloomer, R., Dean, C., 2017. Fine-tuning timing: natural variation informs the mechanistic basis of the switch to flowering in *Arabidopsis thaliana*. J. Exp. Bot. 68, 5439–5452.

Borah, P., Das, A., Milner, M.J., Ali, A., Bentley, A.R., Pandey, R., 2018. Long non-coding RNAs as endogenous target mimics and exploration of their role in low nutrient stress tolerance in plants. Genes 9, 459.

Burleigh, S.H., Harrison, M.J., 1999. The down-regulation of Mt4-like genes by phosphate fertilization occurs systemically and involves phosphate translocation to the shoots. Plant Physiol. 119, 241–248.

Chekanova, J.A., 2015. Long non-coding RNAs and their functions in plants. J. Curr. Opin. Plant Biol. 27, 207–216.

Chen, M., Chory, J., Fankhauser, C., 2004. Light signal transduction in higher plants. Annu. Rev Genet. 38, 87–117.

Chung, P.J., Jung, H., Jeong, D.-H., Ha, S.-H., Do Choi, Y., Kim, J.-K., 2016. Transcriptome profiling of drought responsive noncoding RNAs and their target genes in rice. BMC Genomics 17, 563.

Crespi, M., 2017. CORRECTION: Enod40, a short open reading frame-containing mRNA, induces cytoplasmic localization of a nuclear RNA binding protein in *Medicago truncatula*. Plant Cell. https://doi.org/10.1105/tpc.17.00245.

Ding, J., Lu, Q., Ouyang, Y., Mao, H., Zhang, P., Yao, J., Xu, C., et al., 2012. A long noncoding RNA regulates photoperiod-sensitive male sterility, an essential component of hybrid rice. Proc. Natl. Acad. Sci. 109, 2654–2659.

Efroni, I., Birnbaum, K.D., 2016. The potential of single-cell profiling in plants. Genome Biol. 17, 65.

Fedak, H., Palusinska, M., Krzyczmonik, K., Brzezniak, L., Yatusevich, R., Pietras, Z., Kaczanowski, S., et al., 2016. Control of seed dormancy in Arabidopsis by a cis-acting noncoding antisense transcript. Proc. Natl. Acad. Sci. 113, E7846–E7855.

Fedorenko, O., Topchieva, L., Zaretskaya, M., Lebedeva, O., 2019. Changes in FLC and VIN3 expression during vernalization of *Arabidopsis thaliana* plants from northern natural populations. Russ. J. Genet. 55, 865–871.

Flórez-Zapata, N.M., Reyes-Valdés, M.H., Martínez, O., 2016. Long non-coding RNAs are major contributors to transcriptome changes in sunflower meiocytes with different recombination rates. BMC Genomics 17, 490.

Franco-Zorrilla, J.M., Valli, A., Todesco, M., Mateos, I., Puga, M.I., Rubio-Somoza, I., Leyva, A., et al., 2007. Target mimicry provides a new mechanism for regulation of microRNA activity. Nat. Genet. 39, 1033.

Golicz, A.A., Singh, M.B., Bhalla, P.L., 2018. The long intergenic noncoding RNA (LincRNA) landscape of the soybean genome. Plant Physiol. 176, 2133–2147.

Greb, T., Mylne, J.S., Crevillen, P., Geraldo, N., An, H., Gendall, A.R., Dean, C., 2007. The PHD finger protein VRN5 functions in the epigenetic silencing of Arabidopsis FLC. Curr. Biol. 17, 73–78.

Harvey, J.J., Lewsey, M.G., Patel, K., Westwood, J., Heimstädt, S., Carr, J.P., Baulcombe, D.C., 2011. An antiviral defense role of AGO2 in plants. PLoS One 6, e14639.

Held, M.A., Penning, B., Brandt, A.S., Kessans, S.A., Yong, W., Scofield, S.R., Carpita, N.C., 2008. Small-interfering RNAs from natural antisense transcripts derived from a cellulose synthase gene modulate cell wall biosynthesis in barley. Proc. Natl. Acad. Sci. 105, 20534–20539.

Henriques, R., Wang, H., Liu, J., Boix, M., Huang, L.F., Chua, N.H., 2017. The antiphasic regulatory module comprising CDF5 and its antisense RNA FLORE links the circadian clock to photoperiodic flowering. New Phytol. 216, 854–867.

Heo, J.B., Sung, S., 2011. Vernalization-mediated epigenetic silencing by a long intronic noncoding RNA. Science 331, 76–79.

Hornyik, C., Terzi, L.C., Simpson, G.G., 2010. The spen family protein FPA controls alternative cleavage and polyadenylation of RNA. Dev. Cell 18, 203–213.

Ietswaart, R., Wu, Z., Dean, C., 2012. Flowering time control: another window to the connection between antisense RNA and chromatin. Trends Genet. 28, 445–453.

Jabnoune, M., Secco, D., Lecampion, C., Robaglia, C., Shu, Q., Poirier, Y., 2013. A rice cis-natural antisense RNA acts as a translational enhancer for its cognate mRNA and contributes to phosphate homeostasis and plant fitness. Plant Cell 25, 4166–4182.

Jiao, Y., Lau, O.S., Deng, X.W., 2007. Light-regulated transcriptional networks in higher plants. Nat. Rev. Genet. 8, 217.

Kim, D.-H., Sung, S., 2017. Vernalization-triggered intragenic chromatin loop formation by long noncoding RNAs. Dev. Cell 40, 302–312. e4.

Kim, D.-H., Doyle, M.R., Sung, S., Amasino, R.M., 2009. Vernalization: winter and the timing of flowering in plants. Annu. Rev. Cell Dev. 25, 277–299.

Kim, D.-H., Xi, Y., Sung, S., 2017a. Modular function of long noncoding RNA, COLDAIR, in the vernalization response. PLoS Genet. 13, e1006939.

Kim, E.-D., Xiong, Y., Pyo, Y., Kim, D.-H., Kang, B.-H., Sung, S., 2017b. Spatio-temporal analysis of coding and long noncoding transcripts during maize endosperm development. Sci. Rep. 7, 3838.

Kouchi, H., Takane, K.i., So, R.B., Ladha, J.K., Reddy, P.M., 1999. Rice ENOD40: isolation and expression analysis in rice and transgenic soybean root nodules. Plant J. 18, 121–129.

Lee, C., Kikyo, N., 2012. Strategies to identify long noncoding RNAs involved in gene regulation. Cell Biosci. 2, 37.

Levin, J.Z., Yassour, M., Adiconis, X., Nusbaum, C., Thompson, D.A., Friedman, N., Gnirke, A., et al., 2010. Comprehensive comparative analysis of strand-specific RNA sequencing methods. Nat. Methods 7, 709.

Li, L., Eichten, S.R., Shimizu, R., Petsch, K., Yeh, C.-T., Wu, W., Chettoor, A.M., et al., 2014. Genome-wide discovery and characterization of maize long non-coding RNAs. Genome Biol. 15, R40.

Liu, C., Muchhal, U.S., Raghothama, K., 1997. Differential expression of TPS11, a phosphate starvation-induced gene in tomato. Plant Mol. Biol. 33, 867–874.

Liu, F., Marquardt, S., Lister, C., Swiezewski, S., Dean, C., 2010. Targeted 3′ processing of antisense transcripts triggers Arabidopsis FLC chromatin silencing. Science 327, 94–97.

Liu, J., Jung, C., Xu, J., Wang, H., Deng, S., Bernad, L., Arenas-Huertero, C., et al., 2012. Genome-wide analysis uncovers regulation of long intergenic noncoding RNAs in Arabidopsis. Plant Cell 24, 4333–4345.

Liu, X., Hao, L., Li, D., Zhu, L., Hu, S., 2015. Long non-coding RNAs and their biological roles in plants. Genom. Proteom. Bioinform. 13, 137–147.

Liu, W., Cheng, C., Lin, Y., XuHan, X., Lai, Z., 2018. Genome-wide identification and characterization of mRNAs and lncRNAs involved in cold stress in the wild banana (*Musa itinerans*). PLoS One 13, e0200002.

Lu, X., Chen, X., Mu, M., Wang, J., Wang, X., Wang, D., Yin, Z., et al., 2016. Genome-wide analysis of long noncoding RNAs and their responses to drought stress in cotton (*Gossypium hirsutum* L.). PLoS One 11, e0156723.

Lu, Z., Xia, X., Jiang, B., Ma, K., Zhu, L., Wang, L., Jin, B., 2017. Identification and characterization of novel lncRNAs in *Arabidopsis thaliana*. Biochem. Biophys. Res. 488, 348–354.

Luo, C., Sidote, D.J., Zhang, Y., Kerstetter, R.A., Michael, T.P., Lam, E., 2013. Integrative analysis of chromatin states in Arabidopsis identified potential regulatory mechanisms for natural antisense transcript production. Plant J. 73, 77–90.

Ma, L., Bajic, V.B., Zhang, Z., 2013. On the classification of long non-coding RNAs. RNA Biol. 10, 924–933.

Mach, J., 2017. The long-noncoding RNA ELENA1 functions in plant immunity. Plant Cell 29, 916.

Michaels, S.D., Amasino, R.M., 1999. FLOWERING LOCUS C encodes a novel MADS domain protein that acts as a repressor of flowering. Plant Cell 11, 949–956.

Mockler, T.C., Ecker, J.R., 2005. Applications of DNA tiling arrays for whole-genome analysis. Genomics 85, 1–15.

Mongelli, A., Gaetano, C., Farsetti, A., Martelli, F., 2019. The dark that matters: long non-coding RNAs as master regulators of cellular metabolism in non-communicable diseases. J. Front. Physiol. 10, 369.

Musselman, C.A., Kutateladze, T.G., 2011. Handpicking epigenetic marks with PHD fingers. Nucleic Acids Res. 39, 9061–9071.

Narnoliya, L.K., Kaushal, G., Singh, S.P., Sangwan, R.S., 2017. De novo transcriptome analysis of rose-scented geranium provides insights into the metabolic specificity of terpene and tartaric acid biosynthesis. BMC Genomics 18, 74.

Narnoliya, L.K., Kaushal, G., Singh, S.P., 2019. Long non-coding RNAs and miRNAs regulating terpene and tartaric acid biosynthesis in rose-scented geranium. FEBS Lett. 593, 2235–2249.

Nejat, N., Mantri, N., 2018. Emerging roles of long non-coding RNAs in plant response to biotic and abiotic stresses. Crit. Rev. Biotechnol. 38, 93–105.

Prasad, A., Sharma, N., Muthamilarasan, M., Rana, S., Prasad, M., 2019. Recent advances in small RNA mediated plant-virus interactions. Crit. Rev. Biotechnol. 39, 587–601.

Qi, X., Xie, S., Liu, Y., Yi, F., Yu, J., 2013. Genome-wide annotation of genes and noncoding RNAs of foxtail millet in response to simulated drought stress by deep sequencing. Plant Mol. Biol. 83, 459–473.

Qin, T., Zhao, H., Cui, P., Albesher, N., Xiong, L., 2017. A nucleus-localized long non-coding RNA enhances drought and salt stress tolerance. Plant Physiol. 175, 1321–1336.

Rai, M.I., Alam, M., Lightfoot, D.A., Gurha, P., Afzal, A.J., 2019. Classification and experimental identification of plant long non-coding RNAs. Genomics 111, 997–1005.

Ramirez-Prado, J.S., Abulfaraj, A.A., Rayapuram, N., Benhamed, M., Hirt, H., 2018. Plant immunity: from signaling to epigenetic control of defense. Trends Plant Sci. 23, 833–844.

Rinn, J.L., Kertesz, M., Wang, J.K., Squazzo, S.L., Xu, X., Brugmann, S.A., Goodnough, L.H., et al., 2007. Functional demarcation of active and silent chromatin domains in human HOX loci by noncoding RNAs. Cell 129, 1311–1323.

Seo, J.S., Sun, H.-X., Park, B.S., Huang, C.-H., Yeh, S.-D., Jung, C., Chua, N.-H., 2017. ELF18-INDUCED LONG-NONCODING RNA associates with mediator to enhance expression of innate immune response genes in Arabidopsis. Plant Cell 29, 1024–1038.

Shafiq, S., Li, J., Sun, Q., 2016. Functions of plants long non-coding RNAs. Biochim. Biophys. Acta Gene Regul. Mech. 1859, 155–162.

Sheldon, C.C., Rouse, D.T., Finnegan, E.J., Peacock, W.J., Dennis, E.S., 2000. The molecular basis of vernalization: the central role of FLOWERING LOCUS C (FLC). Proc. Natl. Acad. Sci. 97, 3753–3758.

Shuai, P., Liang, D., Tang, S., Zhang, Z., Ye, C.-Y., Su, Y., Xia, X., et al., 2014. Genome-wide identification and functional prediction of novel and drought-responsive lincRNAs in *Populus trichocarpa*. J. Exp. Bot. 65, 4975–4983.

Shumayla, Sharma, S., Taneja, M., Tyagi, S., Singh, K., Upadhyay, S.K., 2017. Survey of high throughput RNA-seq data reveals potential roles for lncRNAs during development and stress response in bread wheat. Front. Plant Sci. 8, 1019.

Song, J., Angel, A., Howard, M., Dean, C., 2012. Vernalization—a cold-induced epigenetic switch. J. Cell Sci. 125, 3723–3731.

Sousa, C., Johansson, C., Charon, C., Manyani, H., Sautter, C., Kondorosi, A., Crespi, M., 2001. Translational and structural requirements of the early nodulin gene enod40, a short-open reading frame-containing RNA, for elicitation of a cell-specific growth response in the alfalfa root cortex. Mol. Cell. Biol. 21, 354–366.

Srikanth, A., Schmid, M., 2011. Regulation of flowering time: all roads lead to Rome. Cell Mol. Life Sci. 68, 2013–2037.

Sung, S., Schmitz, R.J., Amasino, R.M., 2006. A PHD finger protein involved in both the vernalization and photoperiod pathways in Arabidopsis. Genes Dev. 20, 3244–3248.

Swiezewski, S., Liu, F., Magusin, A., Dean, C., 2009. Cold-induced silencing by long antisense transcripts of an Arabidopsis Polycomb target. Nature 462, 799.

Tian, Y., Zheng, H., Zhang, F., Wang, S., Ji, X., Xu, C., He, Y., et al., 2019. PRC2 recruitment and H3K27me3 deposition at FLC require FCA binding of COOLAIR. Sci. Adv. 5, eaau7246.

Tyagi, S., Sharma, A., Upadhyay, S.K., 2018. Role of Next-Generation RNA-Seq Data in Discovery and Characterization of Long Non-Coding RNA in Plants. In: Next Generation Plant Breeding, Yelda Özden Çiftçi, IntechOpen. https://doi.org/10.5772/intechopen.72773.

Wang, Z., Gerstein, M., Snyder, M., 2009. RNA-Seq: a revolutionary tool for transcriptomics. Nat. Rev. Genet. 10, 57.

Wang, L., Zhao, S., Gu, C., Zhou, Y., Zhou, H., Ma, J., Cheng, J., et al., 2013. Deep RNA-Seq uncovers the peach transcriptome landscape. Plant Mol. Biol. 83, 365–377.

Wang, Y., Fan, X., Lin, F., He, G., Terzaghi, W., Zhu, D., Deng, X.W., 2014. Arabidopsis noncoding RNA mediates control of photomorphogenesis by red light. Proc. Natl. Acad. Sci. 111, 10359–10364.

Wang, H., Niu, Q.W., Wu, H.W., Liu, J., Ye, J., Yu, N., Chua, N.H., 2015a. Analysis of non-coding transcriptome in rice and maize uncovers roles of conserved lnc RNA s associated with agriculture traits. Plant J. 84, 404–416.

Wang, J., Yu, W., Yang, Y., Li, X., Chen, T., Liu, T., Ma, N., et al., 2015b. Genome-wide analysis of tomato long non-coding RNAs and identification as endogenous target mimic for microRNA in response to TYLCV infection. Sci. Rep. 5, 16946.

Wang, T.-Z., Liu, M., Zhao, M.-G., Chen, R., Zhang, W.-H., 2015c. Identification and characterization of long non-coding RNAs involved in osmotic and salt stress in *Medicago truncatula* using genome-wide high-throughput sequencing. BMC Plant Biol. 15, 131.

Wang, C.-Y., Liu, S.-R., Zhang, X.-Y., Ma, Y.-J., Hu, C.-G., Zhang, J.-Z., 2017a. Genome-wide screening and characterization of long non-coding RNAs involved in flowering development of trifoliate orange (*Poncirus trifoliata* L. Raf.). Sci. Rep. 7, 43226.

Wang, J., Meng, X., Dobrovolskaya, O.B., Orlov, Y.L., Chen, M., 2017b. Non-coding RNAs and their roles in stress response in plants. Genom. Proteom. Bioinform. 15, 301–312.

Wang, T., Zhao, M., Zhang, X., Liu, M., Yang, C., Chen, Y., Chen, R., et al., 2017c. Novel phosphate deficiency-responsive long non-coding RNAs in the legume model plant *Medicago truncatula*. J. Exp. Bot. 68, 5937–5948.

Wang, Z., Liu, Y., Li, L., Li, D., Zhang, Q., Guo, Y., Wang, S., et al., 2017d. Whole transcriptome sequencing of *Pseudomonas syringae* pv. actinidiae-infected kiwifruit plants reveals species-specific interaction between long non-coding RNA and coding genes. Sci. Rep. 7, 4910.

Wang, P., Dai, L., Ai, J., Wang, Y., Ren, F., 2019. Identification and functional prediction of cold-related long non-coding RNA (lncRNA) in grapevine. Sci. Rep. 9, 6638.

Wasaki, J., Yonetani, R., Shinano, T., Kai, M., Osaki, M., 2003. Expression of the OsPI1 gene, cloned from rice roots using cDNA microarray, rapidly responds to phosphorus status. New Phytol. 158, 239–248.

Wunderlich, M., Groß-Hardt, R., Schöffl, F., 2014. Heat shock factor HSFB2a involved in gametophyte development of *Arabidopsis thaliana* and its expression is controlled by a heat-inducible long non-coding antisense RNA. Plant Mol. Biol. 85, 541–550.

Xin, M., Wang, Y., Yao, Y., Song, N., Hu, Z., Qin, D., Xie, C., et al., 2011. Identification and characterization of wheat long non-protein coding RNAs responsive to powdery mildew infection and heat stress by using microarray analysis and SBS sequencing. BMC Plant Biol. 11, 61.

Yamaguchi, A., Abe, M., 2012. Regulation of reproductive development by non-coding RNA in *Arabidopsis*: to flower or not to flower. J. Plant Res. 125, 693–704.

Yang, W.C., Katinakis, P., Hendriks, P., Smolders, A., de Vries, F., Spee, J., van Kammen, A., et al., 1993. Characterization of GmENOD40, a gene showing novel patterns of cell-specific expression during soybean nodule development. Plant J. 3, 573–585.

Yao, Y., Ni, Z., Peng, H., Sun, F., Xin, M., Sunkar, R., Zhu, J.-K., et al., 2010. Non-coding small RNAs responsive to abiotic stress in wheat (*Triticum aestivum* L.). Funct. Integr. Genom. 10, 187–190.

Yotsukura, S., duVerle, D., Hancock, T., Natsume-Kitatani, Y., Mamitsuka, H., 2016. Computational recognition for long non-coding RNA (lncRNA): software and databases. Brief. Bioinform. 18, 9–27.

Yu, X., Yang, J., Li, X., Liu, X., Sun, C., Wu, F., He, Y., 2013. Global analysis of cis-natural antisense transcripts and their heat-responsive nat-siRNAs in *Brassica rapa*. BMC Plant Biol. 13, 208.

Zhang, W., Han, Z., Guo, Q., Liu, Y., Zheng, Y., Wu, F., Jin, W., 2014a. Identification of maize long non-coding RNAs responsive to drought stress. PLoS One 9, e98958.

Zhang, Y.-C., Liao, J.-Y., Li, Z.-Y., Yu, Y., Zhang, J.-P., Li, Q.-F., Qu, L.-H., et al., 2014b. Genome-wide screening and functional analysis identify a large number of long noncoding RNAs involved in the sexual reproduction of rice. Genome Biol. 15, 512.

Zhang, G., Duan, A., Zhang, J., He, C.J.G., 2017. Genome-wide analysis of long non-coding RNAs at the mature stage of sea buckthorn (*Hippophae rhamnoides* Linn) fruit. Gene 596, 130–136.

Zhao, J., He, Q., Chen, G., Wang, L., Jin, B., 2016. Regulation of non-coding RNAs in heat stress responses of plants. Front. Plant Sci. **7**, 1213.

Zhu, B., Yang, Y., Li, R., Fu, D., Wen, L., Luo, Y., Zhu, H., 2015. RNA sequencing and functional analysis implicate the regulatory role of long non-coding RNAs in tomato fruit ripening. J. Exp. Bot. 66, 4483–4495.

Zhu, Y., Chen, L., Zhang, C., Hao, P., Jing, X., Li, X., 2017. Global transcriptome analysis reveals extensive gene remodeling, alternative splicing and differential transcription profiles in non-seed vascular plant *Selaginella moellendorffii*. BMC Genomics 18, 1042.

Zubko, E., Meyer, P., 2007. A natural antisense transcript of the *Petunia hybrida* Sho gene suggests a role for an antisense mechanism in cytokinin regulation. Plant J. 52, 1131–1139.

Chapter 21

Applications of lncRNAs in plant improvement

Amandeep Kaur, Madhu, and Santosh Kumar Upadhyay
Department of Botany, Panjab University, Chandigarh, India

21.1 Introduction

At present, the world's population stands at 7.7 billion and estimated to reach over 9 billion by 2050 (Roser, 2013). There is a great demand for a balanced and healthy diet to feed the world's increasing population. Plants are the major source of food for both humans and animals. They also serve as raw materials for numerous industrial products like textile, furniture, bakery, beverages, pharmaceuticals, etc. These demands can be fulfilled by improving the yield and productivity of plants. However, major factors that limit the productivity of plants are nutrient deficiencies and biotic and abiotic stress conditions which suppress the growth and development of plants. The incorporation of resistance traits is an efficient way to stabilize and improve plant production under stress conditions. The traditional plant breeding practices have developed cultivars with high vigor, but these were very time-consuming and arduous. Moreover, the appearance of some undesired traits along with the desired ones and inbreeding depression due to homozygosity are the major drawbacks of this practice.

In this era of advanced technologies, genetic engineering makes these breeding practices faster and more accurate as compared to traditional plant breeding. Tremendous work has been done in this field to improve plant varieties in terms of stress tolerance. By using biotechnological approaches, scientists create transgenics of economically important plants by altering the genetic constitution. These transgenic plants have a better regulating response to various stresses. Several factors are responsible for constraining the research in plants at the molecular level like the multifunctional nature of genes, multifaceted defense signaling, etc. Various stress-resistant genes have been identified and characterized which diminished the negative impact of stresses on the growth and reproduction of plants. Recent advances in sequencing technology show that the noncoding part of the genome has a great role in the gene functioning. Noncoding RNAs (ncRNAs) are special transcripts with either very low or without any coding potential. These ncRNAs regulate the gene expression at three levels: epigenetic, transcriptional,

Long Noncoding RNAs in Plants: Roles in development and stress
https://doi.org/10.1016/B978-0-12-821452-7.00021-0

and posttranscriptional levels (Kaikkonen et al., 2011). Based on their function, ncRNAs are classified into two categories: housekeeping and regulatory ncRNAs. Housekeeping ncRNAs are constitutively expressed in cells such as ribosomal RNAs (rRNAs), small nuclear RNAs (snRNAs), and transfer RNAs (tRNAs) ((Liu et al., 2011). However, regulatory ncRNAs are expressed in specific conditions, for example, during cell differentiation or in response to external stimuli in specific cell types. Regulatory ncRNAs are composed of short noncoding RNAs and long noncoding RNAs. Short noncoding RNAs have the length of fewer than 200 nucleotides such as microRNAs (miRNAs), Piwi-interacting RNAs (piRNAs), enhancer RNAs (eRNAs), promoter-associated RNAs (PARs), and small interfering RNAs (siRNAs) (Zhang et al., 2019a). The nucleotide length of long noncoding RNAs (lncRNAs) is more than 200 bp and its open reading frame length is less than 100 amino acids. Based on genomic locations, lncRNAs are classified into six types: (1) Long intergenic ncRNAs (lincRNAs) are transcribed from a region between two genes; (2) Intronic ncRNAs (lincRNAs) derived from introns; (3) Sense lncRNAs resulting from the overlapping region of exons on sense strand; (4) Antisense lncRNAs produced from the overlapping region of exons on the antisense strand; (5) Natural antisense transcripts (NATs), which are transcribed from the opposite DNA strand of the related protein-coding genes; (6) Bidirectional lncRNAs located on the antisense strand in the promoter region of protein-coding genes, within the distance less than 1000 bp from the transcription start site (Ponting et al., 2009). LncRNAs can be located either in the nucleus or cytoplasm (Karlik et al., 2019). Generally, lncRNAs are transcribed by RNA polymerase II, but in plants, they need plant-specific RNA polymerases IV and V which evolved from RNA polymerase II (Xin et al., 2011).

LncRNAs may be polyadenylated and nonpolyadenylated (Yang et al., 2011; Di et al., 2014). The nonpolyadenylated lncRNAs are shorter with 50–300 nucleotides in length and have low expression, but during stress conditions, they respond more specifically in comparison to polyadenylated lncRNAs (Yang et al., 2011; Di et al., 2014). Furthermore, nonpolyadenylated lncRNAs (poly(A)-) need RNA polymerase III for transcription (Derrien et al., 2012; Di et al., 2014). LncRNAs share similarities with mRNA in many aspects like splicing, polyadenylation, structure configuration, developmental stage, and tissue- and cell-specific expression (Ward et al., 2015; Hirsch et al., 2006). However, lncRNAs have lower expression, as compared to mRNAs (Derrien et al., 2012; Wang et al., 2015b).

Extensive research has been done on lncRNAs in the case of animals, but the studies are limited in plant species. *Early nodulin 40* (*Endo40*) was the first discovered lncRNA in *Glycine max*, having a role in nodulation (Kouchi and Hata, 1993). As per the available findings, lncRNAs have a role in both plant growth and developmental processes and stress responses (Shumayla et al., 2017; Tyagi et al., 2018). For instance, they regulate photomorphogenesis, auxin-controlled development, reproductive processes, root formation, biotic and abiotic stress response, etc. These lncRNAs can be utilized as a potential genetic source for the betterment of plant growth and yield. In this chapter, we have focused on numerous applications of lncRNAs and their diverse mechanisms of gene regulation, which can be targeted for plant improvement.

21.2 Applications of lncRNAs for abiotic stress tolerance

In plants, abiotic stresses like heat, drought, cold, salinity, high light intensity, etc. are the main factors affecting their morphology and physiology. The stresses lead to several molecular and biochemical changes, which limit the plant growth, yield, and also affect the quality of plant-derived products. To reduce the damage caused by abiotic stresses such as the reduction in photosynthesis and respiration, disruption of membrane integrity, wilting, etc., plants have developed complex adaptive mechanisms to reprogram the expression of genes at the transcriptional, posttranscriptional, and posttranslational levels (Kaikkonen et al., 2011). Researchers uncovered the valuable role of lncRNAs in response to abiotic stress conditions through various approaches, which can be used for the development of abiotic stress-tolerant cultivars.

21.2.1 Drought stress-responsive lncRNAs

Plants reduce the harmful impacts of drought stress on their growth, development, and reproduction by changing their metabolism and physiology (Farooq et al., 2009). Drought stress causes cellular dehydration and directs the activation of the biosynthesis and accumulation of the stress hormone abscisic acid (ABA) (Xiong and Zhu, 2003). Based on expression profiling, several drought stress-responsive lncRNAs were identified in some plants like maize, foxtail millet, alfalfa, populus, etc. In maize and populus, 664 and 504 drought-responsive lncRNAs were identified, respectively (Zhang et al., 2014; Shuai et al., 2014). Differential expression analysis illustrated 19 lncRNAs in foxtail millet, of which 17 were lincRNAs and two NATs after polyethylene glycol (PEG)-induced osmotic stress (Qi et al., 2013). In *Medicago truncatula*, osmotic stress-responsive 7863 and 5561 lncRNAs were recognized from leaf and root samples, respectively (Wang et al., 2015b). The experimental evidence in Arabidopsis indicated the positive impact of *DROUGHT-INDUCED lncRNA (DRIR)* on the plants during drought and salt stress conditions (Qin et al., 2017). The $drir^D$ (a T-DNA insertion mutant) and *DRIR* overexpressing Arabidopsis lines were compared with wild-type under drought and salt stress conditions. This indicates the enhanced tolerance against drought and salt stresses in both $drir^D$ and overexpressing *DRIR* in Arabidopsis. When favorable conditions prevailed, the expression of *DRIR* decreased, but it showed higher expression during stress conditions. The transcriptome analysis of mutant and overexpressing lines revealed the altered expression of a large set of genes that might be involved in stress-responsive processes like ABA signaling, stomatal movements, water transport, seedling growth, etc. in comparison to the wild type (Qin et al., 2017). The study suggests the regulatory role of *DRIR*, which can be further explored in other agronomically important plant species encountering the drought and salt stresses.

21.2.2 Heat stress-responsive lncRNAs

At the molecular level, lncRNAs regulate the expression of genes that are involved in heat stress tolerance. Several heat-responsive lncRNAs have been identified in numerous plant species based on expression profiling (Shumayla et al., 2017; Wang et al., 2019a). Using the computational approaches, 4594 heat-responsive lncRNAs were identified in nonheading Chinese cabbage (NHCC) at the genomic level (Wang et al., 2019a). Further, the co-expression based interaction among the lncRNAs, miRNAs, and mRNAs indicates the involvement of various phytohormones such as salicylic acid and brassinosteroid in heat stress tolerance. Thirty-nine lncRNAs in NHCC act as an endogenous target mimic for miRNA, out of which five were confirmed to be involved in heat stress tolerance. In the case of wheat, a large number of heat-responsive lncRNAs were characterized using microarray and sequencing-by-synthesis techniques (Xin et al., 2011). The upregulated expression of *TahlnRNA27* and *TalnRNA5* was also confirmed through qRT PCR and suggested their role in heat stress response in TAM107 and Chinese Spring genotypes (Xin et al., 2011). With the help of high-throughput RNA-sequencing and microarray techniques, we can forecast the existence of lncRNAs under heat stress in other plant species. There is a wide scope of experimental validation of the biological roles of lncRNAs, which can provide deep information about their functional novelty in plants. Additionally, the detailed study of interacting partners and miRNA targets of lncRNAs would be helpful in revealing the stress-responsive gene-regulatory mechanisms. We can positively modulate these lncRNAs for better performance of plant varieties under stress conditions.

21.2.3 Cold stress-responsive lncRNAs

Cold stress can reduce plant productivity to a greater extent. The indications of cold stress are poor germination, stunted seedlings, yellowing of leaves, reduced leaf expansion, wilting, and necrosis (Yadav, 2010). Cold stress also obstructs the reproductive development of plants (Thakur et al., 2010). The cold stress response is sensed by plants via a signal transduction which leads to the activation of transcription factors and cold-responsive genes (Miura and Furumoto, 2013). Such transcription factors and genes are mostly regulated by lncRNAs which help in providing tolerance to plants during cold stress. For example, in Arabidopsis, *Cold-induced long antisense intragenic RNAs (COOLAIR)* and *Cold-assisted intronic noncoding RNA (COLDAIR)* are two lncRNAs that regulate the expression of *FLC* gene. This gene acts as a repressor of flowering during cold temperature (Heo and Sung, 2011). During cold stress, a large number of lncRNAs were identified in some crops such as *Brassica rapa*, *Vitis vinifera* L., and *Manihot esculenta*, which shows their involvement in stress tolerance (Song et al., 2016; Wang et al., 2019b; Li et al., 2017).

21.2.4 LncRNAs in response to high light

Light is an important factor for plant development and growth. Various light-responsive lncRNAs have been discovered in different plants. The transcription of these lncRNAs occurs in the same or opposite direction with their sense transcripts (Wang et al., 2014a). These lncRNAs gave a response to light in the spatiotemporal manner. In Arabidopsis, 37,238 sense-antisense transcript pairs have been found in response to high light (Wang et al., 2014a). These light-responsive lncNAT expression has been validated by qRT PCR. The regulation of about half of light-responsive NAT pairs was associated with histone modifications such as histone acetylation (Wang et al., 2014a; Luo et al., 2013). In the future, high light-responsive lncRNAs in other plant species can also be identified which can pave a way to combat this stress.

21.2.5 Salt-stress-responsive lncRNAs

Like other abiotic stresses, salt stress also has an adverse effect on plant growth and development, resulting in poor quality and low yield and even plant death. Several studies have reported the role of lncRNAs in salt stress tolerance. For instance, some lncRNAs were found to be differentially expressed in roots and leaves of *Arabidopsis thaliana* ecotype Columbia (Col-0) under salt stress (Ben Amor et al., 2009). In *M. truncatula*, 7361 and 7874 lncRNAs were identified in salt-treated leaf and root samples, respectively (Wang et al., 2015b). Moreover, in *Gossypium hirsutum*, 44 lincRNAs were found to be differentially expressed under salt stress (Deng et al., 2018). For instance, *lncRNA973* was involved in regulation of salt stress-responsive gene's expression in cotton (Zhang et al., 2019b). According to earlier studies, a total of 4199 and 6048 lncRNAs were identified that responded to salt stress in *Populus euphratica* and *Populus alba* var. *pyramidalis*, respectively (Wan et al., 2019). Various lncRNAs showed differential expression which suggested their involvement in salt stress response.

21.3 Applications of lncRNAs in biotic stress tolerance

Plant response to biotic stresses such as bacterial, fungal, viral, and pest infections has various defense mechanisms to minimize the damage. LncRNAs are also known to be involved in defense responses including biotic stress. Hence, this can be a key player in improving the crops under such stresses.

For instance, powdery mildew is a common disease caused by the fungus *Blumeria graminis* f. sp. *tritici* (*Bgt*) in wheat. It is one of the most destructive diseases of wheat resulting in significant yield losses (Griffey et al., 1993). A total of 48 long nonprotein-coding RNAs (lnpcRNAs) were reported to be specifically responsive to powdery mildew infection (Xin et al., 2011). Moreover, *TapmlnRNA19* and *TapmlnRNA30* were found to be differentially expressed in response to powdery mildew infection (Xin et al., 2011). In another study, a total of 283 fungal-responsive lncRNAs were identified in *Triticum aestivum* (Zhang et al., 2016). Further, it was observed that 254 and 52 lincRNAs were expressed in response to *Blumeria graminis* f. sp. *tritici* (*Bgt*) and *Puccinia striiformis* f. sp. *Tritici* (*Pst*), respectively (Zhang et al., 2016). In another case, *Sclerotinia sclerotiorum* is a fungal pathogen affecting the global yield of *Brassica napus* (rapeseed). Rapeseed is known to be an important oil crop. *S. sclerotiorum* infects rapeseed to cause a severe fungal disease, Sclerotinia stem rot (Zhao et al., 2016). A total of 3181 lncRNAs have been found in *B. napus* which showed a dynamic response to the *S. sclerotiorum* infection (Joshi et al., 2016).

Phytophthora infestans is another fungus affecting the tomato crop by causing the late blight disease. Cui et al. (2017) reported that *lncRNA16397* decreases reactive oxygen species (ROS) accumulation by inducing *SlGRX*, suggesting its involvement in defense mechanism. Similarly, several lncRNAs are involved in defense-responsive mechanisms against fungus including *Fusarium oxysporum* and *Verticillium dahliae* in *A. thaliana* and *Gossypium* spp., respectively (Zhu et al., 2014; Zhang et al., 2018).

There have been several studies suggesting the roles of lncRNAs against bacterial infections. A total of 1113 lincRNAs are involved in potato defense responses against bacteria, *Pectobacterium carotovorum* subsp. *brasiliense* (Kwenda et al., 2016). In the year 2015 (a), Wang et al. suggested the roles of tomato lncRNAs in *Tomato yellow leaf curl virus* (TYLCV) resistance. Further, TYLCV-resistant tomato line CLN2777A demonstrated the expression of several lncRNAs in response to TYLCV. Moreover, lincRNAs including *slylnc0048*, *slylnc0049*, *slylnc0483*, *slylnc0531*, and *slyinc0934* were found to be upregulated, while some lincRNAs including *slylnc0475*, *slylnc0476*, *slylnc0673*, and *slylnc1052* were observed to be downregulated after TYLCV infection (Wang et al., 2015a). Increasing research in the area of lncRNAs is opening new avenues to exploit the beneficial roles of the lncRNAs for crop improvement programs. Biotic stresses have been one of the biggest threats for the global plant cultivation. Advancements in crop sciences, plant sciences, and bioinformatics collectively give an intriguing idea about the application of lncRNAs in the development of stress-tolerant plants.

21.4 Applications of lncRNAs in nutrient homeostasis

Nutrients including macronutrients and micronutrients are an essential part of the growth and development of plants. Macronutrients including nitrogen (N) and phosphorus (P) are primary elements required for sustainable growth and yield of plants. Micronutrients including magnesium (Mg), iron (Fe), Zinc (Zn), etc. are required in a lesser amount. The involvement of several lncRNAs has been reported to be involved in nitrogen and phosphate homeostasis (Lv et al., 2016; Chiou et al., 2006; Bari et al., 2006). Furthermore, the expression of the target gene, *PHOSPHATE2* (*PHO2*), is repressed by miR399 (Chiou et al., 2006). *PHO2* encodes a ubiquitin-conjugating E2 enzyme that can collaborate with *PHOSPHATE1 (PHO1)*, which further helps in phosphate loading in xylem, and hence, regulates phosphate homeostasis (Chiou et al., 2006; Bari et al., 2006). In Arabidopsis, lncRNA *IPS1* works as a target mimic for miR399 during phosphate deficiency (Franco-Zorrilla et al., 2007). Moreover, other lncRNAs such as *TOMATO PHOSPHATE STARVATION-INDUCED 1* (*TPSI1*) and *Mt4* genes were identified in tomato (Liu et al., 1997) and medicago (Burleigh and Harrison, 1998), respectively, which comprise a region of complementarity with miR399. In rice, the *cis*-natural antisense RNA, $cis\text{-}NAT_{PHO1;2}$, enhances the expression of its sense gene, *PHOSPHATE1;2* (Jabnoune et al., 2013), which is a functional ortholog of *PHO1* in *Arabidopsis* (Secco et al., 2010).

In *Zea mays*, 6211 intergenic and 1034 intronic lncRNAs have been identified under N stress; however, only 637 lncRNAs have shown differential expression (Lv et al., 2016). In another study, during nitrogen deficiency, 126 lncRNAs of *Populus tomentosa* were found to be expressed differentially. Of these 126 lncRNAs, 118 were upregulated and 8 were downregulated during N stress (Chen et al., 2016). Nitrogen and phosphorous, apart from being the nutrients, play important roles in various physiological processes of plants. Therefore, the exploration of lncRNAs is involved in combating the deficiency of nutrients. Although there are fewer studies in this area, the knowledge about these lncRNAs will lead us to the applied approaches.

21.5 Application of lncRNAs in developmental processes

Along with their roles in combating the various stress conditions, lncRNAs are also found to be involved in various developmental processes (Kim and Sung, 2012). They are important components of plant transcriptome, which are entailed in the regulation of various biological processes. These biological processes include reproductive processes, photomorphogenesis, nodule development, auxin regulation, vernalization, fertility, fiber development, etc. (Yu and Zhu, 2019).

21.5.1 LncRNAs in the regulation of photomorphogenesis

Photomorphogenesis is an important phenomenon in plants that depend on light. The complicated regulatory processes of photomorphogenesis have been meticulously explicated. In Arabidopsis, the key regulator involved in this pathway is *Phytochrome-interacting factor 3* (*PIF3*) (Jiao et al., 2007). *PIF3* encodes a transcription factor that is a member of the basic helix-loop-helix family (bHLH) protein and acts as a repressor of photomorphogenesis in *Arabidopsis* (Pham et al., 2018). Furthermore, Deng et al. identified lncRNAs *HID1* (HIDDEN *TREASURE 1*) in Arabidopsis, which plays an important role in seedling photomorphogenesis by modulating *PIF3* expression (Wang et al., 2014b). Further, *HID1*-mediated photomorphogenesis response is changed in red light. In Arabidopsis, it has been found that *HID1* is a positive regulator of photomorphogenesis and represses the expression of *PIF3* (Wang et al., 2014b). In this way, *HID1* lncRNA promotes plant growth by modulating the expression of *PIF3*.

Moreover, Wang et al. (2014a) identified 626 concordant and 766 discordant NAT pairs in *A. thaliana* in response to light. The limited knowledge of lncRNAs in photomorphogenesis provides us an opportunity to study and exploit the role of various photomorphogenesis-related lncRNAs in different plant groups.

21.5.2 LncRNAs and plant fertility

A study suggests the role of lncRNAs in fertility of rice plant (Zhou et al., 2012). Regulatory genes and molecular mechanisms underlying the plant male sterility are not studied extensively. A natural mutant showing photoperiod-sensitive male sterility (PSMS) was found in 1973 in a japonica rice (*Oryza sativa* ssp. *japonica*) variety, Nongken 58 (referred as 58N) in Hubei Province, China (Zhou et al., 2012). Genetic studies explained that PSMS is regulated by a 1236 bp lncRNA called Long-Day-Specific Male fertility-Associated RNA (LDMAR), which is responsible for male fertility when present in sufficient amount under long-day conditions. It causes male sterility under long-day conditions in rice (Ding et al., 2012a) and the main reason for male sterility is programmed cell death (PCD) during anther development. The PCD is caused by a particularly low level of LDMAR transcription under long-day conditions (Ding et al., 2012a). LDMAR differential expression under long-day conditions occurs due to single nucleotide polymorphism (SNP). In this polymorphism, there is a conversion from C to G in Nongken 58S mutant. This C to G SNP change can amend the secondary structure of LDMAR in Nongken 58S and repress its expression by changing the methylation levels in the promoter (Ding et al., 2012b). There is another study that indicates the presence of lncRNAs in cotton, which is responsible for cytoplasmic male sterility and restoration of fertility during anther development (Zhang et al., 2019c). There is a huge scope for such studies in different plants. These lncRNAs have possible applications in either increasing the fertility or development of male sterile lines in various crop plants.

21.5.3 LncRNAs and vernalization

Reproduction is a significant event of the plant's life cycle which is marked by the transition of the vegetative phase into the flowering phase. Vernalization controls the flowering in plants that grow vegetatively in winter seasons and initiate flowering in warmer spring (Song et al., 2012). The *FLOWERING LOCUS C* (*FLC*) is a major regulatory gene that is involved in the transition of the vegetative stage to the reproductive stage (Irish, 2010). *FLC* encodes a MADS-box transcription factor that represses the expression of *SUPPRESSION OF OVEREXPRESSION OF CO1 (SOC1)* and *FLOWERING LOCUS T (FT)*. The flowering is controlled by FT proteins which promote the growth of floral organs in the apical meristem (Corbesier et al., 2007). *FLC* gene locus showed the occurrence of *COOLAIR* and *COLDAIR*. These lncRNAs modulate the expression of *FLC* at the transcriptional level and through histone modification (Heo and Sung, 2011; Sun et al., 2013). The transcription of *COOLAIR* occurs in antisense direction with *FLC*, while *COLDAIR* is transcribed from the intronic region of *FLC* in sense direction. PHD-PRC2 (plant homeodomain-Polycomb Repressive Complex 2) complex is recruited by these lncRNAs and this complex allows histone modification of *FLC* by epigenetic regulation in Arabidopsis (Gitschier, 2013). RNA-sequencing analysis shows that lncRNAs play important role in flowering in other plant species such as *Fragaria vesca*, *Poncirus trifoliata*, *Cicer arietinum*, etc. In *F. vesca*, lncRNAs have been identified from flower and fruit tissues, suggesting their role in reproductive tissue development (Kang and Liu, 2015). Various lncRNAs have been identified from trifoliate orange, *P. trifoliata*, which also regulates flowering (Wang et al., 2017). Furthermore, RNA-sequencing data of eight consecutive stages of floral development and vegetative stages of *C. arietinum* showed a probable regulatory role of lincRNAs in flowering (Khemka et al., 2016).

The endogenous target mimicry (eTM) is another important dimension for the in vitro study of lncRNAs and associated mRNAs. We can use this approach for reducing the flowering period by altering the expression of particular miRNAs (Karakülah et al., 2016), for instance, in rice, this method has efficiently shorten the harvest time which is cost-effective for cultivators.

21.5.4 LncRNAs in fiber development

Fibers have many applications in a wide range of industries such as textile and food industries. The yield and quality of fiber are affected at the initiation and elongation stages of its development. To get a good quality of fiber, various genetic studies have been performed. In 2016, Zou et al. revealed the importance of lncRNAs in fiber development of cotton. By using computational approaches, 3510 lincRNAs and 2486 lncNATs were identified; however, 51.9% of lincRNAs and 54.5% of lncNATs transcripts of cotton were differentially expressed during fiber development (Zou et al., 2016). LncRNAs also showed dynamic changes during elongation stages and indicated their involvement in fiber development in cotton (Zou et al., 2016). By pursuing these approaches, we can determine the involvement of lncRNAs in fiber development in other crops such as coconut, jute, flax, hemp, etc.

21.5.5 LncRNAs in fruit ripening

LncRNAs play a very significant role in fruit ripening of some plants. In earlier studies, 3679 lncRNAs were identified during tomato fruit ripening, out of which 490 lncRNAs were highly expressed and 187 lncRNAs showed lower expression (Zhu et al., 2015). When *lncRNA1459* and *lncRNA1840* were silenced, wild-type fruit ripening got delayed. Hence, lncRNAs revealed their involvement as a regulator in fruit ripening (Zhu et al., 2015). The silencing of related lncRNAs can be done for the investigation of their contribution in late fruit ripening in other crops which can be beneficial for farmers.

21.5.6 The function of lncRNAs in root development

The growth and development of roots mainly depend on the auxin concentration, which is maintained by its transport and auxin-signaling pathways. Earlier studies suggested that lncRNAs regulate the auxin transport in Arabidopsis roots. Auxin-induced *AUXIN-REGULATED PROMOTER LOOP RNA* (*APOLO*) regulates the dynamics of a chromatin loop at *PINOID* (*PID*) promoter, which acts as a key regulator of auxin transport (Ariel et al., 2014). Moreover, in Arabidopsis, Auxin-dependent lateral root formation is controlled by splicing regulators Nuclear Speckle RNA-binding proteins (NSRs), which are further regulated by alternative splicing competitor long noncoding RNA (*ASCO*). It competes with NSRs for pre-mRNA binding and acts as a splicing regulator (Bardou et al., 2014). The previous studies also indicate the involvement of lncRNAs in the root growth of populus. Expression patterns of *lncWOX5* and *lncWOX11* and their target genes at different developmental stages of adventitious roots revealed that lncRNAs played key regulatory roles in populus adventitious root development (Liu et al., 2018).

21.5.7 Role of lncRNAs in nodule formation

Early nodulin (*Enod40*) is induced at specific stages of plant development and functions in nodulation (Kouchi and Hata, 1993). *Enod40* is conserved in legumes (Matvienko et al., 1994; Yang et al., 1993) and is found in several nonlegume species, as well (Kouchi et al., 1999). The *Enod40* gene regulates the symbiotic relations among soil bacteria and leguminous plants. In some plants such as *M. truncatula*, *G. max*, and *O. sativa*, *MtENOD40*, *GmENOD40*, and *OsENOD4* help in nodulation, respectively (Sousa et al., 2001; Yang et al., 1993; Kouchi et al., 1999).

In *M. truncatula*, RNA-binding protein 1 (MtRBP1) showed interaction with *ENOD40* which helps in the relocalization of MtRBP1 from nuclear speckles to cytoplasmic granules during nodulation (Campalans et al., 2004). Hence, *ENOD40* plays a very important role in nodule formation of various plants. We can also search for the presence of *ENOD40* in other legume species.

21.6 Conclusions and future perspective

The advancement of bioinformatics and sequencing technologies such as high-throughput RNA-sequencing provides innovative ways for the discovery of novel lncRNAs in various plant species. Although many lncRNAs were identified during plant development processes and stress conditions, their functions are still not known. The functional studies of lncRNAs are limited to a few plant species. To investigate their functions in response to different stress conditions, few genome editing technologies such as Transcription Activator-Like Effector Nucleases (TALENs), Clustered Regulatory Interspaced Short Palindromic Repeats (CRISPR)/Cas9-induced, CRISPR interference (CRISPRi), etc. can be employed. However, more investigation of gene expression regulation by lncRNAs during developmental stages and stress conditions in plant species will give additional knowledge that will further be useful in improving plant performance.

Acknowledgments

Authors are thankful to Panjab University, Chandigarh, India, for providing the research facility and infrastructure. AK and M are thankful to CSIR and UGC for junior research fellowship, respectively. SKU is thankful to SERB, Government of India, for Early Career Research Award (ECR/2016/001270) and CSIR for research grant (38 (1489)/19/EMR-II). We are thankful to the Department of Science and Technology, Government of India, for partial financial support under Promotion of University Research and Scientific Excellence (PURSE) grant scheme.

References

Ariel, F., Jegu, T., Latrasse, D., Romero-Barrios, N., Christ, A., Benhamed, M., et al., 2014. Non-coding transcription by alternative RNA polymerases dynamically regulates an auxin-driven chromatin loop. Mol. Cell 55 (3), 383–396.

Bardou, F., Ariel, F., Simpson, C.G., Romero-Barrios, N., Laporte, P., Balzergue, S., et al., 2014. Long noncoding RNA modulates alternative splicing regulators in Arabidopsis. Dev. Cell 30 (2), 166–176.

Bari, R., Pant, B.D., Stitt, M., Scheible, W.R., 2006. PHO2, microRNA399, and PHR1 define a phosphate-signaling pathway in plants. Plant Physiol. 141 (3), 988–999.

Ben Amor, B., Wirth, S., Merchan, F., Laporte, P., d'Aubenton-Carafa, Y., Hirsch, J., et al., 2009. Novel long non-protein coding RNAs involved in Arabidopsis differentiation and stress responses. Genome Res. 19 (1), 57–69.

Burleigh, S.M., Harrison, M.J., 1998. Characterization of the Mt4 gene from *Medicago truncatula*. Gene 216 (1), 47–53.

Campalans, A., Kondorosi, A., Crespi, M., 2004. enod40, a short open reading frame-containing mRNA, induces cytoplasmic localization of a nuclear RNA binding protein in *Medicago truncatula*. Plant Cell 16 (4), 1047–1059.

Chen, M., Wang, C., Bao, H., Chen, H., Wang, Y., 2016. Genome-wide identification and characterization of novel lncRNAs in Populus under nitrogen deficiency. Mol. Gen. Genomics. 291 (4), 1663–1680.

Chiou, T.J., Aung, K., Lin, S.I., Wu, C.C., Chiang, S.F., Su, C.L., 2006. Regulation of phosphate homeostasis by microRNA in Arabidopsis. Plant Cell 18 (2), 412–421.

Corbesier, L., Vincent, C., Jang, S., Fornara, F., Fan, Q., Searle, I., et al., 2007. FT protein movement contributes to long-distance signaling in floral induction of Arabidopsis. Science 316 (5827), 1030–1033.

Cui, J., Luan, Y., Jiang, N., Bao, H., Meng, J., 2017. Comparative transcriptome analysis between resistant and susceptible tomato allows the identification of lncRNA 16397 conferring resistance to *Phytophthora infestans* by co-expressing glutaredoxin. Plant J. 89 (3), 577–589.

Deng, F., Zhang, X., Wang, W., Yuan, R., Shen, F., 2018. Identification of *Gossypium hirsutum* long non-coding RNAs (lncRNAs) under salt stress. BMC Plant Biol. 18 (1), 23.

Derrien, T., Johnson, R., Bussotti, G., Tanzer, A., Djebali, S., Tilgner, H., et al., 2012. The GENCODE v7 catalog of human long noncoding RNAs: analysis of their gene structure, evolution, and expression. Genome Res. 22 (9), 1775–1789.

Di, C., Yuan, J., Wu, Y., Li, J., Lin, H., Hu, L., et al., 2014. Characterization of stress-responsive lnc RNAs in *Arabidopsis thaliana* by integrating expression, epigenetic and structural features. Plant J. 80 (5), 848–861.

Ding, J., Lu, Q., Ouyang, Y., Mao, H., Zhang, P., Yao, J., et al., 2012a. A long noncoding RNA regulates photoperiod-sensitive male sterility, an essential component of hybrid rice. Proc. Natl. Acad. Sci. U. S. A. 109, 2654–2659.

Ding, J., Shen, J., Mao, H., Xie, W., Li, X., Zhang, Q., 2012b. RNA-directed DNA methylation is involved in regulating photoperiod-sensitive male sterility in rice. Mol. Plant 5 (6), 1210–1216.

Farooq, M., Wahid, A., Kobayashi, N., Fujita, D.B.S.M.A., Basra, S.M.A., 2009. Plant drought stress: effects, mechanisms and management. In: Sustainable Agriculture. Springer, Dordrecht, pp. 153–188.

Franco-Zorrilla, J.M., Valli, A., Todesco, M., Mateos, I., Puga, M.I., Rubio-Somoza, I., et al., 2007. Target mimicry provides a new mechanism for regulation of microRNA activity. Nat. Genet. 39, 1033–1037.

Gitschier, J., 2013. How cool is that: an interview with Caroline Dean. PLoS Genet. 9, e1003593.

Griffey, C.A., Das, M.K., Stromberg, E.L., 1993. Effectiveness of adult-plant resistance in reducing grain yield loss to powdery mildew in winter wheat. Plant Dis. 77 (6), 618–622.

Heo, J.B., Sung, S., 2011. Vernalization-mediated epigenetic silencing by a long intronic noncoding RNA. Science 331 (6013), 76–79.

Hirsch, J., Lefort, V., Vankersschaver, M., Boualem, A., Lucas, A., Thermes, C., et al., 2006. Characterization of 43 non-protein-coding mRNA genes in Arabidopsis, including the MIR162a-derived transcripts. Plant Physiol. 140 (4), 1192–1204.

Irish, V.F., 2010. The flowering of *Arabidopsis* flower development. Plant J. 61, 1014–1028.

Jabnoune, M., Secco, D., Lecampion, C., Robaglia, C., Shu, Q., Poirier, Y., 2013. A rice cis-natural antisense RNA acts as a translational enhancer for its cognate mRNA and contributes to phosphate homeostasis and plant fitness. Plant Cell 25 (10), 4166–4182.

Jiao, Y., Lau, O.S., Deng, X.W., 2007. Light-regulated transcriptional networks in higher plants. Nat. Rev. Genet. 8 (3), 217.

Joshi, R.K., Megha, S., Basu, U., Rahman, M.H., Kav, N.N., 2016. Genome wide identification and functional prediction of long non-coding RNAs responsive to *Sclerotinia sclerotiorum* infection in *Brassica napus*. PLoS One 11 (7), e0158784.

Kaikkonen, M.U., Lam, M.T., Glass, C.K., 2011. Non-coding RNAs as regulators of gene expression and epigenetics. Cardiovasc. Res. 90 (3), 430–440.

Kang, C., Liu, Z., 2015. Global identification and analysis of long non-coding RNAs in diploid strawberry *Fragaria vesca* during flower and fruit development. BMC Genomics 16, 815.

Karakülah, G., Kurtoğlu, K.Y., Unver, T., 2016. PeTMbase: a database of plant endogenous target mimics (eTMs). PLoS One 11 (12), e0167698.

Karlik, E., Ari, S., Gozukirmizi, N., 2019. LncRNAs: genetic and epigenetic effects in plants. Biotechnol. Biotechnol. Equip. 33 (1), 429–439.

Kim, E.D., Sung, S., 2012. Long noncoding RNA: unveiling hidden layer of gene regulatory networks. Trends Plant Sci 17 (1), 16–21.

Khemka, N., Singh, V.K., Garg, R., Jain, M., 2016. Genome-wide analysis of long intergenic noncoding RNAs in chickpea and their potential role in flower development. Sci. Rep. 6, 33297.

Kouchi, H., Hata, S., 1993. Isolation and characterization of novel nodulin cDNAs representing genes expressed at early stages of soybean nodule development. Mol. Gen. Genet. 238 (1–2), 106–119.

Kouchi, H., Takane, K.I., So, R.B., Ladha, J.K., Reddy, P.M., 1999. Rice ENOD40: isolation and expression analysis in rice and transgenic soybean root nodules. Plant J. 18 (2), 121–129.

Kwenda, S., Birch, P.R., Moleleki, L.N., 2016. Genome-wide identification of potato long intergenic noncoding RNAs responsive to *Pectobacterium carotovorum* subspecies *brasiliense* infection. BMC Genomics 17 (1), 614.

Li, S., Yu, X., Lei, N., Cheng, Z., Zhao, P., He, Y., Wang, W., Peng, M., 2017. Genome-wide identification and functional prediction of cold and/or drought-responsive lncRNAs in cassava. Sci. Rep. 7, 45981.

Liu, C., Muchhal, U.S., Raghothama, K.G., 1997. Differential expression of *TPS11*, a phosphate starvation-induced gene in tomato. Plant Mol. Biol. 33 (5), 867–874.

Liu, J., Jung, C., Xu, J., Wang, H., Deng, S., Bernad, L., Arenas-Huertero, C., Chua, N.H., 2011. Genome-wide analysis uncovers regulation of long intergenic noncoding RNAs in Arabidopsis. Plant Cell 24 (11), 4333–4345.

Liu, S., Sun, Z., Xu, M., 2018. Identification and characterization of long non-coding RNAs involved in the formation and development of poplar adventitious roots. Ind. Crops Prod. 118, 334–346.

Luo, C., Sidote, D.J., Zhang, Y., Kerstetter, R.A., Michael, T.P., Lam, E., 2013. Integrative analysis of chromatin states in Arabidopsis identified potential regulatory mechanisms for natural antisense transcript production. Plant J. 73 (1), 77–90.

Lv, Y., Liang, Z., Ge, M., Qi, W., Zhang, T., Lin, F., et al., 2016. Genome-wide identification and functional prediction of nitrogen-responsive intergenic and intronic long non-coding RNAs in maize (*Zea mays* L.). BMC Genomics 17 (1), 350.

Matvienko, M., Van De Sande, K., Yang, W.C., Van Kammen, A., Bisseling, T., Franssen, H., 1994. Comparison of soybean and pea ENOD40 cDNA clones representing genes expressed during both early and late stages of nodule development. Plant Mol. Biol. 26 (1), 487–493.

Miura, K., Furumoto, T., 2013. Cold signaling and cold response in plants. Int. J. Mol. Sci. 14 (3), 5312–5337.

Pham, V.N., Kathare, P.K., Huq, E., 2018. Phytochromes and phytochrome interacting factors. Plant Physiol. 176 (2), 1025–1038.

Ponting, C.P., Oliver, P.L., Reik, W., 2009. Evolution and functions of long noncoding RNAs. Cell 136, 629–641.

Qi, X., Xie, S., Liu, Y., Yi, F., Yu, J., 2013. Genome-wide annotation of genes and noncoding RNAs of foxtail millet in response to simulated drought stress by deep sequencing. Plant Mol. Biol. 83, 459–473.

Qin, T., Zhao, H., Cui, P., Albesher, N., Xiong, L., 2017. A nucleus-localized long non-coding RNA enhances drought and salt stress tolerance. Plant Physiol. 175 (3), 1321–1336.

Roser, M., 2013. Future population growth. In: Our World in Data.

Secco, D., Baumann, A., Poirier, Y., 2010. Characterization of the rice PHO1 gene family reveals a key role for OsPHO1; 2 in phosphate homeostasis and the evolution of a distinct clade in dicotyledons. Plant Physiol. 152 (3), 1693–1704.

Shumayla, Sharma, S., Taneja, M., Tyagi, S., Singh, K., Upadhyay, S.K., 2017. Survey of high throughput RNA-Seq data reveals potential roles for lncRNAs during development and stress response in bread wheat. Front. Plant Sci. 8, 1019.

Shuai, P., Liang, D., Tang, S., Zhang, Z., Ye, C.Y., Su, Y., et al., 2014. Genome-wide identification and functional prediction of novel and drought-responsive lincRNAs in *Populus trichocarpa*. J. Exp. Bot. 65, 4975–4983.

Song, J., Angel, A., Howard, M., Dean, C., 2012. Vernalization—a cold induced epigenetic switch. J. Cell Sci. 125, 3723–3731.

Song, X., Liu, G., Huang, Z., Duan, W., Tan, H., Li, Y., et al., 2016. Temperature expression patterns of genes and their coexpression with LncRNAs revealed by RNA-Seq in non-heading Chinese cabbage. BMC Genomics 17 (1), 2s97.

Sousa, C., Johansson, C., Charon, C., Manyani, H., Sautter, C., Kondorosi, A., et al., 2001. Translational and structural requirements of the early nodulin gene enod40, a short-open reading frame-containing RNA, for elicitation of a cell-specific growth response in the alfalfa root cortex. Mol. Cell. Biol. 21 (1), 354–366.

Sun, Q., Csorba, T., Skourti-Stathaki, K., Proudfoot, N.J., Dean, C., 2013. R-loop stabilization represses antisense transcription at the *Arabidopsis* FLC locus. Science 340, 619–621.

Thakur, P., Kumar, S., Malik, J.A., Berger, J.D., Nayyar, H., 2010. Cold stress effects on reproductive development in grain crops: an overview. Environ. Exp. Bot. 67 (3), 429–443.

Tyagi, S., Sharma, A., Upadhyay, S.K, 2018. Role of Next-Generation RNA-Seq Data in Discovery and Characterization of Long Non-Coding RNA in Plants. In: Next Generation Plant Breeding. Yelda Özden Çiftçi, IntechOpen. https://doi.org/10.5772/intechopen.72773.

Wan, D., Ma, J., Bai, X., Luo, W., Shao, X., Bai, Q., et al., 2019. Genome-wide identification of long noncoding RNAs (lncRNAs) and their responses to salt stress in two closely related poplars. Front. Genet. 10, 777.

Wang, H., Chung, P.J., Liu, J., Jang, I.C., Kean, M.J., Xu, J., et al., 2014a. Genome-wide identification of long noncoding natural antisense transcripts and their responses to light in Arabidopsis. Genome Res. 24 (3), 444–453.

Wang, Y., Fan, X., Lin, F., He, G., Terzaghi, W., Zhu, D., et al., 2014b. Arabidopsis noncoding RNA mediates control of photomorphogenesis by red light. Proc. Natl. Acad. Sci. U. S. A. 111 (28), 10359–10364.

Wang, J., Yu, W., Yang, Y., Li, X., Chen, T., Liu, T., et al., 2015a. Genome-wide analysis of tomato long non-coding RNAs and identification as endogenous target mimic for microRNA in response to TYLCV infection. Sci. Rep. 5, 16946.

Wang, T.Z., Liu, M., Zhao, M.G., Chen, R., Zhang, W.H., 2015b. Identification and characterization of long non-coding RNAs involved in osmotic and salt stress in *Medicago truncatula* using genome-wide high-throughput sequencing. BMC Plant Biol. 15, 131.

Wang, C.-Y., Liu, S.-R., Zhang, X.-Y., Ma, Y.-Z., Hu, C.-G., Zhang, J.-Z., 2017. Genome-wide screening and characterization of long non-coding RNAs involved in flowering development of trifoliate orange (*Poncirus trifoliata* L. Raf.). Sci. Rep. 7, 43226.

Wang, A., Hu, J., Gao, C., Chen, G., Wang, B., Lin, C., et al., 2019a. Genome-wide analysis of long non-coding RNAs unveils the regulatory roles in the heat tolerance of Chinese cabbage (*Brassica rapa* ssp. *chinensis*). Sci. Rep. 9 (1), 5002.

Wang, P., Dai, L., Ai, J., Wang, Y., Ren, F., 2019b. Identification and functional prediction of cold-related long non-coding RNA (lncRNA) in grapevine. Sci. Rep. 9 (1), 1–15.

Ward, M., McEwan, C., Mills, J.D., Janitz, M., 2015. Conservation and tissue-specific transcription patterns of long noncoding RNAs. J. Hum. Transcr. 1 (1), 2–9.

Xin, M., Wang, Y., Yao, Y., Song, N., Hu, Z., Qin, D., et al., 2011. Identification and characterization of wheat long non-protein coding RNAs responsive to powdery mildew infection and heat stress by using microarray analysis and SBS sequencing. BMC Plant Biol. 11, 61.

Xiong, L., Zhu, J.K., 2003. Regulation of abscisic acid biosynthesis. Plant Physiol. 133, 29–36.

Yadav, S.K., 2010. Cold stress tolerance mechanisms in plants. A review. Agron. Sustain. Dev. 30 (3), 515–527.

Yang, L., Duff, M.O., Graveley, B.R., Carmichael, G.G., Chen, L.L., 2011. Genome wide characterization of non-polyadenylated RNAs. Genome Biol. 12(2), R16.

Yang, W.C., Katinakis, P., Hendriks, P., Smolders, A., de Vries, F., Spee, J., et al., 1993. Characterization of *GmENOD40*, a gene showing novel patterns of cell-specific expression during soybean nodule development. Plant J. 3 (4), 573–585.

Yu, T., Zhu, H., 2019. Long non-coding RNAs: rising regulators of plant reproductive development. Agronomy 9, 53.

Zhang, W., Han, Z., Guo, Q., Liu, Y., Zheng, Y., Wu, F., et al., 2014. Identification of maize long non-coding RNAs responsive to drought stress. PLoS One 9 (6), e98958.

Zhang, H., Hu, W., Hao, J., Lv, S., Wang, C., Tong, W., et al., 2016. Genome-wide identification and functional prediction of novel and fungi-responsive lincRNAs in *Triticum aestivum*. BMC Genomics 17 (1), 238.

Zhang, L., Wang, M., Li, N., Wang, H., Qiu, P., Pei, L., et al., 2018. Long noncoding RNA s involve in resistance to *Verticillium dahliae*, a fungal disease in cotton. Plant Biotechnol. J. 16 (6), 1172–1185.

Zhang, P., Wu, W., Chen, Q., Chen, M., 2019a. Non-coding RNAs and their integrated networks. J. Integr. Bioinform. 16 (3).

Zhang, X., Dong, J., Deng, F., Wang, W., Cheng, Y., Song, L., et al., 2019b. The long non-coding RNA lncRNA973 is involved in cotton response to salt stress. BMC Plant Biol. 19 (1), 459.

Zhang, B., Zhang, X., Zhang, M., Guo, L., Qi, T., Wang, H., et al., 2019c. Transcriptome analysis implicates involvement of long noncoding RNAs in cytoplasmic male sterility and fertility restoration in cotton. Int. J. Mol. Sci. 20 (22), 5530.

Zhao, Y.R., Yu, K.Q., Li, X., He, Y., 2016. Detection of fungus infection on petals of rapeseed (*Brassica napus* L.) using NIR hyperspectral imaging. Sci. Rep. 6, 38878.

Zhou, H., Liu, Q., Li, J., Jiang, D., Zhou, L., Wu, P., et al., 2012. Photoperiod- and thermo-sensitive genic male sterility in rice are caused by a point mutation in a novel noncoding RNA that produces a small RNA. Cell Res. 22, 649–660.

Zhu, Q.H., Stephen, S., Taylor, J., Helliwell, C.A., Wang, M.B., 2014. Long noncoding RNAs responsive to *Fusarium oxysporum* infection in *Arabidopsis thaliana*. New Phytol. 201, 574–584.

Zhu, B., Yang, Y., Li, R., Fu, D., Wen, L., Luo, Y., et al., 2015. RNA sequencing and functional analysis implicate the regulatory role of long non-coding RNAs in tomato fruit ripening. J. Exp. Bot. 66 (15), 4483–4495.

Zou, C., Wang, Q., Lu, C., Yang, W., Zhang, Y., Cheng, H., et al., 2016. Transcriptome analysis reveals long noncoding RNAs involved in fiber development in cotton (*Gossypium arboreum*). Sci. China Life Sci. 59 (2), 164–171.

Index

Note: Page numbers followed by *f* indicate figures, and *t* indicate tables.

D

Kolibri 360 GmbH
Ferdinand-Jühlke-Straße 7
99095 Erfurt, Deutschland
produktsicherheit@kolibri360.de